Laser Cooling

Laser Cooling

Fundamental Properties and Applications

edited by

Galina Nemova

Pan Stanford Publishing

Published by

Pan Stanford Publishing Pte. Ltd.
Penthouse Level, Suntec Tower 3
8 Temasek Boulevard
Singapore 038988

Email: editorial@panstanford.com
Web: www.panstanford.com

British Library Cataloguing-in-Publication Data
A catalogue record for this book is available from the British Library.

Laser Cooling: Fundamental Properties and Applications

ISBN 978-981-4745-04-8 (Hardcover)
ISBN 978-981-4745-05-5 (eBook)

Printed in the USA

Contents

Preface

Refrigeration or cooling is a process in which a substance is maintained at a temperature below that of its surroundings. It is based on controlled heat removal from one location to another and dates back thousands of years when people tried to find different ways of preserving their food. At that time ice or snow, which provided refrigeration through structures as simple as holes in the ground or cold cellars, was the only available cooling technique. In more recent history the efforts of scientists of the nineteenth century to liquefy the permanent gases brought *cryogenics* into existence. Cryogenics is the science of low temperatures, which are below the point at which permanent gases begin to liquefy. The invention of a laser was one of the groundbreaking scientific achievements of the twentieth century, which further revolutionized cryogenics. The rapid development of cryogenic techniques led to a number of fundamental discoveries such as superconductivity (Heike Kamerlingh Onnes, 1908), superfluidity (Pyotr Leonidovich Kapitsa, John Frank Allen, and Don Misener, 1937), and Bose–Einstein condensation (Eric Allin Cornell, Carl Edwin Wieman, and Wolfgang Ketterle, 1995).

Today cryogenics continues to be a burning problem from fundamental and applied points of view. Cooling with light, also known as *laser cooling* or *optical refrigeration*, is one of promising cryogenics techniques. Laser cooling is a physical process by which a system loses thermal energy as a result of its interaction with laser light. This process can be based on the Doppler effect, which can be used to trap atoms and slow their motion to cool a substance. This method along with further evaporative cooling has permitted to realize Bose–Einstein condensation of atomic and molecular species. Optical refrigeration can also be based on redistribution of radiation in dense gas mixtures. Both these techniques permit to remove the thermal energy contained in translational degrees of freedom. They are suitable for gases because at low temperatures the thermal energy in gases is primarily contained in translational degrees of freedom. In some systems, for example, in solids, the thermal energy is primarily contained in vibrations (phonons). Such systems can be laser cooled with anti-Stokes fluorescence. In that case a system illuminated with laser light at long wavelength tail of its absorption spectrum emits photons of higher energy than the absorbed ones as a result of phonon absorption in the system. Laser cooling of solids was first experimentally demonstrated in 1995 with ytterbium-doped fluoride glass. Since the first experimental observation laser cooling of solids has been realized with Yb^{3+}, Er^{3+}, and Tm^{3+} ions doped in a wide variety of low phonon glasses and crystals. Today's record temperature achieved in a Yb^{3+}:YLF sample is 91 K, which is far below the temperature that can be achieved with any other solid state cooling techniques. New materials as well as old ones with extremely low levels of impurities can permit to reach even lower temperatures in the nearest future. Direct band-gap semiconductors are very promising candidates for laser cooling. Laser cooling is progressing rapidly. The latest achievements have motivated me to pull together a team of researchers who are doing frontier research in the field of laser cooling to discuss the fundamental principles and applications of this promising field of physics. Their research and experience have been organized into the 13 chapters of the book.

In **Chapter 1** *Anne Saß, Stavros Christopoulos,* and *Martin Weitz* review their recent results on laser cooling by redistribution of radiation in dense gas mixtures. The achieved cooling powers in these experiments are several orders of magnitude larger than in the Doppler cooling experiments for dilute atomic vapors, and the typically achieved cooling efficiencies are in the order of several per cent. The cooling results are presented in the form of thermal deflection spectroscopy measurements on both rubidium– and potassium–argon mixtures. They also present the Kennard–Stepanov relation, connecting absorption and emission profiles, and apply it to the dense gaseous ensembles under investigation. The authors of the chapter discuss their future experimental aims that include the study of homogeneous nucleation in saturated vapor as well as the application of this very technique to molecular–rare gas ensembles.

In **Chapter 2** *Galina Nemova* presents a detailed analysis of laser cooling with anti-Stokes fluorescence in a two-level rare earth–doped solid state system. The influence of all cooperative effects in rare earth–doped samples on the laser cooling process is considered. Thermodynamics of laser cooling is discussed in the chapter. The advantages of different host materials with low phonon energy for laser cooling with anti-Stokes fluorescence are considered. In this chapter one can see how the efficiency of the laser cooling process can be enhanced with optical cavities. The author presents an overview of techniques, which can provide a local temperature measurement as well as the bulk temperature of a sample. All these techniques have some advantages and disadvantages, which are discussed in the chapter.

Chapter 3 is intended to give an overview of some theoretical and experimental efforts aimed at developing an optical cooling system based solely on optical fibers. In this chapter *Nai-Hang Kwong, Rolf Binder, Dan Nguyen*, and *Arturo Chavez-Pirson* show that such a cooling system has unique advantages, including the possibility for dumping heat waste at arbitrary remote locations and the light-weight, compact, and robust nature of the system. The theoretical basis includes a rate equation–based cooling model for the active ions in the doped fiber as well as heat transport simulations. Experimental results for luminescence spectra for various pump powers show promising trends when compared with spectra as a function of temperature, indicating the possibility of net fiber cooling.

In **Chapter 4** *Guang-Zong Dong* and *Xin-Lu Zhang* propose a modified excitation cooling scheme based on ground-state absorption of $^5I_8 \rightarrow {}^5I_7$ at around 2.07 μm and excited-state absorption of $^5I_5 \rightarrow {}^5F_5$ at around 2.28 μm in Ho^{3+} ions. In this scheme large amount of heat produced by nonradiative $^5I_5 \rightarrow {}^5I_6$ and $^5I_6 \rightarrow {}^5I_7$ transitions in the conventional anti-Stokes cooling case can be avoided, and both the cooling efficiency and the cooling power can be prompted. They also propose to improve the cooling performance of Ho^{3+} optical refrigeration by co-doping Tm^{3+} and Ho^{3+} ion in the host.

Since the invention of the laser, engineers have searched for new ways to increase the laser's power and brightness. These efforts have largely been focused on improved ways to remove the waste heat generated in the lasing medium. In **Chapter 5** *Steven R. Bowman* reviews a novel approach to the problem of laser power scaling: minimizing the waste heat through the incorporation of anti-Stokes fluorescence cooling. This approach has been shown to significantly reduce heat generation in carefully controlled laser systems. This chapter reviews the theory and practice of this novel mode of laser operation.

In **Chapter 6** *Stephen C. Rand* compares three conceptually different approaches to laser cooling of solids. He offers the unique perspective that in solids where free translational motion is absent there are nevertheless interactions analogous to the Doppler effect that may be used to address the energy and momentum of collective excitations of the medium. Moreover there are reasons to explore them systematically. Two methods of cooling solids in particular, which are reminiscent of interactions with free atoms, are analyzed from this perspective to examine their detailed advantages and disadvantages. While attempting to profit from the earlier history of laser cooling in gases and to provide points of comparison between different techniques, this chapter identifies an approach to cooling that results in a temperature-independent cooling rate.

In the "traditional" model, which describes optical refrigeration of rare earth–doped solids based on anti-Stokes fluorescence, a cooling cycle is provided by resonance transitions between the upper ground and lower excited levels of ions, phonon thermalization, and subsequent radiative relaxation to the ground states. In such a model, the rate of the cooling cycle is low because of the small cross section of forbidden transitions between the levels of the same ion configuration and the low population of the upper ground level. In **Chapter 7** *Andrei Ivanov, Yuriy Rozhdestvensky,* and *Evgeniy Perlin* consider a cooling process with stimulated Raman adiabatic passage for optical pumping of a rare earth–doped crystal taking into account vibronic interaction at the Stark split sublevels of ion manifolds. They show that coherent pumping by stimulated Raman adiabatic passage leads to the deep laser cooling of rare earth–doped crystals that have technologically attainable purity to date.

In **Chapter 8** *Azzurra Volpi, Alberto Di Lieto,* and *Mauro Tonelli* report on cooling efficiency measurements performed on various Yb^{3+}-doped fluoride single crystals grown in their laboratory. They show that the investigation of the cooling process for increasing Yb^{3+} concentrations in $LiYF_4$ (YLF) single crystals results in improved cooling efficiency via significant decrease of detrimental parasitic absorption. Energy transfer processes between Yb^{3+} ions and low concentrations of rare earth ions impurities are investigated in this chapter to analyze the effect of detrimental impurities, which participate in multi-phonon relaxation of excited ions. It is demonstrated that Yb^{3+}–Tm^{3+} controlled co-doping led to observation of energy transfer–enhanced Yb^{3+} anti-Stokes efficiency. The results achieved with Yb^{3+}–Tm^{3+} co-doped samples are presented in this chapter.

In **Chapter 9** *B. Rami Reddy* demonstrates the use of Michelson and Mach–Zehnder interferometers to measure the sample temperature changes due to laser beam heating. He has developed an analytical model to estimate the temperature change that depends on fringe (or wave) counting. The passage of a 300 mW of 915 nm laser beam through 1.54 cm long Yb^{3+}:YAG sample raised its temperature by 7.1 ± 0.8 K. The passage of a 115 mW focused argon-ion beam through a 1 cm long Pr^{3+}:YAG sample raised its temperature by 11.7 ± 1.0 K. Optical interferometric measurement of temperature changes agreed with that measured by a thermocouple. These methods apply equally well to estimate the temperature changes in laser cooling of solids also. He has also developed theoretical formulations for the use of optical heterodyne technique in laser cooling studies. His theory reveals that sample temperature changes as small as 1.4 mK can be measured with the heterodyne technique.

Optical fibers are now critical parts in many high-tech devices. Of course fiber parameters have to be tailored to each application needs. Unfortunately, there is no single fiber material that can fulfill all types of application needs, but there are some important criteria that one has to consider when choosing an optical fiber to be used in a specific application. The first criterion is technical. The second criterion is economical. The technology should be mature enough to provide high-quality optical fibers at effective cost with a reasonable yield. Fluoride glasses are very promising hosts for laser cooling. In **Chapter 10** *Mohammed Saad* presents the latest technological achievements concerning three main glass families as representatives of fluoride glasses in general: fluorozirconate, fluoroaluminate, and fluoroindate.

In the last years, the advances enabled by Yb^{3+}-doped fluorides have tremendously increased the interest in this technology. Cryogenic operations as well as competitive steady-state cooling powers have been achieved in the temperature range below 250 K. Minimum achievable temperatures are approaching 80 K. Efficient cooling performances, however, are firstly related to the optical quality of active materials. Stringent requirements on the purity and structural quality need to be satisfied in order to achieve net bulk cooling. In **Chapter 11** *Azzurra Volpi* and *Mauro Tonelli* consider growth techniques and methods for optical and structural quality assessments of fluoride single crystals for optical refrigeration.

In **Chapter 12** *Rolf Binder* and *Nai-Hang Kwong* give an overview of some theoretical efforts to analyze optical refrigeration of inorganic semiconductor crystals. A fermionic theory, describing the conduction and valence bands, combined with Coulomb interaction between the excited charge carriers yields the absorption and luminescence spectra that include excitonic effects valid for arbitrary excitonic ionization ratios. The microscopic theory used as an input to Sheik-Bahae and Epstein's semiconductor cooling model yields a cooling analysis that has been evaluated for GaAs. At low temperatures excitonic effects are found to be crucial for cooling to be possible. Extensions of the basic cooling model for bulk crystals include light propagation and luminescence re-absorption effects and the effects of passivation layers in doped, spatially inhomogeneous structures. While the numerical results presented here are restricted to GaAs, the general theoretical formulation is valid for any direct-gap crystalline semiconductor.

In **Chapter 13** *Iman Hassani Nia* and *Hooman Mohseni* discuss the mechanisms that can potentially increase the laser cooling efficiency to make it practical for bulk semiconductors. They consider Coulomb interaction in piezoelectric semiconductors. The authors describe both Coulomb interaction and piezoelectricity separately to establish the theoretical bases for understanding Coulomb-assisted laser cooling. Their approach is inspired by the optomechanical cooling of micro- and nano-mechanical systems since each phonon mode acts like a mechanical mode of vibration with a low quality factor.

I would like to sincerely express my gratitude to all the contributing authors for their excellent work and patience during the preparation and editing of this book in the International Year of Light 2015.

Galina Nemova
Montreal, Canada
Summer 2016

Chapter 1

Laser Cooling of Dense Gases by Collisional Redistribution of Radiation

Anne Saβ, Stavros Christopoulos, and Martin Weitz
Institut für Angewandte Physik der Universität Bonn, Wegelerstr. 8, 53115 Bonn, Germany
sass@iap.uni-bonn.de, christopoulos@iap.uni-bonn.de, weitz@iap.uni-bonn.de

1.1 Introduction

Light as a tool for cooling matter was considered already in 1929 by P. Pringsheim [1]. The experimental realization of the laser [2] paved the way for experiments on different laser cooling techniques. Among these, Doppler cooling of dilute atomic gases is probably the most common [3–5], first successfully investigated in the 1980s [6]. As an important step, this technique along with further evaporative cooling allowed for the realization of Bose–Einstein condensation of different atomic [7, 8] and molecular species [9] and has today grown into a robust research area. An alternative approach for cooling is anti-Stokes cooling in multilevel systems, culminating, for example, in the cooling of solids. For the first time suggested in 1950 [10] and

Laser Cooling: Fundamental Properties and Applications
Edited by Galina Nemova

ISBN 978-981-4745-04-8 (Hardcover), 978-981-4745-05-5 (eBook)
www.panstanford.com

experimentally realized in 1981 [11], this technique leads nowadays to a significant cooling of heavy metal fluoride glass doped with, for example, trivalent ytterbium ions [12, 13].

In this chapter, we review results on laser cooling by redistribution of radiation in dense gas mixtures. Redistribution of fluorescence is a well-known effect observed in experiments of magneto-optically trapped atoms, where it acts as the main loss mechanism of the trap [5]. Considering atomic collisions at room temperature, redistribution of fluorescence is a consequence of collisional aided excitation [14]. In a theoretical work in 1978, P. Berman and S. Stenholm proposed a cooling mechanism based on collisionally aided fluorescence and the related energy loss in a two-level system [15]. Corresponding experiments with gases at moderate densities never reached the cooling regime [16]; only heating for blue-detuned excitation was observed for these conditions.

Using a high-pressure environment with a system of rubidium atoms subject to 230 bar of argon buffer gas, our group experimentally demonstrated laser cooling by redistribution of fluorescence in 2009 [17]. At the used buffer gas pressures of a few hundred bars, the optical transitions are broadened to linewidths that are in the same order of magnitude as the thermal energy, $k_B T$, in frequency units, where k_B is the Boltzmann constant and T the temperature of the gas mixture. To obtain an idea of the cooling principle in this regime, assume the formation of transient, alkali–noble gas quasi-molecules during each binary collision of the two species. The atomic resonances are thus perturbed by means of the rising intermolecular potential, allowing for absorption of far red-detuned incident radiation. For typical parameters, the radiative lifetime of the excited state exceeds the time of such a collision by 3 to 4 orders of magnitude, which is a few nanoseconds compared to picoseconds.

Subsequent decay of the excited state occurs mostly at larger interatomic distances where the alkali resonance frequency is close to its unperturbed value. In this manner, the mean emitted fluorescence has a smaller wavelength than the absorbed photon; thus energy in the order of the thermal energy is extracted from the sample, cooling the dense mixture.

In the following sections, we initially discuss the mechanism of redistribution of radiation in a more detailed way (Section 1.2) and subsequently present the experimental setup and the high-pressure chambers in Section 1.3. Experimental results will be reviewed throughout Sections 1.4 and 1.5, before concluding and giving an outlook in Section 1.6.

1.2 Redistribution of Radiation

Before discussing the cooling principle for the case of alkali–noble gas mixtures in more detail, we give a brief overview of the basic principle of redistribution of radiation.

1.2.1 Basic Principle

Collisionally induced redistribution of radiation is based on the interaction between an incident light field with an optically active atom of species A surrounded by perturbing atoms of species B. Binary collisions between atoms of the two species can lead to broadening of atom's A resonances [18, 19]. This enables absorption/emission of radiation at a wavelength nonresonant to the free and unperturbed atom. The effect has been observed for the first time in strontium vapor at low buffer gas pressures [20], while a theoretical prediction can be found in Refs. [21, 22].

Let us consider photons with energy $\hbar\omega_L$. We will further consider that the difference between the transition energy $\hbar\omega_0$ and the photon energy is small compared to the thermal energy $k_B T$ of the perturbers. In this case, the photons are either scattered elastically in terms of Rayleigh scattering, or they are absorbed and reemitted as fluorescence (frequency ω_{Fl}) if the energy gap is overcome by energy transfer from the perturbing atoms. A schematic of the described process is shown in Fig. 1.1. Note that in the case of a multilevel atom A surrounded by perturbing atoms, it is possible to achieve excitation into levels that would be too far off resonant in the undisturbed case.

If we consider the collisional pair (A – B) to form a quasi-molecule at small interatomic distances, we can intuitively understand the redistribution process. When the collision takes place, the atom A absorbs the incident photon. The quasi-molecule is thus excited from the ground state into the excited state. Following the collision, the distance between the collisional partners grows and after its $1/e$ natural lifetime—which is of the same order as in the unperturbed case [23]—the electronically excited state decays under the emission of a fluorescence photon. Since the energy levels at larger interatomic distances approach those of the unperturbed radiating atom A, the frequency of the emitted photon ω_{Fl} will be close to the resonance frequency of atom A.

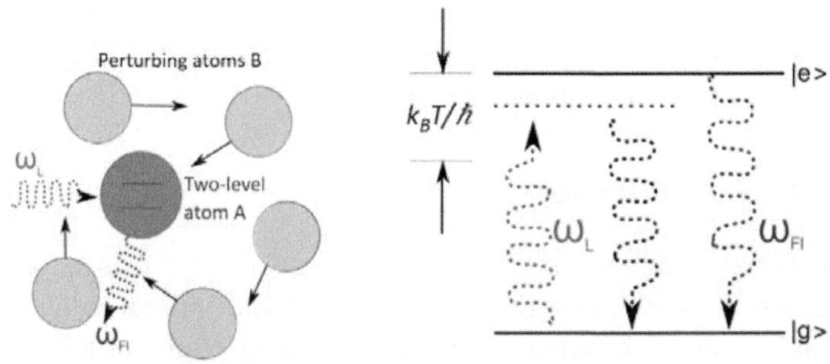

Figure 1.1 Process of redistribution of radiation for a two-level atom A surrounded by perturbing atoms of species B (left). Depending on the degree of elasticity of the collision with the perturbing atoms, resonant Rayleigh scattering ($\omega = \omega_{\mathrm{L}}$) or fluorescence at a different frequency ω_{Fl} can be detected (right).

It can be shown experimentally that the intensity of the redistributed fluorescence is proportional to the intensity of the incident light field and to the number density of species B at a given (far) detuning in case of low intensities and pressures. For larger intensities and number densities, the emitted fluorescence will at some point exhibit saturation [18].

1.2.2 Cooling Principle

We conduct experiments with alkali atoms subject to a high-pressure noble gas. With regard to the nomenclature used above, the alkali atoms are treated as species A, while the noble gas atoms as the perturbing species B. In these experiments, considerable broadening of the alkali resonances occurs due to frequent collisions with the noble gas atoms, making the absorption of far-red-detuned radiation feasible. As already noted above, the radiative lifetime of the electronically excited quasi-molecular state is similar to the lifetime of the excited alkali state. Our main experimental focus lies on rubidium atoms subject to high-pressure argon gas. The corresponding quasi-molecular potentials can be found in Ref. [24], calculated out of [Rb^+] and [e^- – Ar] pseudo-potentials. The resulting cooling principle is sketched in Fig. 1.2 for the case of rubidium atoms subject to argon buffer gas.

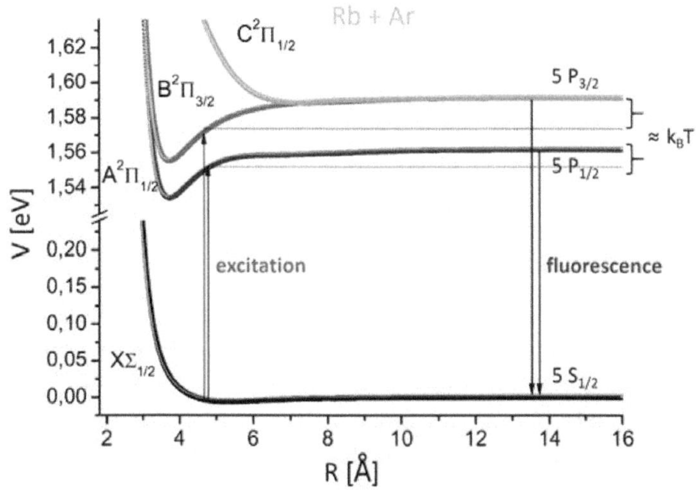

Figure 1.2 Cooling principle for the case of rubidium atoms subject to high-density argon buffer gas, based on the calculated pseudo-potentials [24]. If a rubidium atom collides with an argon atom, absorption of far red-detuned radiation becomes feasible. The formed quasi-molecule is excited, and the excited state subsequently decays under emission of a fluorescence photon mostly at larger interatomic distances, where the energy levels correspond to those of the unperturbed alkali atom. The fluorescence frequency ω_{Fl} is larger than the frequency of the absorbed photon: an amount of energy in the order of the thermal energy, $k_{\mathrm{B}}T$, is extracted from the sample. A cooling of the gas mixture occurs.

For small interatomic distances, a photon can be absorbed also for red detunings of several nanometers. After its $1/e$ natural lifetime, the upper electronic state decays emitting a fluorescence photon. Since the mean frequency of the detected fluorescence is larger than the frequency of the absorbed photon, a cooling of the dense gas mixture is expected as energy is extracted.

Our choice of rubidium atoms as the optically active species becomes now clear. The large elasticity of collisions between the electronically excited states of the rubidium D-lines and noble gas atoms has long been proven [25]. The quasi-molecular potential has a depth of the order of $k_B T$, allowing for a significant part of the atoms' kinetic energy to be extracted. Finally, the cooling transitions are in the near-infrared spectral regime, making them conveniently optically accessible to commercial laser systems.

The cooling power P_{cool} from these experiments can be estimated in a similar way as for anti-Stokes cooling [26, 27] from the mean fluorescence frequency ν_{fl}, the incident power P_{opt}, and the absorption probability of the gas mixture at incident frequency ν, $\alpha(\nu)$.

$$P_{cool} = \alpha(\nu) P_{opt} \frac{\nu_{fl} - \nu}{\nu} \tag{1.1}$$

The typical energy that can be extracted from the sample per cooling cycle is of order of the thermal energy $k_B T$. In the experiments performed so far, we observe cooling powers P_{cool} in the order of up to 100 mW [17, 23, 28, 29]. Note that it is important to choose a detuning with a high-enough absorption probability $\alpha(\nu)$. We are aware that the collisional broadening plays a crucial role here. A measurement of typical fluorescence data recorded with the dense gas mixture will be shown in Section 1.4.1.

Experimentally, it turns out to be crucial that the used buffer gas is of high purity to suppress the influence of concurring heating effects, of which quenching is the most known. In the case of quenching, the excitation energy is transferred to an atom of a third species polluting the dense gas mixture when colliding with the quasi-molecule. Relaxation can then occur via nonradiative decay, and so energy is kept in the system, heating the gas. To reduce possible quenching, an effect known to occur, for example, from an admixture with nitrogen, we use a buffer gas with high purity, namely argon gas preferably with a purity of 99.9999 (argon 6.0).

Residual heating can also be caused by energy pooling. In this scenario, during the collision of two excited alkali atoms one of them gets further excited. Its relaxation to the ground state implicates the emission of a fluorescence photon with higher energy. For the case of rubidium atoms, the reaction equation of that process can be written as [30]:

$$\mathrm{Rb}^* + \mathrm{Rb}^* + \frac{mv_1^2}{2} + \frac{mv_2^2}{2} \rightarrow \mathrm{Rb} + \mathrm{Rb}^{**} + \frac{m\tilde{v}_1^2}{2} + \frac{m\tilde{v}_2^2}{2}. \tag{1.2}$$

The sum of the energies of the single excited rubidium atoms before the collision is higher than the energy of the twice excited atom. The surplus of the energy is thus kept in the system as kinetic energy of the collisional partners [31]. For rubidium atoms subject to a high-pressure noble gas, fluorescence originating from the 6p $\rightarrow$ 5s transition then occurs, and can be detected as blue light at a wavelength $\lambda \approx 420$ nm in the experiments.

At room temperature, the alkali atoms are in the solid phase. To produce a sufficient alkali vapor number density the system is heated to temperatures of up to 680 K, while the buffer gas pressure ranges between 50 bar and 230 bar. Under these conditions, we calculate that the dense gas ensemble consists of approximately 10^{17} cm^{-3} alkali atoms and 10^{21} cm^{-3} noble gas atoms. Then, we can estimate the mean free path length λ_{fp} for a gas of colliding atoms (cross section σ) and number density n

$$\lambda_{\mathrm{fp}} = \frac{1}{n\sigma} \tag{1.3}$$

Let us consider the geometric cross section πd^2, where d denotes the diameter of a rare gas atom, which is twice the van der Waals radius ($d = 2r_{\mathrm{vdW}} = 3.76 \times 10^{-10}$ m for argon [32]). Then, for argon pressure of 200 bar, the mean free path length is in the order of only a few nanometers, $\lambda_{\mathrm{fp}} \approx 2.3$ nm. Following the Boltzmann distribution, still considering a temperature $T = 680$ K, one can also estimate the collisional rate to 2.4×10^{11}/s. Comparing this to the natural lifetime of the first excited alkali atom level (27 ns), we calculate that 10^4 collisions occur within this period of time.

1.3 Experimental Setup and Methods

In the following section, we present our experimental setup in detail, including the techniques used to measure atomic fluorescence and temperature changes inside the gas mixture. We offer insights into the high-pressure cells employed (Section 1.3.1), as well as into the spectroscopic techniques (Section 1.3.2) and the thermal deflection setup (Section 1.3.3).

1.3.1 High-Pressure Cells

For the creation of a high-pressure regime, two different designs of optical cells are employed. Initial experiments were performed with homebuilt cells with volumes of just a few cubic centimeters. The optical access is provided by sapphire windows, which are sealed with Inconel® c-rings into the steel body of the cell and tightly flanged. This design allows for a quick change of the optical windows and, related to this, a change of the inner length in the cell due to the special design we use here. A sketch of the homebuilt high-pressure cell can be seen in Fig. 1.3.

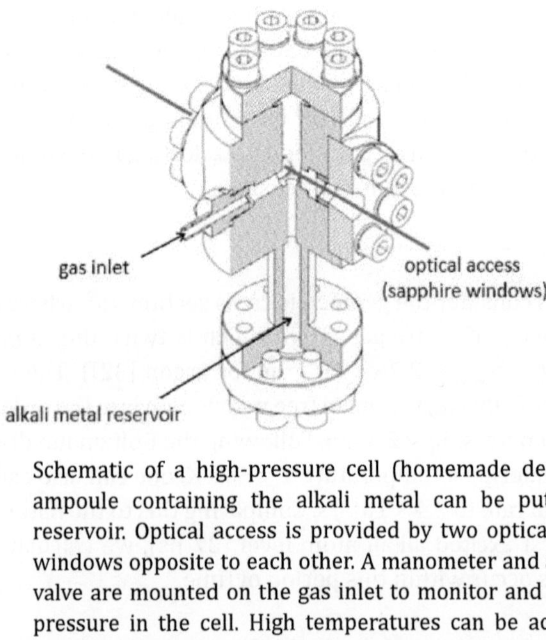

Figure 1.3 Schematic of a high-pressure cell (homemade design). The ampoule containing the alkali metal can be put into the reservoir. Optical access is provided by two optical sapphire windows opposite to each other. A manometer and a pressure valve are mounted on the gas inlet to monitor and adjust the pressure in the cell. High temperatures can be achieved by wrapping heating wires (not sketched for better visibility) around the cell body, enabling the creation of sufficient alkali vapor density.

The second design is commercial, manufactured by SITEC-Sieber Engineering AG. In this case, the sapphire windows are mounted following a Bridgman sealing technique [33]. The different sealing mechanism offers a better long-term stability of several months. Buffer gas pressures of several hundred bar and temperatures of up to 720 K can be achieved using both cell designs. A manometer, placed on the gas inlet of each cell, allows for constant monitoring of the pressure, ensuring that no leaks appear.

A process including ultrasound cleansing and a bake-out period is followed in order to purify the cells, before an ampoule containing 1 g of alkali metal is placed in the reservoir. Subsequently, argon gas is filled in, and the cells are heated by wrapping electrical heating belts around them. In this manner, the alkali atoms are vaporized in a high–buffer gas environment acquiring sufficient optical densities that allow us to begin our experimental efforts.

1.3.2 Spectroscopic Setup

For the excitation of the dense mixture, we employ emission of a continuous-wave Ti:sapphire ring laser that is guided through the cell volume. This type of laser system also allows for an easy change of the incident wavelength, exploited in energy dependence studies of the fluorescence.

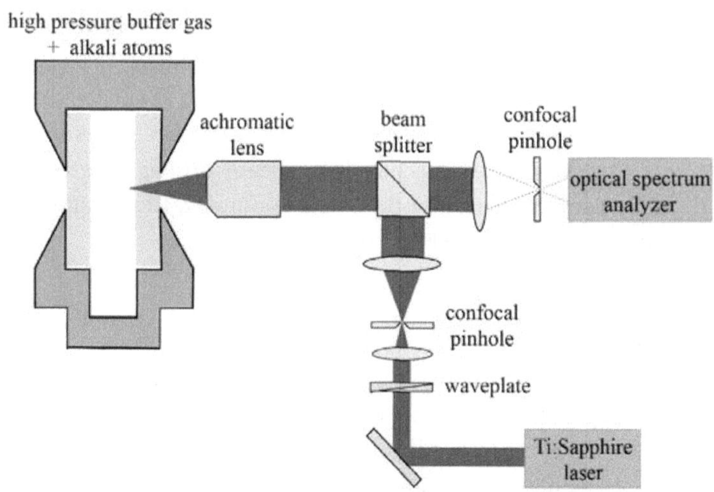

Figure 1.4 Fluorescence detection setup. The incident Ti:sapphire laser beam is spatially filtered and tightly focused into the high-pressure cell. In the back direction, atomic fluorescence is collected and filtered and afterward detected by an optical spectrum analyzer.

For most experiments, both the incident laser beam and the emitted fluorescence were spatially filtered with confocal pinholes. The incoming laser beam was strongly focused to a beam waist of approximately 10 μm. Fluorescence was then collected from that area and analyzed by an optical spectrum analyzer. A sketch of that setup is depicted in Fig. 1.4.

1.3.3 Thermal Deflection Spectroscopy

To determine the temperature change in the dense mixture induced by the cooling laser beam in a quantitative way, thermal deflection spectroscopy [34–36] is employed. The basic idea behind this technique is that a temperature change in the gas produces a corresponding change of density, which in turn induces a variation of the refractive index. This variation is investigated by irradiating a second, nonresonant laser beam collinear to the cooling laser beam into the dense mixture. The cooled gas acts like a graded index lens, causing the probe laser beam to be deflected and consequently to change its propagation direction. By measuring the deflection angle as a function of distance between the two laser beams, the deflection profile can be mapped. In the following, we deduce the relation between the deflection angle and the temperature change according to the work of Ref. [35]. A basic sketch of the concept is shown in Fig. 1.5. Considering a Gaussian laser beam propagating through a spatially varying profile of the refractive index, the change of its propagation direction can be written as follows:

$$\frac{d}{dz}\left(n_0\frac{d\vec{r}}{dz}\right)=\nabla_{\perp}n(\vec{r},t) \tag{1.4}$$

where r is the transversal displacement of the probe beam to its original propagation direction, n_0 denotes the refractive index of the gas mixture at normal conditions, and $\nabla_{\perp}n(\vec{r},t)$ is the gradient of the refractive index orthogonal to the propagation direction [37].

Integration of Eq. 1.4 leads to a relation of the (measurable) deflection angle φ and the spatial change of the refractive index

$$\varphi=\frac{d|\vec{r}|}{dz} \tag{1.5}$$

$$= \frac{1}{n_0} \int \nabla_\perp n(\vec{r},t) dz \tag{1.6}$$

$$= \frac{1}{n_0} \int \frac{dn}{dr} dz \tag{1.7}$$

At this point we consider the deflection angle φ in radial direction r to be small ($dr/dz = \tan\varphi \approx \varphi$). We derive an expression for the temperature change as a function of the deflection angle by reshaping Eq. 1.7:

$$\varphi = \frac{1}{n_0} \int \frac{dn}{dr} dz \tag{1.8}$$

$$\varphi = \frac{1}{n_0} \int \frac{dn}{dT} \frac{dT}{dr} dz \tag{1.9}$$

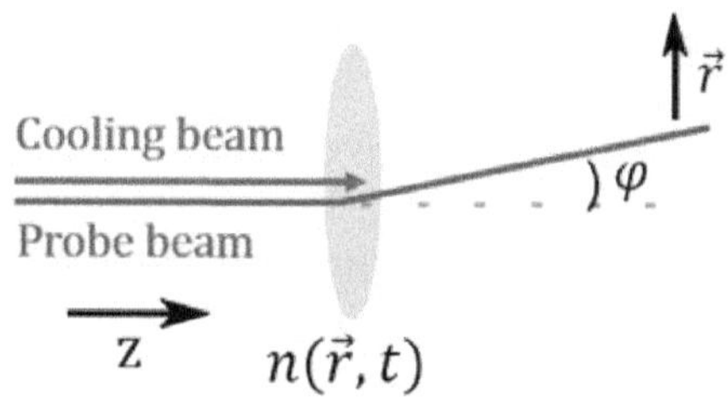

Figure 1.5 Thermal deflection technique. The cooling beam and the so-called probe beam are sent collinearly into the dense gas mixture. The cooling zone induces changes to the gas density and consequently to the refractive index. As a result, the cooling zone acts like a graded index lens for the nonresonant probe beam. By scanning the radial distance between the two beams, the angular deflection φ can be measured as a function of r.

Regarding Eq. 1.9, we can further simplify dn/dT in order to obtain an expression that is only a function of experimental parameters. According to the work of Spear et al. [36], dn/dT is in general determined by two different mechanisms, the thermal expansion and a thermal shift of the absorption bands of the media. The choice of the nonresonant probe laser beam makes it feasible to suppress the influence of absorption bands. We can simplify Eq. 1.9 by considering the Lorentz–Lorenz relation [38]

$$\frac{1}{\rho} = \left(\frac{n^2 - 1}{n^2 + 2}\right) = \text{const.} \tag{1.10}$$

where ρ is the density of the medium. Assuming that the dense gas mixture behaves like an ideal gas, its refractive index n is close to unity. The temperature derivative of Eq. 1.10 can then lead to the following expression for dn/dT:

$$\frac{dn}{dT} \approx -\frac{(n-1)}{T} \tag{1.11}$$

Combining Eqs. 1.9 and 1.11, we obtain

$$\varphi = -\int \frac{(n-1)}{T} \frac{dT}{dr} dz \tag{1.12}$$

To get a rough estimation for the analysis of the measured data, we now consider the fraction $(n - 1)/T$ in Eq. 1.12 to be constant along the propagation direction z

$$\varphi = -\frac{(n-1)}{T} \int \frac{dT}{dr} dz \tag{1.13}$$

This assumption holds for small temperature changes compared to the initial temperature of the dense gas mixture. If we further assume that the cooling laser beam has a Gaussian intensity profile and that the thermal energy is removed with a rate comparable to its intensity distribution, the relative temperature change $\Delta T(r, z)$ can be expressed as follows:

$$\Delta T(r, z) = \Delta T(r, 0) \cdot e^{-\alpha z} \tag{1.14}$$

Hence, as a function of experimental parameters, one derives for ΔT_{exp}:

$$\Delta T_{\text{exp}}(r, z) = \frac{T}{(n-1)} \cdot \frac{\alpha e^{-\alpha z}}{(1 - e^{-\alpha L})} \int_r^{\infty} \varphi(r') dr' \tag{1.15}$$

where r denotes the transverse displacement between cooling and probe laser beam, L is the inner length within the high-pressure cell, T the initial temperature of the dense gas mixture, n is the refractive index of the buffer gas under the particular experimental conditions (p, T) for the wavelength of the probe laser beam, and α denotes the absorption coefficient of the gas and is considered constant for the analysis of the measured deflection data. In our present model, heat transport is considered to occur only in the radial direction because the absorption length, $1/\alpha$, is much longer than the laser focal radius w_0. The heat transfer is described by

$$\nabla(\kappa\nabla T) = -Q(r, z) \tag{1.16}$$

where κ is the thermal conductivity of the buffer gas. The cooling source $Q(r, z)$ of a Gaussian laser beam can be expressed as a function of the cooling power P_{cool}, the cooling beam focal radius w_0, and the absorption coefficient α [39, 40]:

$$Q(r,z) \propto \alpha \cdot P_{\text{cool}} \cdot e^{-2r^2/w_0^2} \cdot e^{-\alpha z} \tag{1.17}$$

Inspection of Eq. 1.16 indicates that the temperature change inside the gas is a function of the thermal conductivity of the buffer gas. Experiments with heavier rare gases such as krypton and xenon are thus supposed to allow for larger temperature changes compared to the so far mainly used argon. In the past, this has often proved problematic as both krypton and xenon are hardly available in purities of 6.0. We have nevertheless achieved first encouraging results with both species as buffer gas [41]. For our experiments, we use a He–Ne laser as a probe beam (wavelength $\lambda \approx 633$ nm). We are positive that this wavelength is far enough nonresonant with the dense gas mixtures of our experiments (rubidium and argon/helium/krypton/xenon as well as potassium–argon mixtures). Furthermore, the choice of a probe laser wavelength in the visible range strongly facilitates the alignment of the setup that can be seen in Fig. 1.6. The Ti:sapphire laser beam is irradiated into the dense gas mixture, while the collinear He–Ne laser beam can be laterally moved by a beam splitter, allowing for a controllable variation of the distance between the two beams. The angular deflection is measured by a position sensitive photodiode behind the cell. To make sure that only the probe beam is detected, an infrared filter is placed directly in front of the photodiode. Finally, a mechanical chopper on the cooling beam path ensures that angular deflection measurements do not entail offsets, while simultaneously allowing for the observation of deflection statistics at each lateral offset position.

In most of the experiments, the width of the refractive index profile is scanned over a few millimeters in 50–100 μm steps. Experiments are conducted with both collimated and focused beams. In Fig. 1.6 the latter case is presented. The beam waist w_0 of the probe beam is set to be smaller than that of the cooling laser beam so that the entire cooling area can be resolved.

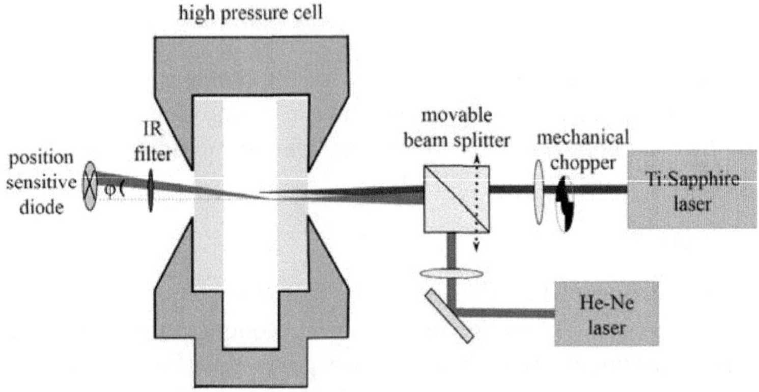

Figure 1.6 Experimental cooling setup incorporated with thermal deflection technique. A movable beam splitter allows for scanning the lateral offset between the cooling beam (Ti:sapphire laser) and the nonresonant probe beam (He-Ne laser). The angular deflection is measured as a function of position by a position sensitive diode. An infrared filter blocks the residual cooling beam behind the cell, while a mechanical chopper allows for measuring offset-free deflection statistics.

We discuss results on thermal deflection spectroscopy measurements in Sections 1.4.2 and 1.4.3 for the case of rubidium and potassium atoms both in dense argon buffer gas.

1.4 Laser Cooling Experiments on Dense Mixtures

In this section, we lay focus on laser cooling experiments of dense atomic mixtures. We begin by presenting a fluorescence spectrum which exhibits a clear redistribution of radiation (Section 1.4.1), before discussing thermal deflection spectroscopy measurements on dense rubidium–argon ensembles (Section 1.4.2). We conclude with the cooling results of dense potassium–argon mixtures (Section 1.4.3).

1.4.1 Fluorescence of a Dense Rubidium–Argon Mixture

For the case of rubidium atoms subject to 160 bar argon atoms at a temperature T = 580 K, fluorescence spectra are recorded for an incident laser power of 100 mW at a wavelength $\lambda \approx 810$ nm using the setup described in Section 1.3.2. The recorded fluorescence as a function of wavelength is presented in Fig. 1.7. A clear redistribution of fluorescence toward the rubidium D resonances can be observed. Residual scattered light at the incident wavelength is also detected by the spectrometer.

Investigation of the atomic resonances leads to the conclusion that they are both pressure and saturation broadened. The linewidths of the two sharper resonances visible on top the broad envelope are $\Delta\nu'(D1)$ = 3.4 THz, $\Delta\nu'(D2)$ = 5.9 THz, respectively, as derived from Lorentzian fits. Early work for pressure broadening of alkali atoms by noble gases at roughly similar noble gas pressures has been performed by Shang-Yi [42].

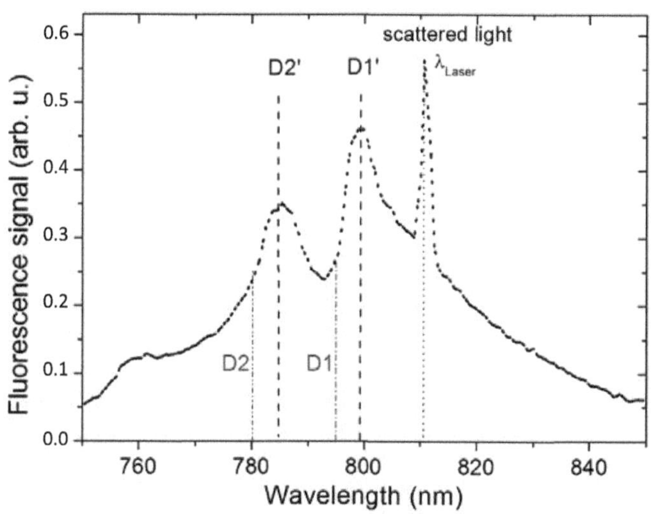

Figure 1.7 Recorded fluorescence of rubidium atoms subject to 160 bar argon buffer gas. Both rubidium D transitions are spectrally resolved. Compared to the unperturbed resonances $D1$, $D2$, a strong shift is observed (center wavelengths labelled with $D2'$, $D1'$, and dashed lines). Due to imperfect filtering, residual scattered light of the incident laser beam at $\lambda \approx 810$ nm is also visible. A clear redistribution toward the D-line resonances is observed.

The resonances are also shifted to higher wavelengths, $\lambda(D1') \approx 800$ nm, $\lambda(D2') \approx 785$ nm, when compared to the undisturbed resonances at $\lambda(D1) = 795$ nm, $\lambda(D2) = 780$ nm [43]. The recorded spectrum is asymmetric toward the center wavelength of the fluorescence. The red wing of the broad envelope, in the region of higher wavelengths, stretches over a significantly higher range than the blue wing. This is in good agreement with a quasi-static approach of the pressure-broadened rubidium argon quasi-molecules, where the red wing is attributed to the $A^2\Pi_{3/2} \rightarrow X^2\Sigma_{1/2}$ transition [44]. The spectrum presented clearly shows redistribution of radiation, which is a prerequisite for the observation of cooling in the dense gas mixture.

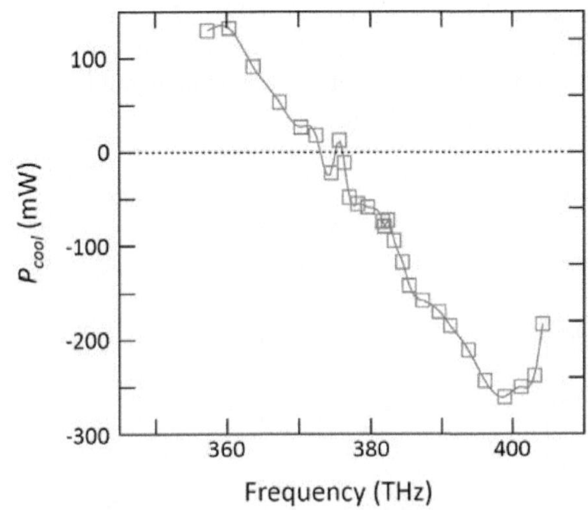

Figure 1.8 Calculated cooling power P_{cool} (represented by squares) as a function of excitation energy, using the differential energy, $(\nu_{\mathrm{fl}} - \nu)/\nu$, extracted directly from fluorescence spectra. The solid line is used as a guide to the eye.

Further investigation involves recording a number of these spectra using different excitation energies. In this manner, the corresponding mean fluorescence ν_{fl} can be determined directly from the spectra, and thus, an estimation of the cooling power P_{cool} becomes feasible through Eq. 1.1. Corresponding results are presented in Fig. 1.8, where the calculated cooling power (squares) is presented as a function of the excitation energy [17]. We observe that the value of zero detuning, corresponding to zero cooling power, is located at 375 THz. Hence, we expect that for lower excitation frequencies (red detunings), the dense gas mixture will be cooled, while for higher (blue detunings) it will be heated. In the subsequent sections we review results on the redistribution laser cooling, which verify this prediction with small deviations attributed to additional heating effects due to quenching.

1.4.2 Cooling of Dense Rubidium–Argon Mixtures

The addressed cooling transitions of rubidium and potassium are spectrally close. As rubidium allows for a higher number vapor density at a certain temperature compared to potassium [45, 46], we initially focus on experiments with the heavier alkali species solved in rare gases.

In the proof of principle experiment [17], a 2 W cooling beam (diameter of the collimated beam was $2w_0 \approx 1$ mm) at a wavelength of $\lambda \approx 820$ nm is irradiated into the high-pressure cell containing rubidium at an initial temperature of 620 K under 230 bar of argon.

Initial evidence of cooling can be obtained by collecting the blackbody radiation of the entrance sapphire window with the use of an infrared camera. In the spectral regime between 7.5 and 13 μm, the sapphire window is opaque and, unlike the gaseous ensemble, its temperature can be measured from its blackbody radiation. The cooling result, following a 30 s optical cooling period, of a $\lambda = 820$ nm laser is presented in Fig. 1.9. The thermal drop induced by the cooling beam in the gas is depicted in the temperature profile of the outer sapphire window surface. The measurement indicates a temperature drop of $\Delta T = -(0.31 \pm 0.03)$ K near the beam center. The high thermal conductivity of sapphire ($\kappa_{\text{sapphire}} = 200$ W cm^{-1} K^{-1}) reduces the here observable cooling. Note that the sapphire window is in contact with the cell's metal body, which is heated due to absorption of scattered fluorescence photons. Nonetheless, the thermocamera image represents a clear qualitative evidence of laser cooling based on collisional redistribution.

Thermal deflection measurements are then performed to accurately determine the temperature change in the gas induced by the cooling beam. The He–Ne laser beam is moved collinearly with respect to the cooling beam in steps of 250 μm with the help of a beam splitter, as shown in Fig. 1.7. The deflection angle is measured at each position. In this way, the temperature profile induced by the Ti:sapphire laser beam is resolved in the radial direction. The probe beam scanning proceeds up to an offset were no deflection is observed.

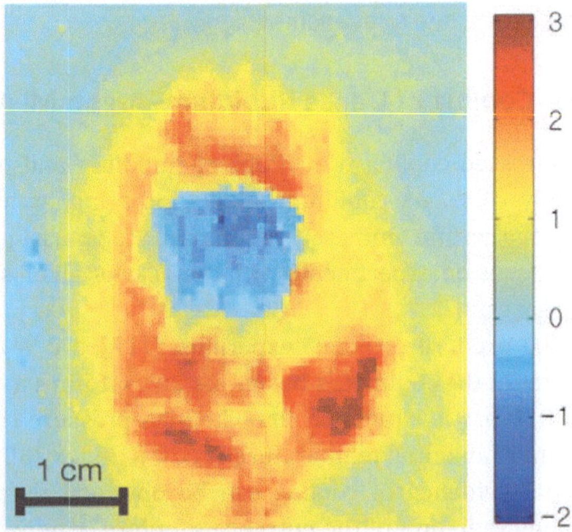

Figure 1.9 Average color-coded temperature change after a 30 s optical cooling period, as measured using an infrared camera directed at the entrance sapphire window. The blue region in the center is the sapphire window, cooled by the underlying gas ensemble. The surrounding metal flange appears red and yellow due to constant heating, attributed to absorption of scattered fluorescence photons.

At the position where the two beams are perfectly superimposed, a symmetric deflection profile is monitored by the position sensitive photodiode. For any other spatial offset, an asymmetric distribution of the measured deflection angles is observed. We therefore fit to the data the heat equation of a temperature gradient as a function of the thermal diffusivity D, the time t, and the beam waist w_0

$$\frac{dT}{dr} \propto \left(\exp\left(-\frac{2r^2}{w_0^2} \right) - \exp\left(-\frac{2r^2}{8Dt + w_0^2} \right) \right) \tag{1.18}$$

similarly to previous work [39]. Note that the constant of proportionality of Eq. 1.18 is determined by the cooling power P_{cool} and the thermal conductivity κ. The actual fitting function $f(r)$ is

$$f(r)=\frac{dT}{dr}=\left(\frac{a}{r-b}\right)\left(\exp\left(-\frac{2(r-b)^2}{w_0^2}\right)-\exp\left(-\frac{2(r-b)^2}{d+w_0^2}\right)\right) \tag{1.19}$$

containing a, b, d, and the beam waist w_0 as fitting parameters. Integration of this devolution leads then to the temperature expression as a function of the angular deflection $\varphi(r)$. As several assumptions were made for deducing this relation, we give an error on the measured values (position of the probe laser beam, photodiode deflection signal) but forego an error estimation of the temperature change. The temperature change calculated out of the experimental deflection data will in the following be denoted as ΔT_{exp}. For this calculation, we assume the absorption α to be constant over the inner cell length.

The results of this measurement are depicted in Fig. 1.10. The experimentally measured deflection angles (red spheres) are numerically integrated. This gives a temperature profile as shown by the solid blue line, with a maximum determined temperature drop of $\Delta T_{\text{exp}} = -66$ K near the cell entrance. For comparison, the deflection data are fitted assuming a temperature profile from a heat transport model, giving a temperature profile as shown by the dashed blue line. Here, the cooling power P_{cool} and the thermal diffusivity of the gas, D, were left as free parameters for the fit.

It is worth noting that temperature changes corresponding to a factor of 2 larger than the one measured in the proof-of-principle experiment have been observed in subsequent experiments [41]. This is attributed to the higher absorption coefficients and smaller beam waists used, corresponding to higher excitation intensity.

Further experiments with rubidium as optically active species include other noble gases as a buffer gas. In the work of Vogl et al. [47], experiments with helium as a buffer gas are described. The observed temperature change is smaller in these experiments. This is in agreement with expectations, since helium has a higher thermal conductivity than argon.

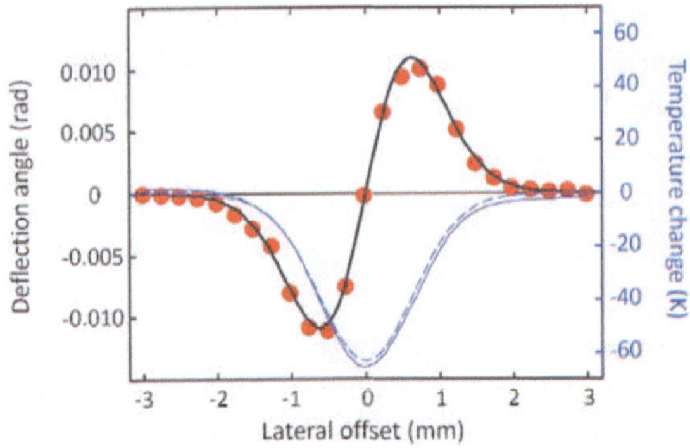

Figure 1.10 Deflection angle (red spheres) of the He–Ne laser beam as a function of its lateral offset to the cooling beam for a dense rubidium-argon mixture (p = 230 bar, T = 620 K). A fit (solid black line) on the measured data is performed following Eq. 1.19. The temperature profile near the cell entrance derived from the deflection data (solid blue line), as well as following a heat transport model (dashed blue line) are in good agreement. The maximum determined temperature drop is 66 K.

A rough estimation based on the ideal gas law indicates an optimum set of parameters for the cooling experiment. When the temperature T decreases, the density $\rho = N/V$ increases, because the pressure p in the cooling region is considered constant, in agreement with the work of Jackson et al. [35].

This affects the fluorescence spectra in a way that the mean fluorescence starts shifting to the red. When its value becomes equal to the incident laser frequency, the cooling process is expected to stop. Therefore, it can be concluded that there is an optimum set of values for both pressure and wavelength. Experiments examining the redistributed fluorescence as function of pressure [18] also give rise to this fact.

We are also currently carrying out further data analysis of measurements in which a higher temperature change than the one reported above is observed [41, 48]. However, the model we presently use for the data analysis neglects possible temperature-dependent changes of the absorption coefficient α. This is expected to become an important factor for temperature changes that are of the same order of magnitude as the absolute temperature of the ensemble.

Another approach for a determination of the temperature change can be reached by monitoring the thermal lens induced by the temperature change using a Shack–Hartmann wavefront detector [49, 50]. Thus, a new high-pressure cell with four optical windows will in the future allow for a determination of a 3D temperature distribution as the cooling region will then be accessible perpendicularly to the laser beam propagation direction.

1.4.3 Experiments on Potassium–Argon Mixtures

We extended our investigation on the redistribution cooling technique by performing measurements using potassium atoms as the optically active species. The fine structure splitting of the excited state of the potassium D resonances is smaller than that of rubidium (3 nm compared to 15 nm [51]). This smaller splitting is expected to better simulate a two-level system, facilitating the interpretation of the cooling results. For these experiments [29], the high-pressure cell has to be heated to its temperature limit (T = 720 K) in order to produce a sufficient particle density. The potential curves for the potassium-argon quasi-molecule are comparable to those of the rubidium-argon system [52, 53].

As a first step for the cooling of potassium-argon mixtures, redistribution of fluorescence has to be proven. Therefore, a setup as the one described in Section 1.3.2 is used to record the fluorescence of the dense mixture. Typical potassium spectra recorded at T = 713 K, under 200 bar of argon, for both red- and blue-detuned excitation wavelengths are depicted in Fig. 1.11. The incident laser power is $P_{\text{Laser}} \approx 600$ mW and the beam is focused down to a waist of $2w_0 \approx 6$ μm. Similarly to the rubidium fluorescence spectrum shown in Section 1.4.1, clear redistribution toward the center of the $D_{1,2}$ resonances, accompanied by strong spectral shift and broadening is observed for both measurements.

More specifically, the resonances are shifted by 3–4 nm to the red compared to the unperturbed case. Residual scattered light at the spectral position of the incident light is also present due to imperfect filtering. On the red wing of the spectrum, the fluorescence level is much higher than on the higher-energy blue wing. We attribute this to satellite resonances of the potassium-argon quasi-molecule in agreement with the work of Kuhn [54] and Kielkopf [55].

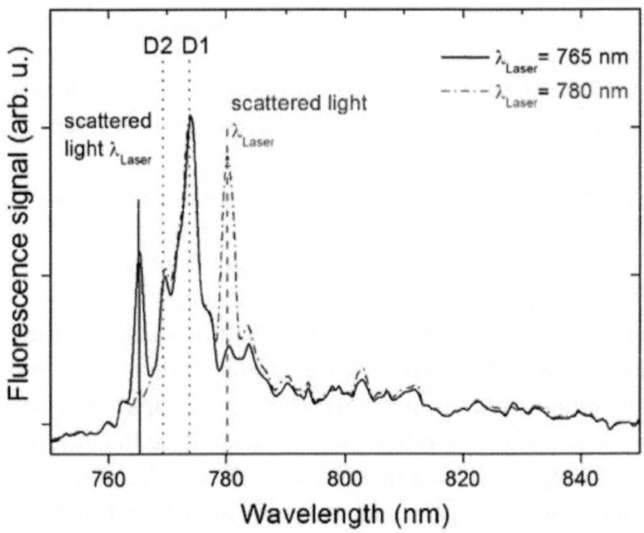

Figure 1.11 Fluorescence of a dense potassium–argon mixture at a pressure of 200 bar and a temperature $T = 713$ K. The dashed line denotes the signal recorded for a red-detuned laser at $\lambda = 780$ nm and the solid line for a blue detuning at $\lambda = 765$ nm. Vertical lines of identical notation are used to indicate the spectral position of the incident light in each case. The $D_{1,2}$ atomic resonances are also depicted (vertical dotted lines).

For different incident wavelengths, red-detuned to the potassium D resonances, thermal deflection spectroscopy measurements were conducted, monitoring the deflection profile in steps of 50 μm. A temperature change of $\Delta T_{exp} \approx -120$ K was observed for an incident wavelength of $\lambda \approx 785$ nm and an incident power of $P_{Laser} \approx 1.4$ W. This measurement is depicted in Fig. 1.12. Here, the cooling laser beam is focused down to a beam diameter of 0.5 mm. The large temperature change observed in this experiment is mainly attributed to the high absorption, producing an absorption length of $l_{abs} = 1.2$ mm.

Using the fluorescence spectrum for the incident wavelength ($\lambda = 785$ nm), we can determine the detuning in Eq. 1.1. In this manner, the cooling power is estimated in the order of 30 mW, which corresponds to a cooling efficiency of 2%.

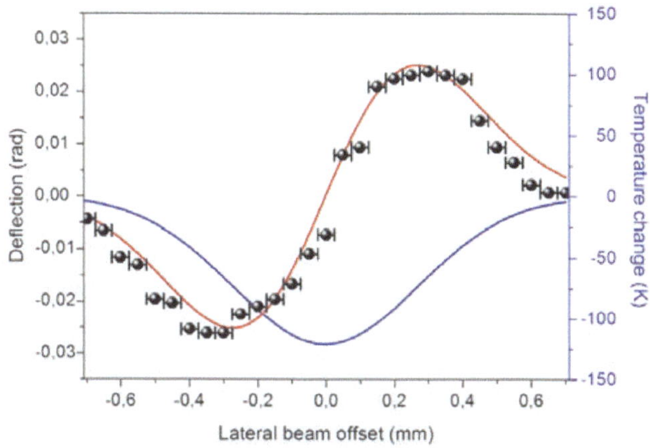

Figure 1.12 Thermal deflection measurement for a dense potassium-argon mixture at an initial temperature T = 713 K and a pressure of 200 bar. The deflection angle (black spheres) of the He–Ne laser beam is shown as a function of its lateral offset to the cooling beam. A fit (red line) on the measured data is performed following Eq. 1.19, while integration of the latter yields the relative temperature change (blue solid line). A temperature drop by 120 K is observed.

1.5 Kennard–Stepanov Experiments

In the following section we present an alternative approach in temperature determination of the dense gaseous ensembles, based purely on spectroscopic data. More specifically, we introduce the Kennard–Stepanov relation, a Boltzmann-type frequency scaling connecting the emission and absorption profiles of the investigated transitions. We begin by offering insight into the theoretical background of the relation and subsequently present experiments focused on temperature determination of the gaseous ensembles, based on the spectral profiles of the rubidium $D_{1,2}$ atomic transitions.

1.5.1 The Kennard–Stepanov Relation

The investigation of an inherent connection between the absorption and emission spectral profiles of transitions in strongly interactive media has been a subject of yearlong research. Following the efforts of Kennard [56] and Stepanov [57], such a universal relation has been derived, applying for systems where efficient relaxation processes lead to thermalization of both the ground and the electronically excited state manifolds of the addressed transition. The so-called Kennard–Stepanov relation can be written as follows:

$$\ln\left(\frac{\alpha(\omega)}{f(\omega)}\right)=\frac{\hbar}{k_{\mathrm{B}}T}\omega+C(T) \tag{1.20}$$

where $\alpha(\omega)$ and $f(\omega)$ are the frequency-dependent absorption and emission line shapes, respectively, T is the so-called spectral temperature and $C(T)$ is a temperature-dependent constant. Its derivation is based on the Boltzmann distribution of the occupation probabilities within both the ground and the electronically excited state manifolds and can be found elsewhere [58]. In the case of fully thermalized state manifolds, the equation predicts a linear behavior of the logarithm versus frequency, with a slope equal to $\hbar/k_{\mathrm{B}}T$. Linearity deviations are mainly attributed to incomplete thermalization of the excited state manifold, internal state conversion, and finite quantum efficiency of the addressed transition. Extraction of the slope using spectroscopic data of absorption and emission profiles can lead to temperature determination of the sample. The method has already been used for studies on dye molecules in liquid solutions [59, 60], solid state inhomogeneous systems [61, 62], and biochemical photosystems [63], as well as thermalization studies of a photon gas [64]. Even though an exact temperature extraction has not been achieved in these experiments, mainly due to inhomogeneous broadening [65, 66], the thermalization of the relevant states has been verified through the linearity of the corresponding Kennard–Stepanov plots. It is however expected that in high-purity gaseous systems this contribution is strongly suppressed and an exact temperature determination becomes possible. To this end, the relation is applied to our dense rubidium–argon gas mixtures.

Before proceeding to the corresponding experiments, we note the particular conditions under which the validity of the relation is ensured. Even though the ground-state manifold of a transition is quite often thermalized, for dilute gases under laser irradiation this is not usually the case for the electronically excited state. Most importantly, the thermalization rate is necessary to be much faster than the radiative lifetime of the excited state, as only in this case a thermal distribution can arise in the upper electronic state manifold before radiative decay takes place. A further requirement for the Kennard–Stepanov relation is the quantum efficiency of the transition, which needs to be near unity.

The universality of the Kennard–Stepanov relation is proven by its independence regarding the relaxation processes that lead to thermalization of the excited state manifold. In solids, interactions with phonons cause such a thermal distribution, whereas in liquid or gaseous samples it is frequent collisions that play that role. In our high-pressure gas ensembles, the number of atomic collisions within the radiative lifetime (27 ns) of the $D_{1,2}$ rubidium transition is of the order of 10^4 [67]. The latter, combined with the extremely elastic nature of the collisions between rubidium and noble atoms [25], leads to efficient thermalization of the electronically excited quasi-molecular manifold.

1.5.2 Kennard–Stepanov Spectroscopy

Our samples are prepared in the same manner as described in Section 1.3.1. In these spectroscopic measurements a three-window cell is utilized, while the rubidium optical density is kept low to suppress reabsorption and preserve quantum efficiency close to unity [25]. We note that here, due to the low optical density requirement, we aim in extracting the ambient temperature of the ensemble. For this reason, the Ti:sapphire laser beam has a diameter of 3 mm, while its power is kept below 1 W, in order to suppress additional heating or cooling effects. The implications of these requirements are discussed in Section 1.5.3.

For a complete Kennard–Stepanov study of the rubidium $D_{1,2}$ transitions [58, 68], pairs of absorption and fluorescence spectra are recorded in order to produce the relevant plots. We collect the laser-induced fluorescence in a perpendicular direction with regard to the laser propagation and analyze it spectrally using a grating spectrometer. The absorption spectrum, however, cannot be accurately obtained using regular white light sources, due to the low optical density of the rubidium vapor. For this reason here, we employ the method of measuring the fluorescence yield. More specifically, the absorption spectrum can be reconstructed by recording the fluorescence power integrated over the emission spectrum and plotting it versus the incident laser wavelength. In this manner, we obtain a signal proportional to the absorption coefficient. The fluorescence yield is additionally tested for different excitation powers in order to verify that the laser intensity is below saturation. Furthermore, fluorescence measurements using different excitation wavelengths allow us to verify that, in the high-pressure regime, the emission's spectral profile is largely independent of the excitation energy. Such a result, also known as Kasha's rule [69], is depicted in Fig. 1.13.

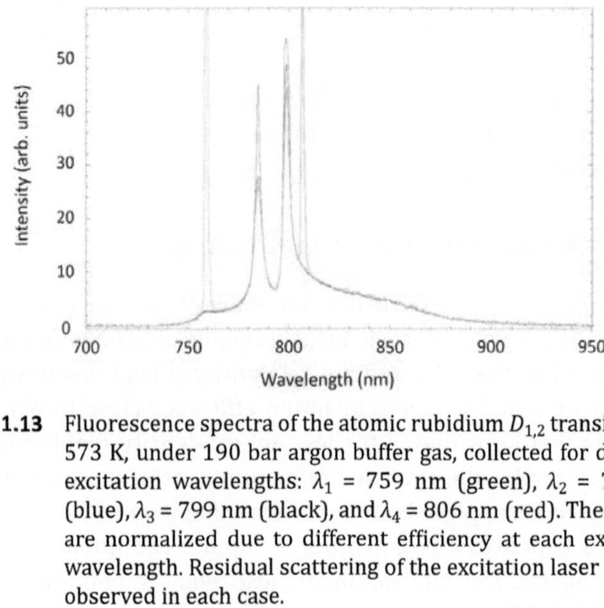

Figure 1.13 Fluorescence spectra of the atomic rubidium $D_{1,2}$ transitions at 573 K, under 190 bar argon buffer gas, collected for different excitation wavelengths: λ_1 = 759 nm (green), λ_2 = 785 nm (blue), λ_3 = 799 nm (black), and λ_4 = 806 nm (red). The spectra are normalized due to different efficiency at each excitation wavelength. Residual scattering of the excitation laser beam is observed in each case.

A measure of the redistribution that occurs within the excited state quasi-molecular manifold can be also given through a pressure dependence of the difference of the observed relative fluorescence signals at the positions of the rubidium D_2 and D_1 lines $f_{\lambda_2}(D_2)/f_{\lambda_2}(D_1)$ – $f_{\lambda_1}(D_2)/f_{\lambda_1}(D_1)$. This ratio is averaged over different excitation wavelength ranges λ_1 = 800 – 820 nm and λ_2 = 760 – 780 nm, which favorably populate the D_1 and D_2 resonances, respectively. Figure 1.14 shows that for high-buffer gas pressures (above 100 bar), the ratio difference takes values below 4%, verifying strong redistribution of the excited state manifold due to the increased number of collisions.

Furthermore, the finer line shape details become wavelength independent in agreement with previous observations of our group [17, 28].

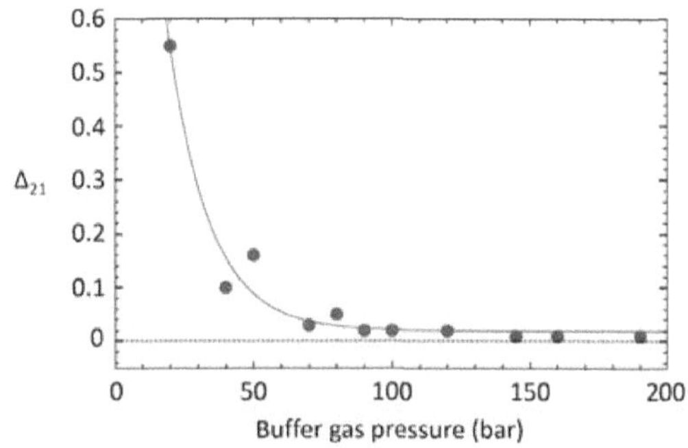

Figure 1.14 Pressure dependence of the difference of the relative collected fluorescence intensity at the position of the rubidium D_1 and D_2 resonances (represented by dots), $\Delta_{21} = f_{\lambda_2}(D_2)/f_{\lambda_2}(D_1) - f_{\lambda_1}(D_2)/f_{\lambda_1}(D_1)$, averaged over excitation wavelength ranges λ_1= 800–820 nm and λ_2= 760–780 nm. An exponential fit is used as a guide for the eye (solid line).

We collect pairs of absorption and fluorescence spectra for a wide range of experimental parameters (p, T) of the $D_{1,2}$ rubidium transitions. Figure 1.15a depicts such a pair obtained at 420 K, under 180 bar of argon buffer gas. In both the pressure-broadened absorption (dashed line) and fluorescence (solid line) spectral profiles the $D_{1,2}$ resonances are clearly resolved, overlapping with a broad spectral wing which extends between 730 nm and 940 nm. A blue satellite (labelled S) is also visible in both cases. Interestingly, the spectral profiles exhibit an exactly opposite asymmetry, with the absorption (fluorescence) having a more pronounced short (long) wing, similar to the so-called mirror rule observed in dye molecule solutions [70]. We plot the corresponding Kennard–Stepanov relation in Fig. 1.15b, where the logarithmic ratio of the absorption over the fluorescence spectra is depicted versus the optical wave number. The experimental data (dots) exhibit a clearly linear behavior, as predicted in Eq. 1.20, for a range that well surpasses $4k_BT$. Small deviations are observed in the vicinity of $D_{1,2}$, attributed to incomplete redistribution, as well as for extreme detunings, where the collected signal is very weak, yielding a low experimental signal-to-noise ratio. We perform a linear fit (solid line) and extract the slope hc/k_BT, with T corresponding to the cell's temperature.

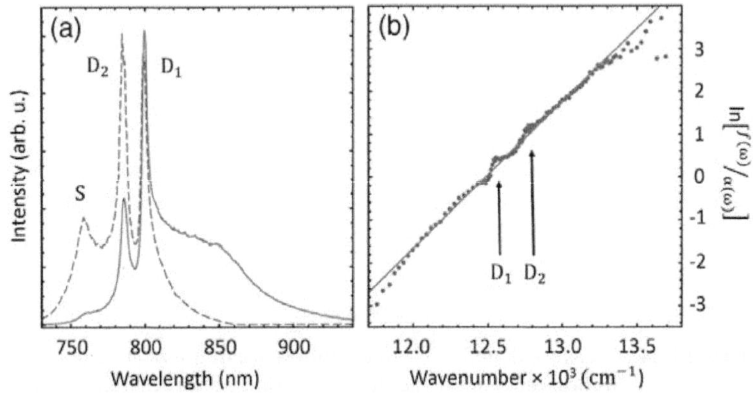

Figure 1.15 (a) Rubidium atomic $D_{1,2}$ experimental absorption (dashed line) and fluorescence (solid line) spectral profiles collected at T = 420 K, under p = 180 bar of argon buffer gas. The atomic resonances ($D_{1,2}$) and the blue satellite (S) are noted. (b) The corresponding Kennard–Stepanov plot depicting the logarithmic ratio of the experimental spectra (dots). A linear fit (solid line) with a slope hc/k_BT is used as guide for the eye.

Experiments performed using lower buffer gas pressures (below 100 bar) exhibit also this linear behavior. However, the observed deviations around the resonances and at the extremes become significantly higher, emphasizing the role of frequent collisions in the thermalization process. More importantly, the requirement for a high collisional rate is compatible with the experimental conditions under which redistributional laser cooling becomes feasible (see Section 1.2.2).

More recently, a theoretical and experimental analysis of the cooling efficiency achieved with the redistribution technique for media that fulfills the Kennard–Stepanov relation has been carried out [71]. The calculations well reproduce an observed asymmetry of the laser cooling and heating efficiencies for red and blue laser detunings from the pressure broadened rubidium resonance, respectively, with the efficiency for laser heating (for blue detuning) being larger than that observed for laser cooling (for red detuning). This agrees with the intuitive expectations that to heat is easier than to cool. The results can also be qualitatively understood from the asymmetry between the absorption and the emission line shapes for the pressure broadened gas predicted by the Kennard–Stepanov relation.

1.5.3 Prospects of the Kennard–Stepanov Technique

The experiments described in Section 1.5.2 highlight the potential of Kennard–Stepanov spectroscopy as a reliable noncontact technique for temperature determination. However, the main issue that arises by the use of the atomic $D_{1,2}$ transitions is that while a high optical density is necessary for the redistributional cooling experiment, this is not compatible with the requirements for a simultaneous Kennard–Stepanov study. More specifically, it is expected that high optical density conditions favor strong reabsorption effects, limiting the quantum efficiency of the transitions. This obstacle may be overcome with the use of auxiliary transitions featuring significantly lower oscillator strengths. Figure 1.16 shows a broadband absorption spectrum in the dense gaseous ensemble with additional atomic and molecular transitions [41]. Here, the absorption coefficient of the $D_{1,2}$ resonances exhibits strong saturation corresponding to an optical density sufficiently high to allow for redistributional cooling. However, the existence of additional atomic and molecular absorption bands featuring significantly lower values fulfills the prerequisite for a simultaneous Kennard–Stepanov investigation. It should be also pointed out that in such dense ensembles energy pooling effects occur [30], resulting also in the emission of fluorescence originating from shorter wavelength transitions when addressing a cooling laser beam to near the D-line spectral resonances. The carriers that populate such weaker oscillator strength transitions (as the atomic $5s \rightarrow 6p$ transition) lie also within the cooling zone, and so it is expected that their spectral temperature should reflect the reduced temperature of the cooling experiment. Ongoing experiments have already produced encouraging results.

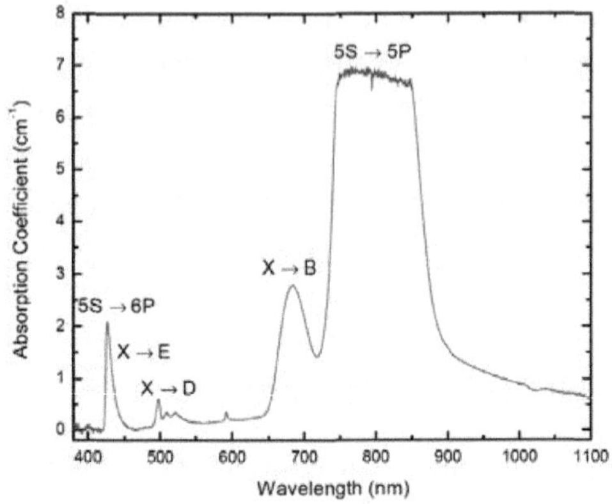

Figure 1.16 Absorption coefficient of rubidium vapor at T = 670 K, under 180 bar argon buffer gas, as measured by spectrally analyzing the transmission of the ensemble using a spectrally broadband halogen lamp. The atomic and molecular absorption bands are noted with respect to Ref. [72]. The second principal series doublet transition, 5s ⟶ 6p at 420 nm, overlaps with the broad rubidium dimer X ⟶ E transition, centered at 430 nm.

1.6 Conclusions

We have reviewed experiments on redistributional laser cooling, a technique currently applicable to dense alkali–rare gas mixtures. The achieved cooling powers here are several orders of magnitude larger than in Doppler cooling experiments for dilute atomic vapors, and the typically achieved cooling efficiencies are in the order of several percent.

The alkali transitions are strongly broadened by the high-pressure gas. The linewidths are in the order of the thermal energy in frequency units, that is, $k_B T/\hbar$. The cooling principle can be described in the frame of a quasi-molecular treatment. Elastic collisions of the optically active species with a surrounding perturber atom enable optical excitation of the former by laser light, red-detuned to the atomic resonances. After the collision, the interatomic distance becomes larger, and the excited state decays under the emission of a fluorescence photon after its natural lifetime. The photon energy is near the difference between the ground and the excited state of an undisturbed alkali atom and is thus higher than the energy of the absorbed photon. Energy in the order of the thermal energy k_BT is thus removed from the dense sample, locally cooling the gas.

The cooling results have been presented in the form of thermal deflection spectroscopy measurements on both rubidium- and potassium-argon mixtures. The technique allows for an estimation of the local temperature induced by the cooling beam. Here, the gas density changes following the temperature change. The former can be investigated by monitoring the local change of the refractive index within the so-called cooling zone. Implementation is achieved by irradiating a second, nonresonant probe laser beam into the high-pressure cell. The cooled gas acts like a graded index lens, causing the probe beam to deflect. Position dependent measurements of the deflection angle, varying the distance between the cooling and probe beams, allow for a mapping of the deflection profile, from which the temperature change can be deduced.

We have further discussed an alternative approach of temperature determination utilizing purely spectroscopic data. To this end, we have presented the Kennard–Stepanov relation, connecting absorption and emission profiles, and applied it in our dense gaseous ensembles. Our investigation shows that the relation is well fulfilled allowing for an accurate temperature determination of the ensemble's ambient temperature. Further studies on auxiliary atomic and molecular transitions are currently in progress.

Future experimental aims include the study of homogeneous nucleation in saturated vapor, as well as the application of this very technique to molecular–rare gas ensembles. Toward this goal, ongoing preparatory experiments seem encouraging [48, 73].

References

1. Pringsheim, P. (1929). Zwei Bemerkungen über den Unterschied von Lumineszenz- und Temperaturstrahlung, *Z. Phys.*, **57**(11–12), pp. 739–746.

2. Maiman, T. H. (1960). Stimulated optical radiation in ruby, *Nature*, **187**(4736), pp. 493–494.

3. Hänsch, T. W., Schawlow, A. L. (1975). Cooling of gases by laser radiation, *Opt. Commun.*, **13**(1), pp. 68–69.

4. Wineland, D. J., Itano, W. M. (1979). Laser cooling of atoms, *Phys. Rev. A,*, **20**(4), 1521–1540.

5. Adams, C. S., Riis, E. (1997). Laser cooling and trapping of neutral atoms, *Prog. Quant. Electron.*, **21**(1), pp. 1–79.
6. Lett, P. D., Watts, R. N., Westbrook, C. I., et al. (1988). Observation of atoms laser cooled below the doppler limit, *Phys. Rev. Lett.*, **61**(2), pp. 169–172.
7. Anderson, M. H., Ensher, J. R., Matthews, M. R., et al. (1995). Observation of Bose-Einstein condensation in a dilute atomic vapor, *Science*, **269**(5221), pp. 198–201.
8. Davis, K. B., Mewes, M., Andrews, M. R., et al. (1995). Bose-Einstein condensation in a gas of sodium atoms, *Phys. Rev. Lett.*, **75**(22), pp. 3969–3973.
9. Shuman, E. S., Barry, J. F., Demille, D. (2010). Laser cooling of a diatomic molecule, *Nature*, **467**(7317), pp. 820–823.
10. Kastler, A. (1950). Quelques suggestions concernant la production optique et la détection optique d'une inégalité de population des niveaux de quantifigation spatiale des atomes. Application à l'expérience de Stern et Gerlach et à la résonance magnétique, *J. Phys. Radium*, **11**(6), pp. 255–265.
11. Djeu, N., Whitney, W. T. (1981). Laser cooling by spontaneous anti-Stokes scattering, *Phys. Rev. Lett.*, **46**(4), pp. 236–239.
12. Epstein, R. I., Buchwald, M. I., Edwards, B. C., et al. (1995). Observation of laser-induced fluorescent cooling of a solid, *Nature*, **377**(6549), pp. 500–503.
13. Seletskiy, D. V., Melgaard, S. D., Bigotta, S., et al. (2010). Laser cooling of solids to cryogenic temperatures, *Nat. Photonics*, **4**(3), pp. 161–164.
14. Yeh, S., Berman, P. R. (1979). Theory of collisionally aided radiative excitation in three-level systems: new interference effect, *Phys. Rev. Lett.*, **43**(12), pp. 848–851.
15. Berman, P. R., Stenholm, S. (1978). Heating or cooling collisionally aided fluorescence, *Opt. Commun.*, **24**(2), pp. 155–157.
16. Giacobino, E., Tawil, M., Berman, P. R., et al. (1983). Production of "hot" excited-state atoms in collisionally aided radiative transitions, *Phys. Rev. A*, **28**(4), pp. 2555–2557.
17. Vogl, U., Weitz, M. (2009). Laser cooling by collisional redistribution of radiation, *Nature*, **461**(7260), pp. 70–73.
18. Carlsten, J. L., Szöke, A., Raymer, M. G. (1977). Collisional redistribution and saturation of near-resonance scattered light, *Phys. Rev. A*, **15**(3), pp. 1029–1045.

19. Burnett, K. (1985). Collisional redistribution of radiation, *Phys. Rep.*, **118**(6), pp. 339–401.

20. Carlsten, J. L., Szöke, A. (1976). Spectral resolution of near-resonant Rayleigh scattering and collision-induced resonance fluorescence, *Phys. Rev. Lett.*, **36**(12), pp. 667–671.

21. Huber, D. L. (1969). Resonant scattering of monochromatic light in gases, *Phys. Rev.*, **178**(1), pp. 93–102.

22. Omont, A., Smith, E. W., Cooper, J. (1972). Redistribution of resonance radiation. I. The effect of collisions, *Astrophys. J.*, **175**, p. 185.

23. Vogl, U. (2010). Kollektive Effekte und stoßinduzierte Redistributionskühlung in dichten atomaren Gasen, PhD thesis.

24. Dhiflaoui, J., Berriche, H., Heaven, M. C., et al. (2011). Electronic Absorption Spectra of the RbAr van der Waals Complex. In *Proceedings of the fifth Saudi Physical Society Conference (SPS5)*, AIP, pp. 234–241.

25. Speller, E., Staudenmayer, B., Kempter, V. (1979). Quenching cross sections for alkali-inert gas collisions, *Z. Phys. A*, **291**(4), pp. 311–318.

26. Mungan, C. E., Gosnell, T. R. (1999). *Laser Cooling of Solids*. Vol. 40, Elsevier, pp. 161–228.

27. Rayner, A., Heckenberg, N. R., Rubinsztein-Dunlop, H. (2003). Condensed-phase optical refrigeration, *J. Opt. Soc. Am. B*, **20**(5), p. 1037.

28. Vogl, U., Saβ, A., Haβelmann, S., et al. (2011). Collisional redistribution laser cooling of a high-pressure atomic gas, *J. Mod. Opt.*, **58**(15), pp. 1300–1309.

29. Saß, A., Vogl, U., Weitz, M. (2011). Laser cooling of a potassium–argon gas mixture using collisional redistribution of radiation, *Appl. Phys. B*, **102**(3), pp. 503–507.

30. Allegrini, M., Gabbanini, C., Moi, L. (1985). Energy pooling processes in laser-excited alkali vapors: an update on experiments, *J. Phys. Colloques*, **46**(C1), pp. C1-61–C1-73.

31. Barbier, L., Cheret, M. (1983). Energy pooling process in rubidium vapour, *J. Phys. B*, **16**(17), pp. 3213–3228.

32. Bondi, A. (1964). van der Waals Volumes and Radii, *J. Phys. Chem.*, **68**(3), pp. 441–451.

33. Bridgman, P. W. (1946). Recent work in the field of high pressures, *Rev. Mod. Phys.*, **18**(1), pp. 1–93.

34. Whinnery, J. R. (1974). Laser measurement of optical absorption in liquids, *Acc. Chem. Res.*, **7**(7), pp. 225–231.

35. Jackson, W. B., Amer, N. M., Boccara, A. C., et al. (1981). Photothermal deflection spectroscopy and detection, *Appl. Opt.*, **20**(8), pp. 1333–1344.

36. Spear, J. D., Russo, R. E., Silva, R. J. (1990). Collinear photothermal deflection spectroscopy with light-scattering samples, *Appl. Opt.*, **29**(28), pp. 4225–4234.

37. Casperson, L. W. (1973). Gaussian light beams in inhomogeneous media, *Appl. Opt.*, **12**(10), pp. 2434–2441.

38. Born, M., Wolf, E. (1999). Principles of optics: electromagnetic theory of propagation, interference and diffraction of light/Max Born and Emil Wolf with contributions by A. B. Bhatia, et al., 7th (expanded) ed. Cambridge University Press, Cambridge.

39. Gordon, J. P., Leite, R. C. C., Moore, R. S., et al. (1965). Long-transient effects in lasers with inserted liquid samples, *J. Appl. Phys.*, **36**(1), p. 3.

40. Zhu, Q., Pan, B., Chen, L., et al. (2010). Analysis of temperature distributions in diode-pumped alkali vapor lasers, *Opt. Commun.*, **283**(11), pp. 2406–2410.

41. Saß, A. (2014). Experimente zur stoßinduzierten Redistributions-laserkühlung dichter atomarer Alkali-Edelgas-Mischungen und Simulationen der zu erwartenden Temperaturverteilung, PhD thesis.

42. Shang-Yi, C. (1940). Broadening, asymmetry and shift of rubidium resonance lines under different pressures of helium and argon up to 100 atmospheres, *Phys. Rev.*, **58**(12), pp. 1051–1058.

43. Daniel A. Steck, "Rubidium 85 D Line Data," available online at http://steck.us/alkalidata (revision 2.1.6, 20 September 2013).

44. Drummond, D. L. (1974). Potentials and continuum spectra of Rb-noble gas molecules, *J. Chem. Phys.*, **60**(9), p. 3426.

45. Alcock, C. B., Itkin, V. P., Horrigan, M. K. (1984). Vapour pressure equations for the metallic elements: 298–2500K, *Can. Metall. Quart.*, **23**(3), pp. 309–313.

46. Mayer, H. (1931). Über eine neue Methode zur Messung kleinster Dampfdrucke. Die Dampfdrucke von Quecksilber und Kalium, *Z. Phys.*, **67**(3–4), pp. 240–263.

47. Vogl, U., Saß, A., Weitz, M. (2012). Laser cooling of dense rubidium-noble gas mixtures via collisional redistribution of radiation. In Epstein, R. I., Sheik-Bahae, M. (eds.) *SPIE OPTO. SPIE*, pp. 827508-1–827508-8.

48. Saß, A., Forge, R., Christopoulos, S., et al. (2014). Laser cooling of dense atomic gases by collisional redistribution of radiation and

spectroscopy of molecular dimers in a dense buffer gas environment. In Epstein, R. I., Seletskiy, D. V., Sheik-Bahae, M. (eds.) *SPIE OPTO. SPIE*, p. 90000A.

49. Hartmann, J. (1900). Bemerkungen über den Bau und die Justierung von Spektrographen, *Z. Instrumentenkd*, **20**, p. 47.
50. Shack, R. V., Platt, B. C. (1971). Production and use of a lenticular Hartmann screen. In Chairman, D. S. (ed.) *Spring Meeting of the Optical Society of America*. Optical Society of America.
51. Tiecke, T. (ed.) (2010). *Properties of Potassium*. PhD Thesis, University of Amsterdam, The Netherlands.
52. Rhouma, M. B. El Hadj, Berriche, H., Lakhdar, Z. B., et al. (2002). One-electron pseudopotential calculations of excited states of LiAr, NaAr, and KAr, *J. Chem. Phys.*, **116**(5), p. 1839.
53. Figl, C. (2004). Optical collisions in crossed beams and Bose-Einstein condensation in a microtrap, PhD thesis.
54. Kuhn, H., Oldenberg, O. (1932). Evidence of space quantization of atoms upon impact, *Phys. Rev.*, **41**(1), pp. 72–78.
55. Kielkopf, J. F. (1968). Semiclassical theory of satellite bands produced in the spectra of alkali metals by interaction with foreign gases, *J. Chem. Phys.*, **48**(12), p. 5570.
56. Kennard, E. H. (1918). On the thermodynamics of fluorescence, *Phys. Rev.*, **11**(1), pp. 29–38.
57. Stepanov, B. I. (1957). Universal relation between the absorption spectra and luminescence spectra of complex molecules, *Dokl. Akad. Nauk SSSR*, **112**, p. 839.
58. Moroshkin, P., Weller, L., Saß, A., et al. (2014). Kennard-Stepanov relation connecting absorption and emission spectra in an atomic gas, *Phys. Rev. Lett.*, **113**(6), p. 063002.
59. Sawicki, D. A., Knox, R. S. (1996). Universal relationship between optical emission and absorption of complex systems: an alternative approach, *Phys. Rev. A*, **54**(6), pp. 4837–4841.
60. Dobek, K. (2011). The influence of temperature on Coumarin 153 fluorescence kinetics, *J. Fluoresc.*, **21**(4), pp. 1547–1557.
61. Band, Y. B., Heller, D. F. (1988). Relationships between the absorption and emission of light in multilevel systems, *Phys. Rev. A*, **38**(4), pp. 1885–1895.

62. Ihara, T., Maruyama, S., Yoshita, M., et al. (2009). Thermal-equilibrium relation between the optical emission and absorption spectra of a doped semiconductor quantum well, *Phys. Rev. B*, **80**(3), p. 033307.
63. Zucchelli, G., Garlaschi, F. M., Croce, R., et al. (1995). A Stepanov relation analysis of steady-state absorption and fluorescence spectra in the isolated D1/D2/cytochrome b-559 complex, *Biochim. Biophys. Acta*, **1229**(1), pp. 59–63.
64. Klaers, J., Schmitt, J., Vewinger, F., et al. (2010). Bose-Einstein condensation of photons in an optical microcavity, *Nature*, **468**(7323), pp. 545–548.
65. van Metter, R., Knox, R. (1976). On the relation between absorption and emission spectra of molecules in solution, *Chem. Phys.*, **12**(3), pp. 333–340.
66. Kawski, A., Bojarski, P., Kukliński, B. (2000). The local temperature dependence of fluorescent centres in PVA films on the excitation wavenumber, *Z. Naturforsch. A*, **55**(6–7), pp. 653–655.
67. Ottinger, C., Scheps, R., York, G. W., et al. (1975). Broadening of the Rb resonance lines by the noble gases, *Phys. Rev. A*, **11**(6), pp. 1815–1828.
68. Christopoulos, S., Saß, A., Moroshkin, P., et al. (2015). A Kennard-Stepanov relation study on redistributional laser cooling in dense gaseous ensembles. In Epstein, R. I., Seletskiy, D. V., Sheik-Bahae, M. (eds.) *SPIE OPTO. SPIE*, p. 93800M.
69. Kasha, M. (1950). Characterization of electronic transitions in complex molecules, *Discuss. Faraday Soc.*, **9**, p. 14.
70. Lakowicz, J. R. (2006). *Principles of Fluorescence Spectroscopy*. Springer US, Boston, MA.
71. Gelbwaser-Klimovsky, D., Szczygielski, K., Vogl, U., et al. (2015). Laser-induced cooling of broadband heat reservoirs, *Phys. Rev. A*, **91**(2), p. 023431.
72. Vdović, S., Sarkisyan, D., Pichler, G. (2006). Absorption spectrum of rubidium and cesium dimers by compact computer operated spectrometer, *Opt. Commun.*, **268**(1), pp. 58–63.
73. Simon, P., Moroshkin, P., Weller, L., et al. (2013). Towards redistribution laser cooling of molecular gases: production of candidate molecules SrH by laser ablation. In Epstein, R. I., Seletskiy, D. V., Sheik-Bahae, M. (eds.) *SPIE OPTO. SPIE*, p. 86380C.

Chapter 2

Laser Cooling in Rare Earth–Doped Glasses and Crystals

Galina Nemova
Department of Engineering Physics, Polytechnique Montréal, C.P. 6079, Succursale Centre-ville, Montréal (Québec) Canada H3C 3A7
galina.nemova@polymtl.ca

2.1 Introduction

Laser cooling is a physical process by which a system loses its thermal energy as a result of interaction with laser light. The term "laser cooling" is most often used to describe cooling and trapping of atoms and ions to extremely low temperatures down to the nano-Kelvin regime. The energy of molecules in gases, especially at low temperatures, is more contained in translational degrees of freedom then in vibrations. Laser cooling in such a system is based on the Doppler's effect. The basic idea is to red-shift the frequency of the light to just below the resonance frequency of the atoms. In that case the atoms moving toward the laser are blue-shifted into resonance. They more strongly absorb photons then the atoms, which move away from the laser and are red-shifted further away from

Laser Cooling: Fundamental Properties and Applications
Edited by Galina Nemova

ISBN 978-981-4745-04-8 (Hardcover), 978-981-4745-05-5 (eBook)
www.panstanford.com

resonance [1]. Adsorbing a photon an atom moving toward the laser loses momentum equal to the photon moment. The excited atom spontaneously relaxes with photon emission in a random direction, on average, without momentum change. The three-dimensional orthogonal setup with counterpropagating laser beams creates an optical molasses, which can be cooled down to the Doppler cooling limit. This area of science has progressed immensely. In 1997, Chu, Cohen-Tannoudji, and Phillips were awarded the Nobel Prize in Physics for developing the methods to cool and trap atoms with light [2]. While the Doppler cooling technique works for atoms in dilute gases it is not suitable for cooling of macroscopic objects like solids. Indeed, in solids translational motion is absent and the thermal energy is contained in vibrational modes known as phonons. The vast majority of luminescent materials emit photons with lower energy than absorbed ones. The associated energy difference is the Stokes shift, named after the physicist George G. Stokes. The Stokes shift is based on electron–phonon coupling, which allows pump light to interact with the vibrational modes of solids. This process is accompanied by heat generation in the system. If the emitted photon has more energy than the absorbed one the energy difference between the emitted and absorbed photons is called the anti-Stokes shift and the emitted photons are termed "anti-Stokes fluorescence." This effect is based on dissipation of thermal phonons in a crystal lattice. In 1928, Lenard showed that anti-Stokes fluorescence does not violate the second law of thermodynamics [3]. In 1929, Pringsheim proposed to use anti-Stokes fluorescence to cool sodium vapor [4]. He proposed to pump sodium vapor in "Dewar 1" with a filtered sodium vapor lamp in order to excite electrons from the $^2S_{1/2}$ ground state to the first $^2P_{1/2}$ excited state. After thermalization, which is based on inelastic collisions in the gas, to the higher $^2P_{3/2}$ excited state the electrons relax to the ground state with resonant ($^2P_{1/2} \rightarrow {}^2S_{1/2}$) and anti-Stokes ($^2P_{3/2} \rightarrow {}^2S_{1/2}$) fluorescence. This fluorescence has to be quenched by nonradiative relaxation with heat generation in "Dewar 2" filled with the mixture of sodium vapor and nitrogen. In this experiment "Dewar 1" is cooled with light. Pringsheim assumed that the emitted photons scatter in all direction and this process is irreversible in principle. Vavilov believed that optical cooling by anti-Stokes fluorescence contradicts the second law of thermodynamics [5, 6]. He argued that the cycle, which includes excitation and fluorescence, is a reversible one and the scattered light can be with sufficient approximation once again converted into a directed beam. Vavilov referred to a Lomonosov's paper published in 1743, in which Lomonosov shows the manner in which the full light flux from a point source can be directed as a parallel beam [7]. In 1946, Landau concluded this controversy [8]. He considered the entropy of the radiation field based on Bose–Einstein statistics applied to a "photon gas." Landau has shown that the entropy of the radiation is a function of the angle of the propagating radiation and its frequency bandwidth. Consequently, laser light has low entropy as compared to higher-entropy anti-Stokes fluorescence. The laser cooling process based on anti-Stokes fluorescence does not violate the second law of thermodynamics. In 1950, Kastler [9] and in 1961 Yatsiv [10] suggested the use of rare earth (RE)-doped solids for laser cooling. The main advantage of RE ions is the optically active 4*f* electrons shielded by the filled 5s and 5p outer shells, which limit the interaction of a RE ion with the surrounding lattice and suppress nonradiative decay, especially in low-phonon hosts. In 1968, Kushida and Geusic tried to cool a neodymium-doped yttrium aluminium garnet (Nd^{3+}:YAG) sample pumped at 1064 nm and placed inside the cavity of the Nd^{3+}:YAG laser [11]. They observed a reduction of heating. In 1981, Djeu and Whitney reached net laser cooling with anti-Stokes fluorescence in low-pressure CO_2 gas with the temperature drop of ~1 K from 600 K using a CO_2 laser as a pump source [12]. The net radiation cooling process based on anti-Stokes fluorescence in solids was observed for the first time only in 1995 by Epstein's research team in Yb^{3+}-doped ZBLANP glass [13]. A 1 wt% Yb^{3+}-doped sample of ZBLANP of volume 2.5 × 2.5 × 6.9 mm^3 was placed in vacuum, pumped in the 870–1050 nm wavelength range and cooled by 0.3 K below room temperature when the pump wavelength was greater the mean fluorescence wavelength. In 2013, laser cooling down to 119 K that is below National Institute of Standards and Technology (NIST) cryogenic temperature (123 K) was achieved in a ytterbium-doped yttrium lithium fluoride (Yb^{3+}:YLF) sample [14]. In the same 2013 a Yb^{3+}:YAG sample of volume 1 × 1 × 10 mm^3 was laser-cooled for the first time in air [15]. Today's record temperature in vacuum is 91 K. It has been achieved in a Yb^{3+}:YLF sample [16, 17].

In RE-doped glasses and crystals the laser cooling process takes place in localized active RE ions; the ground and excited levels of RE ions as well as phonons of the host are involved in the cooling cycle. Semiconductors are an alternative class of materials suitable for laser cooling with anti-Stokes fluorescence. In semiconductors, the cooling cycle involves transitions between the valence and conduction bands of direct-band-gap semiconductors. Unfortunately, the purity of semiconductor samples, surface recombination, high refractive index of semiconductors in comparison with glasses and crystals are the obstacles on the way of laser cooling in bulk semiconductors [18]. In 2013, Zhang et al. observed for the first time laser cooling of CdS nanobelts with a temperature drop of 40 K from room temperature [19]. Laser cooling in bulk semiconductors has not been achieved yet. The progress in laser cooling of solids is summarized in Table 2.1.

Table 2.1 The history of laser cooling with anti-Stokes fluorescence

Year	Achievement	References
1920s	ASF was observed.	
1928	ASF does not violate SLT.	[3]
1929	Cooling with ASF was proposed.	[4]
1946	The entropy of a "photon gas" was estimated.	[8]
1950	RE-doped solids were proposed for cooling.	[9]
1968	Nd^{3+}:YAG—reduced heating with ASF.	[11]
1981	CO_2—net laser cooling with ASF (ΔT = 1 K).	[12]
1995	Yb^{3+}:ZBLANP—net laser cooling with ASF (ΔT = 0.2 K).	[13]
2013	Yb^{3+}:YLF–cooling below NIST cryogenic temperature 123 K.	[14]
2013	Yb^{3+}:YAG—net laser cooling in air (ΔT = 8.8 K).	[15]
2013	CdS nanobelts—net laser cooling (ΔT = 40 K).	[19]
2015	Yb^{3+}:YLF—today's record temperature (T = 91 K).	[17]

ASF, anti-Stokes fluorescence; SLT, the second law of thermodynamics; ΔT, the temperature drop.

2.2 Thermodynamics of Laser Cooling of Solids

For the first time the second law of thermodynamics was applied to luminescence processes by Landau in 1946 [8]. In 1960, Weinstein estimated the limiting efficiency of electroluminescent processes [20]. In 1969, Chukova considered luminescence in general way and obtained the relation for the upper limit of efficiency [21, 22]. Chukova has shown that the limiting efficiency of photoluminescence cannot be larger than the thermodynamic limit on the efficiency of electroluminescence [23, 24].

Let us consider a system consisting of the pump radiation, a luminescent body, which is in thermal equilibrium with an ambient, and the fluorescence radiation. In the steady-state condition, the energy conservation requires that

$$\dot{E}_{\mathrm{p}} + \dot{Q} - \dot{E}_{\mathrm{f}} = 0 \tag{2.1}$$

where $\dot{E}_{\mathrm{p}}$ is the pump laser radiation rate, $\dot{Q}$ is the heat flow power delivered to the body, and $\dot{E}_{\mathrm{f}}$ is the fluorescence radiation rate. The points over the symbols denote the derivative d/dt, where t is time. The second law of thermodynamics requires that the entropy of an isolated system does not decrease. It gives rise to the inequality

$$\dot{S}_{\mathrm{f}} - \frac{\dot{Q}}{T} - \dot{S}_{\mathrm{p}} \geq 0 \tag{2.2}$$

where S_{p} is the entropy of the pump source, S_{f} is the entropy of fluorescence radiation, and T is ambient temperature. The *entropy* of radiation ("photon gas") is determined by the well-known formulae of Bose statistic.

$$S = \kappa_{\mathrm{B}} \iint \frac{l\nu^2}{c^3}[(1+n_\nu)\ln(1+n_\nu) - n_\nu \ln n_\nu] d\nu d\Omega \tag{2.3}$$

where n_ν is the mean photon occupation number per unit volume of the **k**-space, where **k** is the wave vector. In the case of polarized radiation $l = 1$, $l = 2$ specifies unpolarized radiation.

The *entropy flux density* of the radiation in a frequency interval $(\nu, \nu + d\nu)$, which passes the unit area of a surface per unit time into solid angle $d\Omega$ in a direction under an angle θ with the normal to the surface (Fig. 2.1) is

$$\dot{S} = \kappa_{\mathrm{B}} \iint \frac{l\nu^2}{c^2}[(1+n_\nu)\ln(1+n_\nu) - n_\nu \ln n_\nu] \cos\theta d\nu d\Omega \tag{2.4}$$

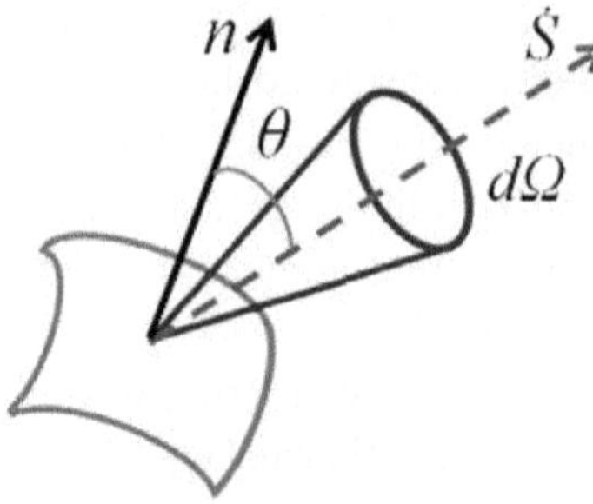

Figure 2.1 Entropy flux density of the radiation emitted from the surface of a sample into a solid angle $d\Omega$. The polar angle θ is measured relative to the surface normal direction n.

The factor $\cos\theta$ projects the unit area of the surface to be normal to radiation. The *energy flux density* of the same radiation can be described as

$$\dot{E} = \iint \frac{l\nu^2}{c^2} h\nu n_\nu \cos\theta d\nu d\Omega \tag{2.5}$$

The *effective flux temperature,* which is the mathematical substitution, can be introduced as the relation between the energy flux density and entropy flow density

$$T^{\text{eff}} = \frac{\dot{E}}{\dot{S}} \tag{2.6}$$

2.2.1 Entropy for Near-Monochromatic Radiation Flow

Suppose that the radiation flow is spread in the frequency interval $\Delta\nu$ around a central frequency ν_0 ($\Delta\nu << \nu_0$) and propagates into the angle $\Delta\Omega$. In this case Eqs. 2.4 and 2.5 can be written as

$$\dot{S} = \kappa_B \frac{l\nu^2}{c^2}[(1+n_\nu)\ln(1+n_\nu) - n_\nu \ln n_\nu]\cos\theta\Delta\Omega\Delta\nu \tag{2.7}$$

$$\dot{E} = \frac{l\nu^2}{c^2} h\nu n_\nu \cos\theta\Delta\Omega\Delta\nu \tag{2.8}$$

respectively. Let us suppose that ether $\Delta\nu$ or $\Delta\Omega$ go to zero. Since the energy flow density is nonzero the relation $n_\nu\Delta\Omega\Delta\nu$ in Eq. 2.8 must be finite, that is, n_ν tends to infinity ($n_\nu \rightarrow \infty$). Consequently, considering Eq. 2.7 for the entropy flow density one can conclude that for monochromatic or unidirectional radiation fluxes the entropy flow density vanishes. For the first time that was proved by Landau in 1945. Using Eq. 2.6 one can conclude that the effective flux temperature for monochromatic or unidirectional radiation fluxes approaches infinity ($T^{\text{eff}} \to \infty$).

2.2.2 Conversion Efficiency

The maximum conversion efficiency, η, of the pump energy flux density, $\dot{E}_{\text{p}}$, into fluorescence energy flux density, $\dot{E}_{\text{f}}$, can be estimated with Eqs. 2.1, 2.2, and 2.6.

$$\eta = \frac{\dot{E}_{\text{f}}}{\dot{E}_{\text{p}}} \leq \frac{T_{\text{f}}^{\text{eff}}}{\left(T_{\text{f}}^{\text{eff}} - T\right)} \left(1 - \frac{T}{T_{\text{p}}^{\text{eff}}}\right) \tag{2.9}$$

where $T_{\text{f}}^{\text{eff}}$ and $T_{\text{p}}^{\text{eff}}$ are the effective flux temperature of the fluorescence and the pump radiation, respectively. In the case of near-monochromatic pump radiation the effective flux temperature $T_{\text{p}}^{\text{eff}} \to \infty$ and, as one can see in Eq. 2.9, the maximum conversion efficiency depends only on the fluorescence effective flux temperature $T_{\text{f}}^{\text{eff}}$. This result for the first time was obtained by Weinstein in 1960 for electroluminescence [20].

2.3 Fundamentals of Laser Cooling in Rare Earth–Doped Solids

2.3.1 The Two-Level Model of Laser Cooling in Rare Earth–Doped Hosts

RE ions (lanthanides) have a long history of optical applications. REs have unique characteristics that distinguish them from other optical ions. RE ions absorb and emit over narrow frequency ranges. The frequencies of absorption and emission transitions are almost insensitive to host materials. The lifetimes of metastable levels are long (μs to ms). These optical properties of RE ions are dictated by their unique electronic structure. All the RE ions have the same outer electronic structure $5s^25p^66s^2$, which are filled shells. The optical properties of the RE ions are connected with the electrons occupying the inner 4f shielded quite effectively orbital. As a result, the RE ions maintain much of their "free ion" properties when are doped into hosts than other ions. For example, the absorption and emission spectra of the REs are less dependent on external electric field than the absorption and emission spectra of the transition elements, which do not experience electronic shielding. While the crystal field interactions are weak in RE ions, they are suitable for laser cooling. The idea to use the RE ions for cooling was proposed by Kastler in 1950 [9].

Let us consider the laser cooling process with ytterbium ions, Yb^{3+}, in the two-level system. Laser cooling in the four-level system has been considered in Refs. [25, 26]. Ionization of the REs usually results in trivalent states. As an example the energy-level diagram of Yb^{3+} ions doped in the YAG host is presented in Fig. 2.2. The energy gap between the $^2F_{7/2}$ and $^2F_{5/2}$ manifolds is ~10,000 cm^{-1}.

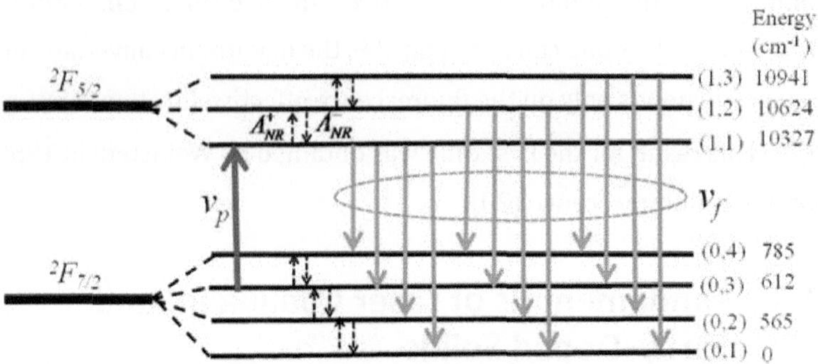

Figure 2.2 Energy levels and major transitions of Yb^{3+} in YAG.

When a free ion is put into a host (in general substitutionally) it is surrounded by a set of nearest-neighbor ligands in some configuration. The host ligands and the dopant ions interact with each other through an electrostatic Coulomb interaction. A ligand field induces a Stark effect, which results in the splitting of the energy levels. The thermalization process, which is based on phonon absorption and creation, occurs within each manifold (Fig. 2.2). It provides Boltzmann's distribution of electrons in Stark split sublevels. The condition of the thermal equilibrium can be presented as

$$A_{NR}^{-}N_{nm} = A_{NR}^{+}N_{n,m-1} \tag{2.12}$$

Here N_{nm} and $N_{n,m-1}$ are the population of each sublevel; n = 0, 1 is the number of the manifold: ground (0) and excited (1); m is the sublevel number; m = 1, 2, 3, and 4 for the ground manifold; and m = 1, 2, and 3 for the excited manifold. A_{NR}^{-} and A_{NR}^{+} are the de-excitation and excitation nonradiative rates, respectively (Fig. 2.2). If $\Delta E_{n,m} = E_{n,m} - E_{n,m-1}$ is the energy difference between sublevels (n, m) and (n, m−1) in the n-manifold, one can obtain the relation

$$\frac{N_{nm}}{N_{n,m-1}} = \frac{A_{NR}^{+}}{A_{NR}^{-}} = \exp\left(-\frac{\Delta E_{n,m}}{k_B T}\right) \tag{2.13}$$

where k_B is Boltzmann's constant. Of course the system has to be free from a "phonon bottleneck," that is, phonons with the energy, which permits to bridge sublevels of manifolds have to be generated in the system. The population of each sublevel can be described with the Boltzmann occupation factor, f_{nm}, which is a function of the temperature, T. The Boltzmann occupation factors of the lower and the upper manifolds can be calculated with relations

$$f_{0i} = \frac{\exp\left(\frac{E_{01} - E_{0i}}{k_B T}\right)}{\sum_{i=1}^{4} \exp\left(\frac{E_{01} - E_{0i}}{k_B T}\right)} \tag{2.14}$$

and

$$f_{1j} = \frac{\exp\left(\frac{E_{11} - E_{1j}}{k_B T}\right)}{\sum_{j=1}^{3} \exp\left(\frac{E_{11} - E_{1j}}{k_B T}\right)} \tag{2.15}$$

respectively. Here index $0i$ (i is integer from 1 to 4) refers to the sublevels of the lower $^2F_{7/2}$ manifold, and index $1j$ (j is integer from 1 to 3) refers to the sublevels of the upper $^2F_{5/2}$ manifold. E_{0i} and E_{1j} are the energies of the sublevels of the lower $^2F_{7/2}$ and the upper $^2F_{5/2}$ manifolds, respectively. The occupation factors obey the equations

$$\sum_{i=1}^{4} f_{0i} = \sum_{j=1}^{3} f_{1j} = 1 \tag{2.16}$$

Each of the Stark sublevels of the $^2F_{5/2}$ manifold can be characterized with the branching ratios, β_{0i}^{1j}, describing the fraction of particles of the sublevel of $1j$ (j = 1, 2, 3) of the upper $^2F_{5/2}$ manifold, which decays down to the sublevel $0i$ (i = 1, 2, 3, 4) of the lower $^2F_{7/2}$ manifold. τ_{1j}^{r} are the lifetimes of the sublevels of the $^2F_{5/2}$ manifold. The difficulty of spectrally resolving and characterizing the individual Stark sublevels and transitions result in uncertainty about these parameters. As a consequence the mean fluorescence frequency of the $^2F_{5/2}$ manifold is more useful parameter, which can be described with the relation

$$\nu_{\mathrm{f}} = \frac{\sum_{j=1}^{3}\sum_{i=1}^{4} \frac{f_{1j}}{\tau_{1j}^{r}} \beta_{0i}^{1j} \nu_{0i}^{1j}}{\frac{1}{\tau_{\mathrm{r}}}} \tag{2.17}$$

where

$$\frac{1}{\tau_{\mathrm{r}}} = \sum_{j=1}^{3} \frac{f_{1j}}{\tau_{1j}^{\mathrm{r}}} \tag{2.18}$$

Here τ_{r} is the average radiative lifetime of the upper $^2F_{5/2}$ manifold commonly known simply as the radiative lifetime of the excited manifold. The mean fluorescence frequency, ν_{f}, does not depend on the pump frequency, ν_{p}, (Vavilov's law) and also can be estimated with the integral relation

$$\nu_{\mathrm{f}} = \frac{\int \Phi(\nu)\nu d\nu}{\int \Phi(\nu) d\nu} \tag{2.19}$$

where $\Phi(\nu)$ is the photon flux density. The population of the lower $^2F_{7/2}$ and the upper $^2F_{5/2}$ manifolds can be calculated with the relations

$$N_0 = \sum_{i=1}^{4} N_{0i} = \sum_{i=1}^{4} f_{0i} N_0 \text{, and} \tag{2.20}$$

$$N_1 = \sum_{j=1}^{3} N_{1j} = \sum_{j=1}^{3} f_{1j} N_1 \tag{2.21}$$

respectively. The laser cooling process in a two-level system can be simulated with the well-known rate equations [27, 28]

$$\begin{cases} \dfrac{dN_1}{dt} = \dfrac{I_p}{h\nu_p}[N_0\sigma_a(\nu_p) - N_1\sigma_e(\nu_p)] - \dfrac{N_1}{\tau_r} - \dfrac{N_1}{\tau_{nr}} \\ N_T = N_0 + N_1 \end{cases} \tag{2.22}$$

where I_p is the pump intensity at the frequency ν_p. $\sigma_{a,e}(\nu_p)$ is the absorption (a) and emission (e) cross sections at the pump frequency ν_p. τ_{nr} is the nonradiative lifetime of the upper $^2F_{5/2}$ manifold. N_T is the RE ions density in the sample. The cooling power density generated in the sample, P_{cool}, can be estimated as a difference between the absorbed and radiated power densities.

$$P_{cool} = -I_p[N_0\sigma_a(\nu_p) - N_1\sigma_e(\nu_p)] + N_1\frac{h\nu_f}{\tau_r} - N_1\frac{h\nu_f}{\tau_{nr}} \tag{2.23a}$$

For a system in the steady-state condition, $dN_0/dt = dN_1/dt = 0$, one can calculate the cooling power using rate equations (Eq. 2.22)

$$P_{cool} = \frac{(h\tilde{\nu}_f - h\nu_p)}{h\nu_p}\frac{I_s\sigma_a(\nu_p)N_T}{\left(1 + \dfrac{\sigma_e(\nu_p)}{\sigma_a(\nu_p)} + \dfrac{I_s}{I_p}\right)} \tag{2.23b}$$

where $\tilde{\nu}_f \approx \eta_q \nu_f$ is the effective mean fluorescence frequency. $I_s = h\nu_p/[\tau_f\sigma_a(\nu_p)]$ is the saturation intensity, where $\tau_f = \eta_q\tau_r$ is the intrinsic fluorescence lifetime of the excited manifold. Here

$$\eta_q = \frac{W_r}{W_r + W_{nr}} \tag{2.24}$$

is the emission quantum efficiency or internal quantum efficiency, which describes the "competition" between the radiative, $W_r = 1/\tau_r$, and nonradiative, $W_{nr} = 1/\tau_{nr}$, decay rates of the upper $^2F_{5/2}$ manifold. In real samples the host material and impurities can absorb the pump photons, too. Denote α_b an average background absorption coefficient. For a real system one has to replace the emission quantum efficiency, η_q, with the relative emission quantum efficiency

$$\tilde{\eta}_q = \eta_q \frac{\alpha_r}{(\alpha_r + \alpha_b)} = \eta_q \eta_{abs} \tag{2.25}$$

where $\alpha_r \approx \sigma_a(\nu_p)N_T$ is the resonant absorption coefficient at the relatively low pump power intensity. η_{abs} is the absorption efficiency, which can be defined as the fraction of pump laser photons engaged in cooling. The cooling efficiency can be estimated as the relation of the cooling power density and absorbed power density

$$\eta_{cool}(I_p) = \frac{P_{cool}}{P_{abs}} = \frac{(h\tilde{\nu}_f - h\nu_p)}{h\nu_p} \frac{I_s}{\left(I_s + \frac{\sigma_e(\nu_p)}{\sigma_a(\nu_p)} I_p\right)} \tag{2.26}$$

As one can see in Eq. 2.26 the cooling efficiency is a function of the pump intensity I_p. As the pump intensity increases the intensity of the stimulated emission, which does not result in the cooling process, increases. For relatively weak pump intensities, I_p, most of the absorbed pump radiation leaves the sample with the spontaneous emission at the mean fluorescence frequency, ν_f, and the cooling efficiency can be estimated as the difference between the energy of the mean fluorescence and pump photon normalized by the energy of the pump photon

$$\eta_{cool} \approx \frac{h\tilde{\nu}_f - h\nu_p}{h\nu_p} = \frac{\lambda_p - \tilde{\lambda}_f}{\tilde{\lambda}_f} \tag{2.27}$$

here $\lambda_p = c/\nu_p$ and $\tilde{\lambda}_f = c/\tilde{\nu}_f$ are the pump wavelength and the mean fluorescence wavelength, respectively. c is the speed of light in vacuum. As one can see in Eq. 2.26 if the pump power intensity reaches saturation $I_p^{Sat} = h\nu_p/(\tau_f\sigma_e(\nu_p))$ the cooling efficiency, η_{cool}, decreases in half of the cooling efficiency corresponding to the weak pump intensity, which is described with Eq. 2.27. The cooling efficiency described with Eq. 2.26 takes place in an ideal unrealistic sample where reabsorption and the total internal reflection (TIR) of radiated photons at the boundaries of the sample are not taken into account. In actual samples used in experiments the process of reabsorption, TIR as well as the cooperative effects, such as energy migration and cooperative luminescence, which become essential at high ion concentrations, influence the fluorescence lifetime of the excited manifold, τ_f, and the cooling efficiency of the system.

Nonradiative decay of the excited level in RE ions with many excited manifolds such as Tm^{3+}, Er^{3+} can include the multiphonon relaxation as well as cross relaxation and energy transfer upconversion, which are a source of heat generation in the system. In the Yb^{3+}-doped samples, which have only two manifolds, energy transfer upconversion is absent. That is the reason why the Yb^{3+} ions are the most suitable dopants for laser cooling of solids.

Let us consider multiphonon relaxation. In this case the relaxation process between upper and lower electronic states includes both an electronic transition and a vibrational transition with the emission of one or several phonons. The combined energy of these phonons has to be equal to the energy gap between the levels involved in the transition. Contrary to high-order multiphoton processes, which are extremely weak, multiphonon processes can be relatively strong since the electron–phonon interaction is stronger than electron–electron interaction and phonons have a density of state that typically are 10 orders of magnitude larger than that of photons. The large variation in vibrational spectra of hosts makes the rate of nonradiative relaxation strongly dependent on the host material. For each host material all levels with the same energy gap have roughly the same nonradiative transition rate, which relatively independent on the nature of the electronic states involved in transition or the RE ions involved in the process, unless a strong selection rule is involved. The theory of multiphonon relaxation for RE ions was developed by Kiel [29] and extended by Riseberg and Moos [30]. The nonradiative decay rate, W_{nr}, is inversely proportional to the exponential function of the energy gap separating the two levels (the energy gap law) [30]

$$W_{nr} = \beta e^{-\alpha M} \tag{2.28}$$

where M is the number of emitted phonons. α and β are vibrational and electronic parameters of the host material, respectively. As one can see in Eq. 2.28 the nonradiative decay rate, W_{nr}, decreases exponentially with an increasing number of phonons, M, participating in the decay process. The vibrations with highest energy, $\hbar\Omega_{max}$, that is, optical phonons, provide the highest multiphonon relaxation rate. The maximum phonon energy depends on the host material, for example, $\hbar\Omega_{max} \approx 203$ cm^{-1} for a KPC crystal and $\hbar\Omega_{max} \approx 1400$ cm^{-1} for borate glasses. Since nonradiative decay is a source of heat generation in the system only RE-doped hosts with $W_{nr} << W_r$ are suitable for laser cooling. The radiative relaxation rate of the first excited state of RE ions is in the range $W_r = 10^2$–10^3 s^{-1} for a wide variety of hosts. Assuming $W_r \approx 10^3$ s^{-1} and using Eqs. 2.24 and 2.28 one can estimate that the emission quantum efficiency $\eta_q > 0.99$ can only be reached in the samples with $\hbar\Omega_{max} \approx < \Delta E_{01}/8$ [31]. Here ΔE_{01} is the energy gap between the manifolds involved into the cooling cycle.

As one can see in Eq. 2.27, if the system is pumped at longer wavelengths the cooling efficiency increases. However, as one moves toward longer wavelengths the absorption cross section decreases and the pump power absorbed by the sample decreases too. For this reason, a figure of merit, $F_{cool}(\lambda)$, for the cooling process has to be considered.

$$F_{cool}(\lambda) = \sigma_{abs}(\lambda)\eta_{cool}(\lambda) \tag{2.29}$$

The maximum $F_{cool}(\lambda)$ can be used to evaluate the host material for the laser cooling application.

2.3.2 Reabsorption or Fluorescence Trapping and Total Internal Reflection

The ions in a sample can interact with each other. One of these mutual interactions is reabsorption, when energy transfer between the ions is radiative (Fig. 2.3a). In this case real photons are emitted by an excited ion are absorbed by another unexcited ions within a photon travel distance. This cycle can repeat itself several times resulting in long-range energy diffusion between identical ions and gives rise to the radiation trapping effect. The probability of this process is proportional to R^{-2}, where R is the distance between ions [32]. Let us assume that an emitted photon undergoes N reabsorption/reemission events and reaches the boundary of the sample at the angle smaller than the critical angle for TIR $\theta_c = \sin^{-1}(1/n)$, where n is the refractive index of the sample. Following Snell's law this photon leaves the sample and do not undergo to TIR at the boundary. The probability of reabsorption for this photon is $r = 1 - \exp(-\alpha_r L)$. Here L is the average distance that photons travel to reach the boundary of the sample. $\alpha_r = \int \alpha_{abs}(\nu)F_0(\nu)d\nu$ is the average reabsorption coefficient, which can be defined as the integrated product of the absorption spectrum, $\alpha_{abs}(\nu)$, and the normalized fluorescence intensity, $F_0(\nu)$, here $\int F_0(\nu)d\nu = 1$. The fraction of emitted photons that after reabsorptions leave sample is $\eta_1 = 1 + \exp(\alpha_r L)$ [33]. Thus in the trapping conditions provided reabsorption the fluorescence lifetime increases and becomes equal to

$$\tilde{\tau}_f = \tau_f(1+\alpha_r L) \tag{2.30}$$

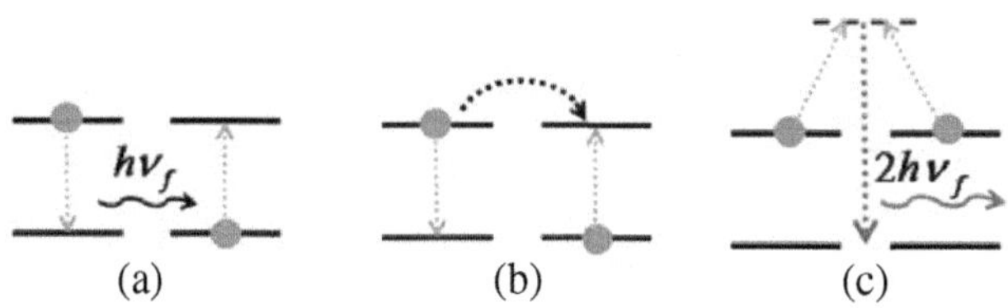

Figure 2.3 Energy transfer processes between two ions: (a) radiative transfer, (b) nonradiative transfer, and (c) cooperative emission.

As one can see in Eq. 2.30 the fluorescence lifetime, $\tilde{\tau}_f$, increases with the increase in the ion concentration and the size of the sample. The fraction of photons, which leaves the sample and do not undergo TIR is $P_{free} = 1 - \cos(\theta_c) = 1 - [1 - (1/n)^2]^{1/2}$. For example, for the Yb^{3+}:ZBLAN sample with the refractive index $n = 1.49$ $P_{free} = 25.9\%$, and for the Yb^{3+}:YAG sample with the refractive index $n = 1.833$ $P_{free} = 16.29\%$. Most of the anti-Stokes emitted photons undergo TIR. The influence of TIR and reabsorption on solid-state optical cooling has been investigated in details in Ref. [33] with a random walk model developed in Refs. [34, 35]. The total escape efficiency or external quantum efficiency in the system, where reabsorption and TIR are taken into account, is

$$\eta_{ext} = \frac{(\tilde{\eta}_q)^N\{\exp(-\alpha_r L)/n - [1-(1/n)^2]^{1/2} + 1\}}{1 - \{[1-(1/n)^2]^{1/2} - \exp(-\alpha_r L)/n\}(\tilde{\eta}_q)^N} \tag{2.31}$$

where $N = G(L/l_{ave})^2$ is the number of emission effects, l_{ave} is the step size or mean free path, that is, the distance corresponding to a probability for reabsorption of one half, that is, $\exp(-l_{ave} \cdot \alpha_r) = 0.5$.

Here $G = 0.308 \cdot (l_{ave}/L) + 1.078 \cdot (l_{ave}/L)^2$ is the dimensionless scaling factor estimated in Ref. [33].

The fluorescence lifetime described with Eq. 2.30 increases with the increase in the ion concentration, but as the ion concentration reaches some critical value, N_0, the fluorescence lifetime begins to decrease. This dependence has been experimentally observed in a number of the Yb^{3+}-doped samples [36]. The decrease in the fluorescence lifetime of the excited Yb^{3+} ions with the increase in the ion concentration indicates that the additional nonradiative processes, which depend on ion concentration, begin to play an important role as the ion concentration exceeds its critical value, N_0, which is different for different host materials. As the ion concentration reaches its critical value two ions can interact nonradiatively with energy transfer. The excitation jumps from one ion to another before one is able to decay with a photon generation and the first ion returns to the ground state (Fig. 2.3b). The mutual interactions of two ions with nearly equal energy gap between the ground and the first excited manifolds are Coulomb interactions of the van der Waals type [37]. Using the quantum mechanical theory Förster considered the dipole–dipole interaction assuming that the interaction between the ions is strongest if for both transitions electric–dipole transitions are allowed. In this case the energy transfer probability is proportional to the inverse of the sixth power of the interaction distance that is R^{-6}. Dexter has developed the theory of ion–ion energy transfer, including higher multipole and exchange interactions [38]. Following Dexter's theory the energy transfer is much more likely in a system in which there is significant overlap between the excited state transitions of the donor ions and the ground-state absorptions of the acceptor ions. At the critical distance between the ions, R_0, energy transfer and spontaneous decay processes have equal probability [32]. Assuming that the energy exchange between ions takes place via dipole–dipole interaction, the total concentration dependent fluorescence lifetime is given by [39, 40]

$$\tilde{\tilde{\tau}}_f(N_T) = \frac{\tau_f(1+\alpha_r L)}{1+\dfrac{9}{2\pi}\left(\dfrac{N_T}{N_0}\right)^2} \tag{2.32}$$

where τ_f is the fluorescence lifetime in absence of excitation migration. $N_0 = \left(\frac{4}{3}\pi R_0^3\right)^{-1}$ is a critical ion concentration. For example, for the Yb^{3+}:YAG sample $N_0 = 2.3 \times 10^{21}$ cm^{-3}. Migrating excitation energy can reach an impurity ion, where it relaxes nonradiatively with phonon (heat) generation. The energy migration process forms an additional loss channel in the system, which is undesirable for the cooling process. It has to be avoided especially in the relatively "big" millimeter- or centimeter-size samples [41].

2.3.3 Cooperative Emission

The cooperative de-excitation of two adjacent excited ions can also take place at high ion concentration in the sample. In this case two simultaneously excited Yb^{3+} ions located nearby to each other relax with emission of a visible photon (Fig. 2.3c). The cooperative emission lifetime is nearly half of that of the infrared (IR) luminescence, τ_f [42]. The contribution into cooling power from cooperative emission can be described as $P_{\text{coop}} = \frac{hc}{\lambda_{\text{coop}}}\frac{1}{N_T}XN_1^2$, where X is the experimental parameter [42]. The intensity of this radiation is very weak in comparison with spontaneous fluorescence [41].

2.3.4 Achievable Temperature

The final equilibrium temperature of the sample depends on the environment. Since most laser cooling experiments have been performed in vacuum chambers, the same conditions are applied here. Hence, there is only a radiative heat load in the system and the final equilibrium temperature of the sample can be estimated following the well-known Stefan–Boltzmann law

$$V_{\text{Samp}}P_{\text{cool}} = \varepsilon\sigma_b S_{\text{Samp}}\left(T_r^4 - T_s^4\right) \tag{2.33}$$

where T_r is room temperature, T_s is the equilibrium temperature of the sample, ε is the hemispherical emissivity of the sample, and $\sigma_B = 5.67 \times 10^{-12}$ W·cm^{-2}·K is the Stefan–Boltzmann constant. V_{Samp} and S_{Samp} are the volume and the total surface area of the sample, respectively. Let us following Ref. [41] consider the *geometrical efficiency factor* (GEF) of the sample as

$$\eta_{\text{Samp}} = V_{\text{Samp}}/S_{\text{Samp}} \tag{2.34}$$

Of course, a sphere is the best choice for cooling due to its high volume-to-surface ratio. Indeed, for a sphere sample $\eta_{\text{Samp}} = R_{\text{Samp}}/3$, where R_{Samp} is the radius of the sphere. For the cube, $\eta_{\text{Samp}} = L_{\text{Samp}}/6$. Using Eqs. 2.33, 2.23b, 2.24, and 2.25 one can obtain the equation

$$T_{\text{r}}^4 - T_{\text{s}}^4 = \eta_{\text{Samp}} \frac{1}{\varepsilon\sigma_{\text{B}}} \left[\eta_{\text{ext}} \eta_{\text{abs}} \frac{\nu_{\text{f}}}{\nu_{\text{p}}} - 1 \right] (\alpha_{\text{r}} + \alpha_{\text{b}}) I_{\text{p}} \tag{2.35}$$

Following Eq. 2.35 we can conclude that if one wants to reach some predetermined temperature, T_{s}, in samples with equal ion concentrations but different sizes, the pump intensity, I_{p}, has to be chosen carefully. It should be noted that not all predetermined temperatures can be reached using a sample with an arbitrary size and ion concentration. The threshold GEF of the sample for any predetermined final equilibrium sample temperature, T_{s}, can be easily estimated. Let us assume that the pump intensity is equal to the saturation pump intensity. In this case the absorbed power density in the sample is $P_{\text{abs}} \approx N_{\text{T}} \sigma_{\text{a}}(\nu_{\text{p}}) I_{\text{p}}^{\text{Sat}}$ and the threshold GEF of the sample can be estimated as

$$\eta_{\text{Samp}}^{\min} \approx \frac{\varepsilon\sigma_{\text{B}} \left(T_{\text{r}}^4 - T_{\text{s}}^4 \right)}{\eta_{\text{cool}} \sigma_{\text{a}}(\nu_{\text{p}}) N_{\text{T}} I_{\text{p}}^{\text{Sat}}} \tag{2.36}$$

Figure 2.4 illustrates the dependence of the threshold GEF of the sample, $\eta_{\text{Samp}}^{\min}$, simulated with Eq. 2.36 on the ion concentration if the final equilibrium temperature of the sample is $T_{\text{s}} = 80$ K or 100 K.

2.4 Optical Cavities

2.4.1 Intracavity Cooling

Laser cooling of solids is the only solid-state technology, which can cool to cryogenic temperatures. Unfortunately, the efficiency of this cooling process is very low. Indeed, in RE-based optical refrigeration the cooling power is only several percent of the absorbed pump power. As one can see in Eq. 2.33 the temperature of the sample, T_s, decreases as the cooling power, P_{cool}, increases. The cooling power $P_{cool} = -\eta_{cool} \cdot P_{abs}$ is proportional to the absorbed pump power. The pump power absorbed in the sample on the length L_{abs} is

$$P_{abs} = P_p(1 - e^{-\alpha_{abs} L_{abs}}) \tag{2.37}$$

where $\alpha_{abs} = \alpha_r(\nu, T) + \alpha_b$ is the absorption coefficient. Due to Boltzmann distribution of electrons in the ground and excited manifolds, resonant absorption decreases with the decreasing temperature. The decrease in the resonant absorption, $\alpha_r(\nu,T)$, can be compensated by the increase in the path length L_{abs}. In 1968, Kushida and Geusic placed the Nd^{3+}:YAG cooling sample inside the cavity of the Nd^{3+}:YAG laser [11]. In an attempt to cool a Nd^{3+}:YAG crystal for the first time they observed a reduction of heating. The intracavity technique has the advantage of very high intracavity circulating power (Fig. 2.5). In 2004, this technique was used by Heeg and his colleagues to realize the neat laser cooling process in a Yb^{3+}:ZBLAN sample with the temperature drop 6 K from room temperature.

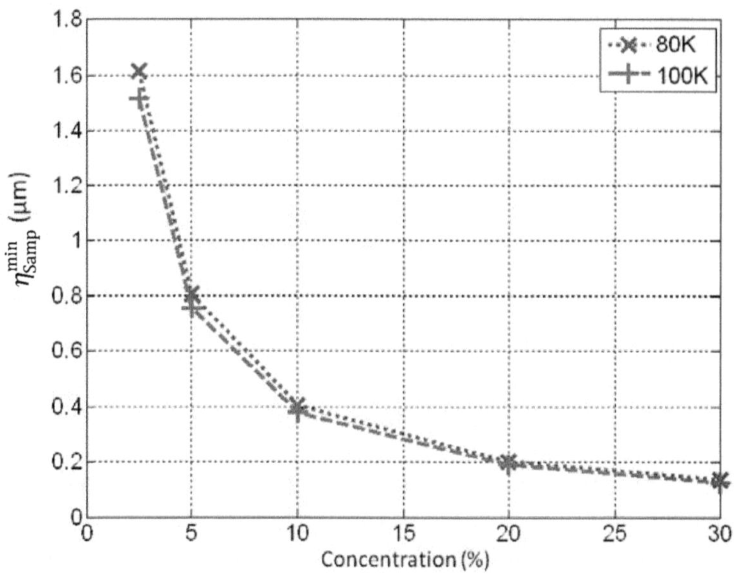

Figure 2.4 The dependence of ion concentration on the threshold GEF, η_{Samp}^{min}, of Yb^{3+}:YAG if T_s = 80 K and 100 K.

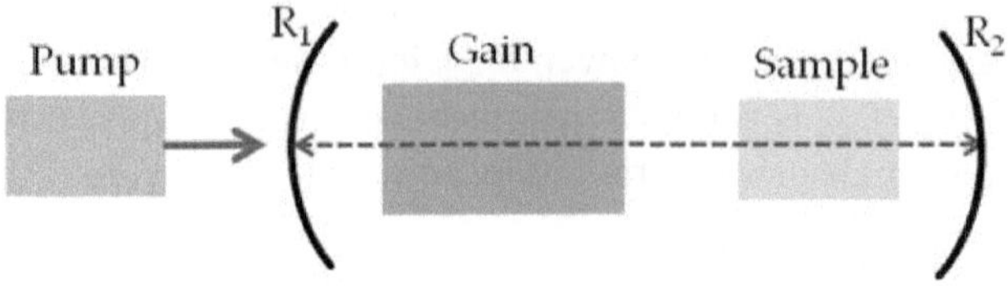

Figure 2.5 An intracavity cooling scheme: a cooling sample is placed inside a laser cavity.

In 2014, this method was applied to a vertical external cavity surface-emitting laser (VECSEL) [43]. In about 6 min a 7% Yb^{3+}:YLF sample placed in the cavity of VECSEL was cooled from room temperature down to 131 K.

2.4.2 Nonresonant Cavity

In 1998, Edwards et al. proposed to use the nonresonant cavity technique to enhance the absorption in the system [44, 45]. In the nonresonant cavity technique the pump beam travels through the cooling sample placed in the resonator at an angle slightly off normal to the mirrors (Fig. 2.6). In Edwards's experiment a Yb^{3+}:ZBLAN cylinder with 12 mm diameter and 10 mm long was mirrored on both ends (Fig. 2.6a). One mirror had a pinhole with diameter 150 μm for admitting the pump radiation.

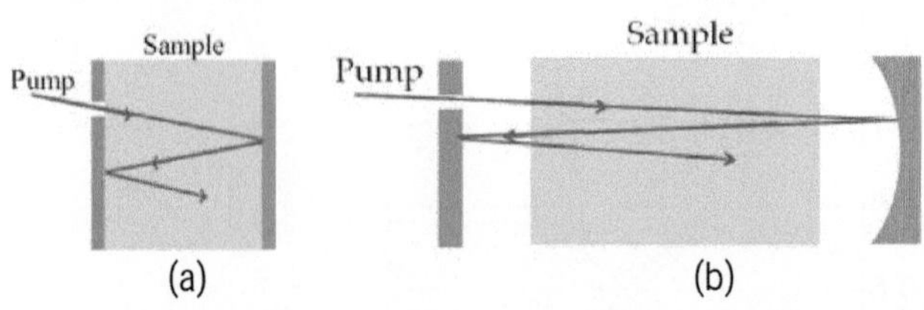

Figure 2.6 Nonresonant cavities formed by dielectric mirrors deposited on the sample (a) and by external mirrors (b).

A maximum temperature drop 48 K starting from room temperature was achieved. This technique has a number of drawbacks. Indeed, besides pump tapping the mirrors trap the fluorescence increasing the reabsorption in the system. Mirror heating is undesirable for the cooling process. In 2003, Hoyt et al. proposed to use external mirrors (Fig. 2.6b) [46]. The improved technique permits to reduce fluorescence trapping and eliminate heat conduction between mirrors and a sample. In this experiment a 1 wt% Tm^{3+}:ZIBLANP sample of volume $4 \times 4 \times 8$ mm^3 was placed in the nonresonant cavity with external dielectric mirrors, which had reflectivity ~99.9%. A planar mirror had a ~0.4 mm hole to admit the pump beam. A second mirror had a 1 m radius of curvature. A temperature change of 19 K from room temperature is achieved in this technique. The nonresonant external cavity technique was used in the experiment where a record temperature drop (ΔT = 185 K) was achieved in a Yb^{3+}:YLF sample [17].

2.4.3 Resonant Cavity

In 2003, a resonant cavity was proposed to enhance pump light absorption [46]. It has been shown that theoretically pump light absorption can reach unity for the mirrors with the reflectivity 1 and $\exp(-2\alpha_{abs}L_R)$, where $\alpha_{abs} = \alpha_r(\nu,T) + \alpha_b$. Since the resonant absorption $\alpha_r(\nu,T)$ is a function of temperature the length L_R has to be properly arranged with temperature. A piezoelectric stabilization can be a solution to the problem. As one can see in Refs. [47–49] the absorption experimentally achieved with the resonant cavity is comparable with the absorption achieved with the nonresonant cavity technique. It reaches ~90%.

2.5 Laser Cooling in Rare Earth–Doped Glasses and Crystals

The properties of radiative transitions can be investigated with absorption and emission spectra. All spectral lines can be divided in three major types of line shapes: Lorentzian, Gaussian, and Voigt. Any process, which has the same probability for all ions of the system, is a source of a Lorentzian line shape, g_L. All physical processes with a random distribution of occurrence for each ion are a source of a Gaussian line shape, g_G. The Voigt line shape, g_V, is a result of the convolution of two broadening mechanisms: one of which results in a Gaussian broadening and the other produces a Lorentzian profile. Lorentzian broadening is referred to as lifetime broadening since the processes responsible for this type of spectral line shapes are generally the processes that shorten the lifetimes of the energy levels involved in the transition, as an example, the Heisenberg uncertainty relationship relating time and energy. This contribution can be considered as the natural linewidth for a transition. The homogeneous spectral line shape can be described with the area-normalized Lorentzian profile

$$g_L(\nu,T)=\frac{2\Delta\nu_L}{\pi\left[(\Delta\nu_L)^2+4(\nu-\nu_0)^2\right]} \tag{2.38}$$

where ν_0 is the mean transition frequency (Fig. 2.7a). $\Delta\nu_L$ is the full-width at half-maximum (FWHM) given by the relation [50, 51]

$$\Delta\nu_L=\frac{1}{2\pi\tau_r}+\frac{1}{\pi T_2} \tag{2.39}$$

where T_2 is the dephasing time. As the temperature decreases the homogeneous transition linewidth $\Delta\nu_L$ is observed to narrow [53]. At very low temperatures (up to 40 K), the homogeneous linewidth is observed to follow a linear increase law with temperature (i.e., $\Delta\nu_L \approx T$) at higher temperatures (up to 300 K), the increase is nearly quadratic (i.e., $\Delta\nu_L \approx T^2$) [51, 52]. The one-phonon direct broadening is associated with liner temperature dependence; the two-phonon Raman scattering is associated with a T^2 low at higher temperatures [54]. Each dopant ion occupies a unique site in the host material and experiences a unique Stark splitting. The structural disorder in a host is a source of inhomogeneous broadening, which is generally well described by a Gaussian distribution (Fig. 2.7b). In a perfect (unrealistic) crystal all dopant ions are subjected to an identical crystal field and as a result are free from inhomogeneous broadening.

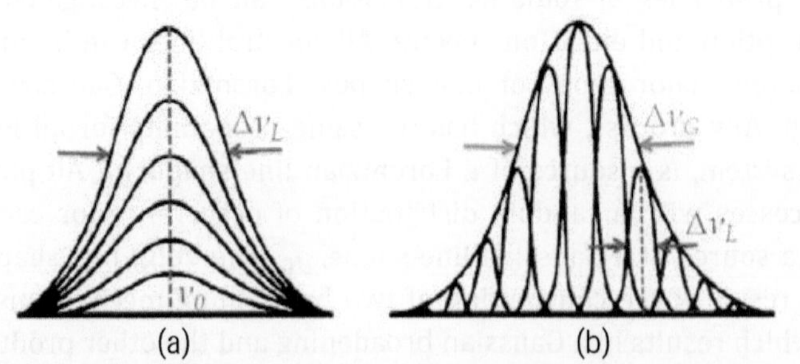

Figure 2.7 Homogeneous broadening (a) and inhomogeneous broadening (b).

On the other hand glass hosts with their structural disorder provide different coordination environment for dopant ions, which results in inhomogeneous broadening. Figure 2.7a illustrates homogeneous absorption at several ion concentrations and Fig. 2.7b illustrates inhomogeneous absorption at one concentration by a number of subpopulations. As one can see in Fig. 2.7 if one pumps a sample with the homogeneous broadening at the long wavelength tail of the absorption spectrum all ions participate in cooling. If a sample with inhomogeneous broadening is pumped at the long wavelength tail of the absorption spectrum some ions can participate in the heat generation process. Inhomogeneous linewidths are essentially independent on temperature. The temperature dependence of the homogeneous and inhomogeneous broadening in the Yb^{3+}:ZBLANP glass sample has been investigated in Ref. [55], where the Voigt absorption and emission profiles have been deconvoluted. The inhomogeneous linewidth for the Yb^{3+}:ZBLANP sample is $\Delta\nu_G \approx 990 \pm 40$ GHz. It is almost independent on temperature. The homogeneous linewidth for the Yb^{3+}:ZBLANP sample changes with temperature as $\Delta\nu_L \approx (12 \pm 2) \times 10^{-3} T^{(1.9\pm0.1)}$ in the temperature range of 10–300 K.

It is well known that the resonant absorption coefficient, $\alpha_r(\nu,T)$, can be estimated as a product of the absorption cross section $\sigma_a(\nu,T)$ and the thermal population density, $N_0(E_{0i},T)$, of the initial state with the energy E_{0i} that is $\alpha_r(\nu_p,T) = \sigma_a(\nu_p,T)\cdot N_0(E_{0i},T)$. Following the Boltzmann distribution described with Eqs. 2.14 and 2.15 the thermal population density of the initial state $N_0(E_{0i},T) \approx \exp(-E_{0i}/k_BT)$. The absorption cross section is proportional to the Voigt spectral line that is $\sigma_a(\nu_p,T) \approx g_V(\nu,T)$. Consequently, the resonant absorption coefficient can be presented as

$$\alpha_r \approx g_V(\nu,T)\cdot\exp(-E_{0i}/k_BT) \tag{2.40}$$

As one can see in Ref. [55] the inhomogeneous broadening is dominant for glass hosts. It almost do not change with temperature that is $g_V(\nu,T) \approx$ constant. As one can see in Eq. 2.40 the peak absorption coefficient decreases exponentially with temperature in glass hosts. In crystal hosts homogeneous broadening is dominant. The homogeneous linewidth scales as $\Delta\nu_L \approx T^2$ in the temperature range 40–300 K, which is interesting for optical refrigeration. As one can see in Eq. 2.38 the peak value $g_L(\nu_0,T) \approx T^{-2}$ at $\nu = \nu_0$. In this case from Eq. 2.40 the resonant absorption coefficient $\alpha_r \approx$

$T^{-2}\cdot\exp(-E_{0i}/k_BT)$. That is in crystal hosts temperature partially offsets the decrease in thermal population of the level. Thus one can conclude that with decreasing temperature a peak absorption coefficient $\alpha_r(\nu_0,T)$ in crystal hosts exceeds one in glass hosts. The absorption efficiency described with Eq. 2.25 can be considered as a "competition" between the resonant absorption coefficient α_r and the background absorption coefficient α_b. Indeed, $\eta_{abs} = \alpha_r/(\alpha_r + \alpha_b)$, that is, the higher peak absorption in crystal hosts, α_r, permits to tolerate the higher background absorption coefficient α_b in the host material. That makes crystal hosts more suitable for laser cooling of solids.

Tremendous progress in purification of fluoride glasses in 1990s caused by development of long-haul fiber-optic telecommunication links permitted to achieve the net laser cooling process in a Yb^{3+}:ZBLANP glass sample for the first time in 1995 [13]. The advantages of RE-doped crystals over RE-doped glasses for laser cooling to low temperatures for the first time was demonstrated by Bowman and Mungan in 2000 [57]. In this work local laser cooling in the Yb^{3+}-doped crystal samples (Yb^{3+}:KGW, Yb^{3+}:YLF, and Yb^{3+}:YAG) has been demonstrated for the first time. Net laser cooling in a Yb^{3+}:YAG crystal sample for the first time was demonstrated in 2001 by Epstein's research team [58]. All latest record temperatures have been achieved in the Yb^{3+}:YLF crystal samples [14, 16]. Let us briefly consider the most well-known experiments in which net laser cooling of solids has been demonstrated.

2.5.1 Laser Cooling in Rare Earth–Doped Glasses

2.5.1.1 Rare earth–doped fluoride glasses

2.5.1.1.1 *Yb^{3+}:ZBLAN and Yb^{3+}:ZBLANP fiber and bulk samples*

Heavy metal fluoride glasses including a family of glasses with a composition **ZrF_4-BaF_2-LaF_3-AlF_3-NaF** (**ZBLAN)** were developed by Poulain and Lucas at the University of Rennes in France in 1975 [59]. ZBLAN has several undisputable advantages over other glasses: low intrinsic losses (theoretically 0.001–0.01 dB/km), the wide transparency window (0.35–4 μm), and the small maximum phonon energy (~580 cm^{-1}) [60–62]. Extrinsic absorption originates from impurities such as transition metals (TMs) (Fe, Cu, Co, Ni), rare earths (Nd, Tb, Er), OH and H_2O, and other molecular species (CO, CO_2, NH_4^+). Extrinsic scattering occurs as a result of bubbles, crystals inclusions, geometrical defects, and other imperfections. The refractive index of ZBLAN is $n \approx 1.5$. It has negative thermo-optic coefficient $dn/dT = -14.75 \times 10^{-6}$ (K^{-1}). Its thermal conductivity is ~0.9 W/(mK) and expansion coefficient is 17.2×10^{-6} (K^{-1}). ZBLAN has a relatively low glass transition temperature $T_g \approx 260$°C. Fluoride glasses have a number of drawbacks: fragility and sensitivity to acids. In a number of experiments devoted to laser cooling of solids a glass composition 53%ZrF_4-18%BaF_2-3%LaF_3-3%AlF_3-20%NaF-2%PbF_2-1%YbF_3 (Yb^{3+}:ZBLANP) have been used. Heavy Pb permits to reduce the maximum phonon energy down to ~506 cm^{-1} and consequently to increase the emission quantum efficiency as one can see in Eq. 2.24. A number of successful laser cooling experiments, in which net laser cooling has been achieved, have been done in bulk as well as in fiber RE-doped ZBLAN samples. In the very first proof of concept experiment a bulk $2.5 \times 2.5 \times 6.9$ mm^3 ZBLANP sample doped with 1 wt% of Yb ions was cooled down by 0.3 K below room temperature [13]. The pump beam at $\lambda_p = 1030$ nm from a continuous-wave titanium–sapphire laser was directed along the long axis of the sample in a single pass geometry. The mean fluorescence wavelength was $\lambda_f = 995$ nm. The experimental cooling efficiency was ~2%. The temperature of the sample was monitored with a liquid-nitrogen InSb IR camera [13]. The largest net laser cooling in the bulk geometry in a cylindrical fiber preform with diameter 8 mm and length 8 mm was achieved in 2005 by Thiede [63]. A 2% Yb^{3+}-doped ZBLANP sample was cooled by ~92 K from room temperature. It was pumped at wavelength 1026 nm with a 10 W Yb^{3+}:YAG laser. The most notable advance with Yb^{3+}:ZBLAN fiber samples was achieved by Gosnell in 1999, when a 1 wt% Yb^{3+}-doped ZBALN multimode fiber with diameter 175 μm and length 7 mm was cooled by 65 K from room temperature in a double-pass geometry pumped at wavelength 1015 nm with a 2.2 W pump source [28].

2.5.1.1.2 *Tm^{3+}:ZBLANP bulk samples*

In 2000 and 2003, laser cooling with Tm^{3+} ions doped in ZBLANP was demonstrated. In Tm^{3+}:ZBLANP the ground, $^3H_6$, and the first excited, $^3F_4$, manifolds of Tm^{3+} ions were involved in the cooling cycle [46, 56]. The energy gap between the $^3H_6$ and $^3F_4$ manifolds is ~6000 cm^{-1}. In Ref. [56] several Tm^{3+}-doped glass and crystal samples such as CaF2, BaF2, YAlO, LuAG, and ZBLANP have been investigated on the subject of cooling with pump excitation at 1.82 μm < λ_p < 1.97 μm. But only with a Tm^{3+}:ZBLANP sample the net laser cooling effect was achieved. The sample with volume 5 × 5 × 9 mm^3 consisted of high purity 1 wt% Tm^{3+}:ZBLANP was cut from a fiber preform. It was pumped with light from an optical parametric oscillator (OPO) at wavelength 1900 nm with the incident average pump power of ~3 W. A calibrated Mach–Zehnder (MZ) interferometer was used to monitor the temperature of the sample. The mean fluorescence wavelength was λ_f = 1820 nm. The energy gap between the manifolds of Tm^{3+} ions involved into the cooling cycle is less that the energy gap between the manifolds in Yb^{3+} ions. That means the nonradiative decay rate, W_{nr}, in Tm^{3+} ions exceeds the nonradiative decay rate in Yb^{3+} ions (see Eq. 2.28). As seen from Eq. 2.24 the increase in the nonradiative decay rate is a source of the decrease in the emission quantum efficiency, η_q. In Tm^{3+}:ZBLANP the nonradiative decay rate $W_{nr} \approx 0.64$ s^{-1}, the radiative decay rate $W_r \approx 83$ s^{-1}, the emission quantum efficiency η_q = 99.2%. The sample was cooled by 1.2 K from room temperature for a single pass of the pump beam. In spite of the increase in the nonradiative decay rate in Tm^{3+}:ZBLANP samples the cooling efficiency of Tm^{3+}:ZBLANP, which is $\eta_{cool} \approx 4.8\%$, exceeds the cooling efficiency of Yb^{3+}:ZBLANP, which is $\eta_{cool} \approx 3.5\%$. But Tm ions have several manifolds. They are not free from excited-state absorption (ESA), which deteriorates the cooling process. In 2003, a 2 wt% Tm^{3+}:ZBLANP sample of 3 × 3 × 10 mm^3 volume was cooled by 19 K from room temperature in a single-pass geometry. The sample was pumped at λ_p = 1900 nm with a pump of ~2.72 W [46].

2.5.1.1.3 *Tm^{3+}:BIG bulk samples*

The resonant absorption coefficient $\alpha_r \approx \sigma_a(\nu_p)N_T$ is proportional to the ion concentration N_T. As one can see in Eq. 2.25 the increase in α_r permits one to reach higher absorption efficiency η_{abs} in the samples with the identical background absorption coefficient α_b. However, crystallization is an obstacle to fabricate ZBLAN samples with high ion concentration. 10 mol% is the maximum concentration of RE ions in ZBLAN without inducing detrimental crystallization. The composition BaF_2-InF_3-GaF_3-ZnF_2-LuF_3-GdF_3 (BIG) can be considered as a solution to the problem, since it can have higher concentration of Yb^{3+} ions than ZBLAN. In 1999 a Yb^{3+}:BIG sample was investigated by Murtagh et al. [64]. Theoretical cooling efficiencies indicate that BIG hosts outperform ZBLANP hosts below 80 K by a factor of 2. BIG may attain a minimum temperature of 45 K compared to 55 K for ZBLANP [64]. The laser cooling process in Yb^{3+}-doped BIG samples has been investigated by Fernandez et al. [65, 66].

2.5.1.2 Laser cooling in rare earth–doped fluorochloride glasses: Yb^{3+}:CNBZn and Er^{3+}:CNBZn bulk samples

The chloride ions in fluorochloride composition such as CdF_2-$CdCl_2$-NaF-BbF_2-$BaCl_2$-ZnF_2 (CNBZn) permit to reduce the maximum phonon energy down to ~370 cm^{-1} and thus to reduce the multiphonon relaxation rate. The reduced maximum phonon energy of a host is very important for laser cooling with RE ions if the energy gap between the manifolds involved in the cooling cycle reduces, like in the case of Tm^{3+} or Er^{3+} ions. In a number of works nonhygroscopic CNBZn glass samples doped with Yb^{3+} ions have been investigated for laser cooling applications [65, 66, 67, 68]. A CNBZn sample of volume $2 \times 8 \times 10$ mm^3 doped with 1 mol% of Yb^{3+} pumped at $\lambda_p = 1010$ nm demonstrated the cooling efficiency ~25% at 300 K, which falls about 20% at 77 K [65]. In 2006 Fernandez et al. observed a temperature drop 0.5 K in a bulk-cooling experiment in a $10.7 \times 10.7 \times 2.2$ mm^3 sized Er^{3+}: CNBZn sample pumped at 860 nm with a pump of 1.9 W [69]. The mean fluorescence wavelength of a Er^{3+}:CNBZn sample was 852.5 nm. The cooling efficiency was 0.7%. The cooling process was monitored by using a combination of photothermal deflection (PTD) measurements and time evolution of the average temperature of the sample acquired with an IR camera. Table 2.2 summarizes the results of experiments devoted to laser cooling of RE-doped glasses in which net laser cooling of solids has been achieved. Figure 2.8 illustrates progress, which has been made toward laser cooling in Yb^{3+}-doped glasses since the very first successful experiment in 1995.

Table 2.2 Net laser cooling with RE-doped glasses

Host	MPE	N_c	ΔT	λ_p	Sample	Reference
		Yb^{3+}				
FIBER						
ZBLANP	506	1 wt%	16	1015	$D_{co} = 250, L = 10$	[74, 75]
ZBLANP		1 wt%	21	1015	$D_{co} = 175, L = 10$	[27]
ZBLANP		1 wt%	65	1015	$D_{co} = 175, L = 7$	[28]
ZBLAN	580	1 wt%	13	1015	$D_{co} = 175, L = 40$	[76]
ZBLAN		1 wt%	3.7	1013	$D_{co} = 175$	[93]
BULK						
ZBLANP		1 wt%	0.3	1015	$V = 2.5 \times 2.5 \times 6.9$	[13]
ZBLAN			48	1030	$D = 12, L = 10$	[45]
ZBLAN		2%	6.0	1027	$V = 2 \times 2 \times 3$	[77]
ZBLANP		2 wt%	92	1026	$D = 8, L = 8$	[63]
		Tm^{3+}				
ZBLANP		1 wt%	1.2	1900	$V = 5 \times 5 \times 9$	[56]
ZBLANP		1 wt%	19	1900	$V = 4 \times 4 \times 8$	[46]
ZBLANP		2 wt%	2	1900	$V = 3 \times 3 \times 10$	[46]
		Er^{3+}				
CBNZn	370	0.5%	0.5	860	$V = 10.7 \times 10.7 \times 2.2$	[69]

MPE (cm^{-1}), the maximum phonon energy; Nc: the ion concentration; ΔT (K), the temperature drop; λ_p (nm), the pump wavelength; D_{co} (μm), the fiber core diameter; D (mm), the preform diameter; L (mm), the length of the sample; V (mm^3), the volume of the sample.

2.5.2 Laser Cooling in Rare Earth–Doped Crystals

2.5.2.1 Yb^{3+}-doped oxide crystal (Yb^{3+}:YAG)

As already noted, the higher peak absorption cross sections in crystal hosts than in glass hosts can be useful for laser cooling to low temperatures. After a number of successful laser cooling experiments with Yb^{3+}-doped ZBLAN and ZBLANP glass samples, in 2001, net laser cooling was achieved for the first time in a Yb^{3+}:YAG crystal sample [58]. A 2.3% Yb^{3+}:YAG sample was pumped at wavelength

1030 nm with a pump of 1.8 W. YAG is the oxide crystal. It is non-hygroscopic. YAG has greater thermal conductivity (~13 W·m^{-1}K^{-1}) than ZBLAN (~0.9 W·m^{-1}K^{-1}). The refractive index (1.83) and the maximum phonon energy (630 cm^{-1}) of YAG exceed the refractive index (1.5) and the maximum phonon energy (580 cm^{-1}) of ZBLAN. As one can see in Eq. 2.30 the increase in the refractive index is a source of the increase in the fluorescence trapping, which is undesirable for laser cooling, since it results in a decrease in the external quantum efficiency. The increase in the maximum phonon energy results in a decrease in the emission quantum efficiency, which is undesirable for laser cooling too. The coefficient of thermal expansion of YAG is 6.14 × 10^{-6} K^{-1}, and its thermo-optic coefficient is $dn/dt \approx 8.4 \times 10^{-6}$ K^{-1}. In 2001, the first experiment devoted to laser cooling of Yb^{3+}:YAG was done in vacuum therefore only the radiative heat load took place. In 2013, a Yb^{3+}:YAG sample was laser-cooled in air [15]. In this experiment Brewster-cut 3 at.% Yb^{3+}:YAG crystal of 1 × 1 × 10 mm^3 volume was cooled by 8.8 K from room temperature with a pump of 4.2 W at wavelength 1029 nm. The laser cooling process in Yb^{3+}:YAG has been theoretically considered in Refs. [41, 80].

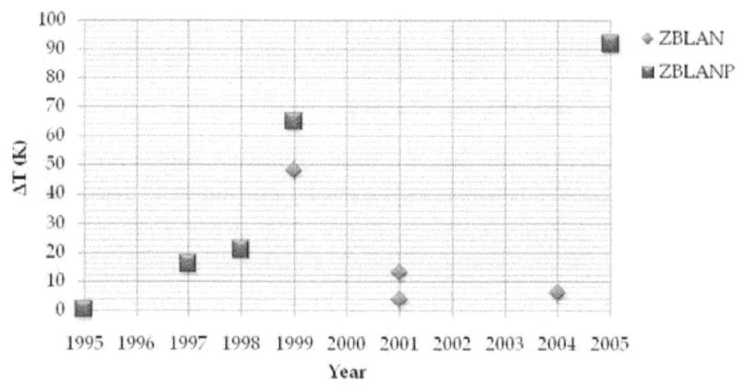

Figure 2.8 The temperature drop, ΔT, in Yb^{3+}-doped glasses achieved in different years.

2.5.2.2 Er^{3+}-doped chloride crystal (Er^{3+}:KPC)

In 2006, net laser cooling in a Er^{3+}-doped chloride KPb_2Cl_5 (KPC) crystal sample with low maximum phonon energy (~203 cm^{-1}) for the first time was observed by Fernandez et al. [69]. KPC is only slightly hygroscopic. A 4.5 × 6.5 × 2.7 mm^3 sized 0.5 mol% Er^{3+}-

doped KPC sample pumped with a pump of 1.9 W at 870 nm was cooled by ~0.7 K starting from room temperature. As one can see in Eqs. 2.28 and 2.24 low phonon energy can provide high emission quantum efficiency even for RE ions with a small energy gap between the manifolds involved in the cooling cycle like Ho^{3+} ($^5I_8$ – $^5I_7$) and Dy^{3+} ($^6H_{15/2}$ - $^6H_{13/2}$). The laser cooling process in Ho^{3+}:KPC and Dy^{3+}:KPC samples has been investigated in Refs. [70, 71].

2.5.2.3 Rare earth–doped fluoride crystals

2.5.2.3.1 *Yb^{3+}- and Tm^{3+}-doped BYF crystals*

In 2004, Patterson et al. observed laser cooling in a Tm^{3+}-doped fluoride BaY_2F_6 (BYF) crystal with low (~350 cm^{-1}) maximum phonon energy with the temperature drop ~1.5 K. A 2 wt% Tm^{3+}:BYF sample was pumped with a OPO (~6 W) at wavelength 1902 nm [72]. Four years later, in 2008, Patterson et al. cooled a 1.2% Tm^{3+}:BYF sample by 3.2 K from room temperature with a pump of 4.4 W at wavelength 1855 nm [73]. In 2006, Bigotta et al. observed laser cooling in a 2.5 mol% Yb^{3+}-doped 3 × 4 × 10 mm^3 BYF sample in single-pass geometry with the temperature drop 4 K starting from room temperature [81, 82].

2.5.2.3.2 *Yb^{3+}-doped YLF crystal*

In 2007, Bigotta et al. for the first time demonstrated laser cooling in a 4 × 4 × 12 mm^3 Yb^{3+}-doped YLiF4 (YLF) sample [83]. They pumped a 5% Yb^{3+}:YLF sample with a pump of 3 W at wavelength 1025 nm and observed cooling by 6.3 K starting from room temperature. YLF is one of widely used laser crystals. Due to the similar size, yttrium ions can be replaced with laser-active RE ions without strongly affecting the lattice structure. The thermal conductivity of undoped YLF crystal is 5.3 $W{\cdot}m^{-1}K^{-1}$ at room temperature. It increases up to 24 $W{\cdot}m^{-1}K^{-1}$ at 100 K. YLF is a low-phonon crystal with the maximum phonon energy ~350 cm^{-1}. YLF is birefringent with tetragonal crystal symmetry. It has 0.1–7.5 μm transition band. YLF has the refractive index $n_o = 1.448$, $n_e = 1.47$, and the negative thermo-optic coefficient $dn_o/dT = -4.6 \times 10^{-6}$ K^{-1}, $dn_e/dT = -6.6 \times 10^{-6}$ K^{-1} at room temperature, which decreases down to $dn_o/dT = -0.5 \times 10^{-6}$ K^{-1}, $dn_e/dT = -1.8 \times 10^{-6}$ K^{-1} at 100 K. Since 2007, several successful laser cooling experiments have been done with Yb^{3+}:YLF samples with Yb^{3+} concentrations in the range 1%–10% (Table 2.3). In 2013, the first laser cooling to a cryogenic temperature of 119 K, which is below the National Institute of Standards and Technology (NIST) defined cryogenic temperature of 123 K, was achieved in a 5 wt% Yb^{3+}:YLF crystal sample pumped at 1020 nm [59]. In 2015, the first laser cooling to double digit temperature 91 K was achieved in a 10% Yb^{3+}:YLF sample pumped at 1020 nm [17]. The source of this success is the improved purity of the host material ($\alpha_b = 10^{-4}$ cm^{-1}) and enhanced absorption of the pump light (42% over the previous experiment). A record temperature 91 K was reached in ~10 min starting from ~255 K. Table 2.3 summarizes the results of experiments devoted to laser cooling of RE-doped crystals in which net laser cooling of solids has been achieved. The progress toward laser cooling in Yb^{3+}-doped crystals is illustrated in Fig. 2.9.

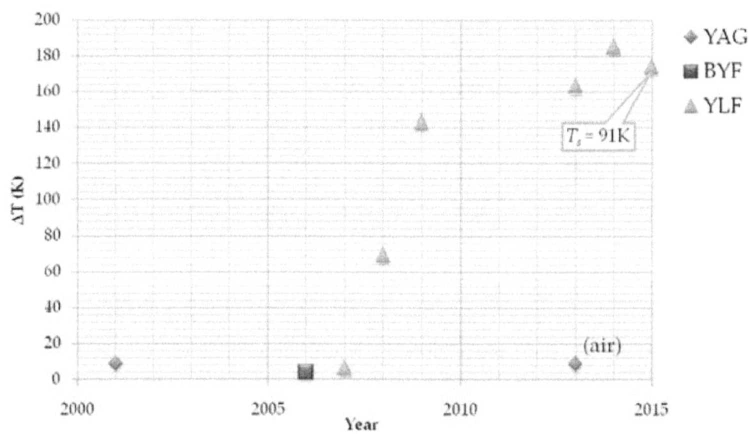

Figure 2.9 The temperature drop, ΔT, in Yb^{3+}-doped crystals achieved in different years.

Figure 2.10 illustrates the progress toward net laser cooling in RE-doped glasses and crystals. As one can see in Fig. 2.10 net laser cooling has been realized with the Yb^{3+}, Tm^{3+}, and Er^{3+} ions doped in a number of low-phonon glass and crystal hosts. Figure 2.10 presents a record temperature drop for each host material. The numbers in the brackets illustrate the year when the result was achieved. All RE-doped crystals presented in Fig. 2.10 are bulk ones with millimeter sizes.

Table 2.3 Net laser cooling with RE-doped crystals

Host	MPE	N_c	ΔT	λ_p	Sample	References
		Yb^{3+}				
YAG	630	2.3 wt%	8.9	1030		[58]
YAG		3%	8.8	1029	$V = 1 \times 1 \times 10$	[15]
BYF	350	2.5%	4	1025	$V = 3 \times 4 \times 10$	[81–83]
YLF	350	5%	6.3	1025	$V = 4 \times 4 \times 12$	[83]
YLF		5%	69	1030		[48]
YLF		5%	143	1022		[49]
YLF		5%	145	1023		[95]
YLF		5%	162	1020	$V = 3 \times 4 \times 12$	[14]
YLF		10%	185	1020		[16]
YLF		10%	174*	1020		[17]
		Tm^{3+}				
BYF		2 wt%	1.5	1902	$V = 4.5 \times 4.5 \times 7$	[72]
BYF		1.2%	3.2	1855		[73]
		Er^{3+}				
KPC	203	0.5%	0.7	870	$V = 4.5 \times 6.5 \times 2.7$	[68, 69]
KPC		3.6×10^{-19} cm^{-3}	0.0136	1567.6	$D = 3, L = 13.6$	[94]

MPE (cm^{-1}), the maximum phonon energy; N_c, the ion concentration; ΔT (K), the temperature drop; λ_p (nm), the pump wavelength; V (mm^3), the volume of the sample; D (mm), the diameter of the sample; L (mm), the length of the sample; *, the double-digit temperature (91 K) has been reached.

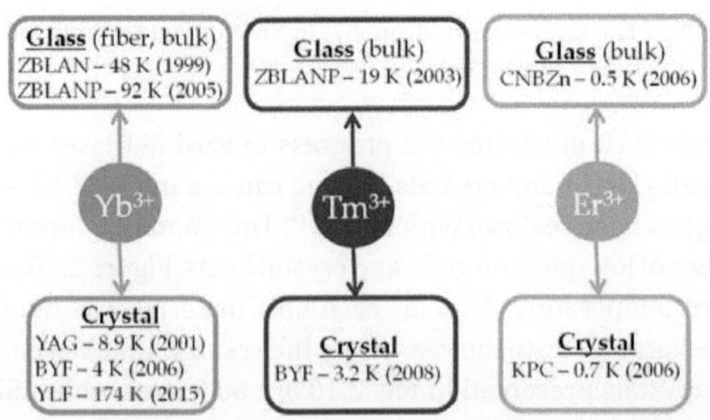

Figure 2.10 The progress toward net laser cooling in RE-doped glasses and crystals with a record temperature drop for each host material.

2.5.3 Laser Cooling in Rare Earth–Doped Oxyfluoride Glass Ceramic

RE-doped glass ceramic (GC) has a unique combination of high chemical and mechanical stability related to the oxide glass and low phonon energy of the fluoride nanocrystals, which trap the majority of RE ions. The effective embedding of RE ions in the crystalline phase with low phonon energy provides high quantum efficiency for the RE ion transitions involved in cooling cycles. On the other hand, GCs are much easier to manufacture on a large scale than are single crystals. Laser cooling with Tm^{3+}-doped oxyfluoride GCs has been considered in Ref. [84]. Yb^{3+}-doped GCs for optical refrigeration have been investigated in Ref. [85]. In this work oxyfluoride glass with chemical composition $30SiO_2$-$15Al_2O_3$-$27CdF_2$-$22PbF_2$-$4YF_3$-$2YbF_3$ was prepared by conventional melt-quenching technique. High-purity (Sigma-Aldrich, 99.99%) precursors were well mixed in an agate mortar to obtain homogeneous powder, and were transferred to a covered platinum (Pt-100%) crucible. The crucible was fired at 1100°C for 2 h in a muffle furnace. The resulting molten glass was poured into a stainless steel mold that was preheated at the glass transition temperature, where the glass was annealed at the same temperature for 5 h to remove internal stresses and strains. The glass samples were annealed at different temperatures and timescales to transform them to GCs. The Yb^{3+} glass and GC samples have been pumped at 1020 nm [85]. Net laser cooling has not been achieved, but a temperature rise of 40 K/W has been observed in the glass samples when pumped at 1020 nm, a level that is not observed in the Yb^{3+}:GC samples even at the pump wavelength 1000 nm, where the resonant absorption is much higher. That indicates a better ratio of anti-Stokes emission to nonradiative decay in the GC samples compared to the glass samples, which is important for cooling.

2.6 Temperature Measurements

The temperature of a sample can be measured with different techniques, which can be divided into contact and noncontact ones. These techniques can provide a local temperature measurement as well as the bulk temperature of a sample. All these techniques have some advantages and disadvantages and have to be properly chosen for each experiment.

2.6.1 Contact Technique

2.6.1.1 Thermocouples

In 1968, Kushida and Geusic used a thermocouple to observe heat mitigation in a Nd^{3+}:YAG sample. It was attached to the sample. Later a direct measurement of the temperature of the entire sample well insulated in an evacuated chamber has been done with a fine wire thermocouple in a number of experiments [45, 63, 76]. Conduction of heat through the thermocouple wires and absorption of the fluorescence are obstacles on a way high-quality measurement, which can be avoided with noncontact techniques.

2.6.1.2 Fiber Bragg grating

In 2013, fiber Bragg grating was used in the first experiment devoted to laser cooling of Yb^{3+}:YAG in air to measure its temperature [15]. The Bragg wavelength shift was 9.9 pm/K. With a Bragg reflection bandwidth of 616 pm, the total measurement range was 62 K. The fiber Bragg grating was in contact with the sample. The fiber Bragg grating was transparent to the fluorescence and the pump radiation. This technique is free from cumbersome calibration and optical alignment. Its dynamic range can be easily switched by changing the interrogation system while keeping the grating on the sample.

2.6.2 Noncontact Technique

2.6.2.1 Photothermal deflection

Photothermal deflection (PTD) is noncontact thermometry technique for analyzing small temperature changes in a sample. PTD was developed in 1980 as a means of measuring optical absorbance down to $\alpha_{abs}l = 10^{-8}$ [86, 87]. In laser cooling of solids PTD is employed to measure the local temperature gradients induced by the pump lasers inside samples. In this case a pump beam is focused into the sample and a probe beam passes through the sample in opposite direction. Angular deflection of the probe beam, which is caused by thermally induced refractive index gradient in the pumped volume of the sample, is measured with a position sensitive photodetector. PTD has been used in a number of works to measure the temperature of laser-cooled samples [57, 64, 65, 69, 74, 75]. Unfortunately, PTD have a number of drawbacks. It can only be utilized to single-pass geometries, since it requires careful alignment of the pump and probe laser beams. It is inherently limited to probing one small region of a sample defined by the probe laser. In this method it is important to separate the thermal contribution to the refractive index change (thermal lensing) from the accompanying contribution caused by pump-induced population changes in both ground and excited states (population lensing).

2.6.2.2 Thermal cameras

In 1995, a liquid-nitrogen-cooled InSb IR camera was used to monitor optical refrigeration in a Yb^{3+}:ZBLANP sample [13]. A thermal camera (TC) has also been used in a number of experiments devoted to laser cooling of solids to measure the temperature of the sample [46, 56, 69]. It is worth to emphasize that TC can be inappropriate for some samples (e.g., KPb_2Cl_5), because the sample can be completely transparent in most of the mid-IR and is therefore invisible to the TC. The detection range of TCs is limited to no lower than 250 K with an absolute resolution of $\sim$0.2 K

2.6.2.2 Differential luminescence thermometry

Differential luminescence thermometry (DLT) is the most widely accepted noncontact technique in recent years. It was used for the first time in 1997 by Mungan et al. to measure the temperature in a 1 wt% Yb^{3+}:ZBLANP fiber sample [74]. The luminescence spectral profile of a RE-doped sample depends on the temperature and does not depend on the pump wavelength. DLT uses temperature-dependent spectral change of a sample to achieve high temperature sensitivity. The temperature of the sample can be determined by direct comparison of sample spectra with a set of reference spectra captured with a closed-cycle cryostat. The sample temperature is then deduced from the reference spectra by a least-squares comparison. DLT has been used in some works [28, 27, 74, 75, 76, 77], but due to acquisition of the full spectrum, this technique is slow and in addition is not immune to inherent noise from CCD dark current and electronic readout. The resolution of ~20–100 mK can be reached with the DLT method.

2.6.2.3 Two-band differential spectral metrology

Two-band differential spectral metrology (2B-DSM) is based on dual-band spectral differencing via a balanced amplified photodetector (BAPD) directly in the spectral domain [78, 79]. Due to the large electronic bandwidth and high common-mode rejection of these detectors, a fast and shot-noise-limited performance can be realized with this method. A monochromator or color filters can be used to sample fluorescence spectra. This technique is sensitive to any changes in the slope of the fluorescence spectrum caused by temperature variations. 2B-DSM permits one to observe temperature on the millisecond time scale.

2.6.2.4 Differential infrared thermography

Differential infrared thermography (DIT) overcomes some of the limitations of the standard IR thermography setup. In IR thermography background absorption is not eliminated and the optical cooling channel competes with background absorption, which is present even in undoped materials resulting in heat generation. In the DIT technique two samples of the same host material are located side by side. Both samples are irradiated simultaneously. The laser spot has to be equally distributed among them. In this way, the undoped sample can be used as a reference to subtract any contributions to heating such as background absorption processes, power fluctuations, etc. not related to laser cooling with the optically active ions. DIT has been used in several experiments devoted to laser cooling of solids [88, 89].

2.6.2.5 Polarization-resolved thermometry

Noncontact and sensitive techniques based on a change of reflectivity [90] and thermally induced stress birefringence [91] in a balanced detector arrangement can be used to determine the quantum and cooling efficiencies. Polarization-resolved thermometry (PRT) is based on thermally induced stress birefringence in a sample caused by temperature gradients. It has a very high temperature resolution of 5 mK. This technique holds certain advantages over existing techniques at low starting temperatures.

2.6.2.6 Mach–Zehnder interferometry

An MZ interferometer has been proposed to measure the temperature of a cooling sample [92]. The refractive index of the sample changes with temperature over a wide temperature range. This approach has been used by Hoyt et al. to measure the laser-induced temperature change in a Tm^{3+}:ZBLAN sample [46]. This technique can be applied to single-pass and multipass geometries providing the bulk sample temperature.

2.7 Conclusions

This chapter is focused on relatively comprehensive consideration of the laser cooling process in RE-doped glasses and crystals. Laser cooling of solids is interesting not only from the fundamental point of view as a process of light and material interaction but its application in portable space-based sensor systems as well. An understanding of laser cooling of solids is very important for development of an all-solid-state cryocooler, which is a compact, reliable device free from vibrations. It is free from moving parts and fluids. It can be used for cryogenic refrigeration of semiconductor devices in a number of applications that have tight constraints for weight and size, for example, airborne and space-based detector systems. The operation of semiconductor detectors is based on electron excitation. The incident photon excites electron from the valence to conduction band inducing an electric current, which has to be measured. But random thermal excitations (thermal noise) can also excite electrons from the valence to conduction band resulting in the "dark" current that limits the performance of detectors. Cryogenic refrigeration of detectors can sufficiently reduce the thermal noise and increase the detector sensitivity. Liquid helium, methane, and ethylene ammonia can be used for cooling down to 150–50 K. But such coolers have a large weight, a finite supply of refrigerant, and a short lifetime (~1 year). Widely used thermoelectric coolers (TECs) with longer lifetimes can reach only ~175 K. Cryogenic refrigeration free from vibrations is important for high-performance IR cameras and night-vision systems, especially in satellite-borne systems. It is also important for gamma-ray spectroscopy, which can provide the required energy resolution when the Ge crystals are cooled down to ~100 K. Ultrastable optical cavities for metrology are another application of all-solid-state optical cryocoolers.

There are several ways for lowering the final temperature of the RE-doped sample and increasing the cooling efficiency. For example, RE ions with lower energy gap between the manifolds involved into cooling cycles such as Tm^{3+}, Ho^{3+}, and Dy^{3+} doped in low-phonon hosts can provide higher cooling efficiency than Yb^{3+} ions doped in the same host material. It is found that the increase in the ion concentration results in an increase in the absorbed pump power, but the cooperative effects such as reabsorption and energy transfer can deteriorate the cooling process. To reach some predetermine temperature of the sample the pump power, ion concentration, and the size of the sample have to be properly arranged. Purification of the host material is very important for all samples and can be considered as a key parameter for the practical realization of optical coolers for any size range from centimeter to nanometer size. Molecular species, which usually have vibration modes with higher energies than the optical phonons of the host, reduced quantum efficiency of the active RE ions participating in the cooling process. The TM impurities such as Fe^{2+}, Cu^{2+}, Co^{2+}, and Ni^{2+} have transitions with high cross sections in the near-IR. They can deteriorate laser cooling. The RE impurities such as Er^{3+}, Tm^{3+}, and Ho^{3+} in the Yb^{3+}-doped samples are undesirable too, but since the excited states of RE ions have higher quantum efficiency than the excited states of TM ions the RE impurities are not so detrimental for the cooling process as TM impurities and a further decrease of TM impurities is a high priority problem on the way to high-quality solid-state samples for laser cooling.

References

1. Hänsch, T. W., Schawlow, A. L. (1975). Cooling of gases with laser radiation, *Opt. Commun.*, **13**, pp. 68–69.
2. Chu, S., Cohen-Tannoudji, C., Philips, W. D. (1997). For development of methods to cool and trap atoms with laser light, *Nobel Prize in Physics*, http://nobelprize.org.

3. Lenard, P., Schmidt, F., Tomascheck, R. (1928). *Handbuch der Experimentalphysik,* 1. Teil Akademische Verlagsgesellschaft, Leipzig.
4. Pringsheim, P. (1929). Zwei Bemerkungen über den Unterschied von Lumineszenz- und Temperature-strahlung, *Z. Phys.*, **57**, pp. 739–746.
5. Vavilov, S. I. (1945). Some remarks on the Stokes of fluorescent liquids, *J. Phys. USSR*, **9**, pp. 68–73.
6. Vavilov, S. I. (1946). Photoluminescence and thermodynamics, *J. Phys. USSR*, **10**, pp. 499–501.
7. Lomonosov, M. V. (1934) Complete papers VI, Leningrad. Ломоносов, М. В. (1934) Собр. Соч., Ленинград. (in Russian).
8. Landau, L. D. (1946). On the thermodynamics of photoluminescence, *J. Phys. USSR*, **10,** pp. 503–506.
9. Kastler, A. (1950). Some suggestions concerning the production and detection by optical means of inequalities in populations of levels of spatial quantization in atoms, *J. Phys. Radium*, **11**, pp. 255–265.
10. Yatsiv, S. (1961). Anti-Stokes fluorescence as a cooling process, in *Advances in Quantum Electronics* (ed. Singer, J. R.) pp. 200–213 (Columbia Univ. Press, New York).
11. Kushida, T., Geusic, J. E. (1968). Optical refrigeration in Nd-doped yttrium aluminum garnet, *Phys. Rev. Lett.*, **21**, pp. 1172–1175.
12. Djeu, N., Whitney, W. T. (1981). Laser cooling by spontaneous anti-Stokes scattering, *Phys. Rev. Lett.*, **46**, pp. 236–239.
13. Epstein, R. I., Buchwald, M. I., Edwards, B. C., Gosnell, T. R., Mungan, C. E. (1995). Observation of laser-induced fluorescent cooling of a solid, *Nature,* **377**, pp. 500–502.
14. Melgaard, S. D, Seletskiy, D. V., Di Lieto, A., Tonelli, M., Sheik-Bahae, M. (2013). Optical refrigeration to 119 K, below National Institute of Standards and Technology cryogenic temperature, *Opt. Lett.*, **38**, pp. 1588–1590.
15. Soares de Lima Filho, E., Nemova, G., Loranger, S., Kashyap, R. (2013). Laser-induced cooling of a Yb:YAG crystal in air at atmospheric pressure, *Opt. Express*, **21**, pp. 24711–24720.
16. Melgaard, S. D., Seletskiy, D. V., Polyak, V., Asmerom, Y., Sheik-Bahae, M. (2014). Identification of parasitic losses in Yb:YLF and prospects for optical refrigeration down to 80 K, *Opt. Express*, **22**, pp. 7756–7764.
17. Melgaard, S. D., Seletskiy, D. V., Albrecht, A., Sheik-Bahae, M. (2015). First solid-state cooling below 100 K, *SPIE Newsroom* (March 13, 2015). doi:10.1117/2.1201503.005790.

18. Nemova, G., Kashyap, R. (2010). Laser cooling of solids, *Rep. Prog. Phys.*, **73**, pp. 086501-1–20.
19. Zhang, J., Li, D., Chen, R., Xiong, Q. (2013). Laser cooling of a semiconductor by 40 kelvin, *Nature*, **493**, pp. 504–508.
20. Weinstein, M. A. (1960). Thermodynamic limitation on the conversion of heat into light, *J. Opt. Soc. Am.*, **50**, pp. 597–602.
21. Chukova, Yu. P. (1969). Thermodynamic limit of the luminescence efficiency, *JETP Lett.*, **10**, pp. 294–296.
22. Chukova, Yu. P. (1971). Thermodynamic limit to the efficiency of broadband photoluminescence, *Bull. Acad. Sci. USSR Phys. Ser.*, **35**, p. 1349.
23. Chukova, Yu. P. (1974). Influence of excitation-line characteristics on efficiency of spectral conversion of energy by ions of trivalent neodymium in yttrium-aluminum garnet, *Bull. Acad. Sci. USSR Phys. Ser.*, **38**, p. 57.
24. Chukova, Yu. P. (1976). The region of thermodynamic admissibility of light efficiencies larger than unity, *Sov. Phys. JETP*, **41**, pp. 613–616.
25. Sheik-Bahae, M., Epstein, R. I. (2009). Laser cooling of solids, *Laser Photonics Rev.*, **3**, pp. 67–84.
26. Hehlen, M. P., Sheik-Bahae, M., Epstein, R. I. (2014). Solid-state optical refrigeration, in *Handbook on the Physics and Chemistry of Rare Earths*, Vol. 45, pp. 179–260.
27. Luo, X., Eisaman, M. D., Gosnell, T. R., (1998). Laser cooling of a solid by 21 K starting from room temperature, *Opt. Lett.*, **23**, pp. 639–641.
28. Gosnell, T. R. (1999). Laser cooling of a solid by 65 K starting from room temperature, *Opt. Lett.*, **24**, pp. 1041–1043.
29. Kiel, A. (1964). Multi-phonon spontaneous emission in paramagnetic crystals, in *Third International Conference on Quantum Electronics* (Grivet, P., Bloembergen, N., eds.), pp. 765–772 (Columbia University Press, New York).
30. Riseberg, L. A., Moos, H. W. (1968). Multiphonon orbit-lattice relaxation of excited states of rare earth ions in crystals, *Phys. Rev.*, **174**, pp. 429–438.
31. Hehlen, M. P. (2009). Design and fabrication of rare-earth doped laser cooling materials, in *Optical Refrigeration: Science and Applications of Laser Cooling of Solids* (Epstein, R. I., Sheik-Bahae, M., eds.), pp. 37–74 (Wiley, Weinheim, Germany).
32. Auzel, F. (2004). Upconversion and anti-Stokes processes with f and d ions in solids, *Chem. Rev.*, **104**, pp. 139–173.

33. Heeg, B., DeBarber, P. A., Rumbles, G. (2005). Influence of fluorescence reabsorption and trapping on solid-state optical cooling, *Appl. Opt.*, **44**, pp. 3117–3124.

34. Chandrasekhar, S. (1943). Stochastic problems in physics and astronomy, *Rev. Mod. Phys.*, **15**, pp. 1–89.

35. Olson, R. W., Loring, R. F., Fayer M. D. (1981). Luminescent solar concentrators and the reabsorption problem, *Appl. Opt.*, **20**, pp. 2934–2940.

36. Goutaudier, C., Lebbou, K., Guyot, Y., Ito, M., Canibano, H., Hassouni, A. E., Laversenne, L., Cohen-Adad, M. T., Boulon, G. (2003). Advances in fibre crystals: growth and optimization of spectroscopic properties for Yb^{3+}-doped laser crystals, *Ann. Chim. Mater.*, **28**, pp. 73–88.

37. Förster, Th. (1948). Zwischenmolekulare Energiewanderung und Fluoreszenz, *Ann. Phys.*, **6**, pp. 55–75.

38. Dexter, D. L. (1953). A theory of sensitized luminescence in solids, *J. Chem. Phys.*, **21**, pp. 836–850.

39. Auzel, F., Bonfigli, F., Gagliari, S., Baldacchini, G. (2001). The interplay of self-trapping and self quenching for resonant transitions in solids; role of a cavity, *J. Lumin.*, **94–95**, pp. 293–297.

40. Jubera, V., Sablayrolles, J., Guillen, F., Decourt, R., Couzi, M., Garcia, A. (2009). From the infrared to the visible range: spectroscopic studies of ytterbium doped oxyborates, *Opt. Commun.*, **282**, pp. 53–59.

41. Nemova, G., Kashyap, R. (2015). Optimization of optical refrigeration in Yb^{3+}:YAG samples, *J. Lumin.*, **164**, pp. 99–104.

42. Malinowski, M., Kaczkan, M., Piramidowicz, R., Frukacz, Z., Sarnecki, J. (2001). Cooperative emission in Yb3+:YAG planar epitaxial waveguides, *J. Lumin.*, **94–95**, pp. 29–33.

43. Ghasemkhani, M., Albrecht, A. R., Melgaard, S. D., Seletskiy, D. V., Cederberg, J., Sheik-Bahae, M. (2014). Intracavity optical refrigeration to 130 K using high-power vertical external-cavity surface-emitting laser (VECSELs), *Proc. SPIE*, **9000**, pp. 900005-1–7.

44. Edwards, B. C., Buchwald, M. I., Epstein, R. I. (1998). Development of the Los Alamos solid-state optical refrigerators, *Rev. Sci. Instrum.*, **69**, pp. 2050–2055.

45. Edwards, B. C., Anderson, J. E., Epstein, R. I., Mills, G. L., Mord, A. J. (1999). Demonstration of a solid-state optical cooler: an approach to cryogenic refrigeration, *J. Appl. Phys.*, **86**, pp. 6489–6493.

46. Hoyt, C. W., Hasselbeck, M. P., Sheik-Bahae, M., Epstein, R. I., Greenfield, S., Thiede, J., Distel, J., Valencia, J. (2003). Advances in laser cooling of thulium-doped glass, *J. Opt. Soc. Am. B*, **20**, pp. 1066–1075.
47. Seletskiy, D. V., Hasselbeck, M. P., Sheik-Bahae, M. (2010). Resonant cavity-enhanced absorption for optical refrigeration, *Appl. Phys. Lett.*, **96**, pp. 181106-1–3.
48. Seletskiy, D. V., Hasselbeck, M. P., Sheik-Bahae, M., Epstein, R. I., Bigotta, S., Tonelli, M. (2008). Cooling of Yb:YLF using cavity enhanced resonant absorption, *Proc. SPIE*, **6907**, pp. B9070–B9079.
49. Seletskiy, D. V., Melgaard, S. D., Bigotta, S., Di Lieto, A., Tonelli, M., Epstein, R. I., Sheik-Bahae, M. (2009). Demonstration of an optical cryocooler, *CLEO/IQEC 2009*, post deadline submission.
50. Weber, M. J. (1985). Laser spectroscopy of glasses, *Ceram. Bull.*, **64**, p. 1439.
51. MacFarlane, R. M., Shelby, R. M. (1987). Homogeneous line broadening of optical transitions of ions and molecules in glasses, *J. Lumin.*, **36**, pp. 179–207.
52. Yen, W. M., Brundage, R. T. (1987). Fluorescence line narrowing in inorganic glasses: linewidth measurements, *J. Lumin.*, **36**, pp. 209–220.
53. Siegman, A. E. (1986). *Lasers* (University Science Books. Mill Valley, CA).
54. Pellegrino, J. M., Yen. W. M., Weber, M. J. (1980). Composition dependence of Nd^{3+} homogeneous linewidths in glasses, *J. Appl. Phys.*, **51**, pp. 6332–6336.
55. Lei, G., Anderson, J. E., Buchwald, M. I., Edwards, B. C., Epstein, R. I. (1998). Determination of spectral linewidths by Voigt profiles in Yb^{3+}-doped fluorozirconate glasses, *Phys. Rev. B*, **57**, pp. 7673–7678.
56. Hoyt, C. W., Sheik-Bahae, M., Epstein, R. I., Edwards, B. C., Anderson, J. E. (2000). Observation of anti-Stokes fluorescence cooling in Thulium-doped glass, *Phys. Rev. Lett.*, **85**, pp. 3600–3603.
57. Bowman, S. R., Mungan, C. E. (2000). New materials for optical cooling, *Appl. Phys. B*, **71**, pp. 807–811.
58. Epstein, R. I., Brown, J. J., Edwards, B. C., Gibbs, A. (2001). Measurements of optical refrigeration in ytterbium-doped crystals, *J. Appl. Phys.*, **90**, pp. 4815–4819.
59. Poulain, M., Poulain, M., Lucas, J. (1975). Verres fluores au tetrafluorure de zirconium proprietes optiques d'un verre dope au Nd^{3+}, *Mater. Res. Bull.*, **10**, pp. 243–246.

60. Tsoukala, V. G., Schoeder, J., Floudas, G. A., Thompson, D. A. (1987). Intrinsic Rayleigh scattering in fluoride glasses, *Mater. Sci. Forum.*, **19–20**, pp. 637–646.

61. Takahashi, S. (1992). Prospects for ultra-low loss using fluoride glass optical fiber: a review, *J. Non-Cryst. Solids*, **140**, pp. 172–178.

62. Gan, F. (1995). Optical properties of fluoride glasses: a review, *J. Non-Cryst. Solids*, **184**, pp. 9–20.

63. Thiede, J., Distel, J., Greenfield, S. R., Epstein, R. I. (2005). Cooling to 208 K by optical refrigeration, *Appl. Phys. Lett.*, **86**, pp. 154107-1–2.

64. Murtagh, M. T., Sigel, G. H., Fajardo, J. C., Epstein, R. I. (1999). Laser induced fluorescent cooling of rare-earth–doped fluoride glasses, *J. Non-Cryst. Solids*, **253**, pp. 50–57.

65. Fernandez, J., Mendioroz, A., Garcia-Adeva, A. J., Balda, R., Adam, J. L. (2000). Anti-Stokes laser-induced internal cooling of Yb^{3+}-doped glasses, *Phys. Rev. B*, **62**, pp. 3213–3217.

66. Fernandez, J., Mendioroz, A., Garcia, A. J., Balda, R., Adam, J. L. (2001). Laser-induced internal cooling of Yb-doped fluoride-based glasses, *J. Alloys Compd.*, **323–324**, pp. 239–244.

67. Fernandez, J., Mendioroz, A., Garcia, A. J., Balda, R., Adam, J. L., Arriandiaga, M. A. (2001). On the origin of anti-Stokes laser-induced cooling of Yb^{3+}-doped glass, *Opt. Mater.*, **16**, pp. 173–179.

68. Fernandez, J., Garcia-Adeva, A. J., Balda, R. (2012). Anti-Stokes laser-induced cooling in rare-earth doped low phonon materials, *Opt. Mater.*, **34**, pp. 579–590.

69. Fernandez, J., Garcia-Adeva, A. J., Balda, R. (2006). Anti-Stokes laser cooling in bulk erbium-doped materials, *Phys. Rev. Lett.*, **97**, pp. 033001-1–4.

70. Nostrand, M. C., Page, R. H., Payne S. A., Isaenko L. I., Yelisseyev, A. P. (2001). Optical properties of Dy^{3+} - and Nd^{3+}-doped KPb_2Cl_5, *J. Opt. Soc. Am. B*, **18**, pp. 264–276.

71. Brown, Ei E., Hömmerich, U., Kumi-Barimah, E., Bluiett, A., Trivedi, S. B. (2015). Comparative spectroscopic studies of Ho: KPb2Cl5, Ho: KPb2Br5, and Ho: YAG for 2 μm laser cooling applications, *Proc. SPIE*, **9380**, pp. 938000-1–7.

72. Patterson, W., Hasselbeck, M. P., Sheik-Bahae, M., Bigotta, S., Parisi, D., Toncelli, J., Tonelli, M., Epstein, R. I., Thiede, J. (2004). Observation of optical refrigeration in Tm^{3+}:BaY_2F_8, *Lasers and Electro-Optics (CLEO* (San Francisco, CA, USA).

73. Patterson, W., Bigotta, S., Sheik-Bahae, M., Parisi, D., Tonelli, M., Epstein, R. I. (2008). Anti-Stokes luminescence cooling of Tm^{3+}doped BaY_2F_8, *Opt. Express,* **16**, pp. 1704–1710.

74. Mungan, C. E., Buchwald, M. L., Edwards, B. C., Epstein, R. I., Gosnell, T. R. (1997a). Laser cooling of a solid by 16 K starting from room-temperature, *Phys. Rev. Lett.,* **78**, pp. 1030–1033.

75. Mungan, C. E., Buchwald, M. I., Edwards, B. C., Epstein, R. I., Gosnell, T. R., (1997b). Internal laser cooling of Yb^{3+}-doped glass measured between 100 and 300 K, *Appl. Phys. Lett.,* **71**, pp. 1458–1460.

76. Rayner, A., Hirsch, M., Heckenberg, N. R., Rubinsztein-Dunlop, H. (2001). Distributed laser refrigeration, *Appl. Opt.,* **40**, pp. 5423–5429.

77. Heeg, B., Stone, M. D., Khizhnyak, A., Rumbles, G., Mills, G., DeBarber, P. A. (2004). Experimental demonstration of intracavity solid-state laser cooling of Yb^{3+}:ZrF_4-BaF_2-LaF_3-AlF_3-NaF glass, *Phys. Rev. A,* **70**, pp. 021401-1–4.

78. Seletskiy, D. V., Melgaard, S. D., Epstein, R. I., Di Lieto, A., Tonelli, M., Sheik-Bahae, M. (2011). Local laser cooling of Yb:YLF to 110 K, *Opt. Express,* **19**, pp. 18229–18236.

79. Seletskiy, D. V., Melgaard, S. D., Epstein, R. I., Di Lieto, A., Tonelli, M., Sheik-Bahae, M. (2012). Precise determination of minimum achievable temperature for solid-state optical refrigeration, *J. Lumin.,* arXiv:1203.5967

80. Nemova, G., Kashyap, R. (2014). Laser cooling in Yb^{3+}:YAG, *J. Opt. Soc. Am. B,* **3**, pp. 340–348.

81. Bigotta, S., Parisi, D., Bonelli, L., Toncelli, A., Tonelli, M., Di Lieto, A. (2006). Spectroscopic and laser cooling results on Yb^{3+}-doped BaY_2F_8 single crystal, *J. Appl. Phys.,* **100**, pp. 013109-1–7.

82. Bigotta, S., Parisi, D., Bonelli, L., Toncelli, A., Di Lieto, A., Tonelli, M. (2006). Laser cooling of Yb^{3+}-doped BaY_2F_8 single crystal, *Opt. Mater.,* **28**, pp. 1321–1324.

83. Bigotta, S., Di Lieto, A., Parisi, D., Bonelli, L., Toncelli, A., Tonelli, M. (2007). Single fluoride crystals as materials for laser cooling application, *Proc. SPIE,* **6461**, pp. 64610E–646110E.

84. Nemova, G., Kashyap, R. (2012). Laser cooling with Tm^{3+}-doped oxy-fluoride glass ceramic, *J. Opt. Soc. Am. B,* **29**, pp. 3034–3038.

85. Soares de Lima Filho, E., Krishnaiah, K. V., Ledemi, Y., Yu Ye-Jin, Messaddeq, Y., Nemova, G., Kashyap, R. (2015) Ytterbium-doped glass-ceramics for optical refrigeration, *Opt. Express,* **23**, pp. 4630–4640.

86. Boccara, C., Fournier, D., Jackson, W., Amer, N. M. (1980). Sensitive photothermal deflection technique for measuring absorption in optically thin media, *Opt. Lett.*, **5**, pp. 377–379.

87. Jackson, W. B., Amer, N. M., Boccara, A. C., Fournier, D. (1981). Photothermal deflection spectroscopy and detection, *Appl. Opt.*, **20**, pp. 1333–1344.

88. Garcia-Adeva, A. J., Balda, R., Al Saleh M., Garcia-Revilla, S., Fernandez, J. (2011). Local internal and bulk optical cooling in Nd-doped crystals and nanocrystalline powders revisited, *Proc. SPIE*, **7951**, pp. 795106-1–10.

89. Garcia-Adeva, A. J., Balda, R., Al Saleh, M., Garcia-Revilla, S., Sola, D., Fernandez, J. (2012). Optical cooling of Nd-doped solids, *Proc. SPIE*, **8275**, pp. 827502-1–14.

90. Ghasemkhani, M., Seletskiy, D. V., Sheik-Bahae, M. (2011). Sensitive thermal reflectance measurement for laser cooling applications, *Proc. SPIE*, **7951**, pp. 79510J-1–6.

91. Ghasemkhani, M., Seletskiy, D. V., Sheik-Bahae, M. (2012). Polarization-resolved optical metrology for noncontact thermometry, *Proc. SPIE*, **8275**, pp. 82750J-1–5.

92. Farley, C. III, Reddy, B. R. (2011). Mach-Zehnder interferometric measurement of laser heating/cooling in Yb^{3+}:YAG, *Proc. SPIE*, **7951**, pp. 79510I-1–8.

93. Rayner. A., Friese, M. E. J., Truscott, A. G., Heckenberg, N. R., Rubinsztein-Dunlop, H. (2001). Laser cooling of a solid from ambient temperature, *J. Mod. Opt.*, **48**, pp. 103–114.

94. Condon, N. J., Bowman, S. R., O`Connor, P. O., Quimby, R. S., Mungan, C. E. (2009). Optical cooling in Er^{3+}:KPb_2Cl_5, *Opt. Express*, **17**, pp. 5466–5472.

95. Seletskiy, D. V., Melgaard, S. D., Bigotta, S., Di Lieto, A., Tonelli, M., Sheik-Bahae, M. (2010). Laser cooling of solids to cryogenic temperatures, *Nat. Photonics*, **4**, pp. 161–164.

Chapter 3

Progress toward Laser Cooling of Thulium-Doped Fibers

Nai-Hang Kwong,[a] Rolf Binder,[a] Dan Nguyen,[b] and Arturo Chavez-Pirson[b]

[a]*College of Optical Sciences, University of Arizona, Tucson, AZ 85721, USA*
[b]*NP Photonics, 9030 Rita Road, Tucson, AZ 85747, USA*
kwong@optics.arizona.edu

3.1 Introduction

Laser cooling has been successfully demonstrated on rare earth-doped glass [1–3] and CdS semiconductor crystals [4]. (For recent general reviews of progress in solid laser cooling, see, for example, Refs. [5, 6].) These advances prompted considerations on how laser cooling of solids may be implemented on a practical scale for applications. In particular, it was noted (2010 Office of Naval Research STTR Solicitation and private communications with Dr. Deborah Van Vechten, ONR) that the possibility of transferring waste heat as photons to remote locations is a desirable advantage of optical refrigeration. These considerations have motivated us to

Laser Cooling: Fundamental Properties and Applications
Edited by Galina Nemova

ISBN 978-981-4745-04-8 (Hardcover), 978-981-4745-05-5 (eBook)
www.panstanford.com

examine the feasibility of the laser cooling of doped optical fibers. In this chapter, we discuss the progress in our theoretical and experimental efforts in this direction.

An all-fiber cooling approach, if successful, has many practical advantages. A fiber device is typically lightweight, small-sized, vibrationless, and almost maintenance free. Fibers have long operation lifetimes, even under critical (e.g., in outer space) environmental conditions. The doping rare earth atoms have been demonstrated to yield high laser cooling efficiency in small glass host samples. It is also convenient to integrate multiple fiber lasers and fiber coolers in one cooling configuration and to design devices to transfer waste (heat) photons out of the fiber cooler to desired locations. Our experimental cooling fibers are made of thulium-doped germinate or tellurite glass which has a potentially high cooling efficiency. A sketch of a design of a fiber cooler is shown in Fig. 3.1.

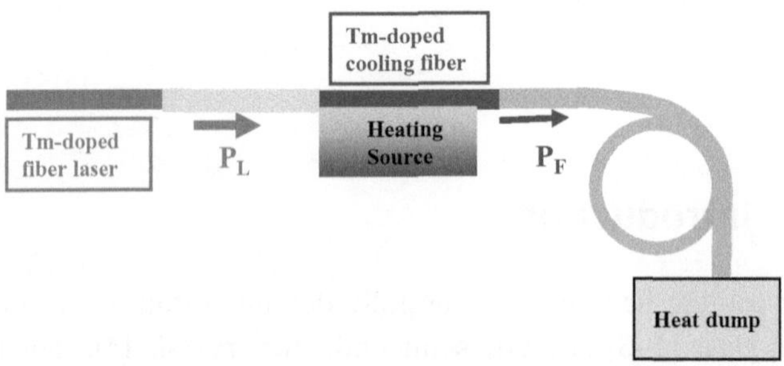

Figure 3.1 Schematic of the fiber cooling geometry. A fiber laser pumps the fiber cooling segment, which is attached to a heat source. The up-converted photoluminescence light can, in principle, be piped arbitrary distances before being dumped and converted to heat.

Expectedly there are challenging issues to be tackled. For example, the purity level of the (undoped) fiber needs to be high to avoid excessive parasitic absorption of photons by impurities. When the fiber cooler and the object to be cooled are macroscopic (centimeter scale or larger), managing the (conductive and convective) heat and photon flows is also a pertinent issue. These and related issues will be discussed below.

In Section 3.2, we describe our theoretical developments in this investigation. We discuss the calculation of the cooling powers of the Tm atoms and of a model isolated fiber and the simulation of heat flow in the cooling of the heat load. In Section 3.3, we discuss experimental progress in fabricating high-purity fiber and temperature measurements under optical pumping conditions.

3.2 Theoretical Developments

3.2.1 Modeling the Cooling of Thulium Atoms

The basic cooling agents in our setup are the thulium (Tm) atoms doping the fiber. We describe in this subsection a photoexcitation model of a Tm atom in a glass host, which we use to estimate its laser cooling efficiency. More detail can be found in [7]. The model, given in Ref. [8], is shown in Fig. 3.2. It consists of four energy levels, each with a degree of degeneracy g_i, as designated in the figure. The lower two levels represent the manifold of states created by the splitting of the 13-fold degenerate ground state ($^3H_6$) of Tm by interactions with the glass host, and the upper two levels represent the split ninefold excited-state ($^3F_4$) manifold. The pump laser energy $h\nu$ is tuned to the energy between |1> and |2>, which is the lowest excitation energy between the two manifolds. The pumped Tm atoms de-excite (incoherently) to a level in the ground-state manifold via luminescence, at rate W_{rad}, and nonradiative transitions, at a rate W_{nr}. If the de-excitation happens sufficiently slowly, the excited atoms have time to thermalize in the excited-state manifold. The luminescent photons are thus at higher energies than that of the pump photons. In each cycle of operation in which a Tm atom absorbs a pump photon and emits a luminescent photon, which then escapes, heat is removed from the system.

The population dynamics of the four-level model obeys the following rate equations [8]:

$$\frac{dN_1}{dt} = -\sigma_{12}\left(N_1 - \frac{g_1}{g_2}N_2\right)\frac{I_{\text{inc}}}{h\nu} + \frac{R}{2}(N_2 + N_3) - w_1\left(N_1 - \frac{g_1}{g_0}N_0 e^{-\delta E_g/k_B T}\right) \tag{3.1}$$

$$\frac{dN_2}{dt} = \sigma_{12}\left(N_1 - \frac{g_1}{g_2}N_2\right)\frac{I_{\text{inc}}}{h\nu} - RN_2 + w_2\left(N_3 - \frac{g_3}{g_2}N_2 e^{-\delta E_u/k_B T}\right) \tag{3.2}$$

$$\frac{dN_3}{dt} = -RN_3 - w_2\left(N_3 - \frac{g_3}{g_2}N_2e^{-\delta E_u/k_BT}\right) \tag{3.3}$$

$$N_{total} = N_0 + N_1 + N_2 + N_3 \tag{3.4}$$

where N_i and g_i are the population density and degree of degeneracy, respectively, of level $|i>$, and N_{total} is the total density of Tm atoms. The pump laser has intensity I_{inc} and frequency v. $R = 2W_{rad} + 2W_{nr}$, and σ_{12} denotes the photoabsorption cross section of the Tm atoms. δE_g and w_1 are the energy difference between the two levels and the intramanifold relaxation rate in the ground-state manifold, and δE_u and w_2 are the corresponding quantities for the excited-state manifold. T is the sample temperature, and k_B is the Boltzmann constant. In Eq. 3.1, the first term on the right side represents stimulated absorption from and emission into the pump laser mode. The second term models the decay of the excited manifold population and the third term the thermalization within a manifold. The terms in Eqs. 3.2 and 3.3 have the same interpretation.

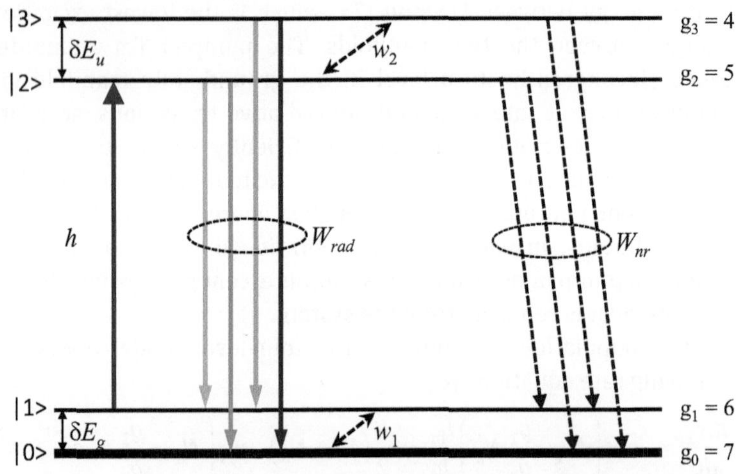

Figure 3.2 Sketch of the four-level model used to simulate cooling of Tm (after Ref. [8]). The ground-state manifold represents the $^3H_6$ state, and the excited-stated manifold $^3F_4$. Reprinted with permission from Ref. [7], Copyright 2011, Society of Photo Optical Instrumentation Engineers.

From the solution of the rate equations, we can obtain the absorbed power density as

$$P_{\text{abs}} = \sigma_{12} N_1 \left(1 - \frac{g_1 N_2}{g_2 N_1}\right) I_{\text{inc}} + \alpha I_{\text{inc}} \tag{3.5}$$

and the luminescence power density as

$$P_{\text{rad}} = W_{\text{rad}}[N_2(E_{21} + E_{20}) + N_3(E_{31} + E_{30})] \tag{3.6}$$

where $E_{ij} = E_i - E_j$ and E_i is the energy of level $|i>$. In Eq. 3.5, apart from the power absorbed by the Tm atoms, we have introduced a parasitic absorption contribution αI_{inc}, which takes into account the photon power absorbed by other excitation modes, such as (unwanted) impurities in the glass host, and converted into heat. $1/\alpha$ represents the spatial scale of this absorption. Assuming all luminescence photons to be able to leave the system, the cooling performance of the Tm-doped glass is characterized by the net power it gains

$$P_{\text{net}} = P_{\text{ads}} - P_{\text{rad}} \tag{3.7}$$

and its cooling efficiency

$$\varepsilon_c = -P_{\text{net}}/P_{\text{abs}} \tag{3.8}$$

A negative P_{net} signifies that cooling is possible.

Figure 3.3 shows some calculated results for the cooling power P_{net} as a function of the (temporally constant) laser pump intensity based on the steady-state solutions of Eqs. 3.1–3.4. The different curves are calculated with different values of the parasitic absorption rate α. The other parameters are set at pump wavelength = 1.94 μm ($h\nu$ = 0.639 eV = 1.024×10^{-19} J), $E_{21} = h\nu$, $\sigma_{12} = 2 \times 10^{-26}$ m^2, W_{rad} = 83 s^{-1}, W_{nr} = 0.64 s^{-1}, $\delta E_u = 1.15 \times 10^{-20}$ J, $\delta E_g = 1.00 \times 10^{-20}$ J, $w_1 = w_2$ = 1000 s^{-1}, $N_{\text{total}} = 3.43 \times 10^{26}$ m^{-3}, and T = 300 K. These parameter values are mostly taken from Ref. [9], where they are validated in an experimental study of laser cooling of Tm-doped ZBLAN glass. Although our glass host is germanate or tellurite, the adopted parameter values can be considered adequate in getting a reasonable estimate of the cooling power.

The trend of the curves in Fig. 3.3 shows the detrimental effect of the parasitic absorption: cooling is not possible for α larger than a value of around 2×10^{-2} m^{-1}. In each curve, the deterioration of the cooling power at high pump intensity is due to the saturation of the absorption/luminescence of the Tm atoms. For steady-state solutions with no parasitic absorption ($\alpha = 0$), the cooling power P_{net} can be shown to approach, as I_{inc} tends to infinity, the limit

$$\lim_{I_{\text{inc}}\to\infty} P_{\text{net}}(\alpha=0) = Rh\nu(1+f_{32})\left(1-\eta_{\text{q}}\frac{h\nu_{\text{m}}}{h\nu}\right)N_2 \tag{3.9}$$

where

$$h\nu_{\text{m}} = h\nu + \frac{\delta E_{\text{g}}}{2} + \frac{f_{32}}{1+f_{32}}\delta E_{\text{u}}, \quad \eta_{\text{q}} = \frac{W_{\text{rad}}}{W_{\text{rad}}+W_{\text{nr}}} \tag{3.10}$$

$$f_{32} = \left(\frac{w_2}{w_2+R}\right)\frac{g_3}{g_2}e^{-\delta E_{\text{u}}/k_{\text{B}}T} \tag{3.11}$$

and the density N_2 approaches the limit

$$\lim_{I_{\text{inc}}\to\infty}\frac{N_2}{N_{\text{total}}} = \left[\frac{g_1}{g_2}\left(\frac{g}{g+1}\right)+\left(1+(1+f_{32})\frac{R}{2gw_1}\right)\right]^{-1}, \quad g = \frac{g_1}{g_0}e^{-\delta E_{\text{g}}/k_{\text{B}}T} \tag{3.12}$$

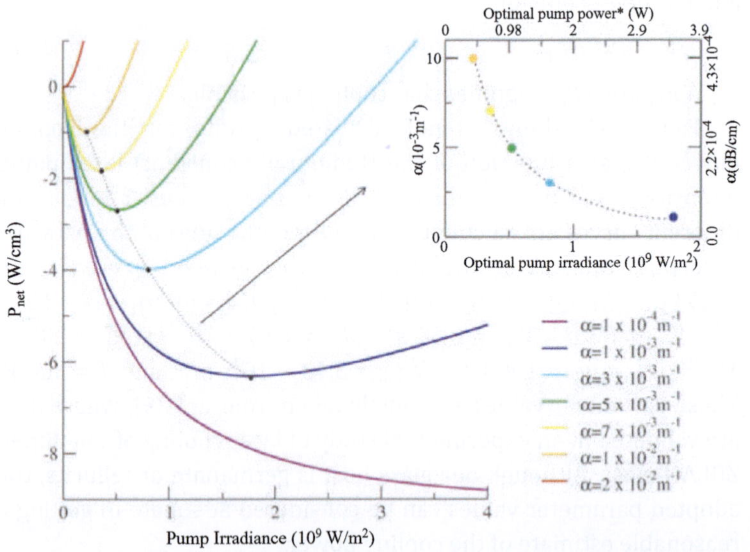

Figure 3.3 Calculated cooling power ($P_{\text{net}} < 0$ signifies cooling) of the fiber versus pump irradiance for a range of values of parasitic absorption rate α. The dotted line and the inset show the optimal cooling power for each α. For each curve, the deterioration of cooling power at high irradiance is due to absorption saturation of the thulium ions.

In Eq. 3.10, $h\nu_{\mathrm{m}}$ is defined as the mean luminescence energy and η_{q} is the quantum efficiency. Equation 3.9 implies that a necessary condition for cooling to occur is

$$\eta_{\mathrm{q}} \frac{h\nu_{\mathrm{m}}}{h\nu} > 1 \tag{3.13}$$

The saturation of the cooling capacity of the Tm atoms at high pump intensities, as implied by Eq. 3.9, allows the parasitic absorption to eventually dominate and disable the cooling of the fiber. The optimal cooling power for each α is marked on the corresponding curve in Fig. 3.3 and plotted in the inset of the figure. We have also repeated the cooling power calculation using a more elaborate model with 13 levels in the ground manifold and 9 levels in the excited manifold, obtaining similar results as in Fig. 3.3.

3.2.2 Cooling Power of a Pumped, Doped Fiber in Free Space

In this subsection, we describe some theory developments in support of our effort to estimate the temperature and infer the cooling power of the irradiated, isolated fiber from its measured emission spectrum. After the fiber is irradiated at low intensity, its emission spectrum shows a systematic dependence on the temperature at which it is set. This dependence allows the emission spectrum to be calibrated for temperature measurements. Then in a laser cooling measurement, when the fiber is pumped at higher intensities, its emission spectrum can be measured, and the fiber's temperature can be determined by comparing this spectrum to the calibration spectra. This strategy was successfully used in Refs. [10, 11]. We have applied it to estimate the temperature of the pumped Tm-doped fiber. The experimental measurement of the spectra is discussed in Section 3.3. The theoretical aspects are treated in this subsection. We show that the temperature dependence of the emission spectrum can be reasonably described by a relation between the absorption and emission spectra. With the estimated temperature of the laser-pumped fiber, we infer from it the fiber's cooling power by performing heat flow simulations in the fiber. The obtained cooling power is compared to the predictions of the microscopic model in the previous subsection.

Figure 3.4 shows a drawing of the cross section of one of the fibers we used in this study. It consists of a core fiber of Tm-doped tellurite glass of radius 52.5 μm surrounded by a cladding with an outer radius of the 62.5 μm. We seek to understand the temperature variations of the fiber's emission spectrum by relating it through general thermal arguments to the photoabsorption spectrum [12]. Photoabsorption and emission are governed by the same quantum transition matrix elements between the levels in the ground ($^3H_6$) manifold and those in the excited ($^3F_4$) manifold of the Tm ions. We assume that the excited Tm atoms have thermalized among the states in the excited manifold before they luminesce. Then their emission spectrum at temperature T, $L(\lambda, T)$, is given by [13]

$$L(\lambda,T)=\frac{C_1}{\lambda^2}\sigma(\lambda,T)\,e^{-\frac{hc}{k_B T}\left(\frac{1}{\lambda}-C_2\right)} \tag{3.14}$$

where λ denotes the wavelength, c is the speed of light in vacuum, and $\sigma(\lambda,T)$ is the single-ion absorption cross section averaged over an ensemble of ions thermally distributed at temperature T within the ground manifold. C_1 and C_2 are parameters that depend on the material but not on T or λ. Knowledge of C_1 and C_2 is not necessary if we are interested only in the spectral distribution but not in the absolute magnitude of $L(\lambda, T)$. If the absorption cross section of the sample is independently measured, then Eq. 3.14 provides a consistency check on the measured emission spectra and can indicate, for example, whether the excited ions have thermalized in the excited manifold. Equation 3.14 differs from two commonly used absorption/emission relations: the Roosbroeck–Shockley relation [14] and the Kubo–Martin–Schwinger (KMS) relation [15] in the conditions under which they are applicable. For the Roosbroeck–Shockley relation to apply, both emission and absorption measurements are performed on a system at thermal equilibrium with the surroundings (the photon bath). In our setup here, where Eq. 3.14 applies, the emission is measured from excited ions that are constrained to thermalize only within the excited ($^3F_4$) manifold and so are not in (complete) thermal equilibrium with the surroundings. The KMS relation has been used in studies of laser cooling of semiconductors [16]. In this context, both absorption and emission are to be measured on a semiconductor sample that is in a metastable quasi-equilibrium state with a fixed density of excited electron–hole pairs.

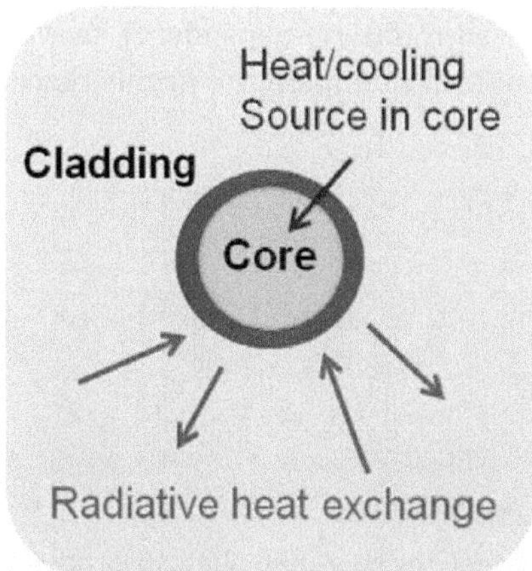

Figure 3.4 Schematic drawing of the cross section of the fiber. The core is a Tm-doped tellurite fiber of radius 52.5 μm, and the outer radius of the cladding is 62.5 μm.

We have measured the absorption cross section of Tm ions in a germinate glass sample at T = 300 K, which is plotted as the black curve in the left panel of Fig. 3.5. Using this cross section, we have calculated through Eq. 3.14 the temperature-dependent emission spectrum. The (scaled) emission spectrum at 300 K is plotted as the red curve in the graph with the absorption cross section. The spectra for a range of temperature are plotted in the right panel of Fig. 3.5, which are to be compared with the experimental calibration spectra. The general trend of the theoretical emission spectra is matched by that in the experiment. Despite the difference in glass hosts in the absorption (germanate) and emission (tellurite) measurements, the spread of the theoretical spectra over 20 K is consistent with that in our experimental calibration spectra (see Section 3.3).

The measured emission spectra of the irradiated fiber give the temperature as a function of the pump laser intensity. We obtain the fiber's cooling power from the temperature by performing heat flow simulations in a thermal model of the fiber. In this model the fiber is infinitely long and has a cross section, as shown in Fig. 3.4, with the dimensions and compositions, as stated there. Heat flows inside the fiber and cladding by diffusion, and the cladding is thermally coupled to the outside by thermal radiation exchange across its outer surface. For each of a range of cooling power densities, we have numerically evolved the (time-dependent) heat diffusion equation inside the fiber until the temperature profile reaches a steady state.

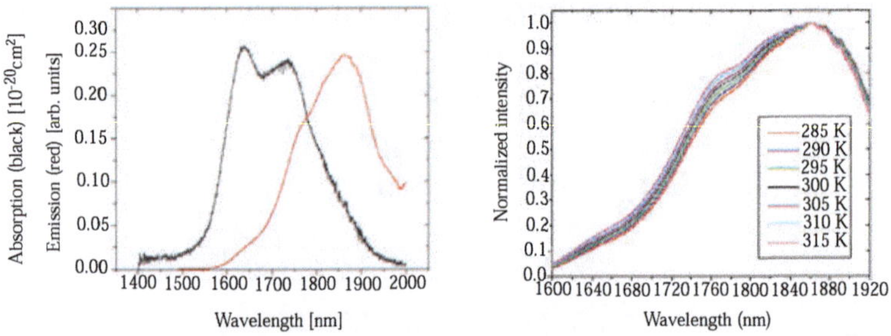

Figure 3.5 Left panel: the measured photoabsorption cross section of the Tm-doped glass (black) and the calculated emission spectrum (red) at 300 K of the pumped fiber according to Eq. 3.14. The absorption cross section has been smoothed before being used in Eq. 3.14. Right panel: the calculated emission spectra for various temperatures. All the curves are normalized to 1 as their peak values. Reprinted with permission from Ref. [12], Copyright 2013, Society of Photo Optical Instrumentation Engineers.

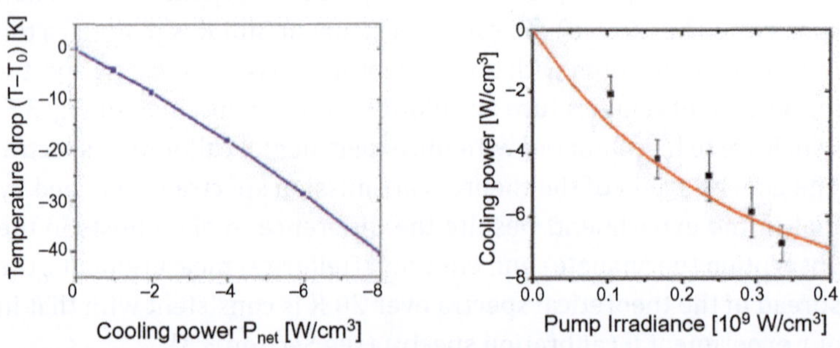

Figure 3.6 Left panel: calculated temperature drop of the fiber as a function of its cooling power P_{net} (T_0 = 300 K). Right panel: cooling power of the pumped fiber as extracted from the experimental emission spectra (black squares) and as calculated by the four-level model in Section 3.2.1 (red curve). Reprinted with permission from Ref. [12], Copyright 2013, Society of Photo Optical Instrumentation Engineers.

The boundary condition of radiative heat exchange with the outside is applied at the outer cladding surface. The outside temperature T_0 and the initial temperature of the fiber-cladding structure are set at 300 K. The calculated steady-state temperature profiles turn out to be essentially uniform across the structure, making it possible to characterize the structure's thermal state by one temperature. The calculated temperature change induced by switching on the cooling power density is plotted as a function of the latter in the left panel of Fig. 3.6. By considering the heat balance across the outer cladding surface, one can also obtain the following analytic estimate of the steady-state temperature T of the structure

$$T^4 = T_0^4 + \frac{r_{\text{core}}^2}{2R\varepsilon\sigma} P_{\text{net}} \tag{3.15}$$

where r_{core} is the core radius, R is the outer cladding radius, ε is the emissivity of the cladding surface, and σ is the Stefan–Boltzmann constant. For the parameter values used here, Eq. 3.15 is a good approximation to the simulation results shown in Fig. 3.6.

Through the relation between temperature and cooling power in (the left panel of) Fig. 3.6, we have converted the experimental temperature-pump-intensity relation into one between the cooling power of the fiber and the pump intensity. In the right panel of Fig. 3.6, this inferred cooling power density (solid squares) is plotted against the pump laser intensity. For comparison, we have also plotted the cooling power (red curve), as calculated in the four-level Tm atom model (see Fig. 3.2), adjusting the parameters to obtain a good fit to the measured data. While most parameters are assigned the values used in Fig. 3.3, we have found that the best fit is obtained by assuming negligible nonradiative decays parasitic absorption, which is unlikely to be the case for our fiber. Even with this and other uncertainties, the comparison shows the microscopic model predictions and the measured data to be qualitatively (to 1 order of magnitude) consistent.

3.2.3 Heat Transport in a Structure Cooled by a Laser-Cooled Fiber

To better understand the prospects of the laser-cooled fiber as a cooler, we have studied possible cooling actions of the fiber on a material structure in contact with it. In this section, we describe our model simulations of heat flow in this cooling device.

3.2.3.1 Model of the cooling device

Figure 3.7 shows a schematic drawing of the heat transport simulation model of the cooling device. The cooling fiber, composed of an active doped inner core and an outer cladding, rests in a V-shaped groove, which is set in a silica glass base. This structure is capped by a metallic heat spreader, which is made up of a nickel-chromium layer on top of a silver layer. The silver reflects the upward-going fluorescent photons from the cooling fiber. A thermistor and heat-load resisters are located on the top of the device. The doped fiber core acts as a heat sink, its cooling power being set as an input to the simulation. The model accounts for two heat sources within the simulation domain: power dissipated by the thermistor and heat-load resistors, and absorption of scattered fluorescence by the metal heat spreader.

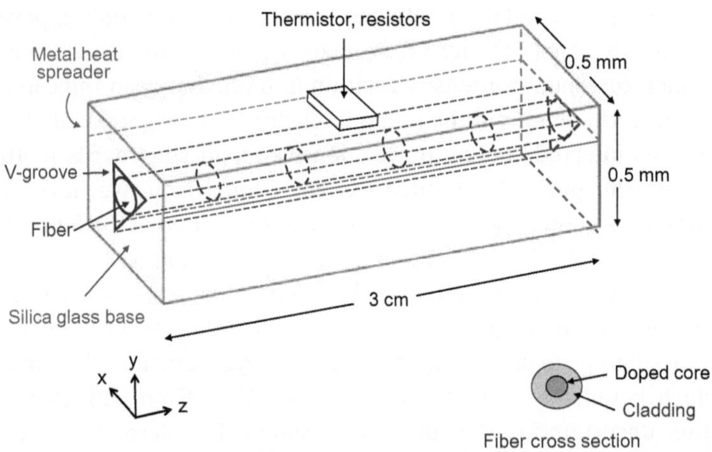

Figure 3.7 3D simulation domain (relative sizes of components not drawn to scale) of the cooling device. Composition details: the fiber is made up of an inner doped core (diameter = 50 μm) and an outer cladding (diameter = 125 μm); the heat spreader is a 100 μm thick silver layer topped by a 50 μm thick nickel-chromium layer; a high conductivity epoxy forms a 20 μm adhesive layer and fills the V-groove around the fiber. The inset (lower right of the figure) shows schematically the fiber cross section with doped core and cladding.

The simulation domain is the interior of the device. The evolution of the temperature field $T(\mathrm{r})$, $r = (x, y, z)$, is governed by the heat diffusion equation with sources

$$\frac{\partial T}{\partial t}=\frac{1}{\rho c_v}\nabla\cdot\left(\kappa\nabla T\right)+\frac{1}{\rho c_v}S \tag{3.16}$$

where ρ is the mass density, c_V is the heat capacity, and κ is the thermal conductivity. $S(r)$ represents the heat sources/sinks inside the simulation domain mentioned above. The boundary condition at the outer surfaces of the cooling device sets the heat flux at the surface J as equal to the radiative exchange with the outside

$$\mathbf{J}=-\kappa\nabla T=\epsilon\sigma\left(T^4-T_{\mathrm{A}}^4\right)\hat{n} \tag{3.17}$$

where ε is the local surface emissivity, σ is the Stefan–Boltzmann constant, $\hat{n}$ is the local unit normal vector, and T_{A} is the outside temperature.

3.2.3.2 Numerical simulation of heat flow in the cooling device model

We have developed a numerical code to solve the heat diffusion Eq. 3.16 in the 3D geometry of the cooling device. The spatial derivatives are discretized by finite differences. The time marching combines a 2D alternating direction implicit (ADI) method [17] for updating the solution across the transverse (x, y) plane and a predictor corrector (PC) method for updating along the longitudinal (z) direction. Since the 3D simulations incur high computational costs, computational efficiency has become an essential requirement in the code's development.

In Figs. 3.8–3.10, we show some simulation results from a sample run of the code. In this simulation, the domain is discretized into 200 cells in each of the two transverse directions (x and y) and 10 cells in the longitudinal (z) direction. The time evolution of the temperature at the thermistor is plotted in Fig. 3.8. An initial steady state of the device is set up with the thermistor (heating power = 1 mW) turned on but the pump laser off. In this state, the heat introduced into the device by the heat load (thermistor) is balanced by the heat radiated out at the device's surfaces. The pump is turned on at $t = 0$. It has two effects on the thermal budget of the device: (i) It supplies the doped fiber core with a cooling power of 8 W/cm^3, and (ii) it generates fluorescent and scattered photons, some of which are absorbed and converted into heat by the metal spreader layer (see Fig. 3.7). As shown in Fig. 3.8, the net effect of all the above processes is that the device is cooled by 1.5 K, reaching a new steady state after several minutes.

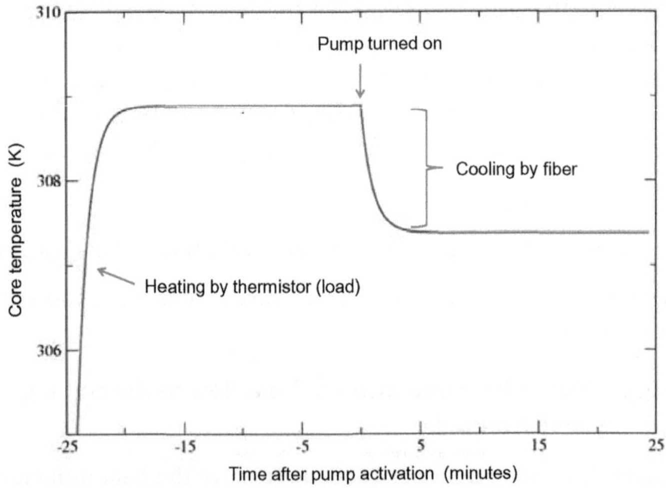

Figure 3.8 Time evolution of the temperature at the thermistor, showing cooling by the laser-pumped fiber. An initial steady state is set up with the thermistor on and the pump laser off. The pump is turned on at $t = 0$, and the temperature drops by 1.5 K several minutes thereafter.

Figure 3.9 shows the temperature profile across a transverse cross section (*x*-*y* plane) of the device at the mid-point in the longitudinal (*z*) direction, at the steady state reached after the pump is turned on. Figure 3.10 shows the temperature profile across a vertical cross section (*y*-*z* plane) at the midpoint in the *x* direction at the same time. The two temperature profiles indicate the flow of heat from the thermistor, the principal heat source, to the fiber core, where the heat is absorbed, and to the boundaries of the device, where it is emitted as thermal radiation. We note that the temperature drop between the middle and the end surfaces in the *z* direction is relatively large (~0.4 K) compared to that between the top and the bottom (<0.1 K).

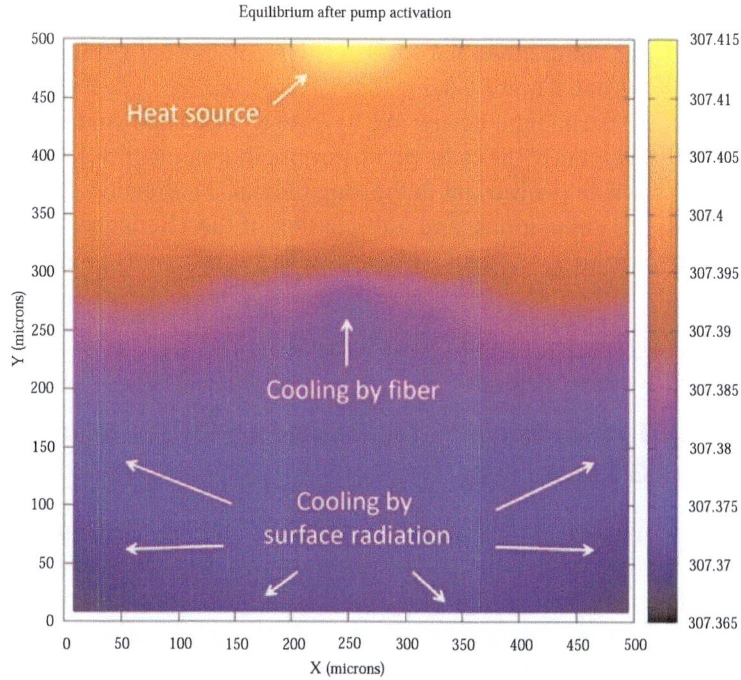

Figure 3.9 Calculated temperature profile in a transverse cross section (*x-y* plane) of the device at the midpoint in the longitudinal (*z*) direction, when the system reaches a steady state after the pump is turned on. The profile indicates the heat flow from the thermistor (top middle) to the cooling fiber (centered at x = 250 μm, y = 275 μm), where the heat is absorbed, and to the device boundaries, where it is radiated. The temperature scale (in K) for the color coding is shown on the right.

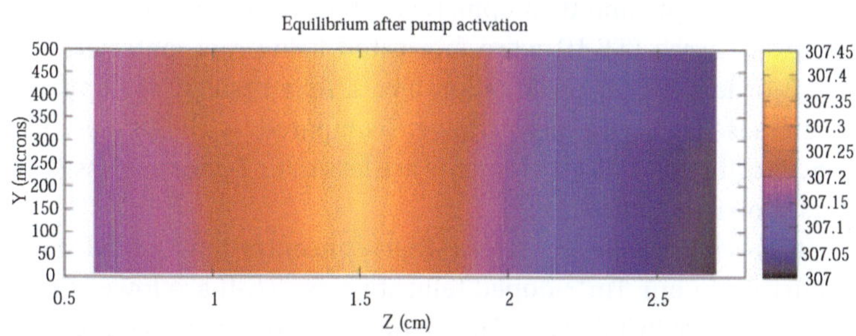

Figure 3.10 Calculated temperature profile in a vertical (*y-z* plane) cross section at the midpoint in the *x* direction. Note the much larger temperature drop between the middle and the sides (~0.4 K) than between the top and the bottom (<0.1 K).

We note that, because of the high computational costs, we have limited our simulations here to a model domain with transverse dimensions of 0.5 mm × 0.5 mm (see Fig. 3.7), which is much smaller than the experimental device. We have also performed simulations with a 2D version of the code, which essentially assumes the device to be translationally invariant in the longitudinal (z) direction. Results of these 2D simulations, and a discussion of the photon reflection and absorption by the metal heat spreader, can be found in Ref. [18].

3.3 Experimental Developments

3.3.1 Fabrication of Highly Purified Tm-Doped Tellurite Glasses

Transition metal (TM) impurities in glasses, especially Fe^{+2} and Cu^{+2} impurities, pose serious challenges to the cooling processes. The TMs Fe^{+2} and Cu^{+2} have strong and broad absorption bands from 1 to 2 μm (peak absorption is ~1 μm). There are two processes caused by the impurities that are negative to the cooling processes in Tm-doped glass: (i) TMs can directly absorb light (both pump and luminescence) and generate heat, and (2) excited Tm ions can transfer energy to surrounding impurities, resulting not only in lower quantum efficiency but also heat generation. We have fabricated Tm-doped tellurite glasses using raw materials having two different purified levels. The first Tm-doped tellurite glasses (TE-I) were fabricated with average Fe^{+2} and Cu^{+2} concentrations of about 3.4 ppm and 0.83 ppm, respectively. The second Tm-doped tellurite glasses (TE-II) were fabricated using raw materials with 1 to 2 orders higher purification levels. As a result, TM impurities in glasses TE-II are much lower compared with glasses TE-I. Concentrations of all TM elements are less than 1 ppm in glasses TE-II, as shown in Table 3.1.

It is worthwhile to stress that concentrations of all metal impurities in our Tm^{+3}-doped tellurite glass TE-II are lower than 1 ppm, as shown in Table 3.1. The metal impurities have been reduced significantly from standard tellurite glasses, in particular Fe^{2+}, and Cu^{2+} has been reduced to ~0.6 ppm and ~0.2 ppm, respectively. As can be seen from Fig. 3.11, the total loss due to metal impurities absorption in the fiber glass is about 0.002 dB/cm around 1.9 μm spectral region where optical cooling processes occur in Tm-doped glass.

Table 3.1 Metal impurities in Tm^{+3}-doped tellurite glass TE-II

Metals	ppm	Metals	ppm	Metals	ppm	Metals	ppm
Ca44(MR)	<1	Er167(MR)	<0.05	Ni58(MR)	0.77	U238(LR)	<0.05
Cd111(LR)	0.16	Eu151(LR)	<0.05	Pb208(MR)	0.12	V51(MR)	<0.05
Co59(MR)	0.19	Fe56(MR)	0.63	Pd105(LR)	<0.05	W182(LR)	<0.05
Cr52(MR)	0.08	Ga69(MR)	<0.05	Pt195(LR)	<0.5	Y89(LR)	<0.05
Cs133(LR)	<0.05	Ge72(MR)	<0.05	Rb85(LR)	<0.2	Yb172(LR)	<0.05
Cu63(MR)	0.18	Hf178(LR)	<0.05	Re187(LR)	<0.05	Zn64(MR)	<0.2

Source: Ref. [12]

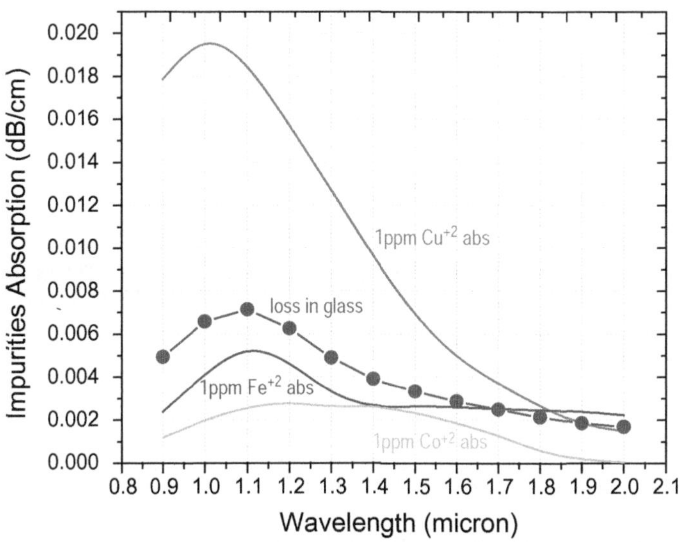

Figure 3.11 Loss due to metal impurities in Tm^{+3}-doped tellurite glass: 1 ppm Cu, 1 ppm Fe, 1 ppm Co, and total loss due to actual impurities in glass. Reprinted with permission from Ref. [12], Copyright 2013, Society of Photo Optical Instrumentation Engineers.

As is well known, the hydroxyl group –OH also has big impact on optical cooling due to the following reasons: (i) Energy transfer from Tm ions in excited levels to –OH proceeds via a nonradiative decay, which generates heat in the system, and (ii) contributions to the fiber loss in general (absorption and scattering, for example) can also generate heat. In this work, we have achieved reducing –OH absorption in glasses TE-II from 75–90 dB/m normal tellurite glass to ~8 dB/m at 3.3 μm, as shown in Fig. 3.12.

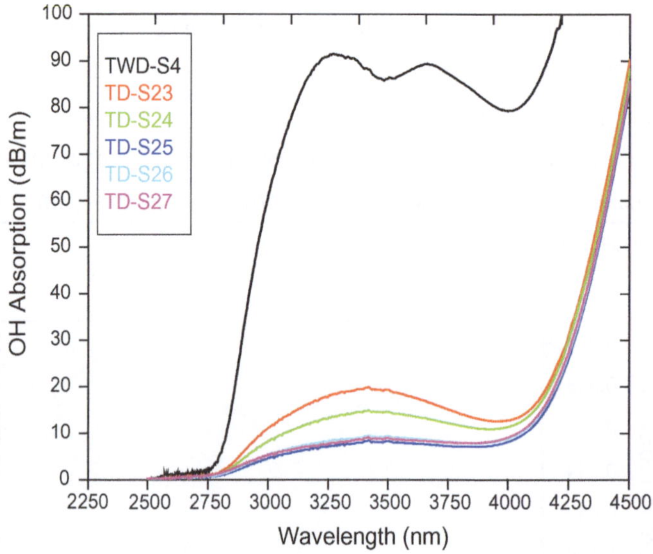

Figure 3.12 –OH absorption in Tm-doped tellurite glasses TD-S23 to TD-S27 compared to standard glass; TDS-25 glass (blue) has been used for fabrication of cooling fiber Tm-CF2. Reprinted with permission from Ref. [12], Copyright 2013, Society of Photo Optical Instrumentation Engineers.

3.3.2 Design and Fabrication of Cooling Fibers

The cooling fiber design requires some special properties that are different compared to fiber lasers and fiber amplifiers. First, the fiber has features that aim at minimizing the reabsorption of the fluorescence, which generates heat in the fiber core. Since the cooling efficiency is on the order of a few percentage points, even 1% of fluorescence reabsorption can have a large negative impact on the overall cooling efficiency. At the same time, the cooling fiber configuration should be able to guide the fluorescence light in the inner cladding. Second, the fiber configuration must satisfy the thermal conditions that can maximize the transfer of heat from the heating sample to the fiber core where cooling effect takes place. We have fabricated three Tm-doped tellurite glass cooling fibers. All fibers have core glasses doped with 2 wt% Tm concentration. The main differences between those fibers are the glasses with the different levels of TM impurities and –OH absorption. To improve the potential cooling performance in the fibers, the different fiber sizes are also designed and fabricated to improve pump absorption and also coupling losses with passive fibers. The specifications of the first cooling fiber (Tm-CF1) are: 2% Tm-doped tellurite glass (TE-I); core/clad size (diameter): 50/125 μm. The glass TE-I has been fabricated with raw materials having a standard level of impurities. However, the –OH absorption is reduced to ~8 dB/m compared to 75–90 dB/m of standard glasses. The second cooling fiber (Tm-CF2) is 2% Tm-doped tellurite glass (TE-II); core/clad size: 105/125 μm. The glass TE-II has been fabricated with highly purified materials having low impurities (all TMs are less than 1 ppm). Figure 3.13 shows the two cooling fibers Tm-CF1 and Tm-CF-2.

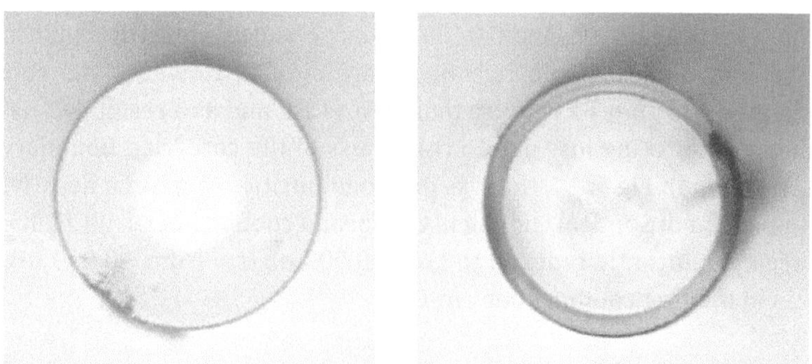

Figure 3.13 Images of two Tm-doped tellurite glass cooling fibers Tm-CF1 (left) and Tm-CF2 (right). Reprinted with permission from Ref. [12], Copyright 2013, Society of Photo Optical Instrumentation Engineers.

Using the data of impurities measured in the glasses and absorption cross sections from Ref. [19] we can calculate the loss due to impurity absorption. Figure 3.14 shows loss due to impurity absorption in Tm-CF1 and Tm-CF2.

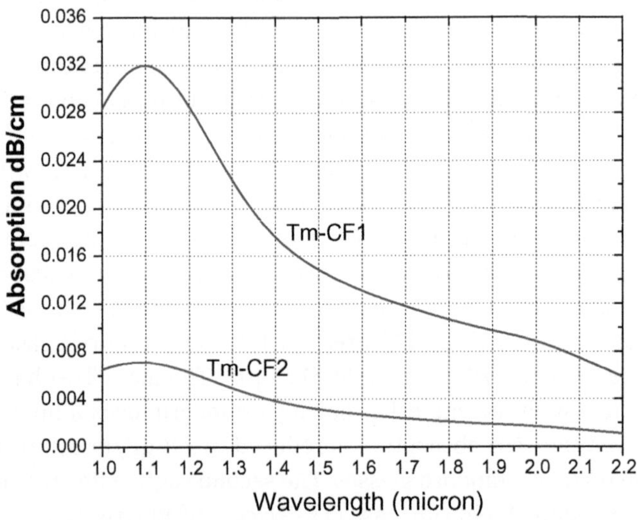

Figure 3.14 Impurities loss in cooling fibers Tm-CF1 and Tm-CF2.

Characterizations of the cooling fibers Tm-CF1 and Tm-CF2 show the total loss of Tm-CF1 and Tm-CF2 at a test wavelength 1310 nm are 0.042 dB/cm and 0.0409 dB/cm, respectively. At first glance, the total losses in the two fibers are similar. However, the core diameter of Tm-CF2 is twice that of Tm-CF1, and as a result CF2 has higher scattering loss due to roughness of the core/clad boundary. As shown in Fig. 3.14, the loss due to impurities that has a negative impact on the optical cooling in the second cooling fiber Tm-CF2 has been significantly reduced to about 0.002 dB/cm from ~0.009 dB/cm in the first cooling fiber Tm-CF1.

3.3.3 Experimental Demonstration of Optical Cooling in Fibers

3.3.3.1 Thermistor measurements

Figure 3.15 shows the sample of the experimental setup with a 3 cm cooling fiber and a thin silver layer that plays the role of a heat spreader for the heat source. The cooling fiber is then put in a V-groove holder with a high conductivity epoxy around it. The epoxy used in the setup is EPO-TEK H20-HC, which is one of the highest thermal-conductivity (9.96 W/mK) epoxies. A high-precision NTC MF51E thermistor (by Cantherm) for extremely accurate temperature measurement is attached to the metal plate. The extremely small size allows the thermistor to respond quickly to tiny changes in temperature.

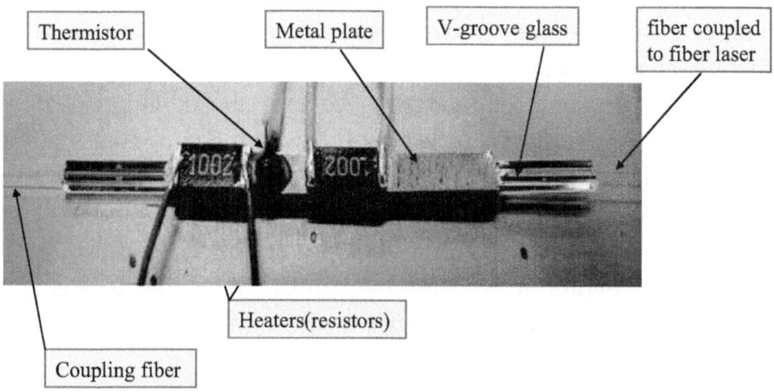

Figure 3.15 The experiment setup and some important parts in the cooling experiment. The cooling fiber is underneath of the metal (silver) plate.

The cooling experiments have been carried out in a vacuum chamber. The experimental setup is shown in Fig. 3.15, including 3 cm of cooling fiber, where one fiber end is fusion-spliced with a pump-coupling fiber of size 25/250 μm that couples to the fiber laser. The other end is fusion-spliced to another coupling fiber of the size 105/125 μm, in which the waste photons can be piped substantial distances and expelled wherever the system designer wishes. The total loss of the chain is 0.31 dB (loss of two coupling + background loss in cooling fiber). Experimental data of temperature versus time in our cooling system in the vacuum chamber are shown in Fig. 3.16. Note that the potential cooling effect depends on pumping power: there is little heat removal if the pump power is too low, but the cooling process could be overwhelmed by heat if the pump power is too high. Therefore we first measured temperature versus time with unchanged pump power in ~1 hour, then increasing power to higher levels. The aim is to find the best range of power for the potential cooling processes in the fiber.

The experiments are repeated with each pump power after about 5 hours. In the experiment, the initial temperature in the vacuum chamber and the sample are set at temperature 26.5°C. When the pump turns on, the temperature measured by the thermistor at the metal plate rises quickly to 29.5°C (P = 1.6 W, @ 1940 nm). Keeping the pump power unchanged, the temperature versus time was measured and recorded from 17:30 to 22:30, as shown in Fig. 3.16. Closer inspection of the results in Fig. 3.16 indicates a temperature difference between the sample and the inside of the vacuum chamber that is consistent with cooling. The temperature in the sample drops more than 0.6°C. We note that the input power is 1.6 W and the pump residue is about 0.25 W, meaning that the fiber absorbs (including loss) 1.35 W of pump power continuously during that time. We have carried out cooling experiments using the second cooling fiber Tm-CF2 and the same configuration as in Fig. 3.15. The temperature evolution for the fiber Tm-CF2 is similar to Fig. 3.16, but the magnitude of temperature in the sample drops more than 1.2°C versus 0.6°C in first fiber Tm-CF1.

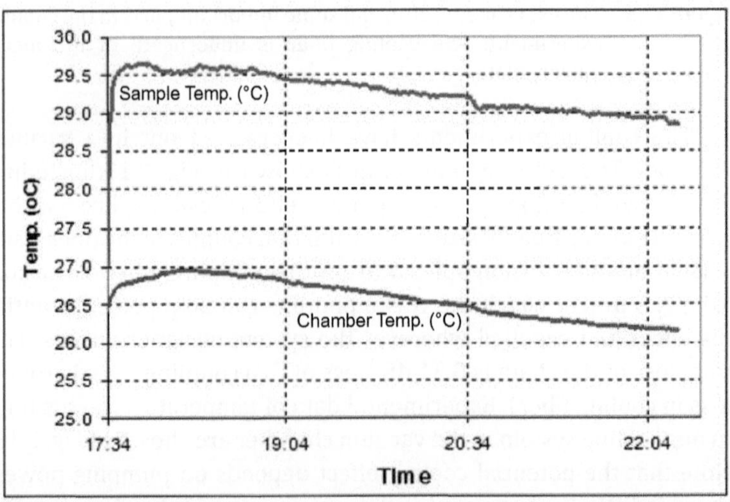

Figure 3.16 Temperature versus time measured in vacuum chamber with pump power 1.6 W at 1940 nm. The curves show temperatures measured at the sample attached to the cooling fiber and inside the vacuum chamber. Reprinted with permission from Ref. [18], Copyright 2012, Society of Photo Optical Instrumentation Engineers.

Qualitatively, the above results of experiments using thermistors to measure temperature in the fiber indicate cooling effects could have occurred in the fiber. It is important to point out here that there are two thermistors in the experiments, and they consume power and therefore generate and contribute heat to the sample. To determine precisely the temperature in the fiber under laser pumping, a noncontact measurement approach has been developed and is presented next.

3.3.3.2 Noncontact measurements

As discussed above, the direct contact measurement approach uses thermistors. The results of direct contact measurements indicate qualitatively that optical cooling could occur in our cooling fiber under laser pumping. To confirm the effect *quantitatively*, we have developed experimentally and theoretically a noncontact measurement approach. The noncontact method investigates the fluorescence spectrum of an isolated fiber under two conditions: (i) The fluorescence spectra of the Tm-doped glass fiber are measured under different pump powers (P-dependent spectra), and (ii) the fluorescence spectra of the Tm-doped glass fiber are measured under low pump power and different temperatures (T-dependent spectra), which are set by a thermoelectric cooler (TEC) attached to the fiber. Figure 3.17 shows configurations of the experiments for measuring P-dependent and T-dependent spectra [10, 11].

Note that the configurations shown in Fig. 3.17 are quite simple in the fiber system using beam combiners. Figure 3.18 shows the power-dependent and temperature-dependent fluorescence spectra under 1950 nm pumping.

We note that the results show similarities between the P-dependent and T-dependent fluorescence spectra from 1600 nm to 1920 nm, which is shorter than the pump wavelength 1950 nm (i.e., the required condition for the optical cooling process). This is consistent with the interpretation that the Tm-doped fiber under different pumping powers undergoes optical cooling processes, and the emitted fluorescence spectra with higher pump power correspond to lower temperature spectra in T-dependent ones. For example, the lowest pump power spectrum (~0.9 W) corresponds to the one at 20°C. Significantly, the results in Fig. 3.19 show that the temperature in the fiber under pump power 2.55 W corresponds to −10°C, and the temperature under power 2.9 W is even lower. Note that, in the present experiment, we only a TEC that can drive the temperature to only −10°C.

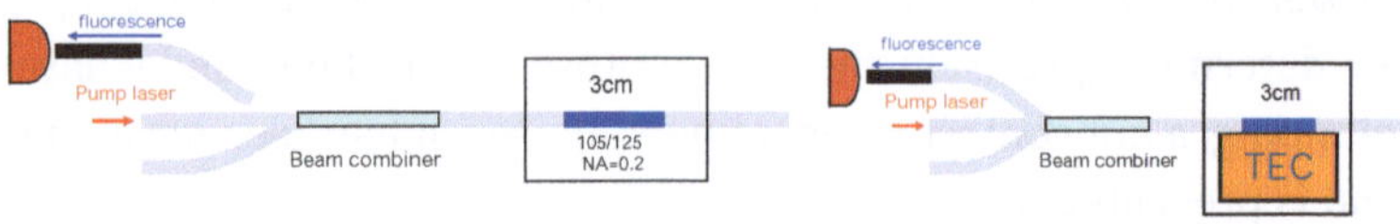

Figure 3.17 Configurations for spectroscopic measurements. Left: backward measurement of pump power-dependent spectra. Right: backward measurement of temperature-dependent spectra of the Tm-doped cooling fiber (blue section).

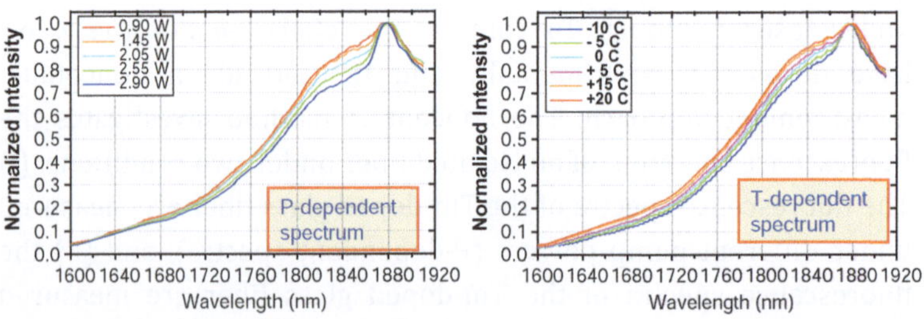

Figure 3.18 Power (*P*)-dependent and temperature (*T*)-dependent spectra under 1950 nm pumping.

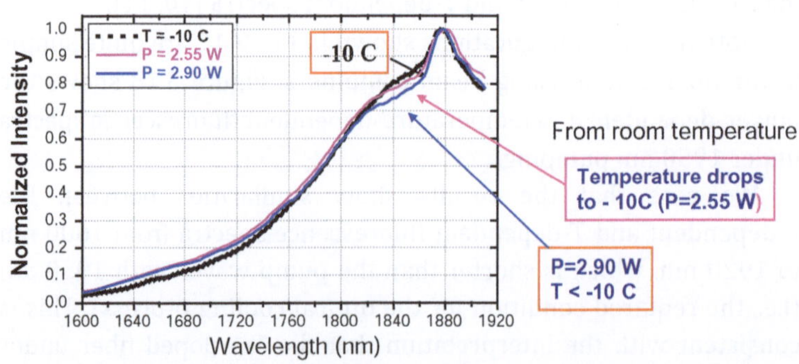

Figure 3.19 Comparison of *P*-dependent and *T*-dependent spectra under 1950 nm pumping. Temperature in the fiber under pumping with $P = 2.55$ W drops from room temperature ~24°C to about −10°C. Temperature under pumping with $P = 2.90$ W is even lower than −10°C.

We also have performed theoretical modeling of the Tm-doped glass and the theoretical results are in good agreement with the experiments (see Fig. 3.20). However, the theoretical calculations are preliminary as several material parameters in the tellurite glass are not well determined. Furthermore, real conditions in the experiments are not fully taken into account in the theoretical modeling. Those issues need to be addressed in the future.

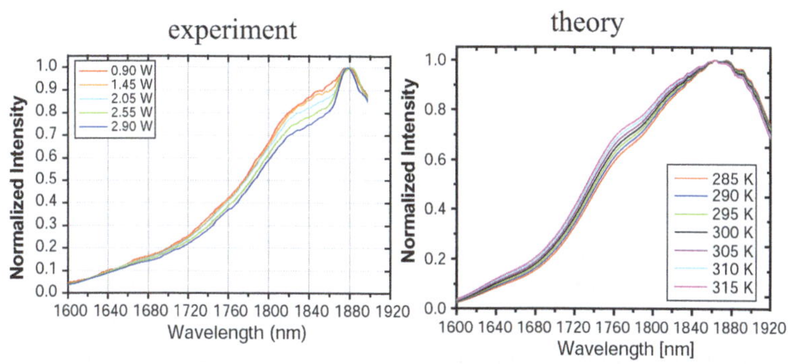

Figure 3.20 We find good agreement between experimental (left) and theoretical spectra (right), comparing power- and temperature-dependent spectra. The results are consistent with the interpretation that the temperature in the fiber drops more than 30°C from room temperature.

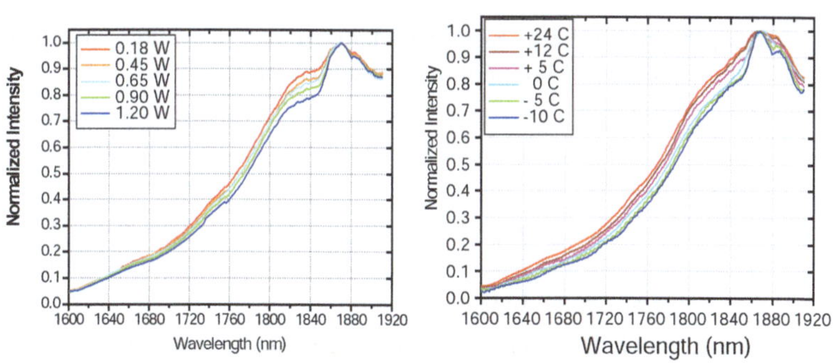

Figure 3.21 *P*-dependent and *T*-dependent spectra under 1940 nm pumping.

Figure 3.21 shows the results for the 1940 nm pump laser. Again, the results are consistent with the interpretation that net optical cooling in the fiber has been achieved. Preliminary analysis shows that the temperature drops about 30°C from room temperature with 1940 nm pump power P = 1.2 W. We believe that this is the first experimental indication of possible optical net cooling in the Tm-doped tellurite glass fiber. In the past, cooling in the Yb-doped ZBLAN fiber has been demonstrated [10, 11]. Note that the ZBLAN glass fiber is fracture-prone compared to tellurite glass. As a result, it is difficult to have an all-fiber integrated system for optical cooling based on the ZBLAN glass fiber. Optical cooling has also been demonstrated experimentally in Tm-doped ZBLAN glass [9] but not in the fiber.

Acknowledgments

Elements of this work were sponsored by the Office of Naval Research (ONR) under contract/grant STTR N00014-11-C-0477, ONR Public Release Distribution Statement A, "Approved for public release; distribution is unlimited." Additional support was provided by the Air Force Office of Scientific Research (AFOSR) under contract/grant FA9550-11-0048 and by TRIF Photonics. We thank G. Hardesty for his contributions to our theoretical effort.

References

1. Pringsheim, P. (1929). Zwei Bemerkungen über den Unterschied von Lumineszenz- und Temperaturstrahlung, *Z. Phys.*, p. 739.
2. Epstein, R. I., Buchwald, M. L., Edwards, B. C., Gosnell, T. R., Mungan, C. E. (1995). Observation of laser-induced fluorescent cooling of a solid, *Nature*, **377**, pp. 500–503.
3. Seletskiy, D. V., Melgaard, S. D., Bigotta, S., Di Lieto, A., Tonelli, M., Sheik-Bahae, M. (2010). Laser cooling of solids to cryogenic temperatures, *Nat. Photonics*, **4**, pp. 161–164.
4. Zhang, J., Li, D., Chen, R., Xiong, Q. (2013). Laser cooling of a semiconductor by 40 kelvin, *Nature*, **493**, pp. 504–508.
5. Sheik-Bahae, M., Epstein, R. I. (2007). Optical refrigeration, *Nat. Photonics*, **1**, pp. 693–699.
6. Nemova, G., Kashyap, R. (2010). Laser cooling of solids, *Rep. Prog. Phys.*, **73**, p. 086501.

7. Nguyen, D. T., Shanor, C., Zong, J., Tian, W., Yao, Z., Wu, J., Weiss, J., Binder, R., Chavez-Pirson, A. (2011). Conceptual study of a fiber-optical approach to solid-state laser cooling, *Proc. SPIE*, **7951**, p. 795109.
8. Epstein, R. I., Sheik-Bahae, M., eds. (2009). *Optical Refrigeration* (WILEY-VCH Verlag, Germany).
9. Hoyt, C. W., Hasselbeck, M., Sheik-Bahae, M., Epstein, R. I., Greenfield, S., Thiede, J., Distel, J., Valencia, J. (2003). Advances in laser cooling of thulium-doped glass, *J. Opt. Soc. Am. B*, **20**, pp. 1066–1074.
10. Mungan, C. E., Buchwald, M. I., Edwards, B. C., Epstein, R. I., Gosnell, T. R. (1997). Laser cooling of a solid by 16 K starting from room temperature, *Phys. Rev. Lett.*, **78**, pp. 1030–1033.
11. Gosnell, T. R. (1998). Laser cooling of a solid by 16 K starting from room temperature, *Opt. Lett.*, **24**, pp. 1041–1043.
12. Nguyen, D. T., Thapa, R., Rhonehouse, D., Zong, J., Miller, A., Kwong, N. H., Binder, R., Chavez-Pirson, A. (2013). Towards all-fiber optical coolers using Tm-doped glass fibers, *Proc. SPIE*, **8638**, p. 86380G.
13. McCumber, D. E. (1964). Einstein relations connecting broadband emission and absorption spectra, *Phys. Rev.*, **136**, pp. A954–A957.
14. van Roosbroeck, W., Shockley, W. (1954). Photon-radiative recombination of electrons and holes in germanium, *Phys. Rev.*, **94**, pp. 1558–1560.
15. Zimmermann, R. (1987). *Many-Particle Theory of Highly Excited Semiconductors* (Teubner, Leipzig).
16. Kwong, N. H., Rupper, G., Binder, R. (2009). Self-consistent T-matrix theory of light-absorption and luminescence, *Phys. Rev. B*, **79**, p. 155205.
17. Richtmyer, R. D., Morton, K. W. (1967). *Difference Methods for Initial-Value Problems*, 2nd Ed. (Interscience Publishers, New York).
18. Nguyen, D. T., Zong, J., Rhonenhouse, D., Yao, Z., Hardesty, G., Kwong, N. H., Binder, R., Chavez-Pirson, A. (2012). All fiber approach to solid-state laser cooling, *Proc. SPIE*, **8275**, p. 827506.
19. Hehlen, M. P., Epstein, R. I., Inoue, H. (2006). Laser cooling in rare-earth doped solids: the next step, *Proc. SPIE*, **6130**, p. 61300N.

Chapter 4

Laser Cooling of Solids around 2.07 Microns: A Theoretical Investigation

Guang-Zong Dong and Xin-Lu Zhang
Key Laboratory of In-Fiber Integrated Optics (Ministry of Education) and College of Science, Harbin Engineering University, Harbin, 150001, P. R. China
zhangxinlu1@aliyun.com

4.1 Introduction

Since the first demonstration provided by Epstein et al. [1] two decades ago, laser cooling of solids or optical refrigeration has been developed systematically [2–4], and recent advances have claimed cryogenic cooling in highly purified Yb^{3+}-doped fluoride crystals [5–8]. The simple energy level structure together with the wide energy gap imparts Yb^{3+} ion–doped sample an extremely high quantum efficiency, making Yb^{3+} ion an ideal choice for cryogenic optical refrigeration. Nevertheless, other trivalent rare earth ions, such as Tm^{3+}, Er^{3+}, and Ho^{3+}, attract increasing interest due to two main reasons. For one thing, the operation wavelength region for optical refrigeration can be expanded by adopting these ions, which could be technically important in practical applications. For another,

Laser Cooling: Fundamental Properties and Applications
Edited by Galina Nemova

ISBN 978-981-4745-04-8 (Hardcover), 978-981-4745-05-5 (eBook)
www.panstanford.com

it is possible to introduce new cooling mechanics to multimanifold cooling systems to improve the cooling performance [9–12]. Since cooling around 1.5 and 2.0 μm has been achieved in Er^{3+}- and Tm^{3+}-doped solids [13–16], it is wondered whether optical refrigeration can be extended beyond 2 μm in Ho^{3+}-doped solids.

In 2013, Bowman et al. experimentally investigated the cooling performances of various crystals and glasses doped with Ho^{3+} ion [17]. Results show that only a 0.3% Ho^{3+}:ZBLANP glass sample tends to exhibit refrigeration when pump wavelength exceeds 2.07 μm. The poor cooling performance of Ho^{3+}-doped solids is attributed to three aspects: multiphonons, background absorption, and fluorescence quenching induced by energy transfer upconversion (ETU). To point out, the multiphonon and background absorption issues can be solved (or partly solved) by adopting highly purified, low-host-phonon materials. Specially, the corresponding energy gap should exceed eight times the maximum host phonon energy to ensure acceptable high quantum efficiency for cooling [18]. The influence of upconversion on optical refrigeration was initially mentioned by Hoyt et al. when Tm^{3+}-doped glass was first used to demonstrate net cooling [13]. The heating or cooling effect caused by upconversion is estimated negligible, and the conventional two-manifold cooling model fits the cooling result well. On the other hand, Garcia-Adeva et al. suggested that excited-state absorption (ESA) and ETU processes in Er^{3+}-doped KPb_2X_5 (X = Cl, Br) crystals could be utilized as additional cooling channels since the nonradiative transitions are largely averted [9].

For optical refrigeration with Ho^{3+}-doped solids, the ETU process is typically strong, and the transitions from the third and second excited states to the lower nearby states are mainly nonradiative, so the heat generated by the fluorescence quenching effect is not negligible [19]. To describe laser cooling of Ho^{3+}-doped solids properly, a cooling model with consideration of upconversion will be presented in Section 4.2. Simulation based on the generalized cooling model will reveal the requirements for achieving net cooling in Ho^{3+}-doped fluoride crystals and the potential cooling performance. Moreover, the role of upconversion in optical refrigeration with multimanifold rare earth ions is comprehensively investigated to obtain a common criterion on whether cooling can be improved through an upconversion channel.

To overcome the ETU-induced heating and realize efficient and enhanced laser cooling in Ho^{3+}-doped solids, in Section 4.3, we will provide a modified excitation scheme based on ground-state absorption (GSA) of $^5I_8 \rightarrow {}^5I_7$ at around 2.07 μm and ESA of $^5I_5 \rightarrow {}^5F_5$ at around 2.28 μm. Compared with the conventional pumping scheme, the co-pumping scheme ameliorates the cooling performance significantly through alleviating the upconversion-quenching-induced heat [12], that is, the heat-producing $^5I_5$ Ho^{3+} ions are pumped to the $^5F_5$ state by the laser at 2.28 μm. Fortunately, the $^5F_5$ Ho^{3+} ions barely transit to the lower nearby state and tend to decay radiatively. Therefore, major heat produced in the conventional laser cooling of Ho^{3+}-doped solids is avoided, and both cooling efficiency and cooling power density can be improved.

In Section 4.4, we broaden the research domain of optical refrigeration to rare earth ion co-doped systems. Specifically, when Ho^{3+}–Tm^{3+} co-doped $YLiF_4$ crystals are pumped at around 2.07 μm, Ho^{3+} ions will populate at the $^5I_7$ manifold. Meanwhile, Tm^{3+} ions will also be sensitized by absorbing energy transferred from $^5I_7$ Ho^{3+} ions. Since the average energy of fluorescence photons generated by Tm^{3+} $^3F_4 \rightarrow {}^3H_6$ radiative transitions is higher than that generated by Ho^{3+} $^5I_7 \rightarrow {}^5I_8$ radiative transitions, energy transfer (ET) from Ho^{3+} to Tm^{3+} ions will induce annihilation of host phonons, and bring enhancement in both the cooling efficiency and the cooling power. Therefore, this cooling mechanism is referred next as *energy transfer–enhanced laser cooling* (ETLC) of solids [11].

4.2 Conventional Laser Cooling of Holmium-Doped Fluoride Crystals

The main upconversion process and the anti-Stokes fluorescence cooling (ASFC) processes of Ho^{3+}-doped fluoride crystals are illustrated in Fig. 4.1. The ground-state ($^5I_8$) Ho^{3+} ions are pumped to the first excited state ($^5I_7$) with laser excitations of energy $h\nu$ (solid red arrow on the left) and then relax thermally among the Stark levels in a timescale of picoseconds with host phonon absorption. Part of the Ho^{3+} $^5I_7$ ions de-excite by emitting fluorescence photons of mean energy $h\nu_{10}$ (solid blue arrow, $\nu_{10} > \nu$), which accounts for the ASFC process. The rest of the $^5I_7$ Ho^{3+} ions de-excited and transfer their energy to the neighboring $^5I_7$ Ho^{3+} ions, causing population accumulation at the third excited state ($^5I_5$) with host phonon absorption (solid green arrows). Comparison between the energy mismatch and the host phonon energy (for $YLiF_4$ crystal host, $\hbar\omega \approx$ 450 cm^{-1}) indicates that about two host phonons are annihilated for each ETU process (see Table 4.1).

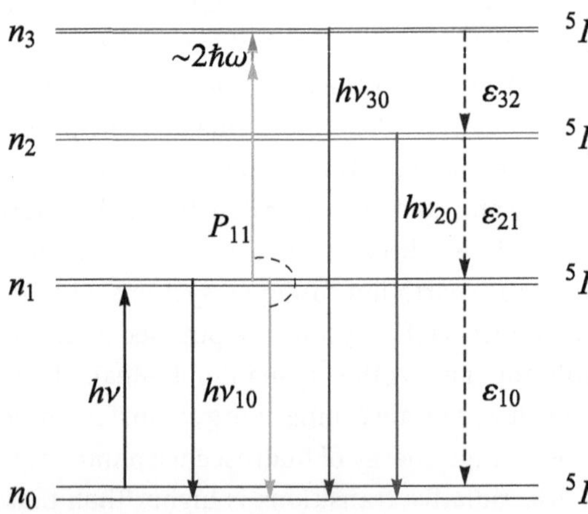

Figure 4.1 Schematic of optical refrigeration in Ho^{3+}-doped fluoride crystal with consideration of ETU. The level labels used in Eq. 4.1 are also indicated.

The ET processes intrinsically appear in pairs, that is, the reverse ET process, $^5I_5 \rightarrow {}^5I_7$; $^5I_8 \rightarrow {}^5I_7$ causes Ho^{3+} ions populating at the $^5I_7$ state. However, this ET process is largely averted since the $^5I_5$ Ho^{3+} ions relax fast (nonradiatively) to the $^5I_6$ state in fluoride crystals [22]. Other ET processes, such as $^5I_7 \rightarrow {}^5I_8$; $^5I_7 \rightarrow {}^5I_6$ and $^5I_6 \rightarrow {}^5I_7$; $^5I_7 \rightarrow {}^5I_8$, are negligible due to the large energy mismatches [21, 22]. It should be noted that $^5I_7$ Ho^{3+} ions can also transit to the $^5I_5$ state by absorbing laser excitations and surrounding phonons (phonon-assisted ESA), whereas the pump-wavelength-dependent ESA process is insignificant compared to the $^5I_7 \rightarrow {}^5I_5$; $^5I_7 \rightarrow {}^5I_8$ ETU process, since more than three phonons are needed in such a transition process when the pump wavelength is tuned to the long-wavelength tail of the $^5I_7$ absorption spectrum (see Fig. 4.2). The $^5I_5$ and $^5I_6$ Ho^{3+} ions de-excite to the lower manifolds either radiatively (solid purple and brown arrows), eliminating heat from the host (upconversion-assisted cooling), or nonradiatively (dashed gray arrows on the right), introducing extra heat to the host.

Table 4.1 Branching ratios, lifetimes, and energy gaps in Ho^{3+}:$YLiF_4$

Transitions	β_{ij}	τ_r (ms)	ε_{ij} (cm^{-1})
$^5I_7 \rightarrow {}^5I_8$	1	13.68	4838
$^5I_6 \rightarrow {}^5I_8$	0.9098	6.01	
$^5I_6 \rightarrow {}^5I_7$	0.0902		3377
$^5I_5 \rightarrow {}^5I_8$	0.4002	7.37	
$^5I_5 \rightarrow {}^5I_7$	0.5649		
$^5I_5 \rightarrow {}^5I_6$	0.0348		2444
$^5F_5 \rightarrow {}^5I_8$	0.7738	0.554	
$^5F_5 \rightarrow {}^5I_7$	0.1813		
$^5F_5 \rightarrow {}^5I_6$	0.0417		
$^5F_5 \rightarrow {}^5I_5$	0.0032		4152
$^5F_5 \rightarrow {}^5I_4$	0.0000		

Source: Refs. 20 and 21.

With consideration of the main ETU process, the population evaluation of each manifold can be described by using the following rate equations:

$$\frac{dn_3}{dt} = -(\beta_{30}W_{30} + \beta_{31}W_{31} + \beta_{32}W_{32})n_3 + p_{11}(C)n_1^2 \tag{4.1a}$$

$$\frac{dn_2}{dt} = -(\beta_{20}W_{20} + \beta_{21}W_{21})n_2 + \beta_{32}W_{32}n_3 \tag{4.1b}$$

$$\frac{dn_1}{dt} = \sigma_{\text{abs}}(\nu)(n_0 - n_1)\frac{I}{h\nu} + \beta_{31}W_{31}n_3 + \beta_{21}W_{21}n_2 \\ -W_{10}n_1 - 2p_{11}(C)n_1^2 \tag{4.1c}$$

$$\sum_{i=0}^{3} n_i = CN \tag{4.1d}$$

where n_i (i = 0, 1, 2, 3) is the population density of the ith level (see Fig. 4.1). I is the intensity of the laser beam at pump frequency ν. σ_{abs} is the absorption cross section of the Ho^{3+} $^5I_7$ manifold. C is the Ho^{3+} dopant concentration and N is the total number density of substitutable sites of the host (for a $YLiF_4$ crystal, $N = 1.396 \times 10^{22}$ ions/cm^3 [23]). β_{ij} ($i > j, i = 1, 2, 3, j = 0, 1, 2$) is the branch ratio for the $i \to j$ transition. The total effective decay rate is expressed as

$$W_{ij} = \eta_{e,ij} W_{ij,r} + W_{ij,nr} \tag{4.2}$$

where $W_{ij,r}$ and $W_{ij,nr}$ are the radiative and nonradiative decay rates, respectively. $\eta_{e,ij}$ is the corresponding fraction of photons escaping from the cooling sample [24]. $p_{11}(C)$ is the dopant-concentration-dependent ETU parameter. Considering a dipole–dipole interaction in the frame of Dexter's ET theory, the microscopic ET rate depends on the separation of the interacting ions to the inverse sixth power [25]. Hence, the ETU parameter is approximately proportional to the second order of the ion concentration. The ETU parameter is also temperature dependent, since host phonons participate in the upconversion process [26]. Therefore,

$$p_{11}(C) \propto C^2 n(T)^2 \tag{4.3}$$

where $n = [\exp(\hbar\omega_{max}/k_B T) - 1]^{-1}$ is the Bose–Einstein occupation number, with $\hbar\omega_{max}$ being the maximum host phonon energy. Take the 2% Ho^{3+}: $YLiF_4$ crystal, for example; the ETU parameter at room temperature is found to be 3.2×10^{-24} m^3·s^{-1} [22].

The total absorption includes the resonant absorption and the background absorption

$$P_{ads} = P_r + P_b = \sigma_{abs}(\nu)(n_{0s} - n_{1s})I + \alpha_b I \tag{4.4}$$

where α_b is the background absorption coefficient that can be treated as a constant [24]. The power density of the fluorescence generated by transitions from the first, second, and third excited states to the lower energy state(s) can be described, respectively, as

$$P_1 = \eta_{e,10} W_{10,r} h\nu_{10} n_{1s} \tag{4.5a}$$

$$P_2 = \eta_{e,20} \beta_{20} W_{20,r} h\nu_{20} n_{2s} \tag{4.5b}$$

$$P_3 = \eta_{e,30} \beta_{30} W_{30,r} h\nu_{30} n_{3s} + \eta_{e,31} \beta_{31} W_{31,r} h_{\nu 31} n_{3s} \tag{4.5c}$$

where n_{is} is the steady-state population of ith energy level that can be calculated from Eq. 4.1 by setting $dn_i/dt = 0$. ν_{ij} is the average fluorescence frequency corresponding to the $i \to j$ transition (as shown in Fig. 4.1). The cooling power density can be expressed by using the energy conservation law

$$P_{\text{cool}} = P_1 + P_2 + P_3 - P_{\text{ads}} \tag{4.6}$$

The cooling efficiency is defined as

$$\eta_{\text{cool}} = P_{\text{cool}}/P_{\text{ads}} \tag{4.7}$$

Analytical expressions of the cooling power density and the cooling efficiency in compact form are hard to present, as quadratic terms emerge in Eq. 4.1, whereas numerical results are easily obtained once detailed spectrum parameters are available.

It is important to mention that a two-manifold treatment of cooling with a Ho^{3+}-doped solid is still valid, given that the second and third excited states recombine predominantly nonradiatively. Neglecting the radiative terms on the right-hand side of Eqs. 4.1a and 4.1b, the population evaluation of the $^5I_7$ manifold can be approximately expressed as

$$\frac{dn_1}{dt} = \sigma_{\text{abs}}(\nu)(n_0 - n_1)\frac{I}{h\nu} - W_{10}n_1 - p_{11}(C)n_1^2 \tag{4.8}$$

Comparing with the traditional two-manifold cooling model that is widely used for describing optical refrigeration with Yb^{3+}-, Tm^{3+}-, and Er^{3+}-doped solids [2], it can be seen that only a dopant-concentration-dependent upconversion term is supplemented in the above equation. The heat generated from ETU can be approximately expressed as

$$p_{11}(C)n_{1s}^2 h\nu = \beta_{21}W_{21,\text{nr}}\varepsilon_{21}n_{2s} + \beta_{32}W_{32,\text{nr}}\varepsilon_{32}n_{3s} \tag{4.9}$$

where ε_{ij} is the energy gap between i and j levels. By using Eqs. 4.4–4.7, the approximate expression of the cooling efficiency can be written as

$$\eta'_{\text{cool}} = \eta'_{\text{ext}}\eta_{\text{abs}}\frac{\nu_{10}}{\nu} - 1 \tag{4.10}$$

where

$$\eta_{\text{abs}} = [1 + \alpha_b/\sigma_{\text{abs}}(n_{0s} - n_{1s})]^{-1} \tag{4.11}$$

is the pump intensity- and frequency-dependent absorption efficiency [5]. In Eq. 4.10, we have introduced the equivalent external quantum efficiency

$$\eta'_{\text{ext}} = (1 + W'_{10,\text{nr}}/W_{10,\text{r}})^{-1} \tag{4.12}$$

where

$$W'_{10,\mathrm{nr}} = W_{10,\mathrm{nr}} + p_{11}(C)n_{1\mathrm{s}} \tag{4.13}$$

is the equivalent nonradiative transition rate that characterizes the quantum defect caused by both nonradiative transition from the first excited state and ETU-induced fluorescence quenching. It can be seen from Eq. 4.3 that the equivalent external quantum efficiency decreases with increasing the dopant concentration. On the contrary, the absorption efficiency increases with increasing the dopant concentration. To achieve net cooling ($\eta'_{\mathrm{cool}} > 0$), the condition

$$\eta'_{\mathrm{ext}}(C)\eta_{\mathrm{abs}}(C) > \frac{\nu}{\nu_{10}} \tag{4.14}$$

needs to be satisfied. By using Eq. 4.14, the appropriate dopant concentration for a Ho^{3+} cooling system should be roughly figured out. So the two-manifold model could still provide approximate prediction on the cooling performance of Ho^{3+}-doped solids, in which the transitions from the higher manifolds are predominantly nonradiatively, and given that the external quantum efficiency is substituted by Eq. 4.12.

The cooling system benefits from the ETU channel only when the overall number of phonons annihilated exceeds that created by the following transitions from the third and second excited states to the lower energy states. Otherwise, the ETU channel harms the cooling performance. For the $^5I_7 \rightarrow {}^5I_8$; $^5I_7 \rightarrow {}^5I_5$ ET process, the $^5I_7$ population is reduced by two, and two laser photons ($E_{\mathrm{u\text{-}a}} = 2h\nu$) originally served as excitations of the conventional cooling channel are absorbed (with host phonon annihilation); as a result, the $^5I_5$ population is increased by one. The $^5I_5$ Ho^{3+} ions de-excite to the ground state through various transitions. If we define the weighting parameters that characterize the probability of each radiative and nonradiative transitions as

$$\eta_{ij} = \eta_{e,ij}\beta_{ij}W_{ij,\mathrm{r}}\left(\sum_{k=0}^{i-1}\beta_{ik}W_{ik}\right)^{-1} \tag{4.15a}$$

$$\xi_{ij} = \beta_{ij}W_{ij,\mathrm{nr}}\left(\sum_{k=0}^{i-1}\beta_{ik}W_{ik}\right)^{-1} \tag{4.15b}$$

where i = 1, 2, 3, j = 0, 1, 2, and $i > j$. The fluorescence energy and the heat generated after absorbing the two laser photons can be described, respectively, as

$$E_{\text{u-f}} = \eta_{30}h\nu_{30} + \eta_{31}h\nu_{31} + \xi_{32}\eta_{20}h\nu_{20} + (\eta_{31} + \xi_{32}\xi_{21})\eta_{10}h\nu_{10} \tag{4.16a}$$

$$\begin{aligned} E_{\text{ETU}} &= \xi_{32}\varepsilon_{32} + \xi_{32}\xi_{21}\varepsilon_{21} + (\eta_{31} + \xi_{32}\xi_{21})\xi_{10}\varepsilon_{10} \\ &\approx \xi_{32}\varepsilon_{32} + \xi_{32}\xi_{21}\varepsilon_{21} \end{aligned} \tag{4.16b}$$

To obtain the last equality on the right-hand side of Eq. 4.16b, the relation $\xi_{10} << \xi_{21(32)}$ (the nonradiative transition rate of the $^5I_7 \rightarrow {}^5I_8$ transition is far smaller than that of the $^5I_5(^5I_6) \rightarrow {}^5I_6(^5I_7)$ transition) has been used. The criterion for cooling with the ETU channel is expressed as

$$E_{\text{u-f}} - E_{\text{u-a}}(2h\nu) > E_{\text{ETU}} \tag{4.17}$$

Substituting Eq. 4.16 into Eq. 4.17, the pump frequency range for cooling with the ETU channel is obtained,

$$\nu < \nu_{\text{u}} \tag{4.18}$$

where the cutoff frequency is written as

$$\begin{aligned} \nu_{\text{u}} &= \eta_{30}\nu_{30} + \eta_{31}\nu_{31} + \xi_{32}\eta_{20}\nu_{20} + (\eta_{31}\eta_{10} + \xi_{32}\xi_{21}\eta_{10})\nu_{10} \\ &-\xi_{32}f_{32} - \xi_{32}\xi_{21}f_{21}, \end{aligned} \tag{4.19}$$

and $f_{ij} = \varepsilon_{ij}/h$ is the equivalent frequency of the corresponding energy gap. Since the absorbed photon energy for each ESA process is $2h\nu$, and the transition rates are almost independent of the pump condition, Eqs. 4.18 and 4.19 also indicate the pump frequency range for cooling through the ESA channel. If the cutoff frequency locals outside the absorption spectrum range of the $^5I_7$ manifold, upconversion cannot assist the cooling process. For cooling samples strongly affected by upconversion quenching, the corresponding nonradiative processes (minus terms on the right-hand side of Eq. 4.19) are stronger than the radiative processes (plus terms), so the value of the cutoff frequency is essentially negative. In such a case, no matter how the pump frequency is red-detuned, the upconversion processes barely obstruct the cooling process. It should be noted that, rare earth ions of a multimanifold exhibit similar transition properties when pumped to their first excited state. Therefore, the criterion is widely applicable in optical refrigeration with rare earth ions of a multimanifold.

To simulate the cooling performance of conventional Ho^{3+} optical refrigeration, we need the spectrum parameters of the Ho^{3+}:$YLiF_4$ crystal, summarized in Table 4.1. The nonradiative transition rates can be obtained by using the energy gap law [27],

$$W_{nr,ij} = \beta(1+n)^m \exp(-\alpha\varepsilon_{ij}) \tag{4.20}$$

where α and β are the host material–dependent characteristic constants. For a $YLiF_4$ crystal, $\alpha = 3.8 \times 10^{-3}$ cm and $\beta = 3.5 \times 10^7\ s^{-1}$ [28]. $m = \varepsilon_{ij}/\hbar\omega_{max}$ is the number of phonons needed to bridge the energy gap of ε_{ij}. At 300 K, the nonradiative transition rates of $^5I_7 \rightarrow {}^5I_8$, $^5I_6 \rightarrow {}^5I_7$, and $^5I_5 \rightarrow {}^5I_6$ transitions are calculated as 1.36, 234.77, and 6308.91 s^{-1}, respectively. Other nonradiative transition rates are comparatively small and are assumed to be zero in the following simulations. The emission cross sections of $^5I_7$, $^5I_6$, and $^5I_5$ manifolds are shown in Fig. 4.2, and the average wavelengths of fluorescence generated by $^5I_7 \rightarrow {}^5I_8$, $^5I_6 \rightarrow {}^5I_8$, and $^5I_5 \rightarrow {}^5I_8$ transitions are found to be 2014, 1185, and 908 nm, respectively. The average wavelength of fluorescence generated by the $^5I_5 \rightarrow {}^5I_7$ transition is estimated to be 1675 nm [21, 29].

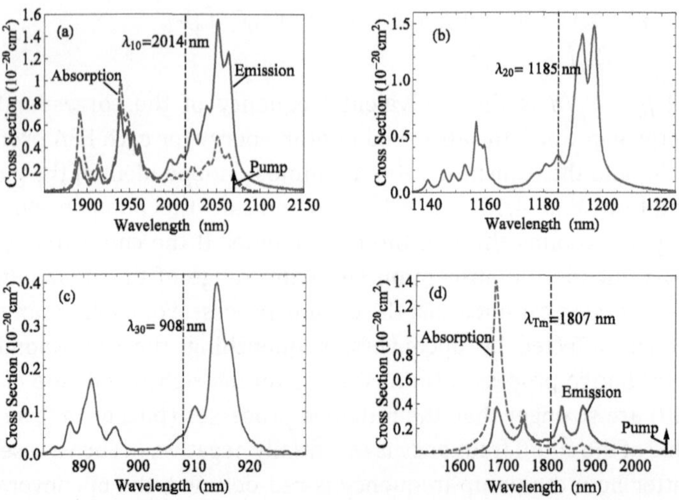

Figure 4.2 (a) Absorption and emission cross section (π-pol.) of Ho^{3+} $^5I_7$ manifold; emission cross section (π-pol.) of Ho^{3+} $^5I_6$ (b) and $^5I_5$ (c) manifolds; and (d) absorption and emission cross section (π-pol.) of Tm^{3+} $^5F_4$ manifold in $YLiF_4$ host matrix at room temperature [20, 23] (the Tm^{3+} spectrum will be needed in Section 4.4). The dashed lines indicate the average fluorescence wavelengths.

Suppose the cooling sample is geometrically optimized, so fluorescence photons totally escape from the cooling sample ($\eta_{e,ij}$ = 1). Calculating the result of Eq. 4.19 shows that for the Ho^{3+}:$YLiF_4$ crystal, the cutoff frequency is negative; therefore, the ETU process of the Ho^{3+}:$YLiF_4$ crystal intrinsically harms its cooling performance. As Ref. [17] indicates, to realize net cooling of Ho^{3+}-doped solids, it is essential to inhibit ETU. For this reason, diluting the dopant could be a feasible solution. To figure out the appropriate dopant concentration region for optical refrigeration with the Ho^{3+}:$YLiF_4$ crystal, the cooling power density and the cooling efficiency that depend on the dopant concentration are simulated below.

From Fig. 4.2a we know that it is rational to select 2070 nm as the pump wavelength. The cooling power density (Eq. 4.6) and the cooling efficiency (Eq. 4.7) as functions of pump intensity and dopant concentration are depicted in Fig. 4.3 (the background absorption is assumed to be 4.0×10^{-4} cm^{-1}, which is currently practical for manufacturing [6]). It can be seen that net cooling would be arduous to achieve if the selected $YLiF_4$ crystal sample is doped with more than 1% Ho^{3+} ions, since the cooling region for pump intensity is considerably narrow and the achievable cooling power density is quite low at such a high concentration level. In view of heating induced by the upconversion quenching, it is rational to limit the dopant concentration within 1%. On the other hand, the radiation center (Ho^{3+} ion) is desired to be densely doped to provide enough cooling, surpassing the background absorption heating. A more practical dopant concentration range for the Ho^{3+}:$YLiF_4$ crystal with a background absorption of 4.0×10^{-4} cm^{-1} is suggested as 0.2%–1%. For each pump wavelength, there exists an optimal dopant concentration from the perspective of maximizing the cooling power density. We figure out that at a pump wavelength of 2070 nm, the optimal dopant concentration for the Ho^{3+}:$YLiF_4$ crystal is 0.54%, the maximum achievable cooling power density is 9.7×10^4 $W{\cdot}m^{-3}$, and the corresponding cooling efficiency is 0.29%.

Figure 4.4 shows the comparison between the cooling results predicted by the conventional two-manifold cooling model without consideration of ETU (M1), the generalized cooling model (M2), and the two-manifold cooling model with consideration of ETU (M3). At low pump intensity ($I < 10^6$ $W{\cdot}m^{-2}$ for a 0.54% Ho^{3+}:$YLiF_4$ crystal), the $^5I_7$ state is sparsely populated, so the ETU process is weak, and all of these models are fit to describe the cooling performance of Ho^{3+}-doped fluoride crystals. However, as the pump intensity increases, M1 tends to overestimate the cooling results, because the heating process induced by upconversion quenching is omitted. On the contrary, M3 tends to underestimate the cooling results because the phonon annihilation induced by phonon-assisted ETU is excluded.

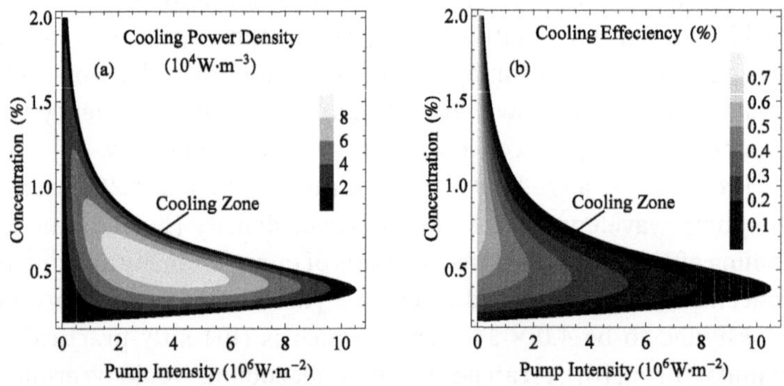

Figure 4.3 (a) The cooling power density and (b) the cooling efficiency of a Ho^{3+}:$YLiF_4$ crystal as functions of pump intensity and dopant concentration. The pump wavelength is 2070 nm, and the background absorption is 4.0×10^{-4} cm^{-1}.

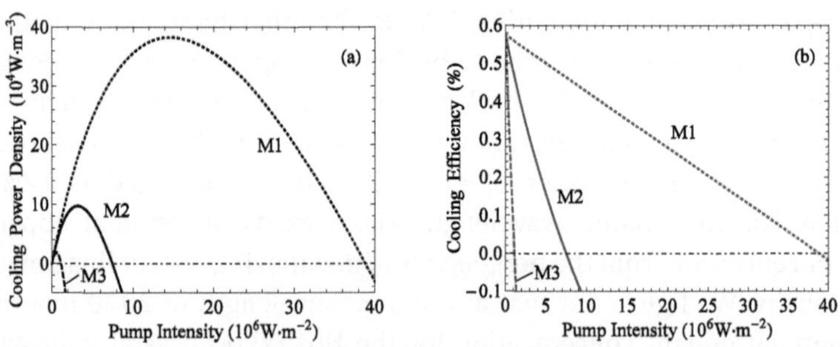

Figure 4.4 Comparison between M1 (dotted curves), M2 (solid curves), and M3 (dashed curves). (a) The cooling power density and (b) the cooling efficiency. The dopant concentration is 0.54%, the pump wavelength is 2070 nm, and the background absorption is 4.0×10^{-4} cm^{-1}.

Figure 4.5 shows the maximum attainable cooling power density and the corresponding optimal dopant concentration as functions of pump wavelength. It can be seen from Fig. 4.5a that for a $YLiF_4$ crystal with a background absorption of 4.0×10^{-4} cm^{-1}, the ideal pump wavelength is around 2067 nm. A cooling power density of about 1.1×10^5 $W \cdot m^{-3}$ is expected in ~0.47% Ho^{3+}-doped samples. Purifying the host is an important way to improve the cooling performance. As it can be seen in Fig. 4.5b, if the background absorption is reduced by an order of magnitude, the maximum cooling power density (MCPD) could exceed 7.5×10^5 $W \cdot m^{-3}$ in ~0.41% Ho^{3+}-doped samples, and the corresponding optimal pump wavelength red-shifts to ~2083 nm. In fact, the optimal dopant concentration is determined by a balancing consideration of heat loads induced by background absorption and fluorescence quenching. As the background absorption decreases, the parasitic heat problem is alleviated and the weight of heating induced by fluorescence quenching effect increases. A reduced dopant concentration essentially weakens the ETU process. Therefore, the optimal dopant concentration for a host with lower background absorption is lower than that for a host with higher background absorption. For optical refrigeration with a highly purified sample, it is possible to tune the pump wavelength to the red-tail part of the absorption spectrum (where the cross section is tiny) to obtain a high cooling power density.

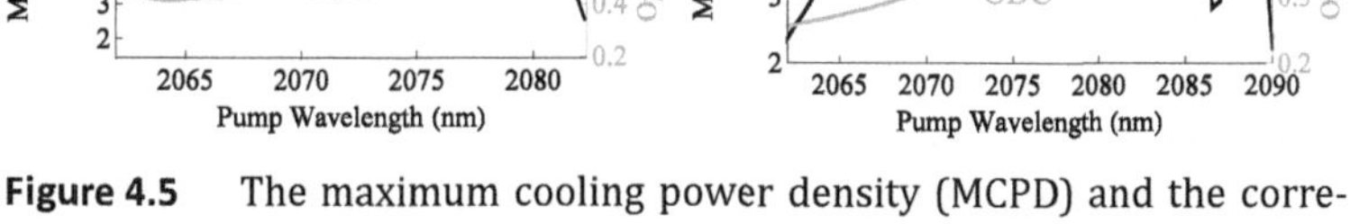

Figure 4.5 The maximum cooling power density (MCPD) and the corresponding optimal dopant concentration (ODC) verses pump wavelength at background absorptions of (a) 4.0×10^{-4} cm^{-1} and (b) 4.0×10^{-5} cm^{-1}.

4.3 Efficient and Enhanced Holmium Optical Refrigeration via Co-Pumping

Upconversion quenching makes the attempt of cooling Ho^{3+}-doped fluoride crystals a challenging task. When pumped at around 2.07

μm for realizing conventional ASFC, Ho^{3+} ions will populate at the $^5I_5$ manifold due to an ETU of $^5I_7 \rightarrow {}^5I_8$; $^5I_7 \rightarrow {}^5I_5$. The following transition of $^5I_5 \rightarrow {}^5I_6$ is predominantly nonradiative, preventing effective cooling. To solve the upconversion-induced heating issue, it is possible to adopt the co-pumping scheme illustrated in Fig. 4.6. First of all, Ho^{3+} ions are pumped from the top of the ground-state manifold ($^5I_8$) to the bottom of the first excited-state manifold ($^5I_7$) by laser at a frequency of v_g to facilitate the ASFC process (solid red and green arrows below). Part of $^5I_7$ Ho^{3+} ions interact through an electromagnetic multipole, resulting in an ETU of $^5I_7 \rightarrow {}^5I_8$; $^5I_7 \rightarrow {}^5I_5$ (solid yellow arrows) with two host phonons' absorption (solid gray arrow). Among all the transitions from $^5I_5$ and $^5I_6$ to the lower manifolds, $^5I_5 \rightarrow {}^5I_6$ and $^5I_6 \rightarrow {}^5I_7$ are predominantly nonradiative due to the small energy gaps, which bring main heat to the system (see Section 4.2).

To cut down the heat induced by upconversion quenching, the laser at a frequency of v_e that pumps $^5I_5$ Ho^{3+} ions to the $^5F_5$ manifold (solid brown arrow above) may be added to the system. The branch ratio of transition from $^5F_5$ to its lower adjacent manifold ($^5I_4$) is approximately zero (see Table 4.1). The branch ratio for the $^5F_5 \rightarrow {}^5I_5$ transition is small, and the corresponding energy gap is much wider than $^5I_5$–$^5I_6$. According to the energy gap law, the $^5F_5$ manifold should be less heat involving than the $^5I_5$ manifold. So, the ESA of $^5I_5 \rightarrow {}^5F_5$ that aims at depopulating the $^5I_5$ manifold may improve the cooling performance. To confirm this guess, a quantitative comparison of heat production through different transition channels is needed.

In assessing the heating process of a particular transition channel, it would be convenient to utilize the weighting parameters (see Section 4.2) that characterize the probability of each radiative and nonradiative transition. The heat that generated after each $^5I_7 \rightarrow {}^5I_8$; $^5I_7 \rightarrow {}^5I_5$ ETU process can be described by Eq. 4.16b. Similarly, the heat that generated after each $^5I_5 \rightarrow {}^5F_5$ ESA transition can be expressed as

$$E_{ESA} \approx \eta_{43}\xi_{32}\varepsilon_{32} + (\eta_{42} + \eta_{43}\xi_{32})\xi_{21}\varepsilon_{21} \quad (4.21)$$

If E_{ESA} is smaller than E_{ETU} in value, we can draw the conclusion that the $^5I_5 \rightarrow {}^5F_5$ ESA process dwindles ETU-induced heat. Indeed, the calculation based on Eqs. 4.16b, 4.20, and 4.21 shows that E_{ESA} is only about 0.3% of E_{ETU} at 300 K. Therefore, the co-pumping scheme could improve the cooling performance of Ho^{3+}-doped solids.

As shown in Fig. 4.6, the rate equations describing the population dynamics of the cooling system are given by

$$\frac{dn_4}{dt} = \sigma_e(\nu_e)(n_3 - n_4)\frac{I_e}{h\nu_e} - A_4 \tag{4.22a}$$

$$\frac{dn_3}{dt} = -\sigma_e(\nu_e)(n_3 - n_4)\frac{I_e}{h\nu_e} - A_3 + B_3 + p_{11}(C)n_1^2 \tag{4.22b}$$

$$\frac{dn_2}{dt} = -A_2 + B_2 \tag{4.22c}$$

$$\frac{dn_1}{dt} = \sigma_g(\nu_g)(n_0 - n_1)\frac{I_g}{h\nu_g} - A_1 + B_1 - 2p_{11}(C)n_1^2 \tag{4.22d}$$

$$\sum_{i=0}^{4} n_i = CN \tag{4.22e}$$

where n_i (i = 0, ..., 4) is the population density of the ith level (see Fig. 4.6). I_g and I_e are the intensities of the incident laser beams at the frequency ν_g and ν_e, respectively. σ_g and σ_e are the temperature-dependent GSA cross section of the $^5I_8 \rightarrow {}^5I_7$ transition and the ESA cross section of the $^5I_5 \rightarrow {}^5F_5$ transition, respectively, and according to the reciprocal relation [30], their temperature dependence can be determined by

$$\sigma_{g(e)}(T) \propto \frac{Z^u_{g(e)}}{Z^l_{g(e)}} \exp\left(\frac{h\nu_{g(e)} - E^{ZL}_{g(e)}}{k_B T}\right) \tag{4.23}$$

where $Z^u_{g(e)}$ and $Z^l_{g(e)}$ are partition functions of $^5F_5(^5I_7)$ and $^5I_5(^5I_8)$ manifolds. $E^{ZL}_{g(e)}$ is the zero-line energy of the $^5F_5(^5I_7)$ manifold.

$$A_i = \sum_{j=0}^{i-1} \beta_{ij} W_{ij} n_i \tag{4.24}$$

is the sum of all decay terms from the ith level (composed of radiative and nonradiative transitions), and

$$B_i = \sum_{j=i+1}^{4} \beta_{ji} W_{ji} n_j \tag{4.25}$$

is the sum of all gain terms to the ith level.

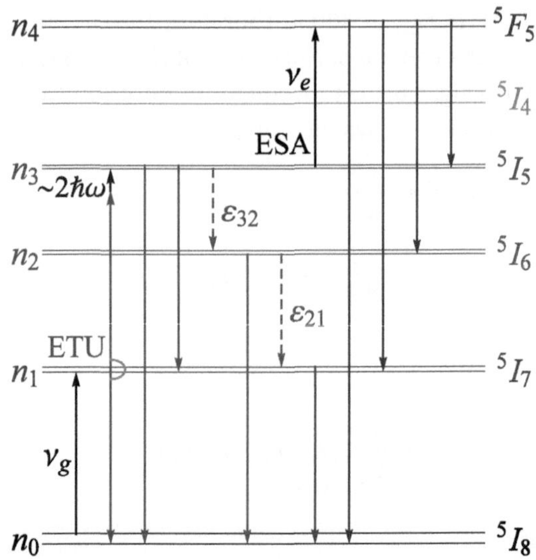

Figure 4.6 Schematic of the co-pumping scheme for efficient optical refrigeration in Ho^{3+}-doped fluoride crystals. The transition of $^5F_5 \rightarrow {}^5I_4$ is omitted due to the tiny branch ratio. The level labels used in Eq. 4.22 are also indicated.

The total absorption includes the resonant absorptions (GSA and ESA) and the background absorption,

$$P_{abs} = \sigma_g(n_{0s} - n_{1s})I_g + \sigma_e(n_{3s} - n_{4s})I_e + \alpha_b(I_g + I_e) \tag{4.26}$$

where n_{is} is the steady-state population of the ith level. The total emitted fluorescence power density can be described as

$$P_f = \sum_{i=1}^{4}\sum_{j=0}^{i-1} \eta_{e,ij}\beta_{ij}W_{ij}n_{is}h\nu_{ij} \tag{4.27}$$

where ν_{ij} is the mean fluorescence frequency for the $i \rightarrow j$ transition. The cooling power density is obtained from the energy conservation law

$$P_{cool} = P_f - P_{ads} \tag{4.28}$$

It is essential to note that while the upconversion quenching–induced heat is alleviated with the ESA transition, the background absorption is increased simultaneously. To achieve cooling enhancement, the ESA-induced heat should be less than the heat

ESA cuts from ETU-induced fluorescence quenching. Calculation shows that the $^5I_5$ manifold is the main source of heating. So the ESA-induced power density change can be approximately expressed as

$$\Delta P_v \approx \beta_{32} W_{32,nr} \varepsilon_{32} \Delta n_{3s} \tag{4.29}$$

where Δn_{3s} is the population density change of the $^5I_5$ manifold after the laser of frequency ν_e is added to the system, assuming the pump intensity for GSA is I_1 and the pump intensity for ESA is I_2. The population density change of the $^5I_5$ manifold can be approximately expressed as (see "Appendix")

$$\Delta n_{3s} \approx \frac{(\eta_{43}-1)p_{11}\sigma_e C^2 N^2}{h\nu_e\left(\sum_{i=0}^{2}\beta_{3i}W_{3i}\right)^2\left(2+\frac{h\nu_g W_{10}}{\sigma_g I_1}\right)^2} I_2 \tag{4.30}$$

The background absorption of the laser for ESA is

$$\Delta P_b = \alpha_b I_2 \tag{4.31}$$

For cooling enhancement to occur

$$\Delta P_b + \Delta P_v < 0 \tag{4.32}$$

needs to be satisfied. By using Eqs. 4.31–4.32, we can obtain

$$\alpha_b < \alpha_0 = \frac{(1-\eta_{43})\xi_{32}\varepsilon_{32}p_{11}\sigma_e C^2 N^2}{h\nu_e\left(\sum_{i=0}^{2}\beta_{3i}W_{3i}\right)^2\left(2+\frac{h\nu_g W_{10}}{\sigma_g I_1}\right)^2} \tag{4.33}$$

where α_0 is defined as the critical background absorption coefficient. Therefore, the co-pumping scheme only improves the cooling performance of a Ho^{3+} cooling system with background absorption lower than α_0. In other words, the sample purity is less demanded to achieve cooling enhancement in a co-pumping cooling system with a larger α_0.

The ETU parameter declines when temperature drops (see Eq. 4.3); as a result, the upconversion quenching becomes less problematic. On the other hand, background absorption consistently brings heat to the sample. Therefore, the co-pumping scheme may become useless or even harmful to the cooling system as the temperature drops to a certain value, which can be determined by using Eq. 4.33. Nevertheless, the co-pumping scheme accelerates the cooling rate at relatively high temperatures, which is helpful in shortening the time needed for reaching the minimum attainable temperature of the cooling system [6].

The cooling power density and cooling efficiency of Ho^{3+}-doped fluoride crystals via the co-pumping scheme can be simulated by using the spectral data found in related references. To do that, the ESA cross section of the $^5I_5 \rightarrow {}^5F_5$ transition is first estimated. In the frame of Judd–Ofelt theory, the ESA cross section is given by [31]

$$\sigma_e = \frac{\lambda_m^4}{8\pi c}\frac{g_f}{g_i}\left(\sum_J A_J\right)\frac{S(\lambda)}{\int S(\lambda)d\lambda} \tag{4.34}$$

where λ_m is the mean absorption wavelength for the $^5I_5 \rightarrow {}^5F_5$ transition. g_i and g_f are the degeneration coefficients of the initial and final states, respectively. A_J is the decay rate from $^5F_5$ to each of the lower manifold ($^5I_4$, $^5I_5$, $^5I_6$, $^5I_7$, and $^5I_8$) for an electric dipole transition. $S(\lambda)$ is the line shape of the ESA spectrum. Here, the ESA spectrum shape is roughly assumed to be rectangular, with a spectrum width of $\Delta\lambda$. The absorption cross section of $^5I_5 \rightarrow {}^5F_5$ can be approximately expressed as

$$\sigma_e(\lambda_m) \approx \frac{\lambda_m^4}{8\pi c\Delta\lambda}\frac{g_f}{g_i}\left(\sum_J A_J\right) \tag{4.35}$$

For Ho^{3+}:$YLiF_4$ crystals, the absorption of the $^5I_5 \rightarrow {}^5F_5$ transition ranges from 2261 to 2408 nm in wavelength, and the mean absorption wavelength is 2282 nm [20, 21]. So the ESA absorption spectrum mainly distributes between 2261 and 2303 nm, and $\Delta\lambda$ is supposed to be 42 nm. The sum of the decay rates from $^5F_5$ to the lower manifolds is 1798.77 s^{-1} [20]. On the basis of the above parametric information, the ESA cross section at 2282 nm is estimated to be 1.54×10^{-18} cm^2.

The spectral data related to $^5I_7$, $^5I_6$, and $^5I_5$ manifolds for a Ho^{3+}:$YLiF_4$ crystal have been provided in Section 4.2. At 300 K, the mean wavelength of fluorescence for the $^5F_5 \rightarrow {}^5I_8$ transition is found to be 649 nm, and the mean wavelengths of fluorescence for $^5F_5 \rightarrow {}^5I_7$, $^5F_5 \rightarrow {}^5I_6$, and $^5F_5 \rightarrow {}^5I_5$ transitions are estimated to be 965, 1457, and 2324 nm, respectively [20]. By using Eq. 4.20, the nonradiative transition rate of the $^5F_5 \rightarrow {}^5I_5$ transition is calculated to be 15.28 s^{-1}. Other nonradiative transitions are insignificant due to the large energy gaps. The pump wavelength for GSA and ESA is selected to be 2070 and 2282 nm, respectively.

Figure 4.7 shows α_0 as a function of dopant concentration for the Ho^{3+}-doped $YLiF_4$ crystal-based cooling system at 300 K. It shows that α_0 increases greatly with the increasing of the dopant concentration, which can be interpreted from two aspects. For one thing, increased dopant concentration feedbacks enlarged ESA absorption ($\sigma_e CN$), so the ratio of the resonant absorption to the background absorption of the laser for ESA is promoted, and the system becomes more endurable to the parasitic heat. For another, the strength of ETU (p_{11}) is proportional to the second order of the dopant concentration. So, in a densely doped system, upconversion quenching rather than parasitic absorption is the dominant source of heating. Besides, the increase in the pump intensity of the laser for GSA promotes the population density at the $^5I_7$ manifold, and as a result, the ETU is strengthened and α_0 gets larger. It can be seen from Fig. 4.7 that, by using the co-pumping scheme, cooling enhancement could be achieved in the Ho^{3+}:$YLiF_4$ crystal with a dopant concentration of more than 0.4%.

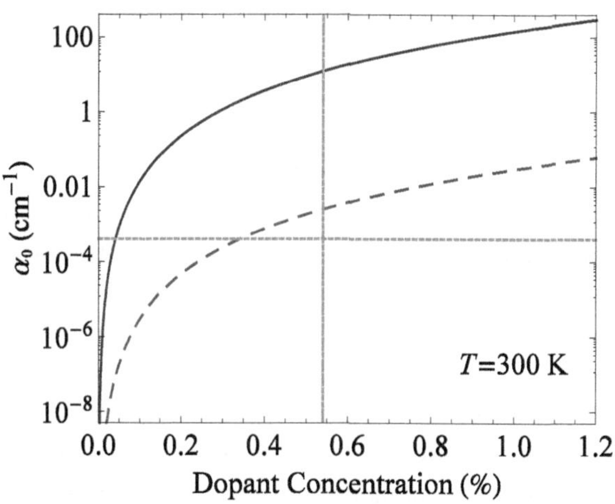

Figure 4.7 Critical background absorption coefficient as a function of dopant concentration at 300 K. The solid curve is for I_1 = 10^5 W·m^{-2}, and the dashed curve is for I_1 = 10^7 W·m^{-2}. The horizontal dotted line implies available background absorption in the $YLiF_4$ host (4.0×10^{-4} cm^{-1}). The vertical dotted line implies the optimum dopant concentration (0.54%) in the conventional cooling case.

To investigate the temperature dependence of GSA (ESA) and ETU transition, we define, respectively, the temperature-dependent coefficients of the GSA (ESA) cross section and ETU parameter as

$$\Theta_{g(e)}(T) = \frac{\sigma_{g(e)}(T)}{\sigma_{g(e)}(T_0)} \tag{4.36a}$$

$$\Theta_u(T) = \frac{p_{11}(T)}{p_{11}(T_0)} = \left[\frac{n(T)}{n(T_0)}\right]^2 \tag{4.36b}$$

where T_0 = 300 K. Using spectrum data found in Ref. [21], the temperature-dependent coefficients are plotted in Fig. 4.8. It can be seen that Θ_u and Θ_g decrease with the decreasing of temperature. When temperature drops to less than 200 K, ETU may become no longer problematic to the cooling system since the ETU parameter is tiny. When the temperature approaches 100 K, the small GSA in the anti-Stokes domain will limit efficient cooling to lower temperatures. On the contrary, Θ_e increases with the decreasing of temperature because the ESA we adopted here is in the Stokes domain. By using Eqs. 4.3, 4.23, and 4.33, the temperature-dependent α_0 is plotted in Fig. 4.9. It can be seen that when the sample temperature drops to around 210 K, $\alpha_0 = \alpha_b$, which means no cooling enhancement occurs. At temperatures lower than 210 K, $\alpha_0 < \alpha_b$, and the co-pumping scheme will introduce extra heat to the cooling system (see Eq. 4.32). Therefore, we conclude that the co-pumping for cooling enhancement only works at relatively high temperatures (for the case discussed here, $T > 210$ K). Although it may not be helpful for cooling at cryogenic temperatures, the co-pumping scheme can shorten the time needed to reach the minimum attainable temperature of a Ho^{3+} cooling system, due to the improved cooling performance at temperatures of 210–300 K. To estimate the cooling improvement, next we simulate the cooling power density and cooling efficiency of the scheme at 300 K.

Generally, the cooling efficiency of an optical refrigeration system dwindles when the input laser power is increased. As the pump intensity approaches the saturation of the resonant absorption, the parasitic absorption will become dominant, and eventually, the cooling power density is lowered with the declining cooling efficiency. To obtain the best cooling effect, the intensity of the pump laser should be deliberately selected. In Ho^{3+} optical refrigeration via the co-pumping scheme, the optimal pump intensities of GSA and ESA for achieving the MCPD appear in pair. Figure 4.10a shows the attainable MCPD in the Ho^{3+}-doped $YLiF_4$ crystal of various dopant concentrations at 300 K. The optimal pump intensities for reaching the MCPD are shown in Fig. 4.10b, and the corresponding cooling efficiency is shown in Fig. 4.10c. It can be seen that an MCPD of 7.5×10^7 W·m^{-3} could be achieved in the 1.2% Ho^{3+}:$YLiF_4$ crystal. The MCPD in the co-pumping scheme is almost 3 orders of magnitude higher than that of the conventional laser cooling with Ho^{3+}:$YLiF_4$ crystals (see Section 4.2). The corresponding cooling efficiency is about 4.2%, which is several times higher than what could be obtained via the conventional pumping scheme. As it has been mentioned, the enhancing effects are due to the cut-down in the population density of the $^5I_5$ manifold with an ESA of $^5I_5 \rightarrow {}^5F_5$. Since the decay from the $^5F_5$ manifold is predominantly radiative, the upconversion quenching is inhibited. Therefore, the dopant concentration, as well as the pump intensity, could be increased, resulting in an immense cooling enhancement.

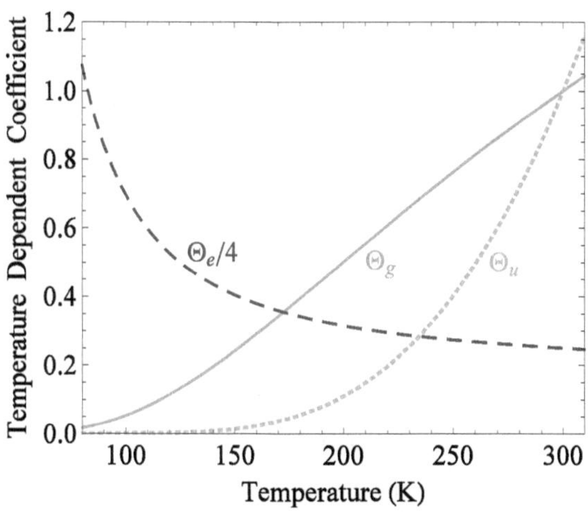

Figure 4.8 Temperature-dependent coefficients of GSA cross section (solid curve), ESA cross section (dashed curve), and ETU parameter (dotted curve).

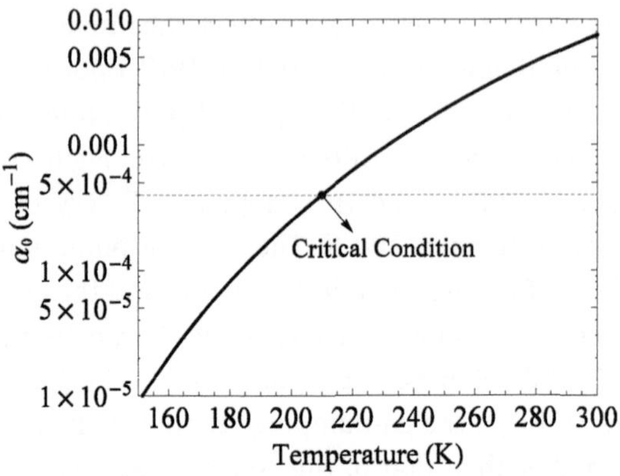

Figure 4.9 Temperature-dependent α_0 of the 1% Ho^{3+}:$YLiF_4$ crystal (I_1 = 10^6 W·m^{-2}). The dashed line indicates the background absorption, α_b = 4.0 × 10^{-4} cm^{-1}. The solid point indicates the critical condition of whether cooling enhancement occurs.

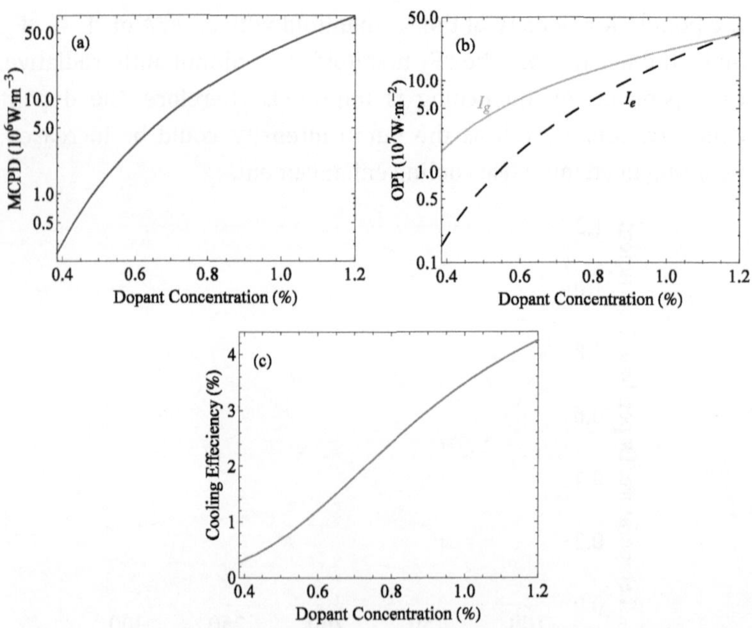

Figure 4.10 At 300 K, (a) the maximum cooling power density (MCPD) as a function of dopant concentration and (b) the corresponding optimum pump intensities (OPIs). The solid curve is for GSA, and the dashed curve is for ESA. (c) The corresponding cooling efficiency. The background absorption is assumed to be 4.0 × 10^{-4} cm^{-1}.

It is essential to point out that the MCPD of the Ho^{3+} cooling system does not always increase with the dopant concentration because high-order upconversion processes, such as $^5I_6 \rightarrow {}^5I_8$; $^5I_6 \rightarrow {}^5S_2$ ($^5F_4$), $^5I_5 \rightarrow {}^5I_8$; $^5I_5 \rightarrow {}^5F_2$ ($^5F_3$), and so on, will eventually emerge when the dopant concentration becomes too high. When these high-order upconversion processes are strongly involved, the population evolution of each manifold no longer obeys Eq. 4.22. The transitions from $^5S_2$, $^5F_4$, $^5F_3$, $^5F_2$, and so on to their adjacent manifolds are predominantly nonradiative, which lower the cooling power density, and even turn the system into a heating one. Anyway, our motivation is to provide a feasible way of realizing efficient and enhanced cooling in Ho^{3+}-doped solids. In the simulations, the dopant concentration is restricted within 1.2%. The population densities of $^5I_5$ and $^5I_6$ manifolds are small enough, so no observable high-order upconversion processes will involve, and the results are fairly reliable.

4.4 Energy Transfer–Enhanced Laser Cooling in Rare Earth–Co-Doped Solids

So far all the discussions of this chapter have been based on the Ho^{3+}-doped cooling system. In this section, we introduce ETLC in rare earth ion co-doped systems. The Ho^{3+}, Tm^{3+}:$YLiF_4$ crystal is selected as the investigated object, in which efficient laser cooling beyond 2 μm could be facilitated. The major ET processes in Ho^{3+}- and Tm^{3+}-co-doped $YLiF_4$ crystals are illustrated in Fig. 4.11. It is important to note that the third and the second excited states of Ho^{3+} and Tm^{3+} ions could also get involved in the transition processes through the ET mechanism. The Ho^{3+} $^5I_7 \rightarrow {}^5I_8$; Ho^{3+} $^5I_7 \rightarrow {}^5I_5$ and the Tm^{3+} $^3F_4 \rightarrow {}^3H_6$; Ho^{3+} $^5I_7 \rightarrow {}^5I_5$ ET processes will induce population accumulation at the $^5I_5$ manifold of Ho^{3+} ions, and the Tm^{3+} $^3F_4 \rightarrow {}^3H_6$; Tm^{3+} $^3F_4 \rightarrow {}^3H_4$ ET processes will induce population accumulation at the $^3H_4$ manifold of Tm^{3+} ions. Due to that the energy gaps of $^5I_5$–$^5I_6$, $^5I_6$–$^5I_7$, $^3H_4$–$^3H_5$, and $^3H_5$–$^3F_4$ are relatively small, and the following $^5I_5 \rightarrow {}^5I_6$, $^5I_6 \rightarrow {}^5I_7$, $^3H_4 \rightarrow {}^3H_5$, and $^3H_5 \rightarrow {}^3F_4$ transitions (dashed arrows) are highly exothermic [23]. Therefore, inhibiting these ETU processes is essential for realizing ETLC.

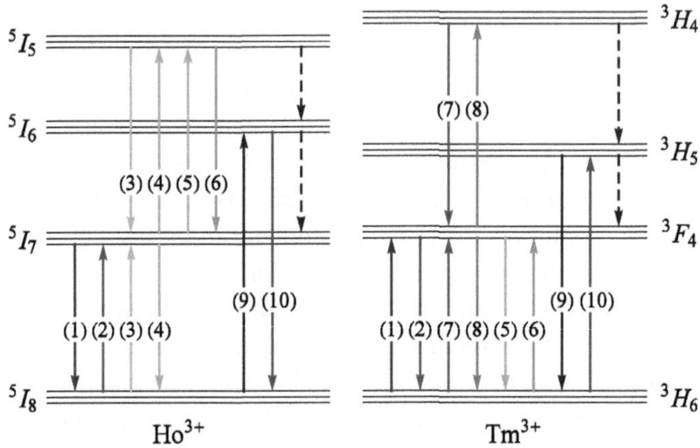

Figure 4.11 Energy level schematic of major Tm^{3+}-Ho^{3+} ET processes in a $LiYF_4$ host matrix [23]. The ETU processes of (4), (5), and (8) result in population accumulation at the third excited electronic energy level of Ho^{3+} and Tm^{3+} ions.

As outcomes of ion–ion interaction, the ET processes depend on the ion concentrations significantly. Analysis of the fluorescence spectrum shows that when the concentrations of Ho^{3+} and Tm^{3+} ions are restricted within relatively low levels (e.g., in the 0.3% Ho^{3+}-, 2% Tm^{3+}-doped $YLiF_4$ crystal), the ETU processes mentioned above could be averted and the Ho^{3+} $^5I_7 \leftrightarrow {}^5I_8$; Tm^{3+} $^3H_6 \leftrightarrow {}^3F_4$ ET processes become dominant [32]. In this case, the Ho-Tm ETLC that is shown schematically in Fig. 4.12 could be realized with proper laser excitation. The incident laser photons are tuned to be resonant for the transitions from the higher $^5I_8$ to the lower $^5I_7$ manifolds (vertical solid arrow on the left). The exited Ho^{3+} ions relax thermally among the manifolds and then decay by emitting fluorescence photons at a mean frequency of v_1 (vertical dashed arrow on the left). For each excitation and de-excitation cycle, an average thermal vibration energy of $hv_1 - hv$ is removed from the host material with extraction of the fluorescence photon. This is the case of conventional ASFC usually discussed.

Meanwhile, part of the ground-state Tm^{3+} ions gets sensitized by absorbing the energy transferred from Ho^{3+} excitations and the surrounding phonon energy (phonon-assisted ET, double-head arrow) [28] and then decays by emitting fluorescence photons at a mean frequency of v_2 (vertical dashed arrow on the right). Since the pump laser is red-edge- tuned for Ho^{3+} ion excitation, the absorption of Tm^{3+} ions is negligible (see Fig. 4.2d). For each excitation and de-excitation cycle of Tm^{3+} ions, an extra thermal vibration energy of $h\nu_2 - h\nu_1$ will be released. According to ET theory, the density population dynamic of the electronic energy levels of the above Ho-Tm cooling system can be described by the following equations:

$$\frac{dn_1}{dt} = \frac{P_r}{h\nu} - \eta_{e,1}W_{1,r}n_1 - W_{1,nr}n_1 - p_{13}n_1n_3 + p_{02}n_0n_2 \tag{4.37a}$$

$$\frac{dn_2}{dt} = -\eta_{e,2}W_{2,r}n_2 - W_{2,nr}n_2 + p_{13}n_1n_3 - p_{02}n_0n_2 \tag{4.37b}$$

$$n_0 + n_1 = C_{Ho}N \tag{4.37c}$$

$$n_2 + n_3 = C_{Tm}N \tag{4.37d}$$

where n_i (i = 0, 1, 2, 3) is the population of the ith level, as shown in Fig. 4.12. P_r is the resonantly absorbed power density at pump frequency ν. $W_{j,r}$ and $W_{j,nr}$ are the radiative decay rate and the nonradiative decay rate, respectively (j = 1 for the Ho^{3+} $^5I_7 \rightarrow {}^5I_8$ transition and j = 2 for the Tm^{3+} $^3F_4 \rightarrow {}^3H_6$ transition). p_{13} and p_{02} are the temperature-dependent macroscopic ET rates [33]. C_{Ho} and C_{Tm} are the percentage concentrations of Ho^{3+} and Tm^{3+} ions, respectively.

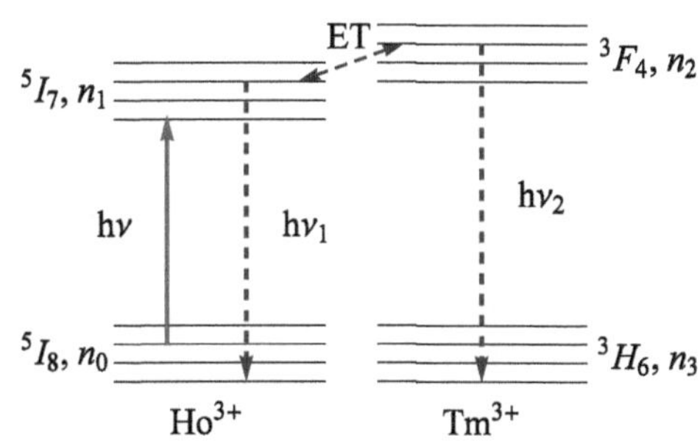

Figure 4.12 Schematic of Ho-Tm ETLC.

The net power density transferred to the sample is obtained by using the energy conservation law

$$P_{net} = -\eta_{e,1}W_{1,r}n_{1,s}h\nu_1 - \eta_{e,2}W_{2,r}n_{2,s}h\nu_2 + P_r + P_b \tag{4.38}$$

where $n_{i,s}$ is the steady-state population of the ith energy level. The first term on the right-hand side of Eq. 4.38 represents the energy that is taken out of the host by the conventional ASFC mechanism of Ho^{3+} ions, and the second term takes into account the cooling contribution of ET processes. The cooling efficiency, which is defined as the ratio of the cooling output to the total absorption, is expressed as

$$\eta_c = \frac{-P_{net}}{P_r + P_b} = \eta_{abs}\frac{\eta_{e,1}W_{1,r}n_{1,s}\nu_1 + \eta_{e,2}W_{2,r}n_{2,s}\nu_2}{(W_1 n_{1,s} + W_2 n_{2,s})\nu} - 1 \tag{4.39}$$

where $W_{1(2)}$ is the total effective decay rate that includes both the radiative decay rate with consideration of florescence reabsorption and nonradiative decay rate. Substituting the analytical results of $n_{i,s}$ into Eq. 4.39, a symmetrical expression is obtained,

$$\eta_c = \eta_{abs}\left(\frac{\eta_{ext,1}\nu_1 + \eta_{ext,2}\nu_2}{2\nu} + \frac{\eta_{ext,1}\nu_1 - \eta_{ext,2}\nu_2}{2\nu_c}\right) - 1 \tag{4.40}$$

where $\eta_{ext,1(2)}$ is the external quantum efficiency for the Ho^{3+} $^5I_7 \rightarrow$ $^5I_8$ (Tm^{3+} $^3F_4 \rightarrow$ $^3H_6$) transition, and

$$\nu_c = P_r/hA \tag{4.41}$$

with

$$A = B - \left[\frac{4p_{13}C_{Tm}W_2 N}{p_{02} - p_{13}}\frac{P_r}{h\nu} + \left(B - \frac{P_r}{h\nu}\right)^2\right]^{1/2} \tag{4.42}$$

and

$$B = \frac{(p_{13}C_{Tm}W_2 + p_{02}C_{Ho}W_1)N + W_1W_2}{p_{02} - p_{13}} \tag{4.43}$$

The appearance of variable ν_c in the cooling efficiency formula (Eq. 4.40) of ETLC is a result of the ET mechanism acting on the co-doped cooling system. As an important and unique parameter of ETLC, ν_c determines the cooling efficiency of a co-doped cooling system. It can be seen from Eqs. 4.41–4.43 that ν_c, which will be referred next as the characteristic frequency, lies on four pairs of parameters: the inputs (P_r and ν), the total effective transition rates (W_1 and W_2) that can be treated as constants, the temperature-dependent macroscopic ET rates (p_{13} and p_{02}), and the percentage concentrations (C_{Ho} and C_{Tm}). The merit of Eq. 4.40 is that knowing the basic property of the characteristic frequency of an ETLC system, one will quickly draw a qualitative conclusion of the cooling efficiency from the expression. For instance, if the concentration of

Tm^{3+} ion equals zero, that is, the Ho^{3+}–Tm^{3+} ET processes do not exist ($p_{13} = 0$), the value of the characteristic frequency then equals that of the pumping frequency ($\nu_c = \nu$). Equation 4.40 can be reduced to

$$\eta_c \Big|_{\nu_c=\nu} = \eta_{\text{ext},1}\eta_{\text{abs}} \frac{\nu_1}{\nu} - 1 \qquad (4.44)$$

This is exactly the cooling efficiency formula of the conventional ASFC system. As a general expression of cooling efficiency for the Ho-Tm ETLC system, Eq. 4.40 shows that the larger the characteristic frequency of an ETLC system is, the larger the corresponding cooling efficiency could be. The above analysis shows that the ETLC cooling scheme is more efficient than the conventional ASFC scheme in principle, so the right-hand side of Eq. 4.40 should be larger than that of Eq. 4.44, indicating that $\nu_c > \nu$. Indeed, for the 0.3% Ho^{3+}, 2% Tm^{3+}:$YLiF_4$ crystal sample, the calculated result on the characteristic frequency (with transition rates and macroscopic ET rates near 300 K found in Refs. [23, 34]) shows that the value of the characteristic frequency is on the order of 2ν.

Next, we predict the cooling performance of an ETLC system on the basis of spectrum data and material parameters of Ho^{3+}, Tm^{3+}:$YLiF_4$ crystals. The energy gap of Tm^{3+} $^3F_4$–$^3H_6$ is 5810 cm^{-1} [23]. By using Eq. 4.20, the nonradiative decay rate of the Tm^{3+} $^3F_4 \rightarrow {}^3H_6$ transition in the $YLiF_4$ crystal is calculated as 0.10 s^{-1}. The radiative decay rate of this transition is 66.67 s^{-1} [23]. The external quantum efficiencies of Ho^{3+} $^5I_7 \rightarrow {}^5I_8$ and Tm^{3+} $^3F_4 \rightarrow {}^3H_6$ transitions in the $YLiF_4$ host are calculated to be 0.9949 and 0.9985, respectively. The average fluorescence wavelength for the Tm^{3+} $^3F_4 \rightarrow {}^3H_6$ transition is found to be 1807 nm (see Fig. 4.2d).

Figure 4.13a shows the cooling efficiencies of the 0.3% Ho, 2% Tm $YLiF_4$ and the 0.3% Ho $YLiF_4$ crystal samples as functions of resonant absorption at a pump wavelength of 2070 nm. To point out, since the dopant concentration of Ho^{3+} ions is relatively low, for the sake of brevity, the model M1 (see Section 4.2) has been applied in simulating Ho ASFC. As the resonant absorption increases, the absorption efficiency will decrease; as a result, the cooling efficiency of Ho ASFC drops. On the contrary, the cooling efficiency of Ho-Tm ETLC first slightly rises with the increasing of the resonant absorption, which is beneficial for boosting the cooling power density.

In essence, as more Ho^{3+} ions are pumped to the excited state with the increased absorption, the average distance between an excited-state Ho^{3+} ion and a ground-state Tm^{3+} ion shrinks. Consequently, the Ho^{3+}–Tm^{3+} ET is accelerated and more Tm^{3+} ions participate in the cooling cycle. At a resonant absorption of around 0.8×10^8 W/m^3, the 0.3% Ho, 2% Tm ETLC system reaches its maximum cooling efficiency (5.08%). However, the cooling efficiency for the 0.3% Ho ASFC system at a resonant absorption of 0.8×10^8 W/m^3 is only about 0.93%. The improving in the cooling efficiency by applying the ET mechanism to a solid-state cooling system is quite remarkable. Moreover, the maximum resonant absorption for cooling ($\eta_c > 0$) of the 0.3% Ho, 2% Tm ETLC system is roughly two times of that of the 0.3% Ho ASFC system, indicating that the co-doped cooling scheme could profoundly enhance the cooling power density. It can be seen from Fig. 4.13b, at a resonant absorption of about 0.7×10^8 W/m^3, the MCPD for the 0.3% Ho ASFC system is 0.8×10^6 W/m^3, whereas at a resonant absorption of about 1.8×10^8 W/m^3, the MCPD for the 0.3% Ho, 2% Tm ETLC system is 8×10^6 W/m^3.

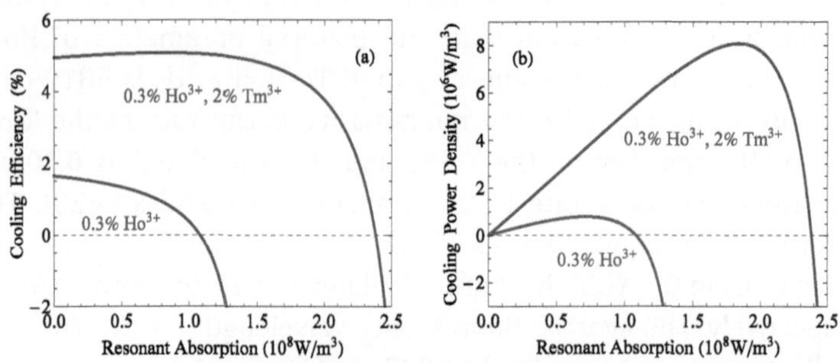

Figure 4.13 The cooling efficiencies (a) and the cooling power densities (b) of Ho-Tm ETLC and Ho ASFC in the $YLiF_4$ crystal as functions of resonant absorption near 300 K. For the 0.3% Ho, 2% Tm: $YLiF_4$ crystal sample, the ET parameters are estimated as $p_{13} = 6 \times 10^{-24}$ cm^3·μs^{-1} and $p_{02} = 8 \times 10^{-23}$ cm^3·μs^{-1} [34]. The background absorption is assumed to be 4.0×10^{-4} cm^{-1}.

The ET-induced cooling enhancement varies with the sample temperature since the macroscopic ET rates contained in the characteristic frequency are temperature dependent. To make a clear statement, we consider the low excitation cases, in which less than 5% of the Ho^{3+} ions are pumped to the excited state for the sample mentioned above. B that is defined by Eq. 4.43 will be more than 20 times larger than $P_r/h\nu$. The right-hand side of Eq. 4.42 can then be expanded as a power series of $P_r/h\nu$. Neglecting the terms that are higher than second order, Eq. 4.42 can be approximately expressed as

$$A \approx \frac{P_r}{h\nu} \frac{p_{02}C_{Ho}W_1 - p_{13}C_{Tm}W_2}{p_{02}C_{Ho}W_1 + p_{13}C_{Tm}W_2} \tag{4.45}$$

where the condition $(p_{02}C_{Ho}W_1 - p_{13}C_{Tm}W_2)N >> W_1W_2$ has been used in the simplification. Substituting Eq. 4.45 into Eq. 4.40, we obtain the cooling efficiency of the ETLC system in the low excitation condition,

$$\eta_{ETA} = \left(\eta_{abs}\eta_{ext,1}\frac{\nu_1}{\nu} - 1\right) + \Gamma\eta_{abs}\frac{\eta_{ext,2}\nu_2 - \eta_{ext,1}\nu_1}{\nu} \tag{4.46}$$

where the enhancing factor that marks the ET-induced cooling enhancement is expressed as

$$\Gamma(T, C_{Tm}/C_{Ho}) = \left(1 + \frac{W_1}{W_2}\frac{C_{Ho}}{C_{Tm}}\frac{p_{02}}{p_{13}}\right)^{-1} \tag{4.47}$$

It can be seen clearly that the terms in the parentheses of Eq. 4.46 represent the conventional ASFC of Ho^{3+} ions, and the last term on the right-hand side accounts for the cooling enhancement induced by the ET mechanism. Equation 4.47 shows that to increase the concentration ratio of the Tm^{3+}–Ho^{3+} ion will bring a larger enhancement in the cooling efficiency. This feature of ETLC is expected since an increased Tm^{3+} concentration promotes the interacting probability of the ground-state Tm^{3+} ions with the excited-state Ho^{3+} ions. It needs to be noted that both of the Ho^{3+} and Tm^{3+} ions should be restricted within relatively low concentration levels to avoid ETU-induced heating (see Section 4.2), whereas to break through the concentration limits and increase the cooling power density, it is possible to adopt alternative low–host phonon materials (e.g., KPb_2Cl_5 crystals), since the multiphonon transitions may be hindered [18].

The ratio of p_{13} to p_{02}, which is usually referred to as the temperature-dependent equilibrium constant, governs the steady-state energy assignment of a Ho-Tm cooling system. By using the approximate expression of the temperature-dependent equilibrium constant derived by Walsh et al. [34], the enhancing factor as a function of temperature is depicted in Fig. 4.14. It can be seen that the enhancing factor decreases with the decreasing of the temperate. Actually, as the temperature drops, more excited-state Ho^{3+} ions will occupy the lower manifolds of the electronic energy level due to the nature of Boltzmann distribution. The enlarged average energy difference between the excited electronic energy levels of the Tm^{3+} and Ho^{3+} ions weakens the Ho^{3+}–Tm^{3+} ET processes, and consequently the enhancing factor decreases. Nevertheless, the ET mechanism assists the cooling processes consistently to the minimum achievable temperature of the cooling system since the value of the enhancing factor is always positive. So the timescale for solid-state laser cooling could be shortened by applying the co-doped cooling scheme.

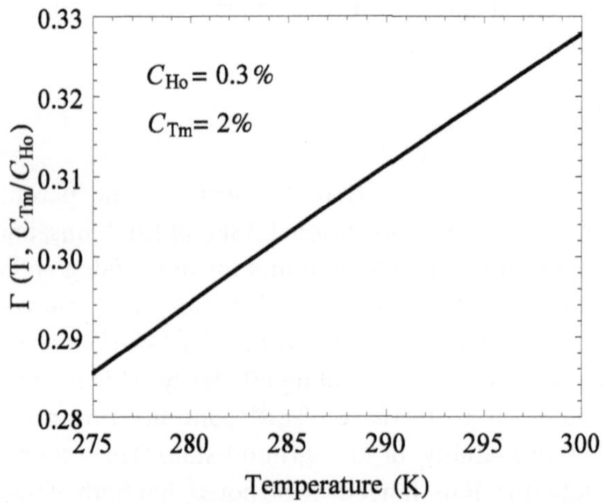

Figure 4.14 The enhancing factor as a function of temperature in a low excitation condition.

It is worth mentioning that the co-doped cooling scheme can also be applied to rare earth–co-doped materials other than the Ho^{3+}, Tm^{3+}:$YLiF_4$ crystals, given that the selected dopants share a similar electronic energy level structure so that ET processes can efficiently occur. For instance, the Tm^{3+}, Er^{3+}-co-doped materials are favorable options for ETLC, since the first excited states ($^3F_4$ and $^4I_{13/2}$) of Tm^{3+} and Er^{3+} ions are close enough in energy. It is important to note that the host material should enable ASFC of each dopant in the first place, and the doping concentrations should be restricted within relatively low levels so that the deleterious ETU processes are averted. Another couple of dopants for ETLC could be Er^{3+} and Yb^{3+}, since the Er^{3+} $^4I_{11/2}$ and Yb^{3+} $^2I_{5/2}$ states are very close in energy. However, the drawback of Er-Yb ETLC is that the intermediate Er^{3+} $^4I_{11/2} \rightarrow {}^4I_{13/2}$ transition usually produces a large amount of heat that may impede the cooling processes. Applying a low–phonon energy (which is small enough compared to the energy gap of Er^{3+} $^4I_{11/2}$ - $^4I_{13/2}$, so the nonradiative transition is negligible) host material could solve the heating issue.

4.5 Conclusions

This chapter theoretically investigates optical refrigeration in Ho^{3+}:$YLiF_4$ and Ho^{3+}, Tm^{3+}:$YLiF_4$ crystals. Due to the small energy mismatch between transitions $^5I_7 \rightarrow {}^5F_8$ and $^5I_7 \rightarrow {}^5I_5$, ETU occurs when the $^5I_5$ Ho^{3+} ion is pumped in the anti-Stokes domain. Therefore, a general cooling model with consideration of upconversion is established to precisely describe laser cooling of Ho^{3+}-doped fluoride crystals. Meanwhile, a guide on selecting a host material for optical refrigeration with multimanifold rare earth ions is provided. Simulation results show that the multiphonon transitions from the second and third excited states of Ho^{3+} ions block effective cooling. To this end, a co-pumping scheme is presented to solve the ETU-induced fluorescence quenching issue. An extra laser is added to the cooling system to pump $^5I_5$ Ho^{3+} ions to a higher excited state ($^5F_5$). The beauty of applying the co-pumping scheme is that the $^5F_5$ Ho^{3+} ions scarcely transit to the lower nearby state and mainly de-excite radiatively. So, the large amount of heat produced by nonradiative $^5I_5 \rightarrow {}^5I_6$ and $^5I_6 \rightarrow {}^5I_7$ transitions in the conventional anti-Stokes cooling case can be avoided, and both cooling efficiency and cooling power can be prompted. Besides, there is another way of improving the cooling performance of Ho^{3+} optical refrigeration: co-doping Tm^{3+} and Ho^{3+} ions in the host. The Ho^{3+} and Tm^{3+} ions share a similar electronic energy level structure, and their first excited states are close in energy. As a result, ET from Ho^{3+} to Tm^{3+} ions arises when Ho^{3+}, Tm^{3+}-co-doped fluoride crystals are pumped at around 2.07 μm. Since the energy gap of Tm^{3+} $^3F_4$ - $^3H_6$ is wider than that of Ho^{3+} $^5I_7$ - $^5I_8$, host phonons will be annihilated when ET from Ho^{3+} to Tm^{3+} ions takes place, bringing enhancement to the cooling effect. However, one weak point of ETLC is that both of the two dopants should be restricted within relatively low concentration levels to avoid ETU-induced heating.

To conclude, ETU-induced fluorescence quenching is the main obstacle to achieve net cooling in Ho^{3+}-doped solids. Nevertheless, by adopting the modified cooling schemes and the optimization procedures presented in this chapter, it is possible to overcome the ETU-induced heating and realize efficient optical refrigeration around 2.07 μm.

Acknowledgments

The authors acknowledge financial support of the Specialized Research Fund for the Program for New Century Excellent Talents in University (grant no. NCET-11-269), the National Natural Science Foundation of China (grant nos. 11204048, 61275138, 10804022, 60878016, and 61378027), the Fundamental Research Funds for the Central Universities, the Special Financial Grant from the China Postdoctoral Science Foundation (grant no. 2013T60345), and the 111 project to the Harbin Engineering University (grant no. B13015).

Appendix

For simplicity, the pump intensities for both GSA and ESA are assumed to be low, so the population densities of $^5I_6$, $^5I_5$, and $^5F_5$ manifolds are far less than the $^5I_7$ manifold. In the conventional cooling case, the steady-state population density (n_{is1}, i = 0, 1, 2, 3) of each manifold can be obtained from the following equations:

$$-(\beta_{30}W_{30}+\beta_{31}W_{31}+\beta_{32}W_{32})n_{3s1}+p_{11}(C)n_{1s1}^2=0 \tag{A.1a}$$

$$\sigma_g(n_{0s1}-n_{1s1})\frac{I_1}{h\nu_g}-W_{10}n_{1s1}\approx 0 \tag{A.1b}$$

$$n_{0s1}+n_{1s1}\approx CN. \tag{A.1c}$$

The approximate expression for the population density of the $^5I_5$ manifold in the conventional pumping scheme reads

$$n_{3s1} \approx \frac{p_{11}C^2N^2}{\left(\sum_{i=0}^{2}\beta_{3i}W_{3i}\right)\left(2+\frac{h\nu_g W_{10}}{\sigma_g I_1}\right)^2} \tag{A.2}$$

In the cooling case via the co-pumping scheme, the equations governing the steady-state population density of $^5I_7$ and $^5I_8$ (n_{1s2} and n_{0s2}) are identical to Eqs. A.1b and A.1c. But the equations governing the steady-state population density of $^5I_5$ and $^5F_5$ (n_{3s2} and n_{4s2}) are written as

$$\sigma_e n_{3s2}\frac{I_2}{h\nu_e} - \left(\sum_{i=0}^{3}\beta_{4i}W_{4i}\right)n_{4s2} \approx 0 \tag{A.3}$$

$$-\sigma_e n_{3s2}\frac{I_2}{h\nu_e} - \left(\sum_{i=0}^{2}\beta_{3i}W_{3i}\right)n_{3s2} + \beta_{43}W_{43}n_{4s2} + p_{11}n_{1s2}^2 \approx 0 \tag{A.4}$$

The approximate expression for the population density of the $^5I_5$ manifold in the co-pumping scheme reads

$$n_{3s2} \approx \frac{p_{11}C^2N^2}{\left[(1-\eta_{43})\frac{\sigma_e I_2}{h\nu_e} + \left(\sum_{i=0}^{2}\beta_{3i}W_{3i}\right)\right]\left(2+\frac{h\nu_g W_{10}}{\sigma_g I_1}\right)^2} \tag{A.5}$$

The population density change of the $^5I_5$ manifold induced by adding the laser for ESA to the system is

$$\begin{aligned}\Delta n_{3s} &= n_{3s2} - n_{3s1} = p_{11}C^2N^2\left(2+\frac{h\nu_g W_{10}}{\sigma_g I_1}\right)^{-2} \\ &\times\left\{\left[(1-\eta_{43})\frac{\sigma_e I_2}{h\nu_e} + \sum_{i=0}^{2}\beta_{3i}W_{3i}\right]^{-1} - \left(\sum_{i=0}^{2}\beta_{3i}W_{3i}\right)^{-1}\right\}\end{aligned} \tag{A.6}$$

Since I_2 is small, the right-hand side of Eq. A.6 can be expanded as a power series of $(1-\eta_{43})\sigma_e I_2(h\nu_e)^{-1}$, and the result formulated in Eq. 4.30 is obtained.

References

1. Epstein, R. I., Buchwald, M. I., Edwards, B. C., Gosnell, T. R., Mungan, C. E. (1995). Observation of laser-induced fluorescent cooling of a solid, *Nature*, **377**, pp. 500–503.

2. Epstein, R. I., Sheik-Bahae, M. (2009). *Optical Refrigeration* (Wiley-VCH, Weinheim).
3. Sheik-Bahaeand, M., Epstein, R. I. (2009). Laser cooling of solids, *Laser Photonics Rev.*, **3**, pp. 67–83.
4. Nemova, G., Kashyap, R. (2010). Laser cooling of solids, *Rep. Prog. Phys.*, **73**, p. 086501.
5. Seletskiy, D. V., Melgaard, S. D., Bigotta, S., Lieto, A. D., Tonelli, M., Sheik-Bahae, M. (2010). Laser cooling of solids to cryogenic temperatures, *Nat. Photonics*, **4**, pp. 161–164.
6. Seletskiy, D. V., Melgaard, S. D., Epstein, R. I., Lieto, A. D., Tonelli, M., Sheik-Bahae, M. (2011). Local laser cooling of Yb:YLF to 110 K, *Opt. Express*, **19**, pp. 18229–18236.
7. Melgaard, S. D., Seletskiy, D. V., Lieto, A. D., Tonelli, M., Sheik-Bahae, M. (2013). Optical refrigeration to 119 K, below National Institute of Standards and Technology cryogenic temperature, *Opt. Lett.*, **38**, pp. 1588–1590.
8. Melgaard, S., Albrecht, A. R., Hehlen, M., Seletskiy, D., Sheik-Bahae, M. (2014). Optical refrigeration cools below 100K, in *CLEO: 2014, OSA Technical Digest* (online) (Optical Society of America), paper FTh4D.4.
9. Garcia-Adeva, A. J., Balda, R., Fernandez, J. (2009). Upconversion cooling of Er-doped low-phonon fluorescent solids, *Phys. Rev. B*, **79**, p. 033110.
10. Nemova, G., Kashyap, R. (2010). Laser cooling of Er^{3+}-doped solids, *Opt. Commun.*, **283**, pp. 3736–3739.
11. Dong, G. Z., Zhang, X. L., Li, L. (2013). Energy transfer enhanced laser cooling in Ho^{3+} and Tm^{3+} codoped lithium yttrium fluoride, *J. Opt. Soc. Am. B*, **30**, pp. 939–944.
12. Dong, G. Z., Zhang, X. L. (2015). Efficient and enhanced optical refrigeration beyond 2 μm in Ho^{3+} doped solids via co-pumping scheme: erratum, *J. Opt. Soc. Am. B*, **32**, pp. 2248–2249.
13. Hoyt, C. W., Sheik-Bahae, M., Epstein, R. I., Edwards, B. C., Anderson, J. E. (2000). Observation of anti-Stokes fluorescence cooling in thulium-doped glass, *Phys. Rev. Lett.*, **85**, pp. 3600–3603.
14. Fernandez, J., Garcia-Adeva, A. J., Balda, R. (2006). Anti-Stokes laser cooling in bulk erbium-doped materials, *Phys. Rev. Lett.*, **97**, p. 033001.
15. Patterson, W., Bigotta, S., Sheik-Bahae, M., Parisi, D., Tonelli, M., Epstein, R. I. (2008). Anti-Stokes luminescence cooling of Tm^{3+}doped BaY_2F_8, *Opt. Express*, **16**, pp. 1704–1710.

16. Condon, N. J., Bowman, S. R., O'Connor, S., Quimby, R. S., Mungan, C. E. (2009). Optical cooling in Er^{3+}:KPb_2Cl_5, *Opt. Express*, **17**, pp. 5446–5472.

17. Bowman, S. R., O'Connor, S., Condon, N. J., Friebele, E. J., Kim, W., Shaw, B., Quimby, R. S. (2013). Non-radiative decay of holmium-doped laser materials, *Proc. SPIE*, **8638**, p. 863803.

18. Hehlen, M. P. (2009). Novel materials for laser refrigeration, *Proc. SPIE*, **7228**, p. 72280E.

19. Dong, G. Z., Zhang, X. L. (2015). Role of upconversion in optical refrigeration: a theoretical study of laser cooling with Ho^{3+} doped fluoride crystals: erratum, *J. Opt. Soc. Am. B*, **32**, pp. 2046–2047.

20. Walsh, B. M., Barnes, N. P., Bartolo, B. D. (1998). Branching ratios, cross sections, and radiative lifetimes of rare earth ions in solids: application to Tm^{3+} and Ho^{3+} ions in $LiYF_4$, *J. Appl. Phys.*, **83**, pp. 2772–2787.

21. Walsh, B. M., Grew, G. W., Barnes, N. P. (2005). Energy levels and intensity parameters of Ho^{3+} ions in GdLiF4, $YLiF_4$ and $LuLiF_4$, *J. Phys.: Condens. Mater.*, **17**, pp. 7643–7665.

22. Barnes, N. P., Walsh, B. M. (2003). Ho:Ho upconversion: applications to Ho lasers, *J. Opt. Soc. Am. B*, **20**, pp. 1212–1219.

23. Walsh, B. M., Barnes, N. P., Petros, M., Yu, J., Singh, U. N. (2004). Spectroscopy and modeling of solid state lanthanide lasers: application to trivalent Tm^{3+} and Ho^{3+} in $YLiF_4$ and $LuLiF_4$, *J. Appl. Phys.*, **95**, pp. 3255–3271.

24. Hoyt, C. W., Hasselbeck, M. P., Sheik-Bahae, M., Epstein, R. I., Greenfield, S., Thiede, J., Distel, J., Valencia, J. (2003). Advances in laser cooling of thulium-doped glass, *J. Opt. Soc. Am. B*, **20**, pp. 1066–1074.

25. Dexter, D. L. (1952). A theory of sensitized luminescence in solids, *J. Chem. Phys.*, **21**, pp. 836–850.

26. Rapaport, A., Milliez, J., Bass, M., Cassanho, A., Jenssen, H. (2005). Numerical model of the temperature dependence of the up-conversion efficiency of fluoride crystals codoped with ytterbium and thulium, *J. Appl. Phys.*, **97**, p. 123527.

27. van Dijk, J. M. F., Schuurmans, M. F. H. (1983). On the nonradiative and radiative decay rates and a modified exponential energy gap law for 4f-4f transitions in rare-earth ions, *J. Chem. Phys.*, **78**, pp. 5317–5323.

28. Powell, R. C. (1998). *Physics of Solid-State Laser Materials* (Springer-Verlag, New York).

29. Lee, T. H., Heo, J. (2005). 1.6 μm emission and gain properties of Ho^{3+} in selenide and chalcohalide glasses, *J. Appl. Phys.*, **98**, p. 113510.

30. Payne, S. A., Chase, L. L., Smith, L. K., Kway, W. L., Krupke, W. F. (1992). Infrared cross section measurements for crystals doped with Er^{3+}, Tm^{3+}, and Ho^{3+}, *IEEE J. Quantum Electron.*, **28**, pp. 2619–2630.

31. Librantz, A. F. H., Jackson, S. D., Gomes, L., Ribeiro, S. J. L., Messaddeq, Y. (2008). Pump excited state absorption in holmium-doped fluoride glass, *J. Appl. Phys.*, **103**, p. 023105.

32. Faiconieri, M., Salvetti, G. (1994). Effects of co-dopant concentrations and excitation conditions on the 2 μm fluorescence dynamics in Tm, Ho: YLF crystals. *Appl. Phys. A*, **59**, pp. 253–258.

33. Walsh, B. M., Barnes, N. P., Bartolo, B. D. (1997). On the distribution of energy between the Tm $^3F_4$ and Ho $^5I_7$ manifolds in Tm-sensitized Ho luminescence, *J. Lumin.*, **75**, pp. 89–98.

34. Walsh, B. M., Barnes, N. P., Bartolo, B. D. (2000). The temperature dependence of energy transfer between the Tm $^3F_4$ and Ho $^5I_7$ manifolds of Tm-sensitized Ho luminescence in YAG and YLF, *J. Lumin.*, **90**, pp. 39–48.

Chapter 5

Optically Cooled Lasers

Steven R. Bowman
Laser Physics Branch, Naval Research Laboratory, Washington, DC 20375-5320, USA
steven.bowman@nrl.navy.mil

Since the invention of the laser, engineers have searched for new ways to increase the laser's power and brightness. These efforts have largely focused on improved ways to remove the waste heat that is generated in the lasing medium. This chapter reviews a novel approach to the problem of laser power scaling: minimizing the waste heat through the incorporation of anti-Stokes fluorescence cooling. This approach has been shown to significantly reduce heat generation in carefully controlled laser systems. This chapter reviews the theory and practice of this novel mode of laser operation.

5.1 Introduction

One of the more useful properties of lasers is their ability to generate extremely high power densities. Take, for example, a commercial fiber laser operating continuously with an average power of 1 kW.

Laser Cooling: Fundamental Properties and Applications
Edited by Galina Nemova

ISBN 978-981-4745-04-8 (Hardcover), 978-981-4745-05-5 (eBook)
www.panstanford.com

Inside the small core of the fiber the beam propagates at intensities as high as 1 GW/cm^2. For comparison, this intensity is 10 billion times higher than terrestrial sunshine. Even the hardest metals are instantly vaporized when placed near the focus of this beam. Yet the laser that generates these intensities is expected to operate reliably for years. The challenge for the laser designers is to cleanly manipulate such high intensities without damaging the laser medium.

What happens to a material when it is exposed to a high-intensity laser depends principally on the amount of power absorbed. Most materials convert absorbed laser power directly into heat. When the absorption is described by a Beer's law coefficient of α and the laser intensity is I_L, then the absorbed power density is simply αI_L. So any material with significant absorption placed in the beam described above would see GW/cm^3 heat loads and be instantly vaporized. Even the most transparent materials, such as a pure silica fiber, have weak background absorption with coefficients of order of 10^{-6} cm^{-1}. So propagation of the 1 kW laser beam described above will still generate 1 kW/cm^3 of waste heat. The core of an optical fiber survives this extreme heat loading by spreading the thermal load over a large surface area. Less transparent materials, such as a laser gain medium, can see even higher thermal loads. As a result, thermal loading remains the primary obstacle to scaling up the average power of most laser systems.

5.2 Laser Thermal Management

The problem of thermal management of the laser is as old as the laser itself. Many groups have contributed to the large body of work and tremendous progress on the problem. This section is intended only as only a brief summary of the major elements of the field. We will start by examining the source and magnitude of the problem, restricting the discussion to optically pumped lasers.

Earlier it was stated that the power density of heat generated in most materials with absorption coefficient α by the passage of a laser beam with intensity I_L would be αI_L. But this is not generally true for laser media. Laser media are designed to reemit absorbed light with high efficiency. The amount of heat generated in a laser depends on the fraction of the pump power that is not reradiated as well as the spectra of the emission. The reradiated light will normally be red-shifted relative to the pump. So even if the excitation and lasing processes were 100% efficient, every laser photon emitted will contribute a fraction of its energy QD = $(1 - \lambda_P/\lambda_L)$ to heating the laser material. Here λ_P and λ_L are the pump and laser wavelengths and QD is the laser quantum defect (QD).

Heat generated in the laser's interior will produce a temperature rise in the center of the gain medium relative to its edge. This temperature rise produces several detrimental effects on the laser beam propagating through it. Firstly, it can reduce the lifetime and increase the background absorption of the lasing ions. These effects combine to reduce net gain. The temperature profile will also modify the refractive index and thickness of the medium distorting the laser beam. The strain generated by the temperature gradient also perturbs the index and can result in depolarization of the beam. If the strain becomes large enough, the laser medium will crack and fail completely. Of course if the temperature rises too high, the laser medium will simply melt or burn. Usually beam quality degradation limits average power well below thermal damage.

To illustrate the detrimental impact of thermal loading consider one of the most common solid-state lasers: neodymium-doped yttrium aluminum garnet (Nd:YAG) operating at 1064 nm. YAG's excellent thermomechanical properties have made it the most popular material for high-average-power solid-state lasers. Nd:YAG is a strong 1 μm emitter with a quantum efficiency near unity. Experiments have shown that when pumped near the optimum absorption at 807 nm, a continuous-wave (cw) Nd:YAG laser converts 25% of the absorbed pump into heat. This is largely the result of the 23% QD between the pump and laser photon energies.

In a simple end-pumped cw Nd:YAG laser, 10 W of pump powers was shown to generate as much as 7 kW/cm^3 of heating and axial temperature rise up to 37 K [56]. This produced a 14-diopter thermal lens in the YAG crystal with a strong spherical aberration. Further, the thermal loading will roughly double when the crystal is not lasing. This is due to strong upconversion and multiphonon quenching. Still, laser systems based on Nd:YAG hold the record for average power (15 kW) with good beam quality [48]. This is a testament to decades of innovations and refinements by the solid-state laser community. Nevertheless, heat generation in Nd:YAG and other conventional bulk laser materials limits their maximum useful average power.

Recently laser architectures based on silica fibers have also made great advances in average power performance. Fiber lasers deal with heat dissipation by virtue of their high aspect ratios, typically of the order of a million to one. This allows them to greatly spread the thermal loading on the laser medium. Also, wave guiding in the fiber's core makes the laser beam much less susceptible to thermo-optic distortions. Fiber lasers based on the 1 μm transition in ytterbium have demonstrated the highest average powers. Currently the highest-power systems produce up to 10 kW with good beam quality [31]. Nevertheless, at the kilowatt average power level, fiber lasers also suffer from internal heating. Longitudinal temperature gradients in the fiber have been shown to generate Raleigh scattering, which destabilizes the transverse mode and spoils the beam quality [61, 64]. As with bulk laser, thermal loading currently limits the maximum average power from a single-fiber laser.

The current best approach to further power scaling is to combine the beams from many parallel lasers. Beam combination can be accomplished through precise control of the phases, wave vectors, or wavelengths of an ensemble of lasers. Each of these approaches has been proven to increase useful power but at the expense of dramatically increasing the complexity of the overall system. Power scaling with this approach rapidly runs into practical limits on the amount of hardware acceptable for the application [58].

Many other strategies have been explored to help manage the thermal loading in high-power lasers. Slab and disk laser geometries that expedite heat dissipation have been fully explored [1, 3, 19, 20, 21, 25, 33, 34, 39, 41]. Wavefront correction techniques such as nonlinear phase conjugation have also been well studied [17]. Pulsed laser approaches have been demonstrated that allow transient heat build-up [2]. New laser materials with higher thermal conductivities are still an active research area [54, 55, 59, 63]. In the early days of lamp-pumped lasers, many systems operated at cryogenic temperatures. Recently cryogenic operation has been revisited to improve material thermo-optic parameters [26]. Finally, laser schemes that minimize the laser QD have been developed in order to reduce heating [16, 18, 43, 55].

The remainder of this chapter explores this last approach: low–quantum defect lasers (LQDL). Specifically we consider the possibility of anti-Stokes fluorescence cooling that can arise in LQDL schemes. Anti-Stokes emission reduces the overall QD further and removes waste heat from the laser medium. The special case of radiation-balanced lasing will be considered in detail. This occurs when fluorescence cooling just cancels out other heating mechanisms. Unique among all the power scaling approaches, a radiation-balanced laser (RBL) can theoretically have zero thermal loading. So an RBL system could potentially break through the average power barrier that has limited the other approaches.

5.3 Theory of Low–Quantum Defect Laser Materials

This section develops a theory for solid-state lasers that operate with minimal excess heating from the laser QD. The goal is to develop a realistic model that can be used to explore the fundamental limitations of laser performance. The approach is to simplify the analysis by just considering the effects inside the lasing medium. By tracking the local power density in the laser medium insights can be obtained about radiative efficiency and waste heat generation. This analysis applies to the most common type of high-average-power lasers: the steady-state quasi-three-level systems.

Several previous studies have developed models for quasi-three-level laser systems [4, 6, 27, 65]. These models have largely focused on predicting gain and power for specific laser architectures, such as an end-pumped resonator or amplifier. As a result, they have focused on the impact of beam profiles and laser geometry. To simplify the analysis these models have largely ignored the contribution of fluorescence to power flow. We will show that this can lead to an underestimate of the gain and an overestimate of the heat loading. The analysis presented below is intended to augment the previous resonant-pumping studies. It also extends and generalizes previous analysis of RBLs [14, 46, 50].

The model present below predicts the local performance of a quasi-three-level laser material. After developing the basic definitions and relations in a lossless idealize material, the analysis will be extended to include the most common processes: quenching, background absorption, and fluorescence reabsorption. Several possible optimization approaches will be considered and compared. The well-characterized laser material Yb:YAG will be used to illustrate these relations and optimization strategies. The techniques used in this analysis can be readily extended to specific gain geometries and beam profiles.

5.3.1 Lossless Laser Materials

LQDLs are typically based on a transition between an energy level just above the ground state and the first excited manifold of a dopant ion. These lasers are often referred to as quasi-three-level systems, but they actually consist of only two well-separated energy bands made up of several nondegenerate levels. Each of these energy manifolds maintains a thermal equilibrium with the host matrix through electron–phonon scattering. The laser transition terminates in a level just high enough above the ground state to have a very small thermal population. So unlike Nd:YAG, lasers based on ions like ytterbium can avoid detrimental energy pooling in the lower level, while still maintaining a low QD.

We begin our analysis by considering an ideal laser system with only radiative losses and no radiative trapping. Here the power flow of the fluorescent field will be included, unlike conventional laser analysis. After deriving some basic results, the model will be extended to handle more realistic cases.

Assume an isotropic laser medium that is doped with active ions at a density of N_T. Optical transitions are presumed to occur between lower and upper manifolds, giving rise to overlapping, thermally broadened absorption and emission spectra. These spectra are characterized by effective absorption and emission cross sections labeled $\sigma_A(\lambda,T)$ and $\sigma_E(\lambda,T)$, respectively, at a wavelength λ and a temperature T. It will be useful to define a fractional cross section as

$$\beta(\lambda,T) \equiv \frac{\sigma_A(\lambda,T)}{\sigma_A(\lambda,T)+\sigma_E(\lambda,T)}. \tag{5.1}$$

Ion densities in the lowest-energy manifold, N_1, are excited to the upper-energy manifold, N_2, by a pump field at wavelength λ_P and irradiance I_P. Ions in the upper manifold radiate with a lifetime τ_R or are stimulated to emit decay by the laser field at wavelength λ_L and irradiance I_L. For simplicity, we assume that the pump and laser fields are directional, monochromatic, and polarization independent. This is not always the case, but the model is easily modified to include broader intensities and nonisotropic media. Our derivation also assumes that ionic dephasing is very rapid compared to τ_R, allowing us to consider only incoherent excitation. With these definitions and mass conservation, we can write the local rate equation as

$$\frac{dN_2}{dt} = \frac{I_P \lambda_P \sigma_A(\lambda_P, T)}{hc} \cdot \left[N_T - N_2/\beta(\lambda_P, T)\right] + \frac{I_L \lambda_L \sigma_A(\lambda_L, T)}{hc} \cdot \left[N_T - N_2/\beta(\lambda_L, T)\right] - N_2/\tau_R. \tag{5.2}$$

The first term on the right describes the pump rate, the second term describes the laser rate, and the last term describes the total spontaneous decay rate of the upper manifold. Applying energy conservation to this laser system, we recognize that the difference between the absorbed power and the emitted power must be the local heat generation rate, H. With the definitions above the power density of waste heat will be

$$H = I_P \sigma_P \cdot \left[N_T - {}^{N_2}\!/_{\beta_P}\right] + I_L \sigma_L \cdot \left[N_T - {}^{N_2}\!/_{\beta_L}\right] - \frac{hc}{\lambda_F} \frac{N_2}{\tau_R}. \tag{5.3}$$

Note that the heat generation depends on the mean fluorescence wavelength, λ_F, defined as the average over the emission band:

$$\lambda_F \equiv \frac{\int_{\text{band}} \lambda \cdot I_F(\lambda) d\lambda}{\int_{\text{band}} I_F(\lambda) d\lambda}. \tag{5.4}$$

We will simplify the derivation using the shorthand notations

$$\sigma_P \equiv \sigma_A(\lambda_P, T), \quad \sigma_L \equiv \sigma_A(\lambda_L, T),$$
$$\beta_P \equiv \beta(\lambda_P, T), \quad \beta_L \equiv \beta(\lambda_L, T), \quad \text{and} \quad \beta_{PL} \equiv \frac{\beta_P \beta_L}{\beta_P - \beta_L}. \tag{5.5}$$

The rate equation above makes it clear that the appropriate saturation intensities for describing the excitation are

$$I_{Psat} \equiv hc\beta_P/\lambda_P \tau_R \sigma_P, \quad \text{and} \quad I_{Lsat} \equiv hc\beta_L/\lambda_L \tau_R \sigma_L. \tag{5.6}$$

For simplicity the results will be cast in terms of dimensionless scaled variables:

$$i_P \equiv I_P/I_{Psat}, \quad i_L \equiv I_L/I_{Lsat}, \quad \text{and} \quad n_2 \equiv N_2/N_T. \tag{5.7}$$

The above material parameters can be evaluated using the well-known approximations of Fuchtbauer–Landenburg and the McCumber reciprocity relations. With these approximations, the temperature-dependent quantities can be computed from the energy-level structure and effective absorption cross sections [8, 29, 49]:

$$\frac{1}{\beta(\lambda,T)} = 1 + \frac{Z_1(T)}{Z_2(T)}\exp\left[\frac{1}{kT}\left(\varepsilon_2 - \frac{hc}{\lambda}\right)\right],$$
$$\frac{1}{\beta_{\mathrm{PL}}} = \frac{Z_1(T)}{Z_2(T)}\exp\left(\frac{\varepsilon_2}{kT}\right)\left[\exp\left(\frac{-hc}{\lambda_{\mathrm{L}}kT}\right) - \exp\left(\frac{-hc}{\lambda_{\mathrm{P}}kT}\right)\right], \tag{5.8}$$
$$\text{and}\quad \lambda_{\mathrm{F}} = \frac{\int_{\mathrm{band}} \sigma_{\mathrm{A}}(\lambda,T)\exp(-hc/\lambda kT)d\lambda/\lambda^4}{\int_{\mathrm{band}} \sigma_{\mathrm{A}}(\lambda,T)\exp(-hc/\lambda kT)d\lambda/\lambda^5}.$$

Here $Z_1(T)$ and $Z_2(T)$ are the Boltzmann thermal partition functions of the lower and upper manifolds and ε_2 is the minimum energy of the upper manifold. The quantity ε_2 is often referred to as the zero-phonon energy.

With the above definitions the steady-state solution to Eq. 5.2 can be written as

$$n_2 = \frac{i_{\mathrm{P}}\beta_{\mathrm{P}} + i_{\mathrm{L}}\beta_{\mathrm{L}}}{1 + i_{\mathrm{P}} + i_{\mathrm{L}}}. \tag{5.9}$$

The steady-state laser gain and pump absorption coefficients are then simply

$$g_{\mathrm{L}} = \sigma_{\mathrm{L}} N_{\mathrm{T}} \cdot \left[\frac{i_{\mathrm{P}}\,\beta_{\mathrm{P}}/\beta_{\mathrm{PL}} - 1}{1 + i_{\mathrm{P}} + i_{\mathrm{L}}}\right] \quad \text{and} \quad \alpha_{\mathrm{P}} = \sigma_{\mathrm{P}} N_{\mathrm{T}} \cdot \left[\frac{1 + i_{\mathrm{L}}\,\beta_{\mathrm{L}}/\beta_{\mathrm{PL}}}{1 + i_{\mathrm{P}} + i_{\mathrm{L}}}\right]. \tag{5.10}$$

The above relations make clear the physical significance of the fractional cross sections. From Eq. 5.2 we can see that $\beta(\lambda, T)$ is simply the excitation fraction at which absorption transitions to gain for wavelength λ. From the expression for laser gain, we can see that $\beta_{\mathrm{PL}}/\beta_{\mathrm{P}}$ is equal to the minimum pump intensity required to produce laser gain. Note that this is true at any laser intensity.

The expression for pump absorption is a bit more subtle. For pump intensities below the gain threshold, pump absorption decreases with laser intensity. Above the threshold, α_{P} increases with increasing laser intensity. This is because stimulated emission competes with absorption and these two have the opposite effect on ground-state populations. The influence of laser intensity on pump absorption clearly scales with the value of β_{PL}/β_L.

To complete the steady-state solution, one should solve for the powers flowing through the optical medium. The optical power densities cycling through the laser medium for the pump and laser fields are identified as

$$P = \alpha_P I_P \quad \text{and} \quad L = g_L I_L. \tag{5.11}$$

The steady-state power conversion directly into heat through the laser's effective QD is found from Eq. 5.3 to be

$$H = \frac{N_T hc}{\lambda_F \tau_R} \frac{\left[\lambda_F(\lambda_L - \lambda_P) i_P i_L \beta_P \beta_L + \lambda_L(\lambda_F - \lambda_P)\beta_P \beta_{PL} i_P - \lambda_P(\lambda_L - \lambda_F)\beta_L \beta_{PL} i_L\right]}{\lambda_P \lambda_L\left[1 + i_P + i_L\right]\beta_{PL}}. \tag{5.12}$$

Unlike the radiative power densities, heat builds up in the lasing medium, leading to temperature gradients and detrimental thermo-optic distortions. For a steady-state laser, temperature gradients must build until equilibrium is established. In a high-power laser system, significant additional power and hardware will be required to manage the thermal effects. It is therefore reasonable to reduce heating, while maintaining laser performance. In the next section we will examine laser optimization strategies using this simple model.

We illustrate the above relations using the radiative properties of Yb:YAG. Figure 5.1 shows the energy structure and cross-sectional spectra at room temperature. The mean fluorescence wavelength and radiative lifetime of Yb:YAG are reported to be 1010 nm and 0.979 ms, respectively, at room temperature [5, 10]. For these calculations the laser wavelength will be chosen at 1050 nm, which corresponds to the lowest-energy transition from the $^2F_{5/2}$ level of Yb:YAG. The peak absorption cross section of the 1050 nm line is only 6.0E-23 cm^2, leading to a saturation intensity of 70 kW/cm^2.

In Fig. 5.2 we illustrate how the cross-sectional parameters defined above vary with the choice of operating conditions in the Yb:YAG system. It shows that the βs can vary by many orders of magnitude for reasonable choices of operating wavelengths and temperature. The single-wavelength parameter $\beta(\lambda, T)$ is plotted using lines with triangles for 100 K, 200 K, and 300 K. At short wavelengths, $\beta(\lambda, T)$ approaches unity for all temperatures, but it falls off rapidly with wavelength and temperature. Since $\beta(\lambda, T)$ is the minimum excitation required for gain, thresholds are significantly reduced for low temperatures.

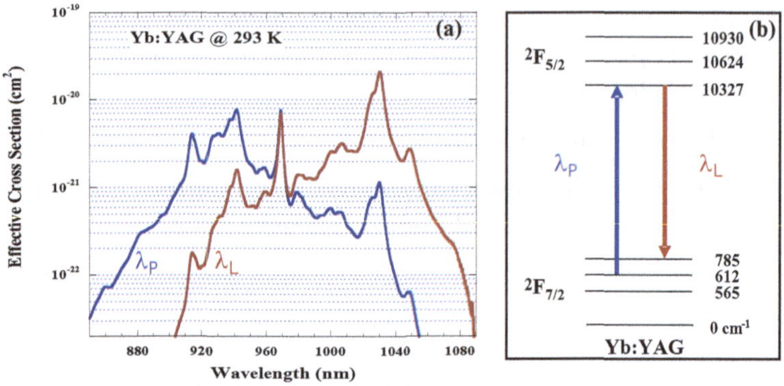

Figure 5.1 (a) Room-temperature absorption (λ_P) and emission (λ_L) cross sections measured in optically thin samples of Yb:YAG. (b) Yb:YAG energy levels at room temperature.

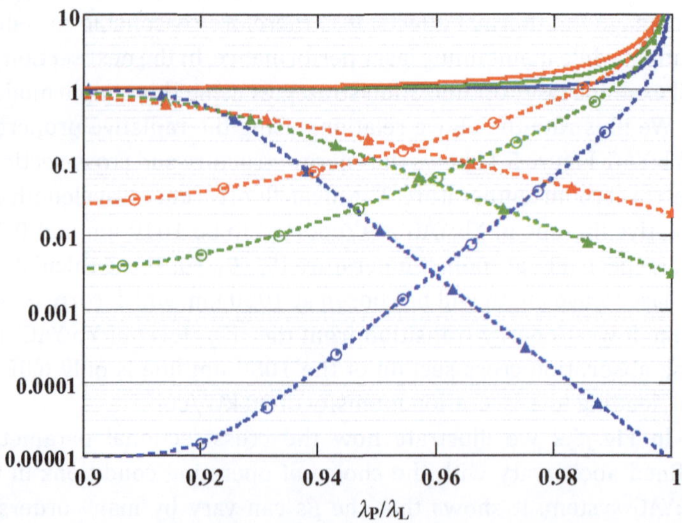

Figure 5.2 Plot of the fractional cross sections discussed in the text evaluated for Yb:YAG. Lines with triangles delineate β_P. Lines with circles are β_{PL}/β_P and solid lines are β_{PL}/β_L. Values are calculated for several temperatures, assuming λ_L = 1050 nm: 300 K (red), 200 K (green), and 100 K (blue).

Room-temperature Yb:YAG requires a minimum excitation of 2% at 1050 nm, but at 100 K the threshold drops by 2000-fold. Note that the curves for all $\beta(\lambda, T)$ cross at 968 nm. This is the zero-phonon transition of Yb:YAG; $\lambda_{00} = hc/\varepsilon_2$.

Figure 5.2 also plots the parameters β_{PL}/β_P and β_{PL}/β_L. As discussed above, β_{PL}/β_P is the pump intensity required to create an inversion and is shown here as dashed lines with open circles. At short pump wavelengths, β_{PL}/β_P approaches β_L, the laser excitation threshold discussed above. However, as the pump wavelength is increased, thresholds grow monotonically and then diverge to infinity as the pump wavelength approaches the laser wavelength. At longer wavelengths higher pump levels are required to offset the effects of pump saturation.

The values of the β_{PL}/β_L parameter in Fig. 5.2 are plotted with solid lines. This parameter controls the influence of laser intensity on pump absorption and laser efficiency. Note that β_{PL}/β_L is always slightly greater than unity. As a result material performance is strongly coupled to laser intensity. Nevertheless, Fig. 5.2 reveals that β_{PL}/β_L is only weakly dependent on pump wavelength and temperature. Only as the pump wavelength approaches that of the laser does β_{PL}/β_L rise sharply. Larger values of β_{PL}/β_L increase pump absorption, but as we will see in the next section, it decreases optical efficiency.

5.3.2 Optimization of Laser Material Performance

The steady-state power densities derived in the previous section allow us to find the optimal performance conditions of the ideal laser medium. For this purpose, we define laser material efficiency as the emitted laser power divided by the absorbed pump power:

$$\eta_m \equiv \frac{L}{P} = \frac{\lambda_P\left[1 - \beta_{PL}/\beta_P i_P\right]}{\lambda_L\left[1 + \beta_{PL}/\beta_L i_L\right]}. \tag{5.13}$$

Note that the dependence of the efficiency on intensity has shifted from that in Eq. 5.10. Now increasing either the pump or the laser intensity increases η_m.

We should be careful to distinguish the laser material efficiency, η_m, from the more commonly used laser optical efficiency. The laser optical efficiency will be proportional to η_m but will also depend on beam propagation through an optical system. Equation 5.13 provides an upper bound to optical efficiency that a well-designed laser system should approach. Of course, the appropriate spatial average must be applied to the local η_m solution.

To compare the magnitude of the laser internal heating, we will define a laser material heating parameter:

$$\Xi_m \equiv \frac{|H|}{L} = \frac{\left\| \left[\lambda_F(\lambda_L - \lambda_P) i_P i_L + \lambda_L(\lambda_F - \lambda_P) i_P \beta_{PL}/\beta_L - \lambda_P(\lambda_L - \lambda_F) i_L \beta_{PL}/\beta_P \right] \right\|}{\lambda_F \lambda_P i_L \left[i_P - \beta_{PL}/\beta_P \right]}. \tag{5.14}$$

Recognizing that the heat generation can be positive or negative, we define Ξ_m in terms of the absolute magnitude. Note that Ξ_m is defined in terms of laser power and not absorbed pump power. The heating parameter more commonly used in the literature would be the product of η_m and Ξ_m. The definitions used here will lead more directly to an optimization procedure.

The above equations describe the performance of a purely radiative cw laser. Laser power can theoretically be optimized by either maximizing η_m or minimizing Ξ_m. This would be accomplished through the appropriate selection of the operating intensities and wavelengths. Note that the heating fraction depends strongly on wavelength separations, while the laser efficiency is independent of the fluorescence wavelength.

Consider first the approach of improving laser performance by maximizing the laser material efficiency. Equation 5.13 reveals that whenever the laser material is pumped far above inversion, the material efficiency will approach λ_P/λ_L. Lower intensities reduce the material efficiency and longer pump wavelength generally increase it. However, we can see from Fig. 5.2 that as λ_P approaches λ_L, β_{PL}/β_P and β_{PL}/β_L increase rapidly, raising the laser's threshold and decreasing efficiency. This optimization approach is illustrated in Fig. 5.3 for the idealized Yb:YAG lasing at 1050 nm. As expected, laser material efficiency initially rises linearly with pump wavelength and then falls to zero as λ_P approaches λ_L. Good efficiencies require that intensities exceed the saturation value. At the highest intensities considered, the room-temperature material efficiency reaches 88% when the pump wavelength is 1005 nm. Cooling is seen to improve material efficiency somewhat at longer wavelengths. For the highest intensity plotted, the 100 K material efficiency is predicted to reach a maximum of 91.5% when pumped at 1028 nm. This simple analysis suggests that for sufficiently high intensities, the material efficiency can approach arbitrarily close to unity.

However, this analysis ignores the increased thermal loading that can occur at very high intensities.

An alternative approach to improving laser performance would be to minimize the heating parameter Ξ_m. This approach will generate the least amount of waste heat per watt of laser output power. Figure 5.4 shows Ξ_m calculated for the room-temperature conditions that were plotted in Fig. 5.3. In contrast to the slowly varying laser efficiency, the heating parameter displays large variations. There is a null in the heat generation that shifts to longer pump wavelengths as the intensities rise. As the pump wavelength is tuned through this point, heat generation switches over to optical cooling. For longer pump wavelengths, Ξ_m increases rapidly exceeding unity above the laser threshold.

The figures above were evaluated for simplicity only at $i_P = i_L$, but in general the pump and laser intensities will vary independently. While the equations are symmetric in the two fields, the large difference in β_P and β_L breaks this symmetry. For pump wavelengths shorter than λ_F, both efficiency and heat parameters depend almost entirely on the laser intensity. For pump wavelengths greater than λ_F, both intensities determine the roll-off and asymptotic values. Nevertheless, the basic shape of both curves remains unchanged at all intensities. The null in the heating efficiency is referred to as the RBL condition.

At the RBL operating point the fluorescence cooling is precisely offset by the laser heating [12, 14]. It is easy to show that when $\Xi_m = 0$, the several simple relations must be true. Firstly, the pump and laser intensities must satisfy a simple relationship determined by the QDs:

$$i_P = \frac{i_{P\min}}{1 - i_{L\min}/i_L} \quad \text{where} \quad i_{P\min} \equiv \frac{\lambda_P(\lambda_L - \lambda_F)}{\lambda_F(\lambda_L - \lambda_P)} \frac{\beta_{PL}}{\beta_P}$$
$$\text{and} \quad i_{L\min} \equiv \frac{\lambda_L(\lambda_P - \lambda_F)}{\lambda_F(\lambda_L - \lambda_P)} \frac{\beta_{PL}}{\beta_L}. \qquad (5.15)$$

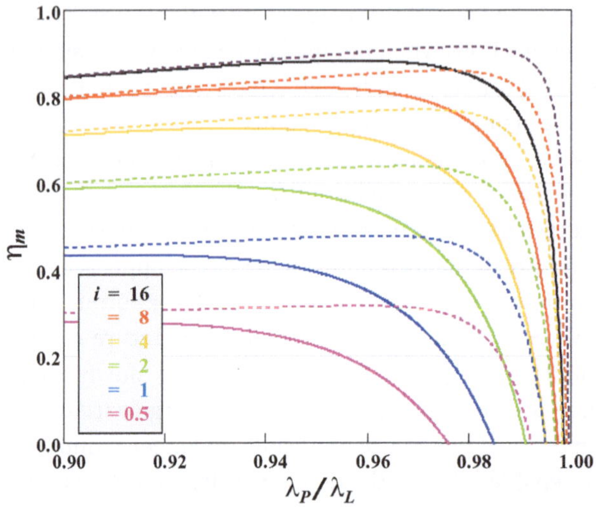

Figure 5.3 Plot of the material efficiency for an ideal Yb:YAG laser with λ_L = 1050 nm and several values of the scaled intensity. For the purpose of this illustration, the scaled pump and laser intensities are set equal. Solid curves are for 300 K and dashed curves are for 100 K.

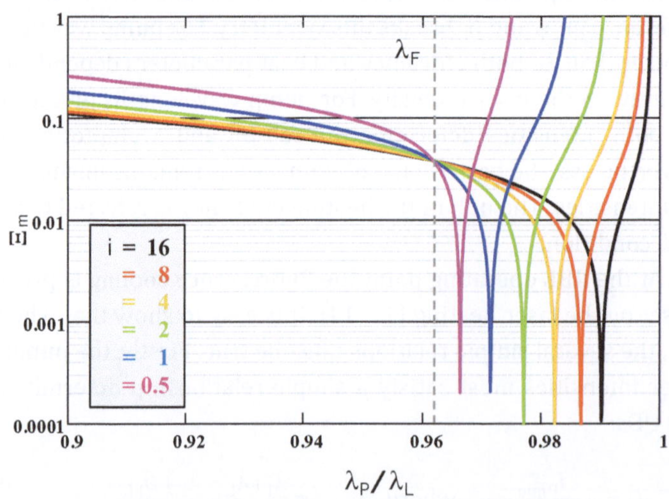

Figure 5.4 Plot of laser material heating parameter for idealized Yb:YAG. Lasing is at 1050 nm and the temperature is 300 K. The vertical line marks the position of the mean fluorescence wavelength. As before, the scaled pump and laser intensities are set equal.

So the RBL pump and laser intensities are inversely related. As the laser intensity increases, the pump intensity approaches a minimum allowed value, i_{Pmin}. Conversely, there is also a minimum laser intensity required to maintain the radiation balance at high pump intensities. Since real RBL systems are likely to be limited by the pump brightness, operation just above i_{Pmin} is likely to be the optimal design.

It is also easy to show that at the RBL point the laser efficiency reduces to

$$\eta_{RBL} = \frac{(\lambda_P - \lambda_F)}{(\lambda_L - \lambda_F)}. \tag{5.16}$$

This simple expression reveals that as the pump wavelength ranges between λ_F and λ_L, η_{RBL} varies from 0 to 1.

To see how the RBL efficiency depends on the pump intensity. Let's assume that the laser intensity is high, so the RBL pump intensity will be close to i_{Pmin}. Figure 5.5 illustrates how η_{RBL} varies with the minimum pump intensity for the example of a room-temperature 1050 nm Yb:YAG laser. From Fig. 5.5 we see that when i_{Pmin} = 10, the RBL efficiency should approach 80%. However, to get η_{RBL} > 90% would require 10 times higher pump intensities.

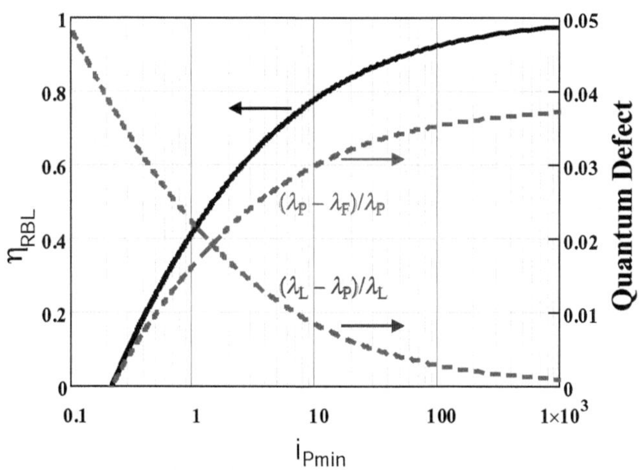

Figure 5.5 The solid curve is the predicted RBL material efficiency as a function of the minimum pump intensity. The dashed lines give the laser quantum defect $(\lambda_L - \lambda_P)/\lambda_L$ and the fluorescence quantum defect $(\lambda_P - \lambda_F)/\lambda_P$. This calculation assumes the conditions of the Yb:YAG laser discussed in the text.

The dashed curves in Fig. 5.5 illustrate how the QDs for the laser and fluorescence emission vary with minimum pump intensity. When the QDs are equal the heat generated by one laser photon is on average offset by the cooling from one fluorescent photon. At this point the RBL efficiency is 50%. As the pump wavelength is increased, laser QD heating decreases and fewer fluorescent photons are required to maintain balance. This requires the laser efficiency increases to maintain the RBL condition. For example, at $i_{\mathrm{Pmin}} = 10$, a single fluorescent photon should balance the defect of three laser photons.

Comparing Figs. 5.3 and 5.4 reveals that optimums for the two approaches do not occur at the same pump wavelength. The RBL points occur at longer wavelengths for the same intensity. Taking the $i_{\mathrm{P}} = i_{\mathrm{L}} = 2$ case, for example, an ideal 1050 nm Yb:YAG laser pumped at 970 nm is predicted to have $\eta_{\mathrm{m}} = 0.593$ and $\Xi_{\mathrm{m}} = 0.102$. The same laser pumped at 1026 nm would see $\eta_{\mathrm{m}} = 0.402$, with no heat generation. At the highest intensities plotted, $i_{\mathrm{P}} = i_{\mathrm{L}} = 16$, the maximum possible efficiency is 0.882, while the RBL efficiency is 0.745. Since thermal loading requires active cooling to maintain temperature and beam quality, these systems' overall efficiencies are likely to be comparable.

A third approach for improving laser performance is motivated by the special case of no lasing. When laser intensities are small, the laser material heating efficiency simplifies to become proportional to $\lambda_{\mathrm{P}} - \lambda_{\mathrm{F}}$. So if the laser medium is pumped at its mean fluorescent wavelength, there will be no heat generated by the pump. This approach might be indicated when the stored energy is not fully extracted from the laser medium, as often occurs in bulk laser geometries or during pulsed mode operation. For a laser operating with the condition of $\lambda_{\mathrm{P}} = \lambda_{\mathrm{F}}$, the fractional laser heating simplifies to the laser's QD:

$$\Xi_{\mathrm{m}} = (1 - \lambda_{\mathrm{L}}/\lambda_{\mathrm{F}}). \tag{5.17}$$

Note that this is the only pump-intensity-independent heat-loading solution for an LQDL. Heat loading would be simply proportional to laser power. In the example plotted in Fig. 5.3 this operating point is labeled by a vertical dashed line. Pumping at 1010 nm and lasing at 1050 nm give a heating parameter of 3.8%. Blue-shifting the laser wavelength could readily reduce the laser heating parameter to below 1%, while maintaining good efficiency.

5.3.3 Nonideal Laser Materials

A more accurate model of laser power flow must include realistic material losses. In the previous section the laser material was assumed to be perfectly radiative, linear, and optically thin. Now we will consider what happens when those assumptions are only mostly correct.

First we will include the possibility of weak quenching of the laser excitation. This could occur through electron–phonon deactivation or through an ion–ion energy transfer mechanism [66]. Direct quenching introduces another linear loss term into Eq. 5.2. This linear loss term is usually handled simply by replacing the radiative lifetime, τ_R, with a fluorescence lifetime, τ_F. Energy transfer losses are effectively handled by adding a quadratic loss term with a characteristic quenching lifetime, τ_Q. For simplicity, we will assume that the quenching process is completely nonradiative; quenched excitations are converted completely into heat.

The second process to be added to the model is background absorption. Background absorption is present at low levels in even the best optical materials [13]. Optical cooling studies have shown that these weak losses become important for low-QD pumping [36]. The background absorption, α_B, is usually assumed to be due to trace impurities, so the excitation model is not directly affected by this process. Here we will also assume that the background absorption is sufficiently broad to affect both pump and laser. Define fractional absorptions as $\delta\alpha_P = \alpha_B(\lambda_P)/\sigma_P N_T$ and $\delta\alpha_L = \alpha_B(\lambda_L)/\sigma_L N_T$. As we will see, even small background absorptions can significantly reduce efficiency and increase heating.

The last process needed for a more realistic laser model is reabsorption of fluorescence inside the laser medium. Often referred to as radiation trapping, this process will occur whenever absorption or reflection prevents some of the spontaneous emission from escaping the laser medium. We will assume that reabsorption is due solely to the active ions. So while trapping can enhance other nonradiative losses by increasing excitation densities, here it is considered a purely radiative phenomenon.

The addition of radiative trapping significantly complicates the modeling of the laser materials. It introduces nonlocal source terms for the fluorescent field that require intense numerical calculation for solution. For this analysis, we will invoke an approximate solution for the trapped fluorescence field [11]. The derivation and general solutions of the revised rate equation are presented in the appendix to this chapter. The approximation assumes that the laser medium traps light uniformly throughout and that the amount of trapping can be described simply by a cavity photon lifetime. This allows the fluorescent field to be removed from the rate equation. The rate equation then simplifies to a quadratic in terms of the excitation density and modified values for fluorescence wavelength, λ'_F, and lifetimes, τ'_F, and τ'_Q. These are the effective values that would be measured external to the optically thick laser medium.

It should be noted here that it is not usually possible to measure the radiative lifetime, τ_R, directly. In the literature it is generally calculated from the absorption or emission spectra. When quenching is weak a more accurate measurement of τ_R is possible by measuring nonradiative losses directly. To obtain the radiative lifetime of Yb:YAG quoted above, we used a low-power tunable laser to perform photothermal deflection on bulk samples [5]. This technique allows an accurate measurement of the pump wavelength at which optical heating crosses over to optical cooling, λ_C. The radiative lifetime can then be computed by correcting the measured fluorescence lifetime for nonradiative losses using the relation $\tau_R = \tau_F\lambda_C/\lambda_F$. The best fluorescence lifetime measurements are acquired using thin powdered samples immersed in an index-matching medium to avoid fluorescence trapping.

Incorporating these loss processes into the revised rate equation allows us to solve for the power densities in a realistic laser material. We will distinguish this solution from that of the ideal material by adding a prime. The revised rate equation now becomes

$$\frac{dN_2'}{dt} = \frac{I_P\lambda_P\sigma_P}{hc}\cdot\left[N_T - N_2'/\beta_P\right] + \frac{I_L\lambda_L\sigma_L}{hc}\cdot\left[N_T - N_2'/\beta_L\right] - \frac{N_2'}{\tau_F'} - \frac{N_2'^2}{N_{Yb}\tau_Q'}. \tag{5.18}$$

While the general solution is presented in the appendix, here we continue with the approximation of weak losses. With that assumption, the steady-state solution can be written in terms of the ideal solution, n_2, given by Eq. 5.9:

$$n_2' = n_2\cdot(1+\delta n') \approx n_2\cdot\left[1+\frac{(1-\tau_R/\tau_F') - n_2\,\tau_R/\tau_Q'}{(i_P + i_L + \tau_R/\tau_F')}\right]. \tag{5.19}$$

This expression makes it is clear that steady-state excitation densities depend principally on the change in the fluorescence lifetime. A decrease in the lifetime from linear quenching will generate a proportional decrease in the excitation density. Conversely, an increase in τ'_F due to radiative trapping will increase excitation densities. For many materials the effect of trapping is more pronounced, leading to longer fluorescence lifetimes and higher excitation densities in optically thick samples. Note that since $n_2 << 1$, the quadratic quenching term will have little effect on the density unless $\tau'_Q \approx \tau_R$.

Using the same fractional definition as before, the laser material efficiency is modified by the revised excitation fraction and by the additional absorption:

$$\eta'_m \equiv \frac{L'_m}{P'_m} = \frac{\lambda_P i_L}{\lambda_L i_P}\left[\frac{n'_2 - \beta_L - \beta_L \delta\alpha_L}{\beta_P - n'_2 + \beta_P \delta\alpha_P}\right]. \tag{5.20}$$

In the approximation of weak losses, it becomes

$$\eta'_m \approx \eta_m \cdot \left[1 + \frac{n_m \delta n'_m - \beta_P \delta\alpha_P}{\beta_P - n_m} + \frac{n_m \delta n'_m - \beta_L \delta\alpha_L}{n_m - \beta_L}\right]. \tag{5.21}$$

Expressed in this way, it is clear that laser material efficiency increases with excitation density and decreases with background absorption. The contribution from background absorption is reduced by β. The impact of these losses on efficiency increases substantially when the material is operated near the laser threshold or pump saturation.

To illustrate the impact of including material losses we again use Yb:YAG as the example. This time we assume small realistic corrections to the material parameters. In the slab laser experiments discussed in the next section, it was found that trapping increased the fluorescence lifetime to 1.25 ms and the mean fluorescence wavelength shifted to 1017 nm. We will also assumed weak quadratic quenching by setting τ'_Q to 20 ms and moderate background absorption with $\delta\alpha_L = \delta\alpha_P = 10^{-3}$. These values are consistent with values observed in samples of high-purity laser materials with low Yb-doping densities [13, 36].

Figure 5.6 shows the result of the evaluation of Eq. 5.21 for the same conditions used in Fig. 5.3. With the more realistic material parameters given above, efficiencies are higher than for the ideal case. This is due to the increased fluorescence lifetime produced by the radiative trapping. The impact is larger at lower intensities where efficiencies increase by roughly 10%. At the highest intensity, room-temperature optimal efficiency rises only slightly to 0.895 at 1008 nm. For operation at 100 K, the optimal efficiency is 0.927 at 1029 nm. Still, the efficiency curves are relatively flat, allowing wide latitude in the selection of the pump wavelength.

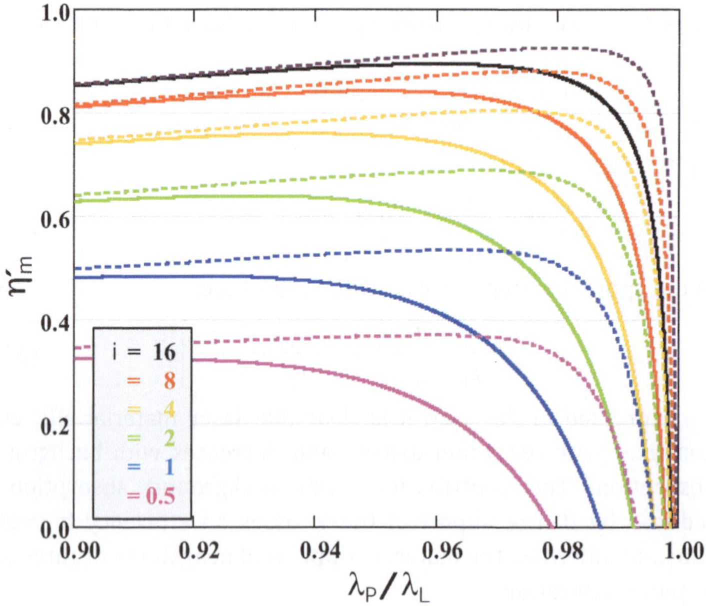

Figure 5.6 Plot of the predicted laser material efficiency for the same conditions used in Fig. 5.3. Material parameters have been modified, as described in the text.

The form of Eq. 5.21 also suggests a strategy to maintain efficiency in the face of higher background absorption. To a limited extent, background absorptive losses will be offset by radiative trapping. Fluorescent trapping significantly increases storage lifetime, excitation density, and laser material efficiency.

In a similar way we can revise the laser heating parameter that incorporates the weak loss processes in the laser material. Heat generation will be modified differently by each of the loss processes. Since radiative trapping is assumed to be purely radiative its contribution to laser heating can be computed directly from the steady-state solution in Eq. 5.14 using the measured effective material properties. The mean fluorescence wavelength, λ'_F, should be measured using the actual laser medium to fully account for the geometrical effects of trapping. The measured fluorescence lifetime should be corrected to remove quenching using the relation, $\tau'_R = \tau'_F\lambda_C/\lambda_F$. Substituting λ'_F and τ'_R into Eq. 5.14 will yield the QD heat generation corrected for the fluorescent reabsorption, H'_F.

In addition to the QD heating, we identify specific contributions for the heat generated by quenching, $\delta H'_Q$, and by background absorption, $\delta H'_A$:

$$\delta H'_Q = \frac{hcN_T n'_2}{\lambda_P}\left(\frac{1}{\tau_F} - \frac{1}{\tau_R} + \frac{n'_2}{\tau'_Q}\right)$$
$$\text{and} \quad \delta H'_A = \frac{hcN_T}{\tau_R}\left(\frac{i_P\beta_P}{\lambda_P}\delta\alpha_P + \frac{i_L\beta_L}{\lambda_L}\delta\alpha_L\right). \tag{5.22}$$

The total heat generation in the realistic laser material now becomes $H' = H'_F + \delta H'_Q + \delta H'_A$. Defining the material heating parameter as before leads to:

$$\Xi'_m \equiv \frac{|H'|}{L'} = \frac{\tau_F}{\tau'_F}\frac{\left\|[\lambda'_F(\lambda_L - \lambda_P)i_P i_L\beta_P/\beta_{PL} + \lambda_L(\lambda'_F - \lambda_P)i_P\beta_P/\beta_L - \lambda_P(\lambda_L - \lambda'_F)i_L]\right\|}{\lambda'_F\lambda_P i_L[i_P\beta_P/\beta_{PL} - 1]}$$
$$+ \frac{\lambda_L}{\lambda_P}\left(\frac{\tau_R}{\tau_F} - 1 + \frac{n_2\tau_R}{\tau'_Q}\right)\left[\frac{i_P\beta_P/i_L\beta_L + 1}{i_P\beta_P/\beta_{PL} - 1}\right] + \left[\frac{i_L\beta_L\delta\alpha_L + i_P\beta_P\delta\alpha_P\lambda_L/\lambda_P}{n_2 - \beta_L}\right]. \tag{5.23}$$

Here the first term on the right accounts for radiative trapping. The second term accounts for quenching and the third accounts for background absorption. Here the approximation, $n'_2 \approx n_2$, found to be valid in Eq. 5.21 has been used. Figure 5.7 plots Ξ'_m for the Yb:YAG example for the same conditions used in Fig. 5.6. Comparing this plot with that of the idealized Yb:YAG in Fig. 5.4 reveals the impact of adding realistic material losses. Linear and quadratic quenching red-shift the crossover wavelength by 5 nm and background absorption adds another 1 nm shift. When radiative trapping is added to the model τ'_F = 1.25 ms, thermal loading decreases and the crossover wavelength red shift increases to 15 nm. This narrows and disrupts the RBL resonance at lower intensities. Background absorption heating is stronger at higher intensities. For the highest intensities, a fractional background absorption of $\delta\alpha_L = \delta\alpha_P = 10^{-3}$ disrupts the

RBL resonance. Increasing the background absorption fraction to 4 × 10^{-3} was found to overwhelm fluorescence cooling, disrupting the RBL resonances at all wavelengths.

Comparing Figs. 5.4 and 5.7 clarifies the impact of weak losses on optimal material performance. The large red shifts in the RBL resonances significantly reduce material efficiency at minimum Ξ'_m. The best efficiency with no heat generation is 0.40 when the pump is 1040 nm with intensities of i = 4. Operation away from the exact RBL resonance allows material efficiencies to recover. If we restrict $\Xi'_m \leq 0.01$, a best efficiency 0.62 occurs for i = 8 with the pump at 1040 nm.

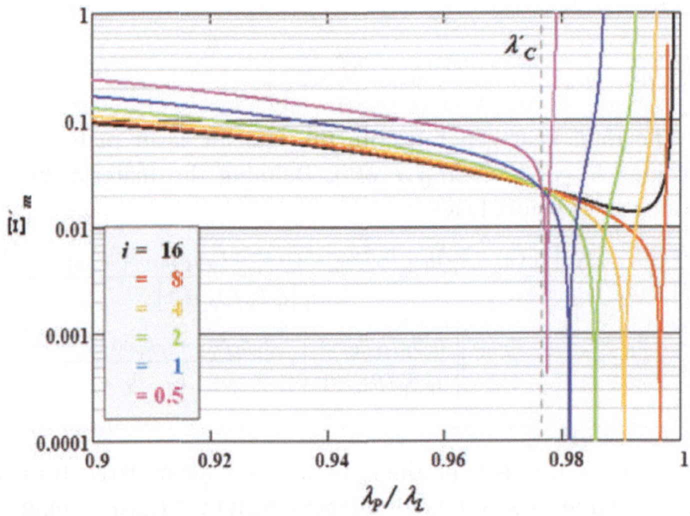

Figure 5.7 Plot of the realistic room-temperature laser material heating parameter for the same conditions used in Fig. 5.4. The vertical dashed line marks the position of the effective crossover from fluorescent heating to fluorescence cooling, λ'_C.

5.4 Low–Quantum Defect Laser Experiments

5.4.1 Introduction to Experiment Review

In the previous section we considered a generic laser material that consisted of two widely separated sets of discreet energy levels.

The discreet energy levels were treated as existing in a thermal equilibrium with their surroundings. The upper and lower manifolds were assumed to be separated sufficiently such that thermal transitions between them were unlikely. It is this condition that allows for the design of laser systems with low QDs. As discussed earlier, these conditions are satisfied in the trivalent ytterbium system. Ytterbium ions have a single pair of 4f states that are roughly 3 kT wide and are separated by approximately 50 kT at room temperature. This energy splitting and the absence of other energy levels make the Yb^{3+} system ideal for LQDL and optical cooling.

In this section we will review a few of the many laser experiments conducted with LQDL systems. The first section will focus on the 1 μm ytterbium system, particularly results approaching radiation balance. Alternative active ions are considered separately. For these ions the presence of additional excited states complicates the simple model developed earlier. Nevertheless many of these systems are also promising candidates for optically cooled lasers.

5.4.2 Ytterbium Laser Experiments

Since its first demonstration, ytterbium has been lased in almost every host material [44]. As mentioned above, the ytterbium 1 μm transition is spectrally broader than most other rare earth transitions. This makes Yb^{3+} ideal for LQDL designs but often leads to lower absorption and emission cross sections. As shown in the previous section, weak absorption and gain tighten the requirements on material losses and increase the need for high-intensity pump sources. Successful LQDL systems are therefore based on the highest–optical quality materials. After decades of process refinement, YAG and fused silica have the highest optical quality and lowest losses. As a result the best LQDLs have so far been based on these materials.

Most ytterbium lasers have utilized GaAs laser diodes as the pump source. Indeed, it was the invention of bright laser diode pumps that made Yb^{3+} a practical laser system. Nevertheless the laser diode's limited brightness and restricted operation wavelength have hampered LQDL design. As a result, most of the experiments have used other pump sources to generate the required intensities and pump wavelength. In this section we will review a few of the more successful LQDL ytterbium lasers.

In 1999 the Naval Research Laboratory began a series of ytterbium LQDL experiments. Early laser designs explored direct-diode pumping of ytterbium-doped $KGd(WO_4)_2$ crystals [11, 13]. This biaxial crystal is attractive for LQDLs because of its somewhat blue-shifted transition and higher cross sections. In 2000 high-power laser diodes with wavelengths approaching 1 μm were just becoming available, so for these experiments we chose λ_P = 995 nm. Several slab and disk laser geometries were tested using the high-cross-section *m* axis transitions for pump absorption and the *p* axis laser transition at 1045 nm. (The principal optical directions are labeled *p*, *m*, and *g* in order of increasing refractive index.) Figure 5.8 shows an edge-pumped 2 × 8 mm Yb:KGW laser disk with half of its 90 radially focused laser diode bars. These Yb:KGW lasers generated up to 490 W at moderate duty factors using only passive cooling. This was the first LQDL system with heat loading below the laser QD. Room-temperature laser saturation intensity for Yb:KGW is only 2.4 kW/cm^2 and a high-Q cavity helped maintain $10 \leq I_L/I_{Lsat} \leq 50$. Background absorption at 1 μm in KGW was measured to be roughly 8×10^{-4} cm^{-1} or $\delta\alpha_L = 4 \times 10^{-4}$. With measured values of Ξ as low as 0.08, these experiments demonstrated the promise of LQDLs for average power scaling.

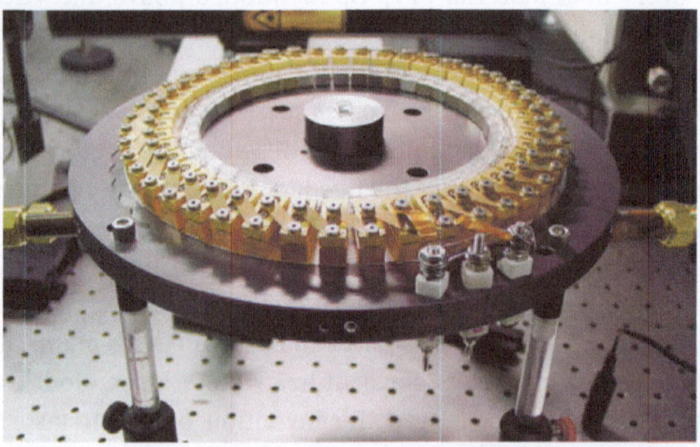

Figure 5.8 Edge-pumped Yb:KGW disk laser with a laser quantum defect of 4.8%.

While direct-diode pumping is attractive from a system perspective, it led to several problems in these prototypes. First high-power diodes were not available at the longer wavelengths needed to generate optical cooling. Further diode coupling was limited to around 55%. For optically thin Yb:KGW, λ_F = 997 nm and the radiative lifetime is 0.275 ms. However, for the 2 × 8 mm laser disk with 2.1E20 Yb density, fluorescent trapping increased these values to λ'_F = 1011 nm and 0.75 ms. Also, the on-axis intensity of the diode pump array limited to 15 kW/cm^2, while the pump saturation intensity of Yb:KGW at 995 nm is I_{Psat} = 14.6 kW/cm^2. So the scaled pump intensity was never much greater than unity. These factors restricted laser optical efficiency that was not better than 17%.

To explore the radiation balance condition, experiments shifted to longitude pumping of Yb:YAG. Ytterbium fiber lasers were selected as pump sources in order to increase brightness and red-shift the pump well beyond the fluorescence line center at 1010 nm. An array of 45 ytterbium-doped double-clad fiber lasers was developed for these experiments. Each fiber generated a near-diffraction-limited 100 W beam at 1030 ± 0.1 nm. The combined output of this fiber array produced pump intensities as high as 140 kW/cm^2, or roughly 10 times higher than the earlier laser diode array. At 1030 nm the pump saturation intensity of room-temperature Yb:YAG is I_{Psat} = 9.05 kW/cm^2. Therefore this fiber array was capable of generating pump intensities $i_P \leq 15$. Figure 5.9 is the steady-state gain for Yb:YAG computed from the spectral data in Fig. 5.1 and Eq. 5.10. This figure assumes $i_L = i_P$, room temperature, and an Yb density of 4.1E20 ions/cm^3. Dashed curves are identical data plotted at the 10 times' scale shown on the right axis. Pump saturation reduces the 1030 nm absorption by eightfold, even in the presence of strong stimulated emission. Note that with increasing intensity, gain at 1049 nm approaches, but can never exceed, $\sigma_L N_{Yb} \beta_{PL}/2\beta_P$, or 0.0176 cm^{-1}. Strong gain clamping is a common characteristic of LQDL systems.

In the first series of experiments the fiber array was used to end-pump a single Yb:YAG rod [10]. The rods used were 3 mm in diameter and roughly 10 cm long. The ends were AR coated and the barrel of the rod was polished. The Yb:YAG laser rod was placed inside a stable optical resonator, as shown in Fig. 5.10. The laser rod was placed on a thermally insulated fixture and a thermal camera was used to measure the surface temperature of the Yb:YAG crystal.

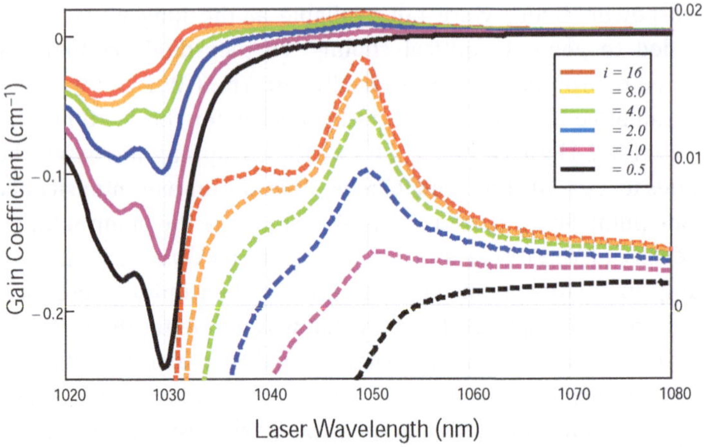

Figure 5.9 Gain coefficient for room-temperature 3% Yb:YAG pumped at 1030 nm. For simplicity the pump and laser intensities are set equal: $i = i_L = i_P$. Solid curves match the left vertical scale. Dashed curves display the same data at the tenfold scale shown on the right axis.

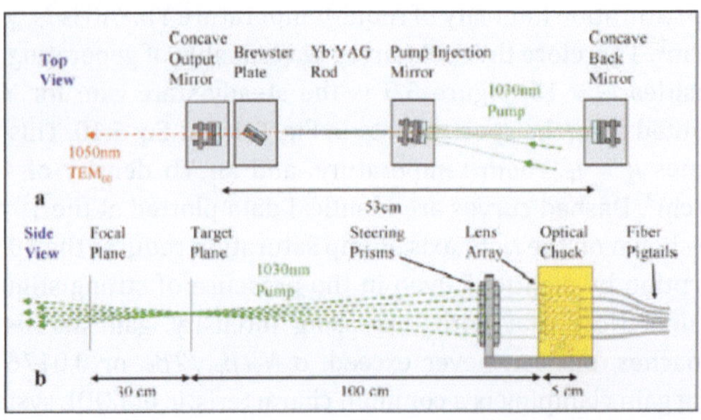

Figure 5.10 Optical configuration of the first radiation-balanced laser. (a) Stable Yb:YAG resonator. (b) Ytterbium fiber laser pump array propagation shown without the target laser rod.

The pump beams were injected into the resonator off-axis using a flat HR mirror with a 3 mm on-axis hole. The broadband HR back-resonator mirror reflected the pump beams back to the center of the rod. The beams crossed near the center of the rod, producing a roughly 2 mm diameter mode volume. The concave output mirror was coated for partial reflection at 1050 nm and low reflectivity at 1030 nm. An 80% reflector was found to generate maximum power. The cavity length was also adjusted for maximum power. A Brewster plate was added to ensure linearly polarized operation.

The resonator above was operated under a variety of conditions. Laser rods with target ytterbium densities of 1%, 2%, 3%, and 5% were tested. Pump power, output coupling, mirror curvature, and cavity length were adjusted while monitoring output power, thermal loading, and beam quality. Typical steady-state laser performance is illustrated in Fig. 5.11. Somewhat higher efficiencies were obtained with the 5% doped crystal, owing to its higher pump absorption. Lower thermal loading and better beam quality were found with the 2% doped crystal. This resonator generated average powers up to 200 W of linearly polarized output with a beam quality of M^2 = 1.2, as illustrated in Fig. 5.12. During laser operation the YAG rod was thermally insulated and monitored with the thermal camera.

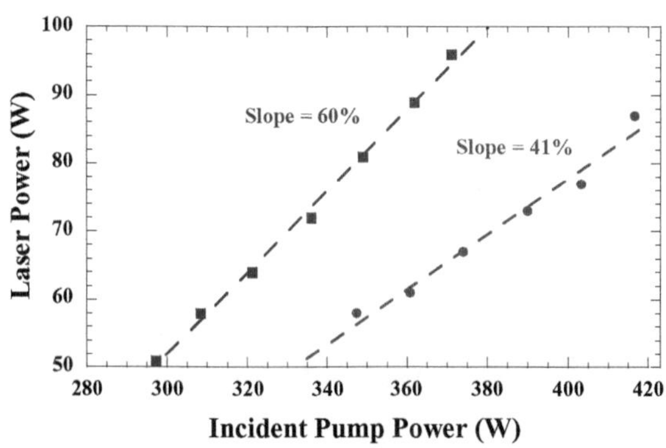

Figure 5.11 Comparison of the laser performance of the Yb:YAG resonator for two doping densities: 3.1E20 ions/cc (dots) and 8.1E20 ion/cc (squares).

For the above laser configuration the minimum heat load condition was found to require an incident pump intensity of i_P = 2.1 and an on-axis intercavity laser intensity of i_L = 1.0. This is a close to the predicted value of the nonideal model developed in Section 5.3.3. This is consistent with the strong radiation trapping observed in these experiments. Fluorescence lifetimes from these laser rods were found to be 2.0 ms, twice as long as the measured value τ_R. Near the RBL resonance, the laser power was 80 W and optical efficiency was

26%. This is consistent with the nonideal LQDL model that predicts the average material efficiency for the rod should be 27%.

Figure 5.13 shows a typical thermal image of the laser rod near the RBL condition. For optimized conditions heat loading in the laser rod was too low to measure with the thermal camera. This corresponds to a material heat parameter $\Xi' < 10^{-4}$. This was the first truly RBL.

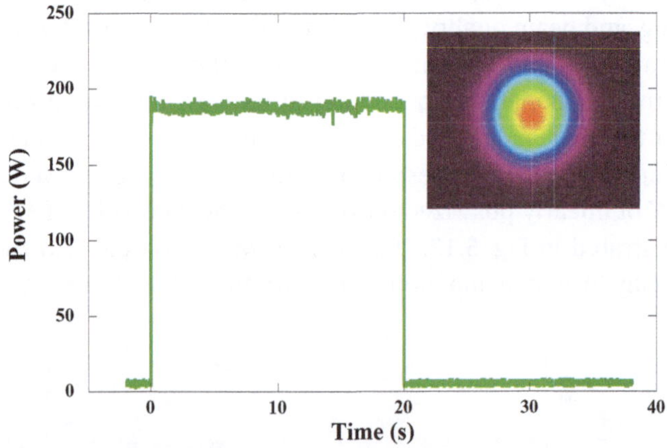

Figure 5.12 Power and near-field beam profile of the 3% Yb:YAG resonator.

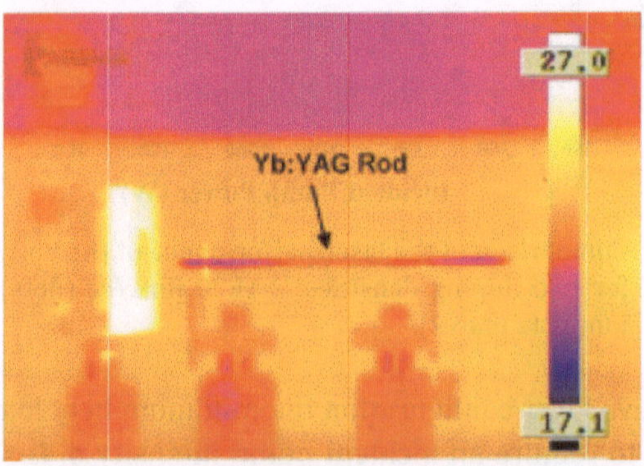

Figure 5.13 A typical thermal image of the 3 × 120 mm Yb:YAG rod inside the radiation-balanced laser. This image was captured after 10 seconds of lasing while being pumped with 372 W at 1030 nm. The false color scale corresponds to ±5°C, as shown on the right. Fluorescence from the laser rod heats the mirror mount on the left, but the thermally insulated laser rod remains very close to the ambient temperature.

A second series of experiments was conducted to improve performance of the Yb:YAG laser system [52]. In these experiments we replaced the laser rod with a composite slab composed of Yb:YAG and sapphire. A single 2 × 110 mm 3% Yb:YAG slab was fusion-bonded to a sapphire plate along its length, as shown in Fig. 5.14. The height of the YAG slab was tapered along its length from 3 mm to 2 mm and an undoped 10 mm end-cap was bonded to the larger end.

Figure 5.14 Radiation-balanced laser gain element constructed from a tapered 110 mm 3% Yb:YAG slab fusion bonded to a 2 × 20 mm sapphire plate. The sapphire plate provides a refractive index mismatch of 0.07. This allows for guiding of the 1030 nm pump lasers, while significantly reducing fluorescent trapping. The gold-plated mount and fused silica pins were designed to avoid radiative heating of the fixture.

The composite slab provided several important improvements for the RBL. It provided a rugged but transparent fixture. The polished edges provided guiding of the fiber pump beams down the length of the YAG. This resulted in more uniform pumping and higher overall pump absorption. The small difference in refractive index between YAG and sapphire produced less trapping of the Yb fluorescence. The sapphire bond reduced the fluorescent trapped lifetime in the slab to 1.25 ms, from the 2 ms measured in the freestanding rods. The tapper helped suppress parasitic modes and smoothed pump intensity.

The slab was end-pumped through the AR-coated end-cap. Pump powers as high as 4.8 kW were incident on the 2 × 3 mm facet, leading to scaled pump intensities of $i_P \leq 8.5$. As with the rod experiments, an 80% and high-reflection mirror created a stable resonator around the slab. Laser powers at 1050 nm up to 1.6 kW were generated under pump-limited conditions, leading to intercavity intensities of $i_L \leq 5.1$.

Minimum heating was observed at a laser power of 600 W, where the measured heating parameter dropped to below 0.001. That corresponds to 1 W of heat for every kilowatt of output laser power. This RBL operating point corresponded to an incident pump intensity of $i_P = 4.7$ and an intercavity laser intensity of $i_L = 1.9$. For the conditions of this experiment the nonideal LQDL model predicts the RBL condition will occur at $i_P = 3$. The measured optical efficiency of the laser was 33% and the model predicts a spatially average efficiency of 31%.

These experiments illustrate the importance of increasing pump intensity and reducing fluorescent trapping. Index matching significantly reduced trapped lifetimes. This shifted the RBL resonance for 1030 nm pumping to higher intensities that increased the RBL power by 7.5 times. They also illustrate the need to scale laser mode volume to maintain the RBL intensities at the highest pump power.

In 2007, Matsubara et al. demonstrated a record-high optical efficiency of 81% in a cw ytterbium laser. Operating at 1048 nm, the lasing medium was a thin disk of 20% Yb:YAG pumped by a high-intensity 940 nm pump. At that pump wavelength the saturation intensity in room-temperature Yb:YAG is 26 kW/cm^2. For this demo a titanium-doped sapphire laser was used to generate pump intensities in the disk as high as 200 kW/cm^2. With values of $\lambda_P/\lambda_L = 0.90$ and $i_P = 7.1$, this demo efficiency matches well the $\eta'_m = 82\%$ calculation shown in Fig. 5.6. The QD heat of 10% limited average power scaling. A thermal lens of only 5 mm was reported at the 800 mW power level.

In 2008 the group at IPG Photonics reported both high efficiency and high power using an LQDL approach in ytterbium-doped silica fibers [31]. They demonstrated 75% optical efficiency in a single-mode, double-clad fiber amplifier that generated 6 kW of average power. This system lased at 1070 nm and was pumped by several ytterbium fibers operating at 1018 nm. Ytterbium cross sections in silica are low, leading to I_{Psat} and I_{Lsat} of 26 and 54 kW/cm^2, respectively [51]. Nevertheless, the reported maximum pump and laser intensities were 20 and 2000 MW/cm^2, respectively. The lossless material model of the previous section predicts that η_m = 95% in this high-intensity amplifier. This system was operated very near the $\lambda_P = \lambda_F$ condition, so the heat loading should be close to the laser QD, 4.8%, when the laser efficiency approaches η_m. In that case, thermal loading would be only 20 W/m, although heat loads were not reported. The overall electrical efficiency of this system was reported to be 25%. Further power scaling was limited by fiber nonlinearity, not heat generation.

5.4.3 Alternative Low–Quantum Defect Laser Materials

The conditions required for LQDL operation are also satisfied by several other rare earth ions: erbium, thulium, and holmium. The first excited states of these ions also decay principally radiatively in most host materials. However, the presence of higher-lying energy levels and lower energy transitions for these ions does give rise to several weak loss processes. Optical cooling has been demonstrated in Er^{3+} and Tm^{3+} but not as yet in holmium [9, 22 , 23, 28, 38, 57]. Efficient resonant-pumped lasers have been demonstrated in all three of these ions.

Resonant pumping of erbium lasers at 1.5 μm has received considerable attention due to the laser's eye-safe nature. Shen et al. demonstrated a 60 W Er:YAG laser end-pumped with an erbium silica fiber laser at 1.532 μm [60]. The 1.645 μm laser generated an optical slope efficiency of 81% with a QD was 7%. Although optical cooling is not possible with this QD, heat loading was reported to be low. Pump intensities in this demo were quite high, 200 kW/cm^2. Higher-power Er:YAG lasers pumped with 1.5 μm diodes have been demonstrated, but weak pump absorption has so far limited their optical efficiencies [32, 45].

Low-QD thulium 2.0 μm lasers have received less attention than erbium. This is principally due to the lack of pump diodes at 1.9 μm and the success of the 2-for-1 pumping scheme using 0.8 μm diodes. Nevertheless, a resonant-pumped thulium laser system has recently been reported with slope efficiencies of 90% [24]. This system used 0.8 μm diodes to pump a thulium fiber laser operating tuned to 1.908 μm. This fiber was then used to pump another thulium fiber lasing at 2.005 μm, leading to a laser QD of 4.8%. As with the erbium demo, fluorescent cooling of thulium was not possible with this pump wavelength.

One of the best-studied midwave LQDL systems is based on the 2.1 μm holmium transition. First demonstrated in 1992, this laser is usually excited with a diode-pumped 2.0 μm thulium laser, as discussed above [16, 18, 62]. This laser scheme provides coherent beam combination of multiple thulium lasers and avoids strong upconversion that occurs in concentrated Tm and Tm/Ho laser materials [15]. The holmium LQDL scheme has received considerable attention due to relatively eye-safe wavelength of 2.1 μm and the superior transmission through the atmosphere.

In 2015 Loiko et al. demonstrated record-high efficiency in a holmium-doped $KLu(WO_4)_2$ microchip laser [47]. They achieved 84% slope efficiency at 2.105 μm pumping with a thulium laser at 1.945 μm at only 4.4 kW/cm^2. Unfortunately optical efficiency was low for this demo due to poor pump absorption and operation near threshold.

A low-power experiment demonstrated 72% slope efficiency from a 2080 nm holmium-doped ZBLAN fiber. The thulium fiber pump source was operated at 2051 nm, leading to a very low laser QD of 1.4% [35]. This LQDL scheme has the potential for optical cooling since λ_F = 2010 nm in Ho:ZBLAN.

Other recent laser demonstrations have explored holmium doped into silica fibers. Average powers up to 400 W have been generated but with lower optical efficiencies [54, 37]. Current work on this material seeks to reduce the hydroxyl-induced background absorption and lower the multiphonon quenching in holmium-doped fibers [30, 42].

5.5 Summary

LQDLs are important because they have the potential to operate with both high efficiency and high power. Optical pumping near the laser wavelength decreases the adverse effects of the laser QD. It also increases the effects of absorption and gain saturation. When the pump wavelength exceeds the mean fluorescence wavelength, optical cooling becomes an important consideration.

In the first part of this chapter a steady-state model was developed that predicts LQDL material performance for an ideal laser medium. However, as the laser QD is decreased further weak loss mechanisms begin to play an important role. Trapping of the fluorescence inside the laser medium impacts both cooling and excitation lifetime. At high intensities, extrinsic background absorption can become a major source of heating. Extensions to the simple model were developed to properly account for weak loss mechanisms.

The steady-state laser model was used to compare optimization schemes. Optical efficiency was found to depend strongly on proper selection of pump wavelength. Higher intensities were shown to improve efficiency. Radiative trapping was found to significantly improve optical efficiency. Optical efficiencies as high as 90% at room temperature appear to be achievable with laser QDs of 4%. Even higher efficiencies are possible at cryogenic temperatures using a 2% QD. Nevertheless optimizing efficiency alone will lead to significant thermal loading in high-power laser systems.

Alternative optimization strategies that minimize thermal loading were also explored. Radiation-balanced lasing was shown to virtually eliminate heat generation at the cost of somewhat reduced optical efficiency. RBL efficiencies of 80% appear to be achievable at room temperature with high-purity materials and careful control of the pump intensity. The potential for higher average power, simplified thermal management, and photovoltaic fluorescent recycling make the RBL an attractive power-scaling approach.

A sample of the LQDL literature was reviewed and performance was compared with steady-state theory. Several ytterbium systems have demonstrated optical efficiencies over 75%. Optically cooled ytterbium lasers have demonstrated heat loads below 0.01% of the output laser power. Record performance of the low–quantum defect ytterbium lasers clearly demonstrates the efficacy of this approach to power scaling. The steady-state model develop here was also found to be in good agreement with the optimum laser performance. Low-QD experiments have been conducted in the longer-wavelength erbium, thulium, and holmium laser systems. These systems have also demonstrated high efficiency but are not yet capable of scaling to high powers.

5.6 Appendix

Radiation trapping significantly impacts the performance of LQDLs. Here we will develop an approximate solution for the excitation density in an optically thick laser medium [10]. Using the notation developed in Section 5.3, the rate equation for the local excitation density will now include a term for quadratic quenching and reabsorption of the local fluorescence:

$$\frac{dN_2}{dt}=\frac{I_\mathrm{P}\sigma_\mathrm{P}}{h\nu_\mathrm{P}}\left(N_\mathrm{T}-\frac{N_2}{\beta_\mathrm{P}}\right)+\frac{I_\mathrm{L}\sigma_\mathrm{L}}{h\nu_\mathrm{L}}\left(N_\mathrm{T}-\frac{N_2}{\beta_\mathrm{L}}\right)-\frac{N_2}{\tau_\mathrm{F}}$$
$$-\frac{N_2{}^2}{N_\mathrm{T}\tau_\mathrm{Q}}+\left\langle\left(\frac{I_\mathrm{F}(\lambda)\cdot\sigma_\mathrm{A}(\lambda)}{h\nu_\mathrm{F}}\right)\left(N_\mathrm{T}-\frac{N_2}{\beta(\lambda)}\right)\right\rangle. \tag{5.24}$$

Here $I_\mathrm{F}(\lambda)$ is the local fluorescence spectral irradiance at wavelength λ. Unlike the pump and laser fields, $I_\mathrm{F}(\lambda)$ will be spectrally broad with complex directionality. The last term in the equation is the rate of fluorescence reabsorption integrated over the entire emission spectrum as well as all directions and polarizations.

For precise evaluation of N_2, the local fluorescence intensity should be the integrated contributions from the entire laser crystal volume. Computing $I_\mathrm{F}(\lambda)$ from first principles requires a detailed knowledge of the excitation state of the entire lasing medium, as well as its surface reflectivity. This approach would require significantly more computation time to build up a self-consistent steady-state spatial solution for all three fields.

To simplify the problem, we can approximate the laser crystal as an optical cavity in which τ_C is the average lifetime of the available cavity modes. This lifetime might be computed from numerical ray traces or treated as an empirical parameter derived from fluorescence decay data. Fluorescence reabsorption reduces the photon lifetime within the crystal cavity according to

$$\frac{1}{\tau_\mathrm{P}}=\frac{1}{\tau_\mathrm{C}}+\frac{c\cdot\alpha(\lambda)}{n}. \tag{5.25}$$

Here c is the speed of light, n is the index of refraction, and $\alpha(\lambda)$ is the absorption coefficient. For simplicity we will assume that τ_C

is independent of wavelength and that the photon lifetime can be approximated using unsaturated absorption.

We can now approximate the fluorescent spectral energy density inside the crystal cavity at wavelength λ as the product of the local emitted power density, the probability of spontaneous emission, and the photon lifetime:

$$U_{\mathrm{F}}(\lambda)=\left(\frac{hc\,N_2}{\lambda\tau_{\mathrm{R}}}\right)\cdot\left(\frac{g(\lambda)}{\int g(\lambda)d\lambda}\right)\cdot\tau_{\mathrm{P}}\,. \tag{5.26}$$

Here the intrinsic fluorescence line shape function, $g(\lambda)$, gives the rate of spontaneous emission of photons with the wavelength between λ and $\lambda + d\lambda$. The line shape function is closely related to the emission cross section,

$$g(\lambda) = 8\pi c\sigma_{\mathrm{E}}(\lambda)/\lambda^4. \tag{5.27}$$

The integral in Eq. 5.26 normalizes $g(\lambda)$ over the entire fluorescence band.

The approximation expressed in Eq. 5.26 is that surface reflections, scattering, and reabsorption effectively smooth the fluorescent energy inside the crystal. The crystal cavity approximation should be valid in a simply shaped laser medium that is roughly uniformly pumped and where amplified spontaneous emission can be ignored.

The fluorescent spectral intensity inside the crystal can be computed using the relation $I_{\mathrm{F}} = cU_{\mathrm{F}}/n$ and substituted back into Eq. 5.24, eliminating the nonlocal dependence on the fluorescence field. The rate equation for the radiation-trapped system is rewritten as

$$\frac{dN_2}{dt}=\frac{I_{\mathrm{P}}\sigma_{\mathrm{P}}}{h\nu_{\mathrm{P}}}\cdot\left(N_{\mathrm{T}}-\frac{N_2}{\beta_{\mathrm{P}}}\right)+\frac{I_{\mathrm{L}}\sigma_{\mathrm{L}}}{h\nu_{\mathrm{L}}}\cdot\left(N_{\mathrm{T}}-\frac{N_2}{\beta_{\mathrm{L}}}\right)-\frac{N_2}{\tau'_{\mathrm{F}}}-\frac{N_2^2}{N_{\mathrm{T}}\tau'_{\mathrm{Q}}} \tag{5.28}$$

which is identical to Eq. 5.19. The definitions of the primed lifetimes fall out of the spectral integral

$$\frac{1}{\tau'_{\mathrm{F}}}\equiv\frac{1}{\tau_{\mathrm{F}}}\cdot\frac{\int g'(\lambda)\,d\lambda}{\int g(\lambda)\,d\lambda}\quad\text{and}$$

$$\frac{1}{\tau'_{\mathrm{Q}}}\equiv\frac{1}{\tau_{\mathrm{Q}}}+\frac{c\tau_{\mathrm{C}}}{n\tau_{\mathrm{F}}}\cdot\frac{\int\frac{\alpha(\lambda)}{\beta(\lambda)}g'(\lambda)\,d\lambda}{\int g(\lambda)\,d\lambda}. \tag{5.29}$$

We recognize that τ_F' is the effective fluorescence lifetime of the radiatively trapped system. As expected, τ_F' increases from τ_R as the fluorescence reabsorption increases. Note that τ_F' is independent of the excited-state density, so it can be measured using fluorescence from the entire crystal. Further, τ_F' depends on the geometry of the crystal cavity only through the parameter τ_C. This means that a measurement of the effective fluorescence lifetime uniquely determines the crystal cavity lifetime. The effective lifetime τ_Q' combines the nonexponential decay from radiation trapping and quadratic quenching. Here increased fluorescence reabsorption decreases τ_Q', leading to faster excitation decay. If the value of τ_C is known, the integral in Eq. 5.29 can be computed and the actual nonradiative lifetime can be extracted from the measurements of τ_Q'.

With these definitions the effective fluorescence line shape function becomes

$$g'(\lambda) = \frac{g(\lambda)}{[1+\alpha(\lambda)c\tau_C/n]}. \tag{5.30}$$

The effective line shape gives the rate of spontaneous emission photons with wavelength λ that will escape from the laser medium.

These expressions make it clear that fluorescence trapping impacts all aspects of laser performance. Trapping increases the energy storage lifetime and excitation densities. This leads to higher gain and more stimulated emission rates. Stimulated emission competes with fluorescence, so optical cooling is decreased. Increasing the quadratic quenching rate also decreases cooling. Trapping diminishes and red-shifts the escaping fluorescence that also reduce cooling efficiency.

In the earlier section we solved for a steady-state solution. For completeness we can also solve for an analytic time-dependent solution. When the pump and laser intensities are turned on and constant after $t = 0$,

$$N_2(t) = N_{SS} \cdot \frac{1-\exp(-A\cdot t)}{1+B\cdot\exp(-A\cdot t)} \tag{5.31}$$

with the following definitions:

$$A \equiv \sqrt{C^2 + 4(i_P\beta_P + i_L\beta_L)/\tau_Q'\tau_R}, \quad C \equiv (i_P + i_L)/\tau_R + 1/\tau_F', \\ B \equiv (A-C)/A+C \quad \text{and} \quad N_{SS} = N_T\cdot(A-C)\cdot\tau_Q'/2. \tag{5.32}$$

This local solution can now be readily extended to the entire laser medium using a numerical grid evaluation.

It is also useful to solve for the special case of $I_P = I_L = 0$. For an initial excited state density, N_0, the solution to Eq. 5.28 is the simple decay function

$$N_2(t) = N_0 \cdot \left[\left(1 + \frac{N_0 \tau'_F}{N_T \tau'_Q} \right) \exp\left(t/\tau'_F \right) - \frac{N_0 \tau'_F}{N_T \tau'_Q} \right]^{-1}. \tag{5.33}$$

This function can be fit to measured fluorescence waveforms from the laser medium in order to determine the effective lifetimes of a specific system. The initial excited-state density needed for the fit can be calculated from the laser pump conditions.

References

1. Abate, J. A., Lund, L., Brown, D., Jacobs, S., Refermat, S., Kelly, J., Gavin, M., Waldbillig, J., Lewis, O. (1981). Active mirror: a large-aperture medium repetition rate Nd:glass amplifier, *Appl. Opt.*, **20**(2), pp. 351–361.
2. Albrecht, G., Sutton, S., George, V., Sooy, W., Krupke, W. (1998). Solid state heat capacity disk laser, *Laser Part. Beams*, **16**, pp. 605–625.
3. Albrecht, G. F., Eggleston, J. M., Ewing, J. J. (1986). Design and characterization of a high average power slab YAG laser, *IEEE J. Quantum Electron.*, **QE-22**, pp. 2099–2106.
4. Beach, R. J. (1996). CW theory of quasi-three level end-pumped laser oscillators, *Opt. Commun.*, **123**, pp. 385–393.
5. Biswal, S., O'Connor, S. P., Bowman, S. R. (2006). Non-radiative losses in Yb:KGd$(WO_4)_2$ and Yb:$Y_3Al_5O1_2$, *Appl. Phys. Lett.*, **89**(9), pp. 0919111–0919113.
6. Bourdet, G. L. (2000). Theoretical investigation of quasi-three-level longitudinally pumped continuous wave lasers, *Appl. Opt.*, **39**(6), pp. 966–971.
7. Bowman, S. R. (2015). Optimizing average power in low quantum defect lasers, *Appl. Opt.*, **54**(31), pp. F78–F84.
8. Bowman, S., Ganem, R. J., Brown, C. G. (2014). Optical cooling in multi-level systems, in *Proceedings of the Laser Refrigeration of Solids VII*, SPIE 90000G, doi:10.1117/12.2040073.
9. Bowman, S. R., O'Connor, S., Condon, N. J., Friebele, E. J., Kim, W., Shaw, B., Quimby, R. S. (2013). Non-radiative decay of holmium-doped laser

materials, in *Proceedings of the Laser Refrigeration of Solids VI,* SPIE 863803, doi:10.1117/12.2007511.

10. Bowman, S. R., O'Connor, S. P., Biswal, S., Condon, N. J., Rosenberg, A. (2010). Minimizing heat generation in solid-state lasers, *IEEE J. Quantum Electron.*, **46**(7), pp. 1076–1085.
11. Bowman, S. R., O'Connor, S., Biswal, S. (2005). Ytterbium laser with reduced thermal heating, *IEEE J. Quantum Electron.,* **41**(12), pp. 1510–1517.
12. Bowman, S. R., Jenkins, N. W., O'Connor, S., Feldman, B. J. (2002). Sensitivity and stability of a radiation balanced laser system, *IEEE J. Quantum Electron.,* **38**(10), pp. 1339–1348.
13. Bowman, S. R., Mungan, C. (2000). New materials for optical cooling, *Appl. Phys. B,* **71**, pp. 807–811.
14. Bowman, S. R. (1999). Lasers without internal heat generation, *IEEE J. Quantum Electron.*, **35**, pp. 115–122.
15. Bowman, S. R., Quarles, G. J., Lynn, J. G., Feldman, B. J. (1993). A comparative study of diode pumped two micron laser materials, *OSA Proc. Adv. Solid-State Lasers,* **15**, pp. 415–418.
16. Bowman, S. R., Feldman, B. J. (1992). Demonstration and analysis of a holmium quasi-two level laser, *Proc. SPIE,* **1627**, p. 46.
17. Brignon, A., Huignard, J. (2004). *Phase Conjugate Optics* (John Wiley & Sons).
18. Budni, P. A., Lemons, M. L., Mosto, J. R., Chicklis, E. P. (2000). High-power/high-brightness diode-pumped 1.9 μm thulium and resonantly pumped 2.1 μm holmium lasers, *IEEE J. Sel. Top. Quantum Electron.*, **6**(4), pp. 649–635.
19. Byer, R. L., Basu, S. (1989). Moving slab laser, US Patent 4833682.
20. Chernoch, J. P., Martin, W. S. (1972). Multiple internal reflection face-pumped laser, US Patent 3633126.
21. Chernoch, J. P., Koenig, H. R. (1969). Disk laser having pumping means in direct optical combination with the disk end faces, US Patent 3423693.
22. Condon, N. J., Bowman, S. R., O'Connor, S. P., Myers, M. J. (2010). Heat loads in erbium-doped laser materials, *Opt. Mater.,* **32**, pp. 1050–1054.
23. Condon, N. J., Bowman, S. R., O'Connor, S. P., Quimby, R. S., Mungan, C. E. (2009). Optical cooling in Er3+:KPb2Cl5, *Opt. Express,* **17**, pp. 5466–5472.

24. Creeden, D., Johnson, B. R., Setzler, S. D., Chicklis, E. P. (2014). Resonantly pumped Tm-doped fiber laser with >90% slope efficiency, *Opt. Lett.*, **39**, pp. 470–473.

25. Eggleston, J., Kane, T., Kuhn, K., Unternahrer, J., Byer, R. L. (1984). The slab geometry laser 1. Theory, *IEEE J. Quantum Electron.*, **20**(3), pp. 289–301.

26. Fan, T. Y., Ripin, D. J., Aggarwal, R. L., Ochoa, J. R., Chann, B. Tilleman, M., Spitzberg, J. (2007). Cryogenic Yb3+-doped solid-state lasers, *IEEE J. Select. Top. Quantum Electron.*, **13**(3), pp. 448–458.

27. Fan, T. Y. (1992). Optimizing the efficiency and stored energy in quasi-three-level lasers, *IEEE J. Quantum Electron.*, **28**, pp. 2692–2697.

28. Fernandez, J., Garcia-Adeva, A. J., Balda, R. (2006). Anti-Stokes laser cooling in bulk erbium-doped materials, *Phys. Rev. Lett.*, **97**, pp. 0330011–0330014.

29. Fowler, W. B., Dexter, D. L. (1962). Relation between absorption and emission probabilities in luminescent centers in ionic solids, *Phys. Rev.*, **128**(5), pp. 2154–2165.

30. Friebele, E. J. , Askins, C. G., Peele, J. R., Wright, B. M., Condon, N. J., O'Connor, S., Brown, C. G., Bowman, S. R. (2014). Ho-doped fiber for high energy laser applications, *Proc. SPIE,* **8961**, pp. 8961201–8961209.

31. Gapontsev, D. (2008). Recent progress on high power fiber lasers, in *11th Annual Directed Energy Symposium,* DEPS 1–27.

32. Garbuzova, D., Kudryashov, D. I., Dubinskii, M. (2005). 110 W, 0.9 J pulsed power from resonantly diode-laser-pumped 1.6-μm Er:YAG laser, *Appl. Phys. Lett.,* **87**, pp. 1211011–1211013.

33. Giesen, A., Speiser, J. (2007). Fifteen years of work on thin-disk lasers: results and scaling laws, *IEEE J. Sel. Top. Quantum Electron.*, **13**(3), p. 598.

34. Giesen, A., Hugel, H., Voss, A., Witting, K., Brauch, U., Opower, H. (1994). Scalable concept for diode-pumped high-power solid-state lasers, *Appl. Phys. B,* **58**, pp. 365–372.

35. Guhur, A., Jackson, S. D. (2010). Efficient holmium-doped fluoride fiber laser emitting 2.1 μm and blue upconversion fluorescence upon excitation at 2 μm, *Opt. Express,* **18**(19), pp. 20164–20169.

36. Hehlen, M. P., Epstein, R. I., Inoue, H. (2007). Model of laser cooling in the Yb3+-doped fluorozirconate glass ZBLAN, *Phys. Rev. B,* **75**, pp. 144302–144313.

37. Hemming, A., Simakov, N., Davidson, A., Bennetts, S., Hughes, M., Carmody, N., Davies, P., Corena, L., Stepanov, P., Haub, J., Swain, R., Carter, A. (2013). Monolithic cladding pumped holmium-doped fibre laser, *IEEE Conference on Lasers and Electro-Optics*, CW1M.1.

38. Hoyt, C. W., Sheik-Bahae, M., Epstein, R. I., Edwards, B. C., Anderson, J. E. (2000). Observation of anti-Stokes fluorescence cooling in thulium-doped glass, *Phys. Rev. Lett.,* **85**, pp. 3600–3603.

39. Injeyan, H., Goodno, R. (2011). *High Power Laser Handbook* (McGraw-Hill).

40. Jauregui, C., Eidam, T., Limpert, J., Tunnermann, A. (2011). The impact of modal interference on the beam quality of high-power fiber amplifiers, *Opt. Express,* **19**, pp. 3258–3271.

41. Kane, T. J., Eggleston, J. M., Byer, R. L. (1985). The slab geometry laser: part II; thermal effects in a finite slab, *IEEE J. Quantum Electron.*, **QE-21**, pp. 1195–1210.

42. Kim, W., Bowman, S. R., Baker, C., Florea, C., Villalobos, G., Shaw, B., Sadowski, B., Hunt, M., Aggarwal, I., Sanghera, J. (2014). Holmium doped laser materials for eye-safe solid state laser application, *Proc. SPIE,* **90814**, pp. 9081051–9081056.

43. Klopp, P., Petrov, V., Griebner, U. (2003). Potassium ytterbium tungstate provides the smallest laser quantum defect, *Jpn. J. Appl. Phys.,* **42**, pp. 246–248.

44. Lacovara, P., Wang, C. A., Choi, H. K., Aggarwal, R. L., Fan, T. Y. (1991). Room-temperature InGaAs diode-pumped Yb-YAG laser, *OSA Proc. Adv. Solid-State Lasers*, **10**, pp. 283–285.

45. Larat, C., Schwarz, M., Lallier, E., Durand, E. (2014). 120 mJ Q-switched Er:YAG laser at 1645 nm, *Opt. Express,* **22**(5), pp. 4861–4866.

46. Li, C., Liu, Q., Gong, M., Chen, G., Yan, P. (2004). Modeling of radiation-balanced continuous-wave laser oscillators, *J. Opt. Soc. Am. B,* **21**(3), pp. 539–542.

47. Loiko, P., Maria Serres, J., Mateos, X., Yumashev, K., Kuleshov, N., Petrov, V., Griebner, U., Aguiló, M., Díaz, F. (2015). In-band-pumped Ho:KLu(WO4)2 microchip laser with 84% slope efficiency, *Opt. Lett.,* **40**(3), pp. 344–347.

48. Mandal, A., Klimek, D. E. (2010). Textron's J-HPSSL 100 kW ThinZag, *IEEE Conference on Lasers and Electro-Optics*, JThH2.

49. Marmo, J., Injeyan, H., Komine, H., McNaught, S., Machan, J., Sollee, J. (2009). Joint high power solid state laser program advancements at Northrop Grumman, *Proc. SPIE,* **7195**, pp. 7195071–7195076.

50. McCumber, D. E. (1964). Einstein Relations connecting broadband emission and absorption spectra, *Phys. Rev.,* 136(4A), pp. 954–957.

51. Nemova G., Kashyap, R. (2009). Athermal continuous-wave fiber amplifier, *Opt. Commun.,* **282**, pp. 2571–2575.

52. Newell, T. C., Peterson, P., Gavrielides, A., Sharma, M. P. (2007). Temperature effects on the emission properties of Yb-doped optical fibers, *Optics Commun.,* **273**, pp. 256–259.

53. O'Connor, S. P., Biswal, S., Condon, N., Bowman, S. R. (2011). Demonstration and analysis of a high power radiation balanced laser, *IEEE Conference on Lasers and Electro-Optics*, CMH4.

54. Oulundsen, G., Carter, A., Samson, B., Bowman, S. R. (2012). Holmium-doped silica fiber designs extend fiber lasers beyond 2 μm, *Laser Focus World,* **48**(8), pp. 47–50.

55. Petermann, K., Huber, G., Fornasiero, L., Kuch, S., Mix, E., Peters, V., Basun, S. A. (2000). Rare earth–doped sesquioxides, *J. Lumin.,* **87**, pp. 973–975.

56. Petit, J., Viana, B., Goldner, P., Vivien, D., Louiseau P., Ferrand, B. (2004). Laser oscillation with low quantum defect in Yb:$GdVO_4$, a crystal with high thermal conductivity, *Opt. Lett.,* **29**(8), pp. 833–835.

57. Pollnau, M., Hardman, P. J., Kern, M. A., Clarkson, W. A., Hanna, D. C. (1998). Upconversion-induced heat generation and thermal lensing in Nd:YLF and Nd:YAG, *Phys. Rev. B,* **58**(24), pp. 16076–16091.

58. Quimby, R. S., Condon, N. J., O'Connor, S. P., Biswal, S., Bowman, S. R. (2008). Upconversion and excited-state absorption in the lower levels of Er:KPb2Cl5, *Opt. Mater.,* **30**(6), pp. 827–834.

59. Richardson, D. J., Nilsson, J., Clarkson, W. A. (2010). High power fiber lasers: current status and future perspectives, *J. Opt. Soc. Am. B,* **27**(11), pp. 63–92.

60. Sanghera, J., Kim, W., Villalobos, G., Shaw, B., Baker, C., Frantz, J., Sadowski, B., Aggarwal, I. (2012). Ceramic laser materials, *Materials,* **5**, pp. 258–277.

61. Shen, D. Y., Sahu, J. K., Clarkson, W. A. (2006). Highly efficient in-band pumped Er:YAG laser with 60 W of output at 1645 nm, *Opt. Lett.,* **31**, pp. 754–756.

62. Smith, A. V., Smith, J. J. (2011). Mode instability in high power fiber amplifiers, *Opt. Express.,* **19**, pp. 10180–10192.

63. Stoneman, R. C., Esterowitz, L. (1992). Intracavity-pumped 2.09-,um Ho:YAG laser, *Opt. Lett.,* **17**(10), pp. 736–738.

64. Takaichi, K., Yagi, H., Shirakawa, A., Ueda, K., Hosokawa, S., Yanagitani, T., Kaminskii, A. A. (2004). Lu2O3:Yb3+ceramics a novel gain material for high-power solid-state lasers, *Phys. Status Solidi A,* **202**, pp. R1–R3.

65. Ward, B., Robin, C., Dajani, I. (2012). Origin of thermal modal instabilities in large mode area fiber amplifiers, *Opt. Express,* **20**, pp. 11407–11422.

66. White, J. O., Dubinskii, M., Merkle, L. D., Kudryashov, I., Garbuzov, D. (2007). Resonant pumping and upconversion in 1.6 μm Er3+ lasers, *J. Opt. Soc. Am. B,* **24**(9), pp. 2454–2460.

67. Wright, J. (1976). Up-conversion and excited state energy transfer in rare-earth doped materials, *Radiationless Processes in Molecules and Condensed Phases, Vol. 15 of Topics in Applied Physics,* Chapter 4, pp. 239–295.

68. Yao, T., Ji, J., Nilsson, J. (2014). Ultra-low quantum-defect heating in ytterbium-doped aluminosilicate fibers, *J. Lightwave Tech.,* **32**(3), pp. 429–434.

Chapter 6

Methods for Laser Cooling of Solids

Stephen C. Rand
MURI Center for Dynamic Magneto-Optics, Appl. Physics & EECS Optics & Photonics Laboratory, 3102 ERB1, 2200 Bonisteel Blvd., University of Michigan, MI, USA
scr@umich.edu

6.1 Introduction

The topic of laser cooling began with the articles of Pringsheim [1] and Hansch [2] long before any laboratory experiments were undertaken. These early papers proposed the use of optical radiation to remove energy from matter for the purpose of refrigeration, but the mere idea that light could transport thermal energy was widely doubted until Landau's thermodynamic analysis showed that radiation fields had entropy that depended on the statistics and mode structure of the field [3]. The influence of various states of the field on solid state cooling was examined in detail later [4], but once it was realized that entropy of fields could be exchanged for entropy of matter, the stage was already set for a concerted worldwide effort to alter the temperature of collections of atoms with light.

Experiments to observe laser cooling of atomic motion in gases originated from investigations of the mechanical effects of

Laser Cooling: Fundamental Properties and Applications
Edited by Galina Nemova

ISBN 978-981-4745-04-8 (Hardcover), 978-981-4745-05-5 (eBook)
www.panstanford.com

radiation pressure on small dielectric particles [5]. Similarly, optical refrigeration of solids was preceded by demonstrations that the amplitude of axial vibrations of mirrors could be reduced in Fabry–Perot resonators [6]. Reductions of this kind were accomplished by monitoring excursions of the cavity from resonance and using the error signal to control mirror position with variations of light pressure applied to the back side of one of the cavity mirrors. However, this type of optomechanical feedback does not lead to overall refrigeration, because it does not reduce the amplitude of internal modes of vibration irreversibly. Some modes are heated, while others are cooled. To refrigerate a solid, light must do more than redistribute vibrational energy among the available modes. Ideally a low-entropy (single-mode) input field couples to one or more vibrational or translational modes, undergoes an increase in entropy during its interaction with the sample, and then carries that entropy away via radiative transport into space.

Laser cooling techniques for solids can build on past experience in atomic vapors. In gases the motion of atoms along a chosen direction can be exploited to shift "hot" atoms preferentially into resonance with light propagating opposite to their motion via the Doppler effect [7]. In this way the momentum of the light wave is transferred to fast-moving atoms selectively, gradually slowing them down through the accumulated effect of many opposing photon interactions. In condensed matter, however, free translational motion is absent. As a consequence, one might think that laser cooling in glasses or crystals has to exploit completely different principles. Yet there are analogies to the Doppler effect in solids and ways to address the energy and momentum of collective excitations of the medium that are reminiscent of interactions with free atoms. Hence this chapter reviews several methods with this perspective in mind, attempting to profit from the earlier history of laser cooling in gases and to provide points of comparison between different techniques.

Researchers face the challenge of developing the most effective means of controlling the temperature of macroscopic objects with light. Success on this front will make possible the thermal management of hot spots in micromechanical or integrated electronic devices, improved operation of high-power lasers through radiation balancing, and low-temperature operation of active devices in vacuum where cooling improves sensitivity but is only possible through radiative transport.

6.2 Cooling with Anti-Stokes Fluorescence

The earliest demonstration of optical refrigeration of a solid relied on inducing anti-Stokes fluorescence by simply tuning the wavelength of incident light to the "red" or low-energy side of an impurity resonance [8]. When vibronic coupling of the medium is weak, the detuning is into the spectral wing of an essentially pure electronic transition. When the vibronic coupling is strong, a condition characterized by a large Huang–Rhys factor [9], the light is detuned into the phonon sideband associated with the electronic transition. Then, consistent with quantum theory [10], the absorption of photons on the low-energy side of resonance is followed on average by photon emission at the center frequency of the transition. In this circumstance an overall loss of energy is imposed even on stationary atoms (Fig. 6.1). Hence this general approach has met with considerable success in demonstrations of cooling in rare earth solids [1], liquids [12], semiconductors [13], and internal acoustic modes of optomechanical resonators [14].

The efficiency of cooling that relies on inducing anti-Stokes fluorescence can easily be estimated. In the two-level system of Fig. 6.1, an amount of energy given by $h\nu$ is added to the system for each absorbed photon. Energy $h\nu_{\mathrm{fi}}$ is removed by each fluorescent photon. Hence if these two steps account for all system dynamics, then an input power of P_{in} produces a cooling power per absorbed photon of

$$P_{\mathrm{c}} = P_{\mathrm{in}}\left(\frac{\nu_{\mathrm{fl}} - \nu_{\mathrm{L}}}{\nu_{\mathrm{L}}}\right) \tag{6.1}$$

and the cooling efficiency would be $\eta = P_{\mathrm{c}}/P_{\mathrm{in}} = (\nu_{\mathrm{fi}} - \nu_{\mathrm{L}})\nu_{\mathrm{L}}$. However, in the presence of nonradiative relaxation, which is commonplace in rare earth solids, the radiative quantum efficiency is lowered. Radiative and nonradiative decay at rates W_{r} and W_{nr} then contribute an external efficiency factor of the form $\eta_{\mathrm{ext}} = W_{\mathrm{r}}/(W_{\mathrm{r}} + W_{\mathrm{nr}})$. When corrected for fluorescence escape efficiency η_{e}, which is the fraction of emitted photons that escape the sample, this factor becomes

$$\eta_{\mathrm{ext}} = \eta_{\mathrm{r}}W_{\mathrm{r}}/(\eta_{\mathrm{e}}W_{\mathrm{r}} + W_{\mathrm{nr}}) \tag{6.2}$$

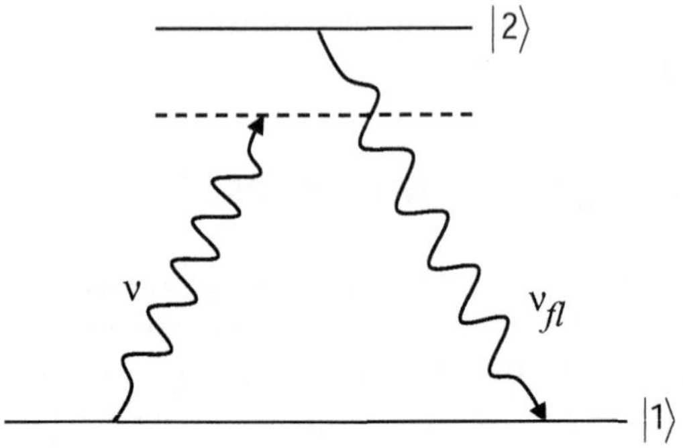

Figure 6.1 Anti-Stokes fluorescence emission in a two-level system.

Additionally, only a fraction of the absorbed photons result in excitation of coolant atoms. The rest are absorbed by background impurities. With the absorption coefficient of the coolant atoms designated by α_c and that of the background by α_b, the proportion of absorbed photons that contribute to cooling is

$$\eta_{abs} = \frac{\alpha_c}{(\alpha_c + \alpha_b)} \tag{6.3}$$

If the assumption is made that phonon-mediated absorption is proportional to the occupation probability of the phonon mode, namely $\langle n \rangle = (\exp[\hbar\Omega/k_B T]\text{-}1)^{-1}$, the efficiency becomes $\eta_{abs} = \alpha_c \langle n \rangle / (\alpha_c \langle n \rangle + \alpha_b)$. Then at high temperatures ($k_B T >> \hbar\Omega$)), η_{abs} varies little with temperature when the background absorption is low ($\alpha_b/\alpha_c << 1$). On the other hand at low temperatures, as $\langle n \rangle \to 0$, η_{abs} decreases in proportion to $\langle n \rangle$. Hence the absorption efficiency drops dramatically through the important cryogenic regime. The overall cooling is determined by the product of the various efficiency factors.

$$\eta_c = \eta_{ext}\eta_{abs}\left(\frac{\nu_{fl} - \nu}{\nu}\right) \tag{6.4}$$

From Eqs. 6.1 and 6.4 it is easy to show that the cooling power density per unit frequency is $P_c/V = \eta_{abs}\eta_{ext}\alpha_{tot}I_\nu(\nu_{fl} - \nu)/\nu$. Here $I_\nu = I_0 g(\nu)$ is the intensity per unit frequency and $g(\nu)$ is a normalized spectral line shape function. V is the interaction volume. If the bandwidth of the light source equals the full linewidth 2Γ of the optical transition, and the density of atoms in the interaction volume is N/V, the energy loss rate per atom is

$$R = 2\Gamma\eta_c\alpha_{tot}I_\nu(V/N) = \eta_c\alpha_{tot}I_0(V/N) \quad (6.5)$$

The condition for net cooling can be expressed in a simple way based on Eq. 6.4, since η_c must exceed zero. By assuming that the laser detuning is optimized at the phonon frequency Ω, we can rewrite the numerator using $h\nu_{ft} - h\nu = \hbar\Omega$. Furthermore the average phonon energy is $\hbar\Omega \approx k_B T$ near the boundary of the classical regime [15]. Then the condition for laser cooling ($\eta_c > 0$) becomes

$$\eta_{ext}\eta_{abs} > 1 - \frac{k_B T}{h\nu} \quad (6.6)$$

An important limitation of refrigeration based on anti-Stokes fluorescence emerges from this condition by determining the temperature at which the cooling efficiency drops to zero ($\eta_c = 0$). By solving Eq. 6.6 as an equality, it is found that a lower bound exists for the attainable temperature. The minimum temperature T_{min} that can be reached with anti-Stokes cooling found from Eq. 6.6 is

$$T_{min} = \frac{h\nu}{k_B}\left(\frac{(\alpha_b/\alpha_c) + (1 - \eta_{ext})}{1 + (\alpha_b/\alpha_c)}\right) \quad (6.7)$$

According to Eq. 6.7, if the quantum efficiency is high ($n_{ext} \approx 1$), the temperature limit for laser cooling by anti-Stokes emission is primarily determined by the frequency of light and the background-to-resonant-absorption ratio. The dependence on these two factors arises because heat input to the system is proportional to incident photon energy and absorptive heating efficiency. Only when the background absorption coefficient vanishes could arbitrarily low temperatures be reached. For the moderate levels of background absorption encountered in practice in solids, this places a severe constraint on temperatures that can be reached by laser cooling. As a simple example of this, the anti-Stokes cooling limit predicted by Eq. 6.7 at a wavelength of 1 micron is T_{min} = 145.3 K when the external quantum efficiency is $\eta_{ext} = 0.90$ and background absorption coefficient is $\alpha_b = \alpha_c/1000$.

The lower bound on attainable temperatures can be interpreted in another way. It arises because the cooling rate depends on phonon occupation, a factor that diminishes as temperature decreases. Eventually the rate of anti-Stokes cooling drops below the rate of heating from background impurity absorption and cooling is no longer possible [16]. For 3D refrigeration of solids to be as effective as laser cooling of gases, techniques capable of maintaining high rates of cooling through the cryogenic temperature range are needed. This is discussed further in what follows.

In the remainder of this chapter momentum is incorporated into the discussion to investigate whether the fundamental temperature limitation of anti-Stokes fluorescence cooling can be evaded when the dynamics are viewed comprehensively. Judging from the successes of laser cooling of gases, cooling schemes that conserve momentum and energy simultaneously during optical interactions are the most efficient. Yet momentum is disregarded in the anti-Stokes fluorescence method. Other techniques may offer the capability of balancing the momentum and energy requirements more consistently to accommodate phonon dispersion (Fig. 6.2).

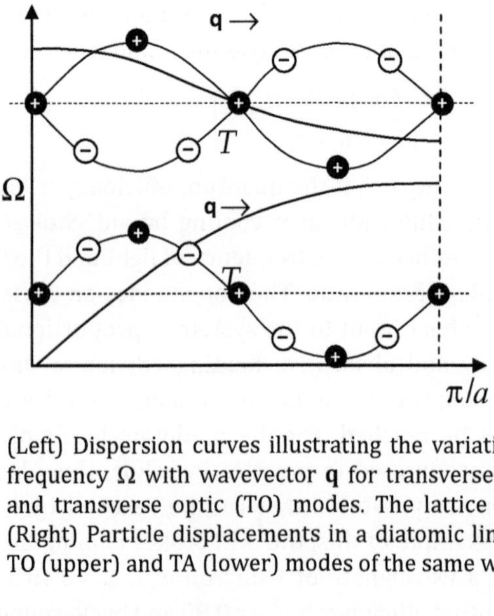

Figure 6.2 (Left) Dispersion curves illustrating the variation of phonon frequency Ω with wavevector **q** for transverse acoustic (TA) and transverse optic (TO) modes. The lattice constant is *a*. (Right) Particle displacements in a diatomic linear lattice for TO (upper) and TA (lower) modes of the same wavelength.

6.3 Brillouin Cooling

In gases the link between linear momentum and the absorption of a photon is simple. For each photon absorbed, the linear momentum of an approaching atom decreases by an average of $\hbar k$. Translational momentum is thereby reduced, provided fluorescence randomizes the direction of reemitted photons. In solids things are not so simple because free translation of atoms cannot take place. Atomic displacements are oscillatory. Motion is conveyed in collective waves and momentum is distributed, as illustrated in Fig. 6.2. Nevertheless, charge motions associated with departures of atoms from their equilibrium positions have Fourier components with a well-defined amplitude, momentum, and energy [11]. Moreover, Doppler-shifted interactions with these phonon modes are well-known to take place in acousto-optic (Brillouin) scattering [15]. So analogies to Doppler cooling methods in gases are useful. At this point we therefore formulate the problem of light interacting with acoustic and optical modes of vibration in terms of interaction Hamiltonians and equations of motion that display momentum explicitly. Two recent developments are reviewed. The most detailed consideration is given to laser cooling by stimulated Raman scattering because in theory it furnishes a cooling rate that is independent of temperature.

The energy of interaction between light and a solid derives from the electron–phonon interaction in which the force of light acts to displace ions from equilibrium by an amount $\delta R^*(t)$. This displacement can be written as a Fourier expansion in terms of normal modes of polarization $^*\varepsilon_{k\lambda}$ ($\lambda = 1,2,3$), amplitude $Q_{q\lambda}$, wavevector $\mathbf{q}$, and frequency $\Omega_{q\lambda}$. For an ion situated at equilibrium position $\bar{R}_0$ in a solid, undergoing motion described by a single-phonon mode, the displacement consists of only one Fourier term.

$$\delta\bar{R}(t) = (NM)^{-1/2} Q(\Omega,t)\hat{\varepsilon}\exp[i\bar{q}\cdot\bar{R}_0] \tag{6.8}$$

$\hat{\varepsilon}$ is the polarization vector, N is the number of particles, and M is the particle mass. The form of the interaction Hamiltonian is given by force × distance.

$$H_{\mathrm{I}} = -\bar{\nabla}V\cdot\delta\bar{R} \tag{6.9}$$

The quantized form of H_{I} depends on the coupling mechanism between light and the vibrational mode of interest, so the cases of acoustic and optical vibrational modes are considered separately below.

For *transverse acoustic modes,* which are propagating density fluctuations, the light–matter interaction is determined by electrostrictive deformations of the medium because transverse optical fields cannot couple directly to longitudinal waves. The energy dependence of acoustic and optical modes on the wavevector is shown in Fig. 6.2. To generate pressure waves, a deformation potential [17] is needed of the form

$$\bar{\nabla} V = -\gamma \bar{\nabla}(\bar{E}_1 \cdot \bar{E}_2^*) = -\gamma q E_1 E_2^* \hat{\varepsilon} \quad (6.10)$$

Here

$$E_1(z,t) = \xi_k(\hat{a}_1^- e^{-i(\omega_1 t - kz)} + \hat{a}_1^+ e^{i(\omega_1 t - kz)}) \quad (6.11)$$

and

$$E_2^*(z,t) = \xi_{k+q}(\hat{a}_2^+ e^{i(\omega_2 t - [k+q]z)} + \hat{a}_2^- e^{-i(\omega_2 t - [k+q]z)}) \quad (6.12)$$

are quantized pump and anti-Stokes fields, $\xi_k = \sqrt{\hbar\omega_k / 2\varepsilon_0 V}$ is the field per photon, and $\gamma = [\rho(\partial\varepsilon / \partial\rho]_{\rho 0}$ is the electrostrictive constant relating changes of permittivity to changes of density [18]. The operator for normalized mode amplitude is related to the phonon creation ($\hat{b}^+$) and annihilation ($\hat{b}^-$) operators by

$$\hat{Q}(\Omega,t) = \left(\frac{\hbar}{2\Omega}\right)^{1/2} (\hat{b}_q^- \exp[-i\Omega t] + \hat{b}_q^+ \exp[i\Omega t]) \quad (6.13)$$

When Eqs. 6.8 and 6.10–6.13 are combined with Eq. 6.9 the interaction Hamiltonian becomes

$$H_I = -(NM)^{-1/2}(\hat{\varepsilon} \cdot \bar{\nabla} V) Q(\Omega,t) \exp[i\bar{q} \cdot \bar{R}_0]$$
$$= \hbar f(\hat{b}_q^- a_{k+q}^+ \hat{a}_k^- + h.c.) \quad (6.14)$$

where the interaction strength is defined as $f^* = (2\hbar NM\Omega)^{-1/2} V\gamma q \xi_k \xi_{k+q}$. V is the volume of interaction. The conservation of momentum in the interaction is illustrated in Fig. 6.3.

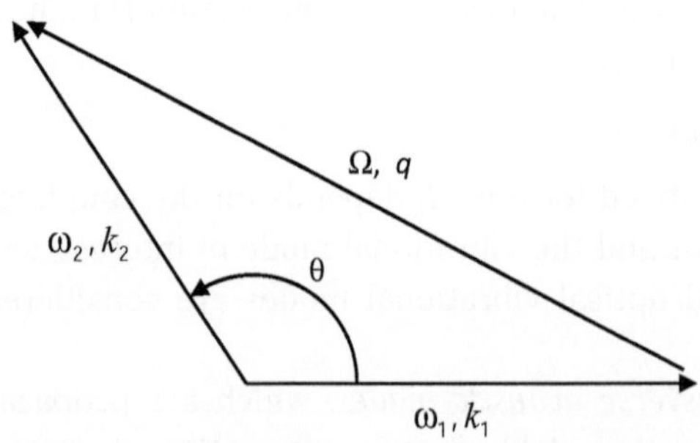

Figure 6.3 Wavevector diagram for phonon scattering through angle θ. Phonon frequency Ω and wavevector $\bar{q} = \bar{k}_2 - \bar{k}_1$ conserve energy and linear momentum in the wave interaction involving optical fields $\bar{E}_1(\omega_1)$ and $\bar{E}_2(\omega_2)$.

The Hamiltonian can be used to calculate temporal changes in the vibrational energy (or temperature) of the medium. To appreciate the connection between H_I and vibrational energy or temperature, note that the conjugate momentum of the vibrational amplitude $\hat{Q}(\Omega,t)$ is

$$\hat{P}(\Omega,t) = -i\left(\frac{\hbar\Omega}{2}\right)^{1/2}(\hat{b}\exp[-i\Omega t] - \hat{b}^{+}\exp[i\Omega t]) \quad (6.15)$$

When $\hat{Q}$ and $\hat{P}$ are defined as above, the single-mode Hamiltonian specifying vibrational energy reduces to a familiar form.

$$H_{q\lambda} = \tfrac{1}{2}(\hat{P}^{+}_{q\lambda}\hat{P}_{q\lambda} + \Omega^{2}_{q\lambda}\hat{Q}^{+}_{q\lambda}\hat{Q}_{q\lambda}) = \tfrac{1}{2}\hbar\Omega_{q\lambda}(\hat{b}^{+}_{q\lambda}\hat{b}^{-}_{q\lambda} + h.c.) \quad (6.16)$$

Taking all possible modes into account, the total energy is therefore found to be

$$H = \sum_{q,\lambda} H_{q\lambda} = \sum_{q,\lambda} \hbar\Omega_{q\lambda}(\hat{b}^{+}_{q\lambda}\hat{b}^{-}_{q\lambda} + \tfrac{1}{2}) \quad (6.17)$$

Since $\left\langle \hat{n}_{q\lambda} \right\rangle = \left\langle \hat{b}^{+}_{q\lambda}\hat{b}^{-}_{q\lambda} \right\rangle$ is the number of phonons in mode q, the summation in Eq. 6.17 describes the total vibrational energy of the medium as a weighted sum of the individual mode energies which in turn determines the sample temperature.

With the interaction Hamiltonian of Eq. 6.14 in hand, optically driven dynamics of two optical waves interacting with an acoustic mode can be analyzed. The static Hamiltonian $H_0 = \hbar\omega_1\hat{a}^{+}_1\hat{a}^{-}_1 + \hbar\omega_2\hat{a}^{+}_2\hat{a}^{-}_2 + \hbar\Omega\hat{b}^{+}\hat{b}^{-}$ plays no role in steady-state dynamics. For example, the change of phonon occupation with time can be calculated using the Heisenberg equation of motion

$$\begin{aligned}\dot{\hat{n}}_b &= (i/\hbar)[H_0 + H_{\text{int}}, \hat{n}] \\ &= (i/\hbar)\left\{\hbar f(\hat{b}^{-}\hat{a}^{+}_2\hat{a}^{-}_1 + h.c.), \hat{b}^{+}\hat{b}^{-}\right\} \\ &= \text{if}\left\{\hat{b}^{-}\hat{a}^{+}_2\hat{a}^{-}_1(\hat{b}^{+}\hat{b}^{-}) - \hat{b}^{+}(\hat{b}^{-}\hat{b}^{+})\hat{a}^{-}_2\hat{a}^{+}_1\right\}\end{aligned} \quad (6.18)$$

The first term on the right determines the rate of annihilation of phonons, while the second term is the rate of creation. We can show this explicitly by solving for the temporal behavior of $\hat{b}^{+}$ and $\hat{b}^{-}$, again using the Heisenberg picture.

$$(d\hat{b}^{+}/dt) = i\Omega\hat{b}^{+} + \text{i}f\hat{b}^{-}\hat{a}_1^{-}\hat{a}_2^{+}\hat{b}^{+} - \text{i}f\hat{b}^{+}\hat{a}_1^{+}\hat{a}_2^{-}\hat{b}^{-} - \Gamma_b\hat{b}^{+} \quad (6.19)$$

$$(d\hat{b}^{-}/dt) = -i\Omega\hat{b}^{-} - \text{i}f\hat{b}^{+}\hat{a}_1^{+}\hat{a}_2^{-}\hat{b}^{-} + \text{i}f\hat{b}^{-}\hat{a}_1^{-}\hat{a}_2^{+}\hat{b}^{+} - \Gamma_b\hat{b}^{-} \quad (6.20)$$

Making use of the slowly varying envelope approximation (SVEA), steady-state solutions for the operator amplitudes are simply

$$\hat{b}^{+} = \left\{\text{i}f\hat{b}^{-}\hat{a}_1^{-}\hat{a}_2^{+}\hat{b}^{+} - \text{i}f\hat{b}^{+}\hat{a}_1^{+}\hat{a}_2^{-}\hat{b}^{-}\right\}/\Gamma_b$$

$$\hat{b}^{-} = \left\{-\text{i}f\hat{b}^{+}\hat{a}_1^{+}\hat{a}_2^{-}\hat{b}^{-} + \text{i}f\hat{b}^{-}\hat{a}_1^{-}\hat{a}_2^{+}\hat{b}^{+}\right\}/\Gamma_b \quad (6.21)$$

With these substitutions one finds that the net rate of change of phonon occupation in the state $|n_1\rangle|n_2\rangle|n_b\rangle$ is determined by the balance of annihilation versus creation.

$$\left\langle \dot{\hat{n}}_b \right\rangle = \text{i}f\langle n_b|\langle n_1|\langle n_2|\left\{-\text{i}f\hat{b}^{+}\hat{a}_1^{+}\hat{a}_2^{-}\hat{b}^{-} + \text{i}f\hat{b}^{-}\hat{a}_1^{-}\hat{b}_2^{+}\hat{b}^{+}\right\}\hat{a}_2^{+}\hat{b}_1^{-}(\hat{b}^{+}\hat{b}^{-})|n_2\rangle|n_1\rangle|n_b\rangle/\Gamma_b$$
$$-\text{i}f\langle n_b|\langle n_1|\langle n_2|\left\{\text{i}f\hat{b}^{-}\hat{a}_1^{-}\hat{a}_2^{+}\hat{b}^{+} - \text{i}f\hat{b}^{+}\hat{a}_1^{+}\hat{a}_2^{-}\hat{b}^{-}\right\}\hat{a}_2^{-}\hat{a}_1^{+}(\hat{b}^{-}\hat{b}^{+})|n_2\rangle|n_1\rangle|n_b\rangle/\Gamma_b \quad (6.22)$$

Upon evaluation of the matrix elements we obtain the rate equation for phonon occupation.

$$\dot{n}_b = -(f^2/\Gamma_b)n_1(n_2+1)n_b + (f^2/\Gamma_b)(n_1+1)n_2(n_b+1) \quad (6.23)$$

The first term on the right side of Eq. 6.23 describes cooling, while the second describes heating. Possible detuning of the modes has been ignored in this analysis. The phonon annihilation rate for resonant Brillouin scattering in the interaction volume is therefore the coefficient of the first term in Eq. 6.23.

$$\gamma_B = (f^2/\Gamma_b)n_1(n_2+1)n_b \quad (6.24)$$

The energy loss rate takes into account the phonon frequency.

$$R = \gamma_B\hbar\Omega = (f^2/\Gamma_b)n_1(n_2+1)n_b\hbar\Omega \quad (6.25)$$

Laser cooling is realized if conversion of an optical pump wave to an anti-Stokes wave prevails over its conversion to a Stokes-shifted wave [19]. Brillouin cooling of a mechanical mode was first demonstrated in a microresonator by simultaneously matching its momentum and energy to the difference of wavevectors and frequencies of the pump and anti-Stokes waves [14]. To do this the pump and anti-Stokes waves were simultaneously enhanced in the resonator, while the Stokes component of Brillouin scattering was suppressed by wavevector mismatch.

6.4 Raman Cooling

In this section stimulated scattering is considered for the cooling of solids. The general theory of stimulated Brillouin and Raman scattering [20, 21] was developed in the 1960s, and despite the complexity of equations coupling the various mode amplitudes, Louisell [22] pointed out that the interaction energy is just the added electromagnetic energy resulting from a change in the permittivity ε, or $H_{\mathrm{I}} = \frac{1}{2}\int \delta\varepsilon E^2$. This perspective facilitates the analysis of both Brillouin and Raman scattering since it is straightforward to write $\delta\varepsilon$ to reflect one interaction or the other. In the last section the dielectric fluctuation in Brillouin scattering was written $\delta\varepsilon = -\mathrm{l}(\partial Q/\partial x)$to describe electrostriction by the incident light. In this section $\delta\varepsilon$ arises from dielectric fluctuations due to transverse charge motion caused directly by the light field itself. We shall proceed as before nevertheless, writing down the interaction Hamiltonian using Eq. 6.9, since the deformation potential approach is equivalent to that of Louisell.

For *transverse optical modes*, the net optical electric field can act directly on charged ions at the difference frequency to create or destroy optical phonons. In this case the potential is just the optical energy density at the mode frequency. Hence the gradient of the potential in Eq. 6.9 is $\bar{\nabla} V = \varepsilon_0 \bar{\nabla}(\bar{E}_1 \cdot \bar{E}_2^*)\hat{\varepsilon}$, and the single-mode interaction Hamiltonian is

$$H_{\mathrm{I}} = \left(\frac{\hbar\varepsilon_0^2}{2\Omega NM}\right)^{1/2} \xi_k \xi_{k+q} (\hat{a}_1^- \hat{a}_2^+ \hat{b}^- + h.c.)$$

$$= \hbar f'(\hat{a}_1^- \hat{a}_2^+ \hat{b}^- + h.c.) \tag{6.26}$$

where $f' = -(2\hbar NM\Omega)^{-1/2}\varepsilon_0 Vq\ \xi_k\ \xi_{k+q}$ is the coupling strength in the interaction volume V. By careful selection of the frequencies and wavevectors of two optical fields, the energy and momentum can be conserved for optical modes just as for acoustic modes (Fig. 6.3).

Two-photon Raman transitions (Fig. 6.4) can couple to optical modes of vibration purely through anti-Stokes scattering. Consequently they can be utilized for efficient cooling. Early analysis and experiments on Raman cooling may be found in Refs. [23–25]. Here we follow Ref. [26] to analyze how the Raman technique can be adapted to cooling solids and to estimate the cooling rate. The potential advantages of applying adiabatic rapid passage to this technique are also considered. The fact that the cooling rate is temperature independent is emphasized.

With pulsed excitation, the two fields at ω_1 and ω_2 in Fig. 6.4 cause the system population to evolve in the direction indicated by wiggly arrows. The light drives population from state 1 to state 3 at a rate determined by the one- and two-photon detunings Δ and δ. The $1 \rightarrow 2$ and $2 \rightarrow 3$ transitions are assumed to be electric-dipole-allowed. The selection rules therefore dictate that state 3 is a long-lived storage state. For prolonged excitation, population builds up in level 3 and the reverse process begins to take place at a comparable rate. For the two-photon detuning shown in the figure, a small energy deficit ($\hbar\delta$) and momentum reduction ($2\hbar k$) is incurred for each Raman cycle, provided the pulses are short enough to avoid substantial buildup of population [27] in state 3. Atoms with a velocity centered at $v = 2\hbar k/m$ lose kinetic energy. By emptying state 3 periodically with a fast, random process like fluorescence, changes in the energy and momentum distributions become irreversible through an increase in the entropy [3, 4] of the outgoing radiation field [28]. Hence the ensemble of atoms can be progressively cooled.

To estimate the stimulated Raman transition rate for the $1 \rightarrow 3$ transition it is instructive to develop a sense of the importance of *forward* and *backward* pumping processes between states 1 and 3 (Figs. 6.4 and 6.5, respectively). It is easily shown that the net rate of change of population is zero in the steady state because of the balance of these two processes, which provide cooling and heating, respectively. One can also show that the transition rate for a single process is given by the square of an off-diagonal (coherence) element of the density matrix that may be evaluated straightforwardly to allow estimation of cooling rates.

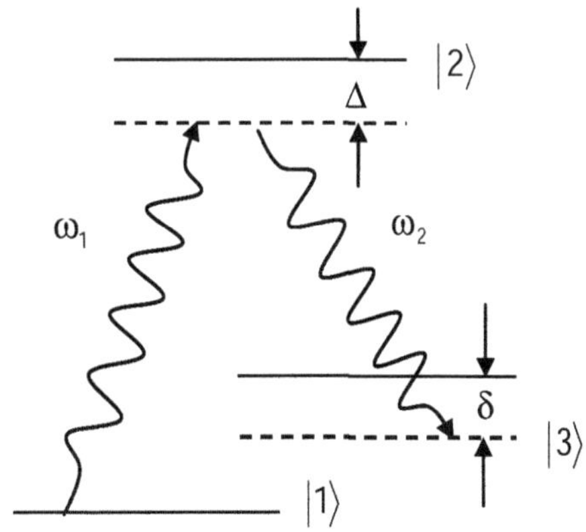

Figure 6.4 A three-level system undergoing a two-photon Raman process. $\Delta = \omega_{12} - \omega$ is the one-photon detuning and $\delta = \omega_{13} - (\omega_1 - \omega_2)$ is the two-photon detuning.

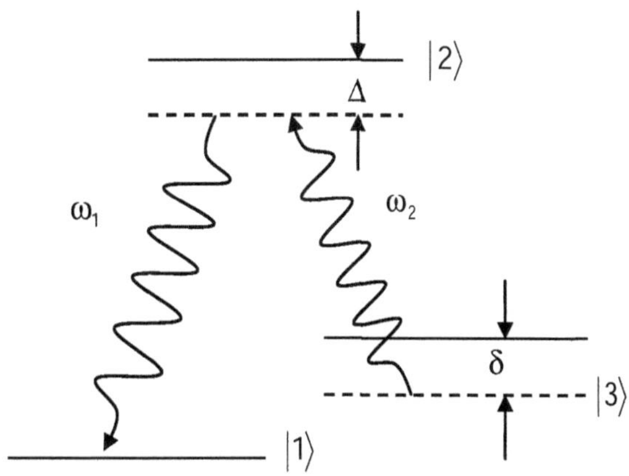

Figure 6.5 Raman interaction of fields $E_1(\omega_1)$ and $E_2(\omega_2)$ in a time-reversed (backward) sequence.

The net rate of population change in excited state 3 is just $\dot{\rho}_{33} \equiv d\rho_{33}/dt$. In the weak field limit the wave function is dominated by its ground-state component, which varies little with time. Hence we may assume $\rho_{11} = c_1 c_1^* \cong 1$ in order to calculate the transition rate

$$\dot{\rho}_{33} = \frac{d}{dt}\left[\langle 3|\psi\rangle\langle\psi|3\rangle\right] \cong \frac{d}{dt}\left[\langle 3|c_3|3\rangle\langle 1|c_1^* c_1|1\rangle\langle 3|c_3^*|3\rangle\right] = \dot{\rho}_{31}\rho_{13} + \rho_{31}\dot{\rho}_{13} \tag{6.27}$$

Since $\rho_{31} = \rho_{13}^*$ and $|\rho 31|$ = $|\rho_{13}|$, one finds $\dot{\rho}_{31} = \dot{\rho}_{13}^* = -i\omega_{13}\rho_{31}$ and $\dot{\rho}_{13} = i\omega_{13}\rho_{13}$. The result is that the two contributions to ρ_{33} given in Eq. 6.27 are equal and opposite. Under steady-state conditions the occupation of state 3 therefore does not change.

$$\dot{\rho}_{33} = -i\omega_{13}|\rho_{13}|^2 + i\omega_{13}|\rho_{13}|^2 = 0 \tag{6.28}$$

The rate at which the level 3 population increases is determined by the positive term in Eq. 6.28, assuming one vibrational quantum occupies the mode. The transition rate per atom per unit (angular frequency) excitation bandwidth that leads to cooling is therefore

$$|\dot{\rho}_{33} / \omega_{13}| = |\rho_{13}|^2 \tag{6.29}$$

To determine the rate itself from the field-driven coherence in Eq. 6.29 the source bandwidth must be specified. Taking this to equal the full transition linewidth 2 Γ_{13} in the discussion below, only the calculation of the matrix element ρ_{13} explicitly remains.

For quasi-, steady-state fields the interaction Hamiltonian can be written as

$$V = (-\tfrac{1}{2}\hbar\Omega_{12}\exp(i\omega_1 t) + c.c.) + (-\tfrac{1}{2}\hbar\Omega_{23}\exp(i\omega_2 t) + c.c.) \tag{6.30}$$

Contributions to state 3 dynamics other than the one depicted in Fig. 6.4 will be ignored, owing to their large detunings from resonance. The first-order coherence is therefore given by the standard first-order perturbation expression [29]:

$$\tilde{\rho}_{12}^{(1)} = \left(\frac{\Omega_{12}/2}{\Delta_1 + i\Gamma_{12}}\right)(\rho_{11}^{(0)} - \rho_{22}^{(0)}) \tag{6.31}$$

The second-order coherence established by the forward Raman process is obtained by solving the second-order transport equation

$$i\hbar\dot{\rho}_{13}^{(2)} = \hbar(\omega_1 - \omega_3)\rho_{13}^{(2)} + [V^{(1)}, \rho^{(1)}]_{13} - i\hbar\Gamma_{13}\rho_{13}^{(2)} \tag{6.32}$$

The two-photon coherence oscillates at the frequency of the (combined) driving fields. So if the slowly varying envelope approximation is adopted, the form of ρ_{13} is

$$\rho_{13}^{(2)}(t) = \tilde{\rho}_{13}^{(2)} \exp[i(\omega_1 - \omega_2)t] \tag{6.33}$$

By substituting Eqs. 6.30, 6.31, and 6.33 into Eq. 6.32 and grouping terms with the same time dependence, one finds the solution for the coherence to be

$$\tilde{\rho}_{13}^{(2)} = \left\{ \left(\frac{\Omega_{12}\Omega_{23}^{*}/4}{[\Delta_1 + i\Gamma_{12}][\Delta_2 + i\Gamma_{13}]} \right) (\rho_{11}^{(0)} - \rho_{22}^{(0)}) \right\} \tag{6.34}$$

The second denominator incorporates the two-photon detuning $\Delta_2 = \omega_{31} - (\omega_1 - \omega_2)$. Using this result in Eq. 6.29, the Raman transition rate per atom is obtained with an appropriate assumption regarding excitation bandwidth. If the excitation matches the bandwidth of the Raman transition ($2\Gamma_{13}$), the number of phonon annihilations per impurity atom per second is

$$\gamma_R = 2\Gamma_{13} < n > \left| \tilde{\rho}_{13}^{(2)} \right|^2 = 2\Gamma_{13} < n > \left| \langle f | M | i \rangle \right|^2 \tag{6.35}$$

where adjustment has been made for the number of phonons occupying the mode using the Planck distribution $\langle n \rangle = (\exp[\hbar\Omega / k_B T] - 1)^{-1}$. In this model the implicit summation over intermediate states in the matrix element $\langle f | M | i \rangle$ reduces to a single term:

$$\langle f | M | i \rangle = \left[\frac{\langle 3 | \mu^{(e)} \cdot \hat{e}_2 | 2 \rangle \langle 2 | \mu^{(e)} \cdot \hat{e}_1 | 1 \rangle}{[\Delta_1 + i\Gamma_{12}][\Delta_2 + i\Gamma_{13}]} \right] \frac{E(\omega_1) E^*(-\omega_2)}{4\hbar^2} \tag{6.36}$$

The rate of energy loss is therefore

$$R = \gamma_R \hbar\Omega = 2\Gamma_{13} \langle n \rangle \left| \langle f | M | i \rangle \right|^2 \hbar\Omega \tag{6.37}$$

This rate applies to each atom in the interaction volume because it was calculated using the density matrix that is normalized to the number of coolant ions. To realize this cooling rate in practice, the optical interaction must be made irreversible. To this point, the dynamics we have described are coherent and proceed as readily in the forward direction as in the backward direction. We must ensure that the second term in Eq. 6.27 describing reverse transitions is negligible. Also the overall efficiency may benefit from interaction with more than one mode at a time. So we briefly consider these aspects of Raman cooling next with the aid of Fig. 6.6.

Refrigeration requires irreversibility. The coherent (reversible) Raman process outlined above must therefore be combined with a random process that increases the entropy of outgoing radiation [4] without decreasing efficiency. For this purpose fluorescence on an allowed transition at frequency ω_4 can be induced by a strong repump beam (at frequency ω_3), as illustrated schematically in Fig. 6.6. To avoid lowering the rate of cooling set by anti-Stokes Raman transitions, the return of the excited-state population to the ground state should be fast. Provided state 3 is emptied at a rate faster than the Raman excitation process that fills it, and no extra heat is generated in the process, efficient cooling will be maintained.

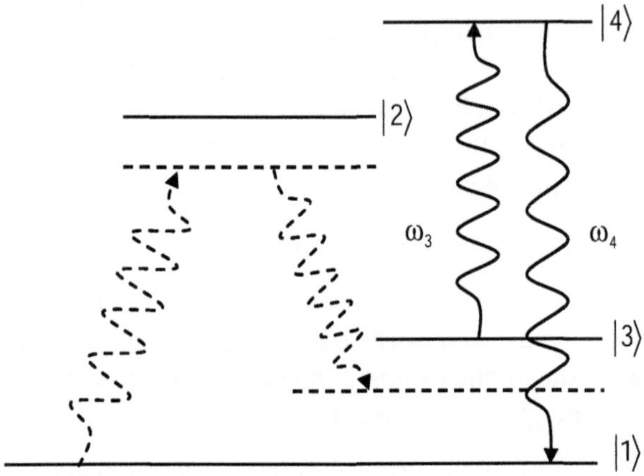

Figure 6.6 The coherent Raman process (dashed arrows) must be followed by random emission for there to be net cooling. Here, spontaneous emission is induced at ω_4 with a pump source at frequency ω_3 that returns the population from state 3 to state 1.

Improvement in the Raman cooling rate can be anticipated by interacting with a range of vibrational modes during each cooling cycle. More than one mode could be addressed, for example, by rapid scanning the two-photon detuning δ in discrete steps [27] or by using some form of adiabatic passage. Reversal of the temporal order of two fixed-frequency pulses can transfer the population between two discrete states with nearly unit efficiency at a fixed detuning. However, for rapid cooling over a broad range of phonon frequencies, rapid adiabatic passage is a superior method in which the excitation detuning is swept from one side of Raman resonance to the other adiabatically. This approach can be implemented very simply by imposing a "chirp" on $\delta(t)$. By sweeping the frequency rapidly from one side of resonance to the other, all phonons with frequencies between the endpoints $\pm\delta_{\max}$ of the chirp are cooled at once [30]. Adiabatic passage is discussed further in Chapter 7.

The need to avoid heat generation during the repump cycle affects how closely the repump beam may be tuned to resonance. The criterion for optimizing one-photon detuning in Raman cooling of solids is different from that for gases because the configuration of neighboring atoms surrounding a coolant atom changes as the dipole forms during transition of a coolant atom from a low- to a high-energy state. That is, the equilibrium positions of neighbors change in response to the absorption process. Phonons are unavoidably generated by relaxation of this nonequilibrium configuration unless the light is detuned on the red side of resonance by more than the phonon energy.

In the illustration of Fig. 6.7, detuning requirements are considered in the context of energy levels of Ce^{3+} ions. Given the electronic structure of Ce^{3+}, the laser must be detuned by an amount equal to or greater than the Stokes shift of the longest-wavelength $4f^{n-1}5d$–$4f^n$ transition. For Ce^{3+} this means

$$\Delta_{\text{Laser}} \geq \Delta_{\text{Stokes}} + \Delta_F \tag{6.38}$$

where Δ_F is the energy separating the $^2F_{5/2}$ and $^2F_{7/2}$ energy levels. This choice ensures that absorbed photons have less than the energy required to generate either a vibrationally excited upper state or a vibrationally excited ground state. In the Ce^{3+} system, cooling ought to maximize when both the Raman and repump beams are detuned by $\Delta_{\text{Laser}} = \Delta_{\text{Stokes}} + \Delta_F$. In practice, smaller detunings might improve the cooling if the transition rate from the 5d state to $^2F_{7/2}$ is less than that to $^2F_{5/2}$, as it is in Ce:YAG [31]. Despite the reduction in cooling rate that large detunings incur in general, high overall cooling rates can be achieved when the excited state is exceptionally broad, as predicted [26] for the 5d state of Ce^{3+}.

To conclude this chapter we emphasize two key advantages of Raman cooling of solids, namely that the cooling rate is temperature independent and that the cooling process is divided into two main steps involving different levels, thereby permitting separate optimization of absorption and fluorescence. At low temperatures, the phonon occupation factor in Eq. 6.27 drops below 1 and varies as $\langle n \rangle = \exp(-\hbar\Omega/k_BT)$. In this nonclassical regime, any cooling rate that is solely proportional to $\langle n \rangle$ falls exponentially as temperature decreases. Indeed γ_R is proportional to $\langle n \rangle$ according to Eq. 6.35, just like the rate for anti-Stokes fluorescence cooling. However, the phonon-broadened linewidth Γ_{13} of the Raman transition is strongly temperature dependent and offsets the decline with a matching proportionality to $<n>$ of its own. In the model below there is complete cancellation of factors of $<n>$ in Eq. 6.35 at two-photon resonance. Since the transition to state 3 is detuned by the phonon frequency below the terminal electronic state (see Fig. 6.4), the relevant resonant condition is $\Delta_2 = 0$.

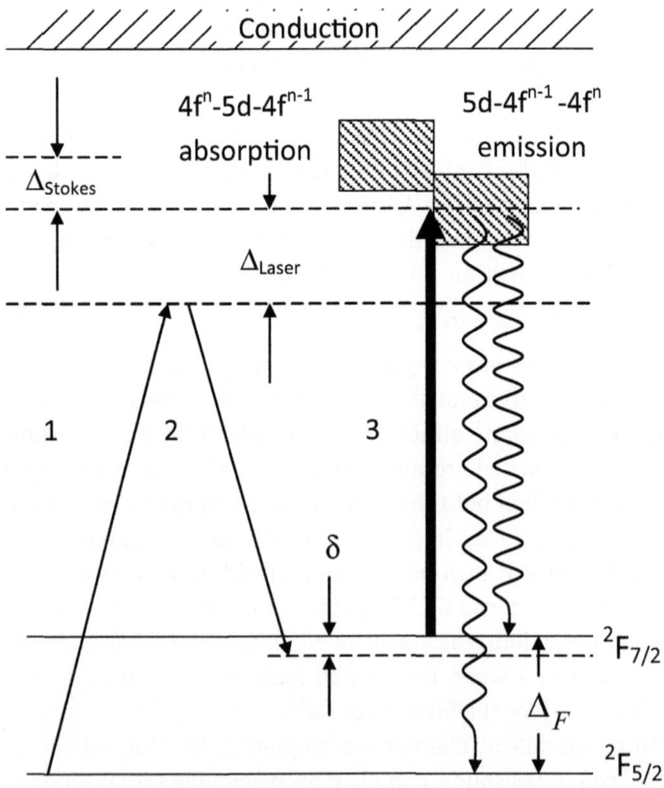

Figure 6.7 Energy levels and detunings involved in Raman laser cooling of a three-level coolant atom in a solid. The example is based on the electronic structure of Ce^{3+}.

While other mechanisms may contribute, we assume here that the Raman linewidth is governed primarily by a two-phonon Orbach process [32]. Thus it can be modeled as $\Gamma_{13}(T) \cong \Gamma_0\left(1+\langle n\rangle[\Gamma_0'/\Gamma_0]\right)$. In this expression Γ_0 is the limiting inhomogeneous linewidth at zero temperature. When $\Gamma'_0 > \Gamma_0$ the linewidth reduces to $\Gamma_{13}(T) \cong <n> \Gamma'_0$. Upon substitution into Eq. 6.35 with zero two-photon detuning ($\Delta_2 = 0$), the rate of phonon annihilation per atom becomes constant.

$$\gamma_{\mathrm{R}} = \frac{2}{\Gamma'_0} \left| \langle f | M' | i \rangle \right|^2 \tag{6.39}$$

In this expression the modified matrix element M' appears because of the cancellation between $<n>$ and two-photon linewidth factors in the Orbach model outlined above. Its definition is

$$\langle f | M' | i \rangle = \left[\frac{\langle 3 | \mu^{(e)} \cdot \hat{e}_2 | 2 \rangle \langle 2 | \mu^{(e)} \cdot \hat{e}_1 | 1 \rangle}{[\Delta_1 + i\Gamma_{12}]} \right] \frac{E(\omega_1) E^*(-\omega_2)}{4\hbar^2} \tag{6.40}$$

The energy loss rate per atom

$$R = \gamma_{\mathrm{R}} \hbar\Omega = \frac{2}{\Gamma'_0} \left| \langle f | M' | i \rangle \right|^2 \hbar\Omega \tag{6.41}$$

is similarly expected to be temperature independent over the entire range $k_B T > \hbar\Gamma_0$. Unlike acoustic modes, optical phonons have very little dispersion (Fig. 6.2). Consequently the frequency of the dominant optical phonon does not vary significantly and R remains constant over a wide range of temperature.

The 1D Raman interaction described above involves a phonon near the center of the Brillouin zone propagating along a single optical axis. However, zone center phonons cannot participate directly in the collisional processes that establish thermal equilibrium within the phonon distribution. Hence it is natural to inquire as to how effectively a mode that is cooled irreversibly can equilibrate with the reservoir of other phonons in the solid. Restoration of equilibrium in solids is governed by Umklapp processes, phonon collisions requiring a reciprocal lattice vector $\bar{G}$ to conserve momentum [33]. Zone center phonons have low wavevectors, so when two phonons interact with a third via an Umklapp process, there is no way to satisfy the relation $\bar{k}_1 + \bar{k}_2 = \bar{k}_3 + \bar{G}$ with a wavevector $\bar{G}$ that spans the entire Brillouin zone. To achieve uniform cooling of even a single mode, phonons propagating along orthogonal axes should be addressed, and this calls for the introduction of two more sets of counterpropagating Raman beam pairs to cover all three orthogonal space axes. Additionally, beam switching along each axis is needed to interrogate both forward- and backward-traveling phonons of the same frequency. Fast directional switching of each beam pair can be implemented with a Pockel's cell that reverses the propagation direction of the ω_1 and ω_2 beams simultaneously within the sample [28] on a timescale much shorter than the lifetime of the $^2F_{7/2}$ shelving state.

While uniform cooling of one mode can be assured by such procedures, thermal equilibration of the sample as a whole relies on Umklapp processes that involve phonons of wavevector $\bar{G}$. Although the occupation probability of such phonons drops exponentially with decreasing temperature, the thermal conductivity that is proportional to the phonon mean free path actually rises through most of the cryogenic range. Hence sample equilibration times drop until temperatures around 10–20 K are reached. Then the mean free path becomes limited by sample dimensions and thermal conductivity drops as T^3 due to its proportionality to specific heat [33]. The consequence of this is that fast thermal equilibration can be anticipated in Raman laser cooling of solids at all but the very lowest ($T < 10$ K) temperatures.

References

1. Pringsheim, P. (1929). Zwei Bemerkungenüber über den Unterschied von Lumineszenz- und Temperaturstrahlung, *Z. Phys. A,* **57**, pp. 739–746.
2. Hansch, T. W., Schawlow, A. L. (1975). Cooling of gases by laser radiation, *Opt. Commun.,* **13**, pp. 68–69.
3. Landau, L. (1946). On the thermodynamics of photoluminescence, *J. Phys. (Moscow),* **10**, pp. 503–506.
4. Ruan, X. L., Rand, S. C., Kaviany, M. (2007). Entropy and efficiency in laser cooling of solids, *Phys. Rev. B,* **75**, pp. 214304(1–9).
5. Ashkin, A., Dziedzic, J. M. (1987). Optical trapping and manipulation of viruses and bacteria, *Science,* **235**, pp. 1517–1520.
6. Cohadon, P. F., Heidmann, A., Pinard, M. (1999). Cooling of a mirror by radiation pressure, *Phys. Rev. Lett.,* **83**, pp. 3174–3177.
7. Metcalf, H., van der Straten, P. (1999). *Laser Cooling and Trapping* (Springer-Verlag, New York).
8. Epstein, R. I., Buchwald, M. I., Edwards, B. C., Gosnell, T. R., Mungan, C. E. (1995). Observation of laser-induced fluorescent cooling of a solid, *Nature,* **377**, pp. 500–503.

9. Powell, R. C. (1998). *Physics of Solid State Laser Materials* (Springer-Verlag, New York), pp. 154.
10. Loudon, R. (2000). *The Quantum Theory of Light,* 3rd Ed. (Oxford University Press, Oxford), p. 349.
11. Donovan, B., Angress, J. F. (1971). *Lattice Vibrations* (Chapman and Hall, London).
12. Clark, J. L., Miller, P. F., Rumbles, G. (1998). Red edge photophysics of ethanolic rhodamine 101 and the observation of laser cooling in the condensed phase, *J. Phys. Chem. A,* **102**, pp. 4428–4437.
13. Zhang, J., Li, D., Chen, R., Xiong, Q. (2013). Laser cooling of a semiconductor by 40 Kelvin, *Nature,* **493**, pp. 504–508.
14. Bahl, G., Tomes, M., Marquardt, F., Carmon, T. (2012). Observation of spontaneous Brillouin cooling, *Nat. Phys.,* **8**, pp. 203–207.
15. Kittel, C. (1971). *Introduction to Solid State Physics*, 4th Ed. (John Wiley & Sons, New York).
16. Sheik-Bahae, M., Epstein, R. I. (2007). Optical refrigeration, *Nat. Photonics,* **1**, pp. 693–699.
17. Pines, D. (1964). *Elementary Excitations in Solids* (W.A. Benjamin, Inc., New York), p. 232.
18. Shen, Y. R. (1984). *The Principles of Nonlinear Optics* (John Wiley & Sons, New York), p. 192.
19. Tomes, M., Marquardt, F., Bahl, G., Carmon, T. (2011). Quantum mechanical theory of optomechanical Brillouin cooling, *Phys. Rev. A,* **84**, pp. 063806(1–5).
20. Shen, Y. R., Bloembergen, N. (1965). Theory of stimulated Brillouin and Raman scattering, *Phys. Rev.,* **137**(6A), pp. 1787–1805.
21. Yariv, A. (1965). Quantum theory for parametric interactions of light and hypersound, *IEEE J. Quantum Electron.,* **1**(1), pp. 28–36.
22. Louisell, W. H. (1960). *Coupled Modes and Parametric Electronics* (Wiley, New York).
23. Toschek, P. E. (1985). Lumino-refrigeration, *Ann. Phys.,* **10**, pp. 761–775.
24. Lindberg, M., Javanainen, J. (1986). Temperature of a laser-cooled trapped three-level ion, *J. Opt. Soc. Am. B,* **3**(7), pp. 1008–1017.
25. Kasevich, M., Chu, S. (1992). Laser cooling below a photon recoil with three-level atoms, *Phys. Rev. Lett.,* **69**, pp. 1741–1744.
26. Rand, S. C. (2013). Raman laser cooling of solids, *J. Lumin.,* **133**, pp. 10–14.

27. Reichel, J., Morice, O., Tino, G. M., Salomon, C. (1994). Subrecoil Raman cooling of Cesium atoms, *Europhys. Lett.*, **28**, p. 477.

28. Reichel, J., Bardou, F., Ben Dahan, M., Peik, E., Rand, S. C., Salomon, C., Cohen-Tannoudji, C. (1995). Raman cooling of Cesium below 3 nK: new approach inspired by Levy flight statistics, *Phys. Rev. Lett.*, **75**, pp. 4575–4578.

29. Rand, S. C. (2015). *Lectures on Light*, 2nd Ed. (Oxford University Press, Oxford), Appendix G.

30. Kuhn, A., Perrin, H., Hansel, W., Salomon, C. (1997). Three dimensional Raman cooling using velocity selective rapid adiabatic passage, in *OSA TOPS on Ultracold Atoms and BEC*, **7**, pp. 58–66.

31. Feofilov, S. P., Kulinkin, A. B., Gacoin, T., Mialon, G., Dantelle, G., Meltzer, R. S., Dujardin, C. (2012). Mechanisms for Ce3+ excitation at energies below the zero-phonon line in YAG crystals and nanocrystals, *J. Lumin.*, **132**, pp. 3082–3088.

32. Abragam, A., Bleaney, B. (1970). *Electron Paramagnetic Resonance of Transition Ions* (Dover, New York), pp. 560–562.

33. Kittel, C. (1971). *Introduction to Solid State Physics*, 4th Ed. (Wiley, New York), p. 228.

Chapter 7

Deep Laser Cooling of Rare Earth–Doped Crystals by Stimulated Raman Adiabatic Passage

Andrei Ivanov, Yuriy Rozhdestvensky, and Evgeniy Perlin
Center of Information Optical Technologies, ITMO University, Birzhevayaliniya, 14, Saint Petersburg, 199034, Russia
avivanov@mail.ifmo.ru

7.1 Introduction

Cooling of solid-state matter by laser radiation has definite advantages over other types of cooling such as thermoelectric, mechanical, and fluid. Optical coolers are compact, they have no vibrations, and they use light as a coolant. Such coolers can be effectively utilized in the case of extremely small solid-state objects that are used in up-to-date optoelectronics. In the list of possible applications of laser cooling in solids there are self-cooled lasers, thermostabilized light-emitting diodes (LEDs), and space sensors.

Although different solid-state materials were proposed for laser cooling in the last few decades, great success has been achieved to

Laser Cooling: Fundamental Properties and Applications
Edited by Galina Nemova

ISBN 978-981-4745-04-8 (Hardcover), 978-981-4745-05-5 (eBook)
www.panstanford.com

date for glasses and crystals doped by the *rare earth* (RE) ions. There are two reasons that ensure the success of laser cooling of such materials. The first reason is a strong coupling between electrons and vibrations in the RE ion. The ion electronic structure consists of levels that are split into sublevels by a crystal field of the matrix due to the Stark effect. The energy gaps between *Stark-split sublevels* (SSSs) are comparable with phonon energies of the crystal matrix. So, there are conditions to observe the *anti-Stokes fluorescence* (ASF) in these solid systems. The second reason is the development of technology that allows producing doped materials of high purity. It means that in such RE ion systems there are low concentrations of unwanted impurities. Reducing the impurity concentrations decreases background absorption and unintended heating by a laser radiation.

Laser cooling of a solid was originally attained in a Yb^{3+}-doped fluorozirconate glass that was cooled down by 0.3°C below room temperature in 1995 [1]. Since that time, laser cooling was observed in various glasses and crystals doped by the RE ions [2–7]. At present, progress of the technology makes it possible to reach the range of cryogenic temperatures for a Yb:YLF crystal [8]. This result is better than the results obtained on cooling with the use of thermoelectric Peltier elements.

The traditional model that describes the optical refrigeration of doped solids is based on a four-level system of 4f multiplet states of an RE ion [6]. In this model, the ASF is provided by (i) resonance transitions between the upper ground and lower excited levels, (ii) phonon absorption between the SSSs, and (iii) subsequent radiative relaxation to the ground states. At the same time, in such a model, the rate of the cooling cycle is very low because of the small cross section of forbidden transitions between the levels of the same ion configuration and because of the low population of the upper ground level. For improving the cooling performance in the traditional model, a nonresonant cavity for maximizing the absorbed laser power is used [6].

In the last few decades, the coherent phenomena stimulated by laser radiation, such as *coherent population trapping* (CPT) and *electromagnetically induced transparency* (EIT), have been studied in atomic physics most intensively [9]. At the same time, the CPT and EIT effects, including the room-temperature case, were demonstrated for solid systems by several groups [10–12]. For the CPT effect to be observed, it is very important to comply with certain relations between the lifetimes of the electron levels involved in the cooling process. For example, for a Λ system, in which there exist three levels (the ground and first and upper excited levels), the presence of one metastable level with a long lifetime (the first excited level) provides optical interference phenomena. Such a situation is possible in RE-doped crystals due to the existence of dipole-allowed 4f–5d and forbidden 4f–4f optical transitions.

To improve the characteristics of the cooling process, such as cooling efficiency, a minimum attainable temperature, and a cooling rate, Raman mechanisms were recently proposed [13–15]. Raman cooling used laser pulse pumping through dipole-allowed *5d* ion levels by the *two-photon Raman* (TPR) scattering proposed for doped materials (see Section 6.4). The mechanism is based on 4f–5d and 5d–4f optical transitions involving longitudinal optical phonons of the crystal matrix. The acceleration of the cooling cycle is achieved by combining the stimulated TPR process with third pulse pumping at the allowed fluorescence transition. Nevertheless, in this model, the cooling efficiency is 0.6%. This is because of the low transition rate of the two-photon process that involves phonons. Moreover, this cooling mechanism stops working after freezing the vibrations of the optical modes. Another Raman mechanism of cooling is based on methods of coherent pumping of doped crystals [14, 15]. The coherent and complete population transfer between the ground and the first excited levels of 4f multiplet through the 5d ion levels was achieved by using the different Raman techniques, namely TPR scattering, *stimulated Raman adiabatic passage* (STIRAP), and the π-pulse method. It was shown in these works that the coherent pumping increases the number of electrons that participate in the cooling cycle, which leads to increasing of the cooling power. At the same time, the results showed that the deformation potential mechanism for description of the interaction between ion electrons and vibrations of the crystal medium failed to explain the establishment of the Boltzmann distribution of the electrons at the SSSs.

In this chapter, we consider a cooling model with STIRAP for optical pumping of an RE-ion-doped crystal, taking into account vibronic interaction at the SSSs of ion levels. In the model, cooling and heating processes, which make substantial contributions to the cooling mechanism, are defined. The cooling processes are attributed to the vibronic coupling of the SSSs. In addition, using vibronic interaction allows one to achieve the Boltzmann distribution of the electrons at the SSSs in a framework of the proposed model. We apply density matrix formalism for determining the time dependence of electron populations at the ion levels under the action of pulse pumping. Then, the level populations are used for estimation of cooling characteristics of the model. The calculations, which are obtained for the Yb^{3+}:CaF_2 system, show that coherent pumping by STIRAP leads to deep laser cooling of doped crystals that have technologically attainable purity to date.

7.2 Vibronic Model of Laser Cooling of Rare Earth Ions

Let us consider a crystal doped by trivalent RE ions. The specific feature of the system under consideration is the existence of 5d ion levels below the conduction band edge of the crystal. In addition, there are only two 4f ion levels (the ground and first excited levels) below the 5d levels. We assume that we are dealing with a five-level energy structure (see Fig. 7.1), which consists of one level of the 5d configuration (denoted by $|3\rangle$) and four levels of the 4f configuration: one ground level (denoted by $|1\rangle$) and three SSSs of the first excited level (denoted by $|2\rangle$,$|4\rangle$, and $|5\rangle$, respectively). In the model, we assume that the 5d ion level is empty. The energy difference between the ground level $|1\rangle$ and the lower SSS $|2\rangle$ is assumed to be large enough for radiative relaxation from the first excited level to occur and, thus, to prevent the system from significant heating [16].

We assume that the laser cooling cycle of the five-level system contains the following steps: (i) optical pumping of the SSS $|2\rangle$, (ii) interaction of the electrons with the ion vibrations, and (iii) transference of the population to the ground level $|1\rangle$ by radiative decay (Fig. 7.1). We denote this cooling process as C1. For laser cooling of the considered system, we use a sequence of two laser pulses (Fig. 7.2b) with electromagnetic field frequencies ω_1 and ω_2, respectively. The pulse sequence is used to implement the STIRAP in which laser pulses are partially overlapped and come in the counterintuitive order when the first pulse initially couples the two empty levels, $|3\rangle$ and $|2\rangle$, and the second pulse couples levels $|1\rangle$ and $|3\rangle$. Such coherent pumping gives rise to the transfer of all the population from level $|1\rangle$ to level $|2\rangle$ without populating level $|3\rangle$ [14]. The establishment of the Boltzmann distribution at levels $|2\rangle$, $|4\rangle$, and $|5\rangle$ is the result of vibronic interaction. This interaction leads to consumption of vibrational energy of the RE ions and consequently to the cooling of the system. Finally, the population returns to the ground level by radiative decay, and the cycle is completed.

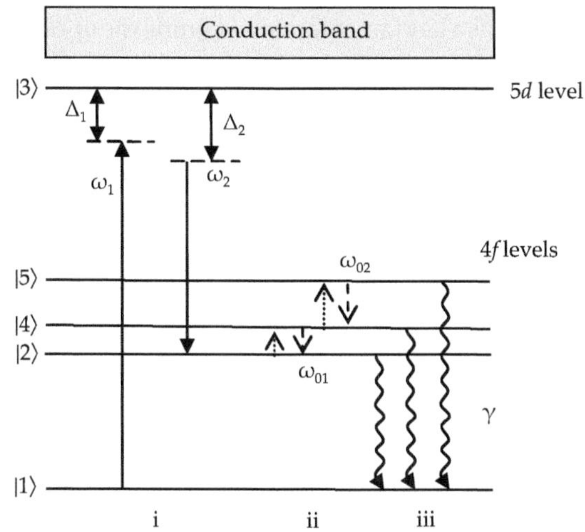

Figure 7.1 Scheme of the laser cooling cycle for the five-level energy structure of an RE ion: (i) two-photon optical pumping, (ii) vibronic interaction, and (iii) radiative decay. Solid arrows show the optical transitions at the frequencies ω_1 and ω_2; dotted arrows show the vibronic interaction with energies ω_{01} and ω_{02}; downward dashed arrows refer to the nonradiative decay; wavy arrows show the radiative decay γ; Δ_1 and Δ_2 are the detunings of optical transitions; the detunings of the transitions under the vibronic interactions are not depicted (see text).

It should be noted that the actual efficiency of the above-proposed cooling cycle is defined by the lack of population of $4f^{n-1}$ 5d levels. This is necessary to avoid heating processes. In addition, the efficiency of the cooling cycle is defined by the creation of a high population of SSSs to stimulate vibronic interaction. Such conditions can be achieved on coherent interaction of radiation with the doped crystal and based on the differences between the energy relaxation times of the 5d and 4f levels. The 5d levels rapidly decay to lower levels, with a lifetime of the order of tens of nanoseconds [17], whereas the lifetime of the 4f configuration levels is of the order of milliseconds [16]. This relationship between the decay times of the excited levels provides the conditions for the cooling process.

7.2.1 Vibronic Interaction

Vibronic coupling is a key factor for determining the optical properties of local centers in a crystal [18, 19]. In case of quasi-degenerate or weak split electronic states, these properties are governed by the *pseudo Jahn–Teller effect* (PJTE) [20]. The small intervals in the electronic spectrum mean a relatively slow electron movement that is comparable with a nuclear movement. So, the movements of the electrons and the nucleus in the RE ion cannot be divorced and a nonadiabaticity of the coupling must be taken into account. It should be emphasized that in case of the RE ion with unfilled d or f electron shells, there is a strong interaction between localized electrons and local or pseudolocal vibrations [21].

Vibronic interaction exerts a strong influence on the laser cooling process. The existence of vibronic coupling leads to decreasing of vibrational energy of the RE ion interacting with laser radiation under cooling conditions. Since the ion is bound with the atoms of the crystal by elastic interaction, laser cooling of the RE ion gives rise to cooling of the crystal matrix as a result of the achievement of thermal equilibrium. The Boltzmann distribution at SSSs is a sequence of the coupling between the vibronic subsystem and the phonon reservoir.

It is worth mentioning that the use of deformation potential formalism to describe the interaction of ion electrons with lattice phonons fails to explain the establishment of a Boltzmann distribution at SSSs of the excited RE ion level [15].

7.2.2 Model Hamiltonian for the Rare Earth Ion System

For a theoretical description of the time evolution of the five-level system of the RE ion, we use strict quantum mechanical treatment that is based on density matrix formalism. We consider the quantum mechanical system that consists of electronic and local, or pseudolocal, vibration subsystems. We assume that the energy spectra of the noninteracting subsystems are known. So, the Hamiltonian of the system without interaction can be written in the following form:

$$H_0 = H_{0\mathrm{e}} + H_{0\mathrm{v}},\ H_{0\mathrm{e}} = \sum_i \varepsilon_\mathrm{i} a_\mathrm{i}^+ a_\mathrm{i},\ H_{0\mathrm{v}} = \sum_n \hbar\omega_{0\mathrm{n}} b_\mathrm{n}^+ b_\mathrm{n} \tag{7.1}$$

Here H_{0e} is the Hamiltonian of the electronic subsystem and H_{0v} is the Hamiltonian of the pseudolocal mode; a_i^+ and a_i are the creation and annihilation operators of an electron at the ion level i of energy ε_i; b_n^+ and b_n are the creation and annihilation operators of a pseudolocal mode n of frequency ω_{0n}. The subsystems interact with each other and with linearly polarized electromagnetic waves. We assume that there is a strong linear vibronic coupling [22, 23] and there are two Jahn–Teller active modes of frequencies ω_{01} and ω_{02} in the system (Fig. 7.1). The interaction Hamiltonian is taken in the following form:

$$\begin{aligned} H_{\mathrm{int}} &= \hbar\widehat{\Omega}_{13,\lambda_1\kappa_1} a_3^+ a_1 \exp(i\Delta_1 t) + a_1^+ a_3 \hbar\widehat{\Omega}^+_{31,\lambda_1\kappa_1} \exp(-i\Delta_1 t) \\ &+ \hbar\widehat{\Omega}_{23,\lambda_2\kappa_2} a_3^+ a_2 \exp(i\Delta_2 t) + a_2^+ a_3 \hbar\widehat{\Omega}^+_{32,\lambda_2\kappa_2} \exp(-i\Delta_2 t) \\ &+ \hbar\widehat{Q}_{24,01} a_4^+ a_2 \exp(i\delta_{01} t) + a_2^+ a_4 \hbar\widehat{Q}^+_{42,01} \exp(-i\delta_{01} t) \\ &+ \hbar\widehat{Q}_{45,02} a_5^+ a_4 \exp(i\delta_{02} t) + a_4^+ a_5 \hbar\widehat{Q}^+_{54,02} \exp(-i\delta_{02} t), \end{aligned} \tag{7.2}$$

$$\widehat{\Omega}_{ij,\lambda_m\kappa_m} = \frac{\mu_{ij}}{\hbar}\widehat{E}_{\lambda_m\kappa_m},\ \widehat{Q}_{ij,0n} = \frac{k_{ij}}{\hbar} b_n, \tag{7.3}$$

Here μ_{ij} are the matrix elements of the dipole moment operator for levels i and j; $\widehat{E}_{\lambda_m\kappa_m}$ ($\widehat{E}^+_{\lambda_m\kappa_m}$) is the electric field operator that contains the annihilation (creation) operator of a photon; λ_m, and κ_m are the polarization and the wavevector of the photon related to the laser pulse m; $\Delta_1 = \omega_{31} - \omega_1$ and $\Delta_2 = \omega_{32} - \omega_2$ are the detunings of optical transitions, where ω_{ji} are the frequencies of optical transitions between the levels i and j; k_{ij} are the matrix elements of vibronic coupling for levels i and j; $\delta_{01} = \omega_{42} - \omega_{01}$ and $\delta_{02} = \omega_{54} - \omega_{02}$ are the detunings of transitions under the vibronic interaction; and an operator A^+ denotes a Hermitian conjugate operator of A.

The coupling of the system with the phonon continuum is taken into account by the introduction of the relaxation constants in a framework of density matrix formalism [23].

7.2.3 Time Evolution of the Five-Level System

To estimate the cooling performance, we use a density matrix method to take into account the coherent interaction of the pulse sequence with the considered system. Since optical pumping transfers the population of the ground level $|1\rangle$ to the lowest excited level $|2\rangle$, cooling energy is determined by a number of the electrons moved to the upper levels $|4\rangle$ and $|5\rangle$ by the vibronic interaction. Hence, our goal is to calculate the populations of levels $|4\rangle$ and $|5\rangle$ using diagonal elements of the density matrix. To achieve this goal we consider the dynamics of the system that is described by a set of equations for the density matrix elements.

We theoretically treat the time evolution of a five-level system using the interaction representation in the rotating-wave approximation. In this case, the set of equations is

$$\dot{\rho}_{11} = i\rho_{13}\hat{\Omega}^{+}_{31,\lambda_1\kappa_1} - i\hat{\Omega}_{13,\lambda_1\kappa_1}\rho_{31} + \gamma_{22}\rho_{22} + \gamma_{33}\rho_{33} + \gamma_{44}\rho_{44} + \gamma_{55}\rho_{55},$$

$$\dot{\rho}_{12} = i(\Delta_1 - \Delta_2)\rho_{12} - i\hat{\Omega}_{13,\lambda_1\kappa_1}\rho_{32} + i\rho_{13}\hat{\Omega}^{+}_{32,\lambda_2\kappa_2} + i\rho_{14}\hat{Q}^{+}_{42,01} - \gamma_{12}\rho_{12},$$

$$\dot{\rho}_{13} = i\rho_{11}\hat{\Omega}_{13,\lambda_1\kappa_1} - i\hat{\Omega}_{13,\lambda_1\kappa_1}\rho_{33} + i\rho_{12}\hat{\Omega}_{23,\lambda_2\kappa_2} + i\Delta_1\rho_{13} - \gamma_{13}\rho_{13},$$

$$\dot{\rho}_{14} = i(\delta_{01} - \Delta_2 + \Delta_1)\rho_{14} - i\hat{\Omega}_{13,\lambda_1\kappa_1}\rho_{34} + i\rho_{12}\hat{Q}_{24,01} + i\rho_{15}\hat{Q}^{+}_{54,02} - \gamma_{14}\rho_{14},$$

$$\dot{\rho}_{15} = i(\delta_{01} + \delta_{02} - \Delta_2 + \Delta_1)\rho_{15} - i\hat{\Omega}_{13,\lambda_1\kappa_1}\rho_{35} + i\rho_{14}\hat{Q}_{45,02} - \gamma_{15}\rho_{15},$$

$$\dot{\rho}_{22} = i\rho_{23}\hat{\Omega}^{+}_{32,\lambda_2\kappa_2} - i\hat{\Omega}_{23,\lambda_2\kappa_2}\rho_{32} + i\rho_{24}\hat{Q}^{+}_{42,01} - i\hat{Q}_{24,01}\rho_{42}$$
$$+\gamma^{\text{nr}}_{44}\rho_{44} + \gamma_{33}\rho_{33} - \gamma_{22}\rho_{22},$$

$$\dot{\rho}_{23} = i\Delta_2\rho_{23} + i\rho_{22}\hat{\Omega}_{23,\lambda_2\kappa_2} - i\hat{\Omega}_{23,\lambda_2\kappa_2}\rho_{33} - i\hat{Q}_{24,01}\rho_{43} + i\rho_{21}\hat{\Omega}_{13,\lambda_1\kappa_1} - \gamma_{23}\rho_{23},$$

$$\dot{\rho}_{24} = i\delta_{01}\rho_{24} + i\rho_{22}\hat{Q}_{24,01} - i\hat{Q}_{24,01}\rho_{44} - i\hat{\Omega}_{23,\lambda_2\kappa_2}\rho_{34} + i\rho_{25}\hat{Q}^{+}_{54,02} - \gamma^{\text{nr}}_{24}\rho_{24},$$

$$\dot{\rho}_{25} = i(\delta_{01} + \delta_{02})\rho_{25} - i\hat{Q}_{24,01}\rho_{45} - i\hat{\Omega}_{23,\lambda_2\kappa_2}\rho_{35} + i\rho_{24}\hat{Q}_{45,02} - \gamma^{\text{nr}}_{25}\rho_{25},$$

$$\dot{\rho}_{33} = i\rho_{31}\hat{\Omega}_{13,\lambda_1\kappa_1} - i\hat{\Omega}^{+}_{31,\lambda_1\kappa_1}\rho_{13} + i\rho_{32}\hat{\Omega}_{23,\lambda_2\kappa_2} - i\hat{\Omega}^{+}_{32,\lambda_2\kappa_2}\rho_{23} - 4\gamma_{33}\rho_{33},$$

$$\dot{\rho}_{34} = i(\delta_{01} - \Delta_2)\rho_{34} - i\hat{\Omega}^{+}_{31,\lambda_1\kappa_1}\rho_{14} - i\hat{\Omega}^{+}_{32,\lambda_2\kappa_2}\rho_{24}$$

$$+i\rho_{32}\hat{Q}_{24,01} + i\rho_{35}\hat{Q}^{+}_{54,02} - \gamma_{34}\rho_{34},$$

$$\dot{\rho}_{35} = i(\delta_{01} + \delta_{02} - \Delta_2)\rho_{35} - i\hat{\Omega}^{+}_{31,\lambda_1\kappa_1}\rho_{15} - i\hat{\Omega}^{+}_{32,\lambda_2\kappa_2}\rho_{25} + i\rho_{34}\hat{Q}_{45,02} - \gamma_{35}\rho_{35},$$

$$\dot{\rho}_{44} = i\rho_{42}\hat{Q}_{24,01} - i\hat{Q}^{+}_{42,01}\rho_{24} + i\rho_{45}\hat{Q}^{+}_{54,02} - i\hat{Q}_{45,02}\rho_{54}$$

$$-(\gamma_{44} + \gamma^{\mathrm{nr}}_{44})\rho_{44} + \gamma^{\mathrm{nr}}_{55}\rho_{55} + \gamma_{33}\rho_{33},$$

$$\dot{\rho}_{45} = i\delta_{02}\rho_{45} - i\hat{Q}^{+}_{42,01}\rho_{25} + i\rho_{44}\hat{Q}_{45,02} - i\hat{Q}_{45,02}\rho_{55} - \gamma_{45}\rho_{45},$$

$$\dot{\rho}_{55} = i\rho_{54}\hat{Q}_{45,02} - i\hat{Q}^{+}_{54,02}\rho_{45} - (\gamma_{55} + \gamma^{\mathrm{nr}}_{55})\rho_{55} + \gamma_{33}\rho_{33}. \tag{7.4}$$

Other matrix elements are determined by the relation $\rho_{ij} = \rho^*_{ji}$. In addition, the particle number conservation condition $\rho_{11} + \rho_{22} + \rho_{33} + \rho_{44} + \rho_{55} = 1$ must be met. In Eq. 7.4, the quantities γ_{ij} describe radiative relaxation of the density matrix elements, and $\gamma_{ij}^{\mathrm{nr}}$ describe the nonradiative relaxation. Equation 7.4 is written not taking into account nonradiative relaxation from the SSSs because of the short duration of the laser cooling cycle compared to the relaxation time. This set of equations can be solved only numerically.

In our calculations, we determine the expectation values of the level populations by statistical averaging over states of the photonic and vibrational subsystems. The averaging procedure is based on the standard statistical approximation in which the photonic and vibrational subsystems can be described by the equilibrium distribution functions. This means that the density matrix of the complete system is replaced by the product of the density matrices of the subsystems

$$\rho(t) = \rho_{\mathrm{e}}(t)\rho_{\mathrm{ph}}(0)\rho_{\mathrm{v}}(0), \tag{7.5}$$

where subscripts e, ph, and v denote the electronic, photonic, and vibrational subsystems, respectively. In case of a photonic subsystem, we use the Glauber–Sudarshan P-representation with a δ-like diagonal weight function for single-mode laser radiation for each pulse [24]:

$$\rho_{\mathrm{ph}}(0) = \int \vartheta(\{\xi_\kappa\}) \prod_\kappa |\xi_\kappa\rangle\langle\xi_\kappa| d^2\xi_\kappa,$$

$$\vartheta(|\xi_\kappa|) = \frac{1}{2\pi\sqrt{\bar{n}_\kappa}}\delta(|\xi_\kappa| - \sqrt{n_\kappa}) \tag{7.6}$$

Here ξ_κ and $|\xi_\kappa\rangle$ are the eigenvalues and eigenstates of the annihilation or the creation operator; n_κ is the average photon number in the mode. And the vibrational subsystem is described by the density matrix operator

$$\rho_v(0) = Z^{-1}\exp\left(-\sum_j \beta_T \hbar\omega_{0j} b_j^+ b_j\right),$$

$$Z = \mathrm{Tr}\left[\exp\left(-\sum_j \beta_T \hbar\omega_{0j} b_j^+ b_j\right)\right], \quad \beta_T^{-1} = k_B T. \tag{7.7}$$

where k_B is the Boltzmann constant and T is the temperature of the system. In the case of a pseudolocal vibrational subsystem, the averaging over the states is performed using the local *density of states* (DOS) that is approximated by the δ-function for each frequency ω_{01} and ω_{02} in our calculations. We assume the resonance vibronic interaction that allows us to use the detunings $\delta_{01} = 0$ and $\delta_{02} = 0$. Thus, taking into account the averaging procedure, we can change the boson annihilation and creation operators by c-numbers $\sqrt{n}\exp(\varphi)$, where n is the average particle number in the mode and φ is the phase of the eigenvalue of the annihilation or the creation operator.

The time dependence of the diagonal elements of the reduced density matrix for the first cooling cycle is shown in Fig. 7.2a. For implementation of laser cooling we use two identical pulses that have a Gaussian shape:

$$E_0(t) = E_0\exp\left(-\frac{(t-t_0)^2}{\sigma^2}\right). \tag{7.8}$$

The width of each pulse is $\sigma = 2.0 \times 10^{-7}$ s and the electric field amplitude of each pulse is $E_0 = 1.2 \times 10^3$ V/cm. The distance between the pulses is 9.8×10^{-8} s (Fig. 7.2b). The first pulse couples the two empty levels $|3\rangle$ and $|2\rangle$, and the second pulse couples levels $|1\rangle$ and $|3\rangle$ (see Fig. 7.1). The shapes and parameters of the pulses are chosen to satisfy the adiabatic conditions: (i) a smooth pulse, (ii) a long interaction time, and (iii) a large Rabi frequency [25]. Hence, the adiabatic passage process can be actualized if the Rabi frequencies are sufficiently larger than the inverse time of the pulse overlap.

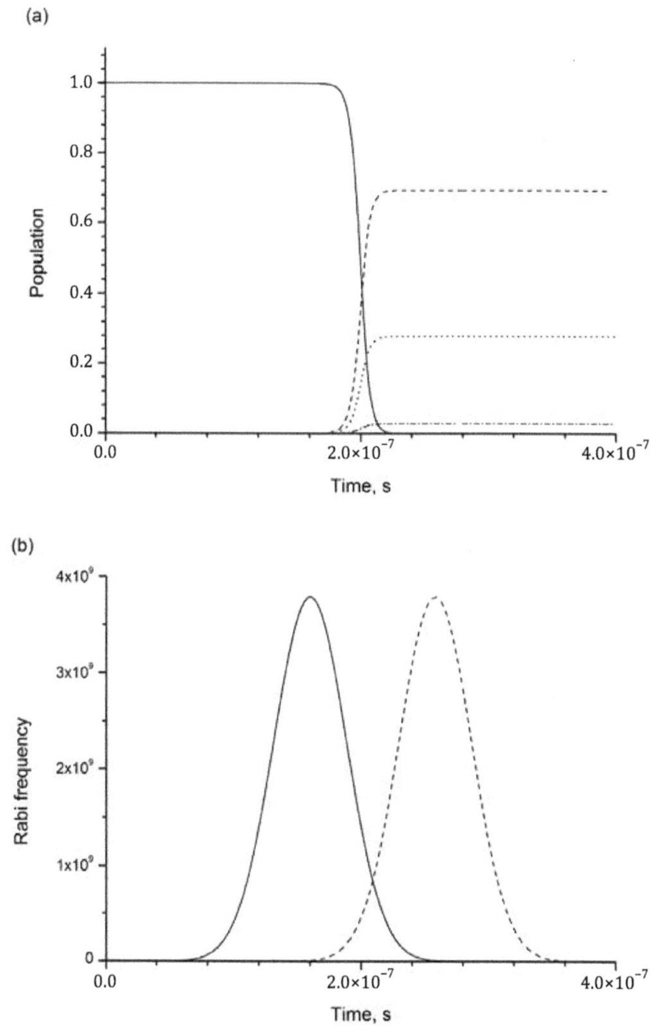

Figure 7.2 (a) Time dependencies of the diagonal elements of the density matrix: ρ_{11} (solid line), ρ_{22} (dashed line), ρ_{44} (dotted line), and ρ_{55} (dash-dotted line). (b) Rabi frequencies of laser beams. Note that the sequence of pulses is counterintuitive: first is Ω_{32} (solid line) and second is Ω_{13} (dashed line). Results are obtained at 300 K (cooling process C1).

The time dependencies of the diagonal elements of the density matrix are shown in Fig. 7.2a. It is shown that the total population of level $|1\rangle$ is transferred to levels $|2\rangle$, $|4\rangle$, and $|5\rangle$ throughout the STIRAP without populating level $|3\rangle$. In the calculations, we use the detunings $\Delta_1 = 0$ and $\Delta_2 = 0$. This method of pumping can also be used if the detunings Δ_1 and Δ_2 are not equal to zero. In the case of substantial detunings, the total population of the ground level can be transferred if the electric field amplitude will be increased.

In the calculations, we use the parameters typical of the solid systems under study: $\mu_{ij} = 3.34 \times 10^{-28}$ C cm; $\hbar\omega_{01} = 191$ cm^{-1}; $\hbar\omega_{02} = 487$ cm^{-1}; $\gamma_{33} = 1.0 \times 10^8$ s^{-1}; $\gamma_{13} = \gamma_{23} = \gamma_{34} = \gamma_{35} = 4\gamma_{33}$; $\gamma = 1.0 \times 10^3$ s^{-1}; $\gamma_{12} = \gamma_{14} = 0.5\,\gamma$; $\gamma_{22} = \gamma_{44} = \gamma$; $\gamma^{nr} = 1.0 \times 10^{12}$ s^{-1}; $\gamma_{44}{}^{nr} = \gamma_{55}{}^{nr} = \gamma^{nr}$; $\gamma_{24}{}^{nr} = \gamma_{45}{}^{nr} = 0.5\,\gamma^{nr}$; $\gamma_{25}{}^{nr} = 0$; and cycle duration $\tau = 4 \times 10^{-7}$ s. The lattice temperature was assumed to be $T = 300$ K for the first cooling cycle. The expectation values of quantities Q_{ij} were determined to satisfy the condition of establishment of the Boltzmann distribution at the levels $|2\rangle$, $|4\rangle$, and $|5\rangle$ at 300 K. Thus, the strengths of vibronic interaction were assumed to be $k_{24}/\hbar = 5.1 \times 10^{11}$ s^{-1} and $k_{45}/\hbar = 4 \times 10^{11}$ s^{-1} in the calculations (see Eq. 7.3). The frequencies of the laser beams and optical transitions for certain system are discussed below.

7.3 Laser Cooling of the Yb^{3+}:CaF_2 System

For the real system that can be described by the model of laser cooling developed in Section 7.2, we can choose the calcium fluoride crystal doped with trivalent ytterbium, Yb^{3+}:CaF_2. This material has the following features that are important in considered case. The lowest 5d level of the Yb^{3+} ion falls within the bandgap of CaF_2. The 4f multiplet of the Yb^{3+} ion involves only two levels, $^2F_{7/2}$ (ground level) and $^2F_{5/2}$ (excited level), that are split into four (E_1, E_2, E_3, E_4) and three (E_5, E_6, E_7) SSSs, respectively. The energy gap between the ground and excited levels is 9523 cm^{-1}. Such a large bandgap allows for not taking into account the nonradiative relaxation to the ground level. The energy diagram of the 4f multiplet is shown in Fig. 7.3 [26]. The optical spectrum of the CaF_2 crystal doped by Yb^{3+} ions manifests a rich vibronic structure [27]. For example, detailed analysis of the phonon side band shows the local breathing-mode oscillations of frequency 450 cm^{-1} that is close to the frequencies of the electronic transitions ω_{23} and ω_{67} of the Yb^{3+} ion (see Fig. 7.3). It is noteworthy that for this specific system, we use the same parameters as those listed in Section 7.2.

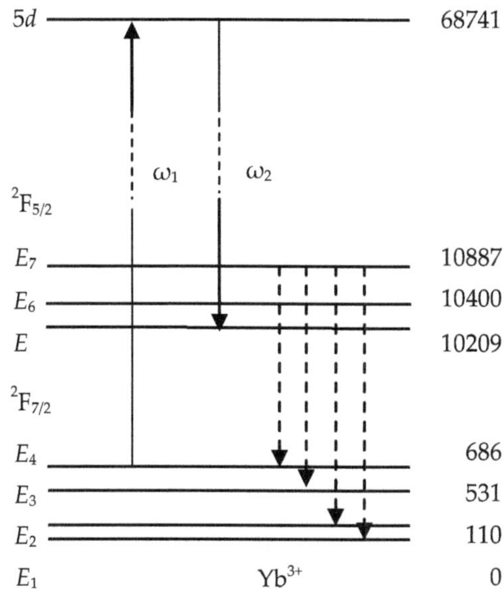

Figure 7.3 Energy diagram of the Yb^{3+} ion and the laser radiation photon energies used in calculations (in cm^{-1}). Downward dashed arrows denote the processes of radiative (nonradiative) relaxation from the excited to ground sublevels. Only one set of relaxation processes from level 7 is depicted (see text).

To implement coherent pumping of the doped system chosen here, vacuum ultraviolet (VUV) lasers must be used [28]. We assume that the photon energy of the radiation for the first laser pulse, ω_1, is 68,055 cm^{-1} and falls in resonance with the electron transition between the highest ground SSS (denoted by E_4) and the lowest 5d level of the Yb^{3+} ion. The photon energy for the second laser pulse, ω_2, is 58,532 cm^{-1} and falls in resonance with the transition between the lowest 5d level and the lowest SSS (denoted by E_5) of the excited level (see Fig. 7.3). The relation between photon energies is chosen to avoid the population of the 4f ^{12}5d levels by optical transitions from the ground SSSs and, so hence, to avoid additional heating of the system.

7.3.1 Number of Electrons Involved in the Cooling Cycle

The efficiency of the proposed cooling model depends on the number of electrons involved in the cooling cycle. For determining the number of electrons that are transferred to the excited level from the ground level, we use a simplified five-level model that takes into account vibronic interaction at the SSSs the both RE ion levels (Fig. 7.4). For this goal, we modify Eq. 7.4 in accordance with the scheme depicted in Fig. 7.4 to a new set of equations:

$$\dot{\rho}_{11} = i\rho_{13}\hat{Q}^{+}_{31,\lambda_1\kappa_1} - i\hat{Q}_{13,\lambda_1\kappa_1}\rho_{31} + \gamma_{33}\rho_{33} + \gamma_{44}\rho_{44} + \gamma_{55}\rho_{55},$$

$$\dot{\rho}_{12} = -i\hat{Q}_{13,\lambda_1\kappa_1}\rho_{32} + i\rho_{13}\hat{Q}_{32,\lambda_2\kappa_2} + i\rho_{14}\hat{\Omega}^{+}_{42,01} - \gamma^{\mathrm{nr}}_{12}\rho_{12},$$

$$\dot{\rho}_{13} = i\rho_{11}\hat{Q}_{13,\lambda_1\kappa_1} - i\hat{Q}_{13,\lambda_1\kappa_1}\rho_{33} + i\rho_{12}\hat{Q}^{+}_{23,\lambda_2\kappa_2} - \gamma^{\mathrm{nr}}_{13}\rho_{13},$$

$$\dot{\rho}_{14} = -i\hat{Q}_{13,\lambda_1\kappa_1}\rho_{34} + i\rho_{12}\Omega_{24,01} + i\rho_{15}\hat{Q}^{+}_{54,02} - \gamma_{14}\rho_{14},$$

$$\dot{\rho}_{15} = -i\hat{Q}_{13,\lambda_1\kappa_1}\rho_{35} + i\rho_{14}\hat{Q}_{45,02} - \gamma_{15}\rho_{15},$$

$$\dot{\rho}_{22} = i\rho_{23}\hat{Q}_{23,\lambda_2\kappa_2} - i\hat{Q}^{+}_{23,\lambda_2\kappa_2}\rho_{32} + i\rho_{24}\hat{\Omega}^{+}_{42,01} - i\hat{\Omega}_{24,01}\rho_{42}$$
$$+\gamma_{55}\rho_{55} + \gamma_{44}\rho_{44} - \gamma^{\mathrm{nr}}_{22}\rho_{22},$$

$$\dot{\rho}_{23} = i\rho_{22}\hat{Q}^{+}_{23,\lambda_2\kappa_2} - i\hat{Q}^{+}_{23,\lambda_2\kappa_2}\rho_{33} - i\hat{\Omega}_{24,01}\rho_{43} + i\rho_{21}\hat{Q}_{13,\lambda_1\kappa_1} - \gamma^{\mathrm{nr}}_{23}\rho_{23},$$

$$\dot{\rho}_{24} = i\rho_{22}\hat{\Omega}_{24,01} - i\hat{\Omega}_{24,01}\rho_{44} - i\hat{Q}^{+}_{23,\lambda_2\kappa_2}\rho_{34} + i\rho_{25}\hat{Q}^{+}_{54,02} - \gamma_{24}\rho_{24},$$

$$\dot{\rho}_{25} = -i\hat{\Omega}_{24,01}\rho_{45} - i\hat{Q}^{+}_{23,\lambda_2\kappa_2}\rho_{35} + i\rho_{24}\hat{Q}_{45,02} - \gamma_{25}\rho_{25},$$

$$\dot{\rho}_{33} = i\rho_{31}\hat{Q}_{13,\lambda_1\kappa_1} - i\hat{Q}^{+}_{31,\lambda_1\kappa_1}\rho_{13} + i\rho_{32}\hat{Q}^{+}_{23,\lambda_2\kappa_2} - i\hat{Q}_{32,\lambda_2\kappa_2}\rho_{23}$$
$$-\gamma^{\mathrm{nr}}_{33}\rho_{33} + \gamma^{\mathrm{nr}}_{22}\rho_{22} + \gamma_{44}\rho_{44} + \gamma_{55}\rho_{55},$$

$$\dot{\rho}_{34} = -i\hat{Q}^{+}_{31,\lambda_1\kappa_1}\rho_{14} - i\hat{Q}_{32,\lambda_2\kappa_2}\rho_{24} + i\rho_{32}\hat{\Omega}_{24,01} + i\rho_{35}\hat{Q}^{+}_{54,02} - \gamma_{34}\rho_{34},$$

$$\dot{\rho}_{35} = -i\hat{Q}^{+}_{31,\lambda_1\kappa_1}\rho_{15} - i\hat{Q}_{32,\lambda_2\kappa_2}\rho_{25} + i\rho_{34}\hat{Q}_{45,02} - \gamma_{35}\rho_{35},$$

$$\dot{\rho}_{44} = i\rho_{42}\hat{\Omega}_{24,01} - i\hat{\Omega}^{+}_{42,01}\rho_{24} + i\rho_{45}\hat{Q}^{+}_{54,02} - i\hat{Q}_{45,02}\rho_{54} - 3\gamma_{44}\rho_{44} + \gamma^{\mathrm{nr}}_{55}\rho_{55},$$

$$\dot{\rho}_{45} = -i\hat{\Omega}^{+}_{42,01}\rho_{25} + i\rho_{44}\hat{Q}_{45,02} - i\hat{Q}_{45,02}\rho_{55} - \gamma^{\mathrm{nr}}_{45}\rho_{45},$$

$$\dot{\rho}_{55} = i\rho_{54}\hat{Q}_{45,02} - i\hat{Q}^{+}_{54,02}\rho_{45} - 3\gamma_{55}\rho_{55} + \gamma^{nr}_{55}\rho_{55}. \quad (7.9)$$

In the equations, we assume that all of the detunings are zero. In Eq. 7.9 we save the notations used in Eq. 7.4.

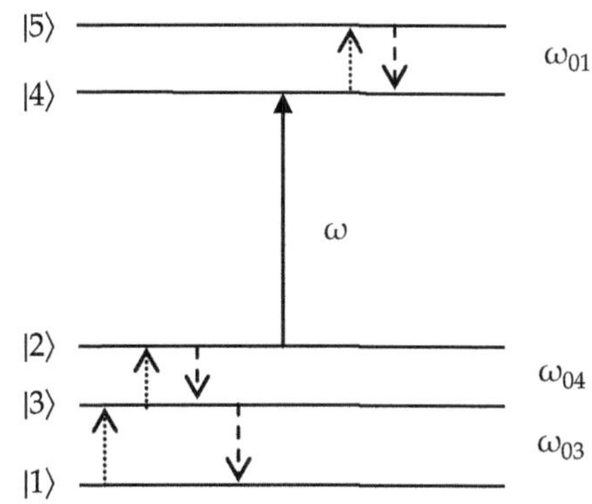

Figure 7.4 Scheme of vibronic interaction at the SSSs of the ground and excited RE ion levels coupled by the electromagnetic field for calculation of the number of electrons involved in the cooling process. The five-level energy structure is used for simplicity of calculations. The solid arrow shows the optical transitions at the frequencies ω; dotted arrows show the vibronic interaction with frequencies ω_{01}, ω_{03}, and ω_{04}; and downward dashed arrows refer to the nonradiative decay.

The changing of the level populations at the SSSs under the action of the Gaussian laser pulse is depicted in Fig. 7.5. It is shown that electrons from all SSSs of the ground level participate in the cooling cycle at 300 K. Moreover, the aggregate population that is transferred to the excited levels $|4\rangle$ and $|5\rangle$ after the pulse action is larger than the population at the upper level $|2\rangle$ before the pulse action (Fig. 7.4). Thus, the existence of vibronic interaction at SSSs leads to an increase of the number of electrons involved in the cooling cycle. Since the electrons initially populated the ground levels $|1\rangle$ and $|3\rangle$ consume energy to transfer to the upper level $|2\rangle$, these electrons participate in another cooling process that is additional to the process described in Section 7.2. We denote this process as C2.

In the calculations, we use the following parameters: $\hbar\omega_{03}$ = 421 cm^{-1}; $\hbar\omega_{04}$ = 155 cm^{-1}; $\gamma_{14} = \gamma_{15} = \gamma_{24} = \gamma_{25} = \gamma_{34} = \gamma_{35} = \gamma/3$; $\gamma_{55} = \gamma_{44} = \gamma$; $\gamma_{22}^{nr} = \gamma_{33}^{nr} = \gamma_{55}^{nr} = \gamma^{nr}$; $\gamma_{12}^{nr} = \gamma_{13}^{nr} = \gamma_{23}^{nr} = \gamma_{45}^{nr} = 0.5\ \gamma^{nr}$; and T

= 300 K. The Boltzmann distribution at the levels is established if the strengths of vibronic interaction are $k_{12}/\hbar$ = 3.3 × 10^{11} s^{-1}, $k_{23}/\hbar$ = 4.3 × 10^{11} s^{-1}, and $k_{45}/\hbar$ = 5 × 10^{11} s^{-1}. Other parameters are the same as in the Section 7.2.

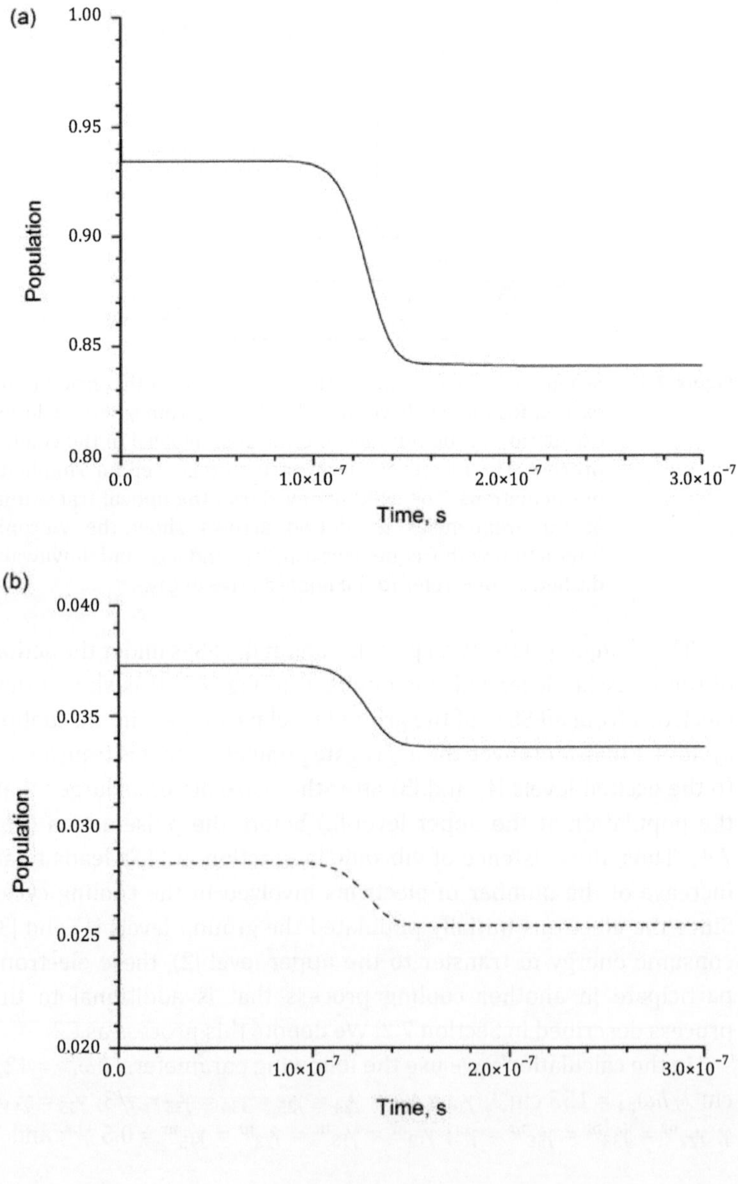

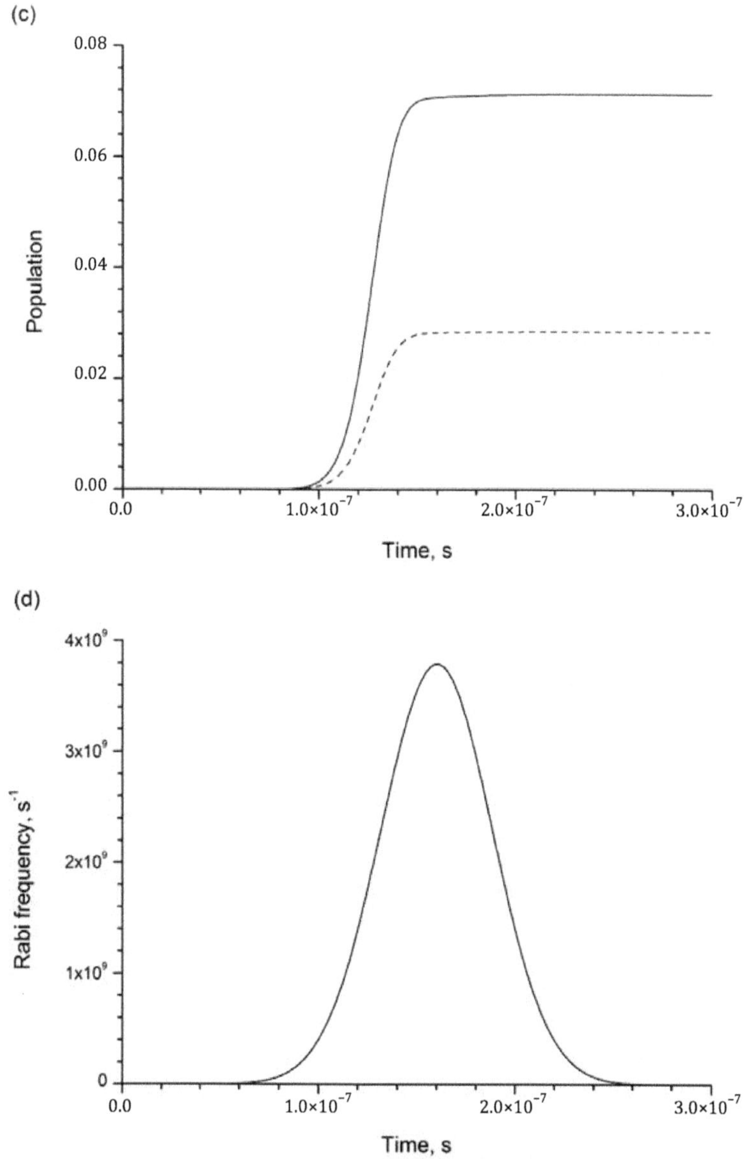

Figure 7.5 Time dependencies of the diagonal elements of the density matrix: (a) ρ_{11}, (b) ρ_{33} (solid line) and ρ_{22} (dashed line), and (c) ρ_{44} (solid line) and ρ_{55} (dashed line). (d) Rabi frequency of laser beam. Results are obtained at 300 K (cooling process C2).

7.3.2 Estimation of Cooling Performance

For determining cooling performance, we must take into account all the processes that contribute to cooling and heating of the considered doped system. The total power left over in the system is determined by the difference between heating (P_{heat}) and cooling (P_{cool}) powers:

$$P_{tot} = P_{heat} - P_{cool}, \; P_{heat} = P_{im} + P_{vib}. \tag{7.10}$$

Here, P_{im} is the power absorbed by impurity ions and P_{vib} is the power released in consequence of electronic transitions that lead to the energy transfer to the vibrational subsystem at the SSSs. The cooling power is determined by two above-considered cooling processes that are denoted by C1 and C2, respectively. Since the average thermal energy per one electron $\bar{\varepsilon}^{C}_{vib}$ is removed from the host matrix by radiative decay from the excited level with the rate γ, the expression for the cooling power can be written as

$$P_{cool} = N_1\gamma\left(\bar{\varepsilon}^{C1}_{vib} + \bar{\varepsilon}^{C2}_{vib}\right), \tag{7.11}$$

$$\bar{\varepsilon}^{C1}_{vib} = \left\langle\rho^1_{22}(\tau)\right\rangle\frac{S}{4} + \left\langle\rho^1_{44}(\tau)\right\rangle\left(E_6 - E_5 + \frac{S}{4}\right) + \left\langle\rho^1_{44}(\tau)\right\rangle\left(E_7 - E_5 + \frac{S}{4}\right), \tag{7.12}$$

$$\bar{\varepsilon}^{C2}_{vib} = \left(\left\langle\rho^2_{11}(0)\right\rangle - \left\langle\rho^2_{11}(\tau)\right\rangle\right)(E_4 - E_1) + \left(\left\langle\rho^2_{33}(0)\right\rangle - \left\langle\rho^2_{33}(\tau)\right\rangle\right)(E_4 - E_3), \tag{7.13}$$

$$S \equiv 3E_4 - \sum_{m=1}^{3} E_m, \; N_1 = n_{ion}N_e\left(\left\langle\rho^2_{44}(\tau)\right\rangle + \left\langle\rho^2_{55}(\tau)\right\rangle\right). \tag{7.14}$$

Here $\left\langle\rho^n_{ii}\right\rangle$ is the averaged value of the population of the level i for the system of differential equations n(n = 1 denotes Eq. 7.4; n = 2 denotes Eq. 7.9); N_1 is the number of electrons involved in the cooling cycle; n_{ion} is the ion concentration in the crystal; and N_e is the number of electrons at the ground level. In Eq. 7.12, we assume that the rates of radiative decay from all of the excited SSSs to all the ground SSSs are equal. One set of relaxation processes from level 7 is depicted in Fig. 7.3. The relaxation is related to two heating processes that we take into account in our calculations. Since the electron has an equal probability to appear at the ground SSSs as a result of the relaxation processes, establishment of the Boltzmann distribution at these SSSs heats the system. We denote this process as H1. Another heating process is nonradiative decay from the excited SSSs. The rate of radiative decay is of 4 orders of magnitude higher than the rate of nonradiative decay [16]. Due to this fact, no account is taken of nonradiative relaxation in Eqs. 7.4 and 7.9. However, the contribution of the nonradiative relaxation in heating power must be included. We denote this process as H2. So, the total contribution to the heating power can be written as

$$P_{\text{vib}} = N_1\left(\gamma\bar{\varepsilon}_{\text{vib}}^{\text{H1}} + W_{\text{nr}}\bar{\varepsilon}_{\text{vib}}^{\text{H2}}\right) \tag{7.15}$$

Here W_{nr} is the nonradiative decay rate. The average thermal energy per electron that is absorbed in the system can be determined for heating process H1 as

$$\begin{aligned}\bar{\varepsilon}_{\text{vib}}^{\text{H1}} = &\frac{S_1}{4} - \left\langle\rho_{11}^2(0)\right\rangle(E_2 - E_1) - \left\langle\rho_{22}^2(0)\right\rangle(E_4 + E_3 - E_1) - \\ &-\left\langle\rho_{33}^2(0)\right\rangle(E_3 - E_1)\end{aligned} \tag{7.16}$$

and for heating process H2 as

$$\begin{aligned}\bar{\varepsilon}_{\text{vib}}^{\text{H2}} = &\left\langle\rho_{22}^1(\tau)\right\rangle\frac{S_2}{4} + \left\langle\rho_{44}^1(\tau)\right\rangle\left(\frac{S_2}{4} - E_6 + E_5\right) \\ &+\left\langle\rho_{44}^1(\tau)\right\rangle\left(\frac{S_2}{4} - E_7 + E_5\right),\end{aligned} \tag{7.17}$$

$$S_1 \equiv \sum_{m=2}^{4} E_m - 3E_1,\; S_2 \equiv 4(E_5 - E_4) - \sum_{m=1}^{3} E_m\,. \tag{7.18}$$

The cooling efficiency is defined as a ratio of the negative total power to the power absorbed in the sample:

$$\eta_{\text{cool}} = -\frac{P_{\text{tot}}}{P_{\text{abs}}}\,. \tag{7.19}$$

The absorbed power can be taken in the form

$$P_{\text{abs}} = N_1\hbar\,(\omega_1 - \omega_2)\,\lambda. \tag{7.20}$$

Finally, we must determine the power absorbed by impurity ions. For this goal we can write the ratio of the power absorbed by impurity ions to the power absorbed in the sample in Eq. 7.19 using the Beer–Lambert law for both cases:

$$\frac{P_{\text{im}}}{P_{\text{abs}}} = \frac{1-\exp(-\beta)}{1-\exp(-1)},\; \beta = \frac{\alpha_{\text{im}}}{\alpha_{\text{rad}}}, \tag{7.21}$$

where α_{rad} is the optical absorption coefficient for the RE ions and α_{im} is the absorption coefficient for the impurity ions. Equation 7.21 defines the absorption of laser radiation on the length of $\alpha_{rad}{}^{-1}$.

It should be noted that the cooling process in the model proposed in this study is basically more efficient than the process involving forbidden optical transitions between the levels of the 4f multiplet in the traditional model. To demonstrate this advantage of the proposed model, we directly calculate the difference between the cooling efficiencies in the above-mentioned two models using the formulas available in Ref. [6]. Since the fundamental distinction between the models is that pumping of the SSSs is implemented at different (allowed and forbidden) optical transitions, the difference between the cooling efficiencies is defined by the difference between the absorption efficiencies only (all other parameters of the system being the same). The absorption efficiency can be expressed as [6]

$$\eta_{abs} = (1 + \beta)^{-1}. \tag{7.22}$$

Then, we use the expression for the absorption coefficient to find quantity β. The expression for the absorption coefficient can be written as

$$\alpha_i = \sigma_i N_i, \tag{7.23}$$

where σ_i is the absorption cross-section for a certain wavelength and N_i is the concentration of absorbing ions. If we assume that the absorption cross sections for the RE and impurity ions are equal, the quantity β is defined as the ratio of the concentrations of the RE and impurity ions. Typical commercial fluoride starting materials have transition metal impurities at the parts per million (parts in 10^6) level [6]. And we assume that the crystal doped to the 5 mol% Yb content. So hence, the ratio of the absorption coefficients is 2.0×10^{-5} in the case of allowed optical transitions. The optical absorption coefficient controlled by dipole-allowed transitions in the Yb^{3+} ion is 2 orders of magnitude larger than the absorption coefficient controlled by dipole-forbidden transitions. Hence, by determining the absorption efficiencies for the two models by Eqs. 7.22 and 7.23, we find that the cooling efficiency in the proposed model with coherent pumping is higher by 0.005 (or by 14%) than the cooling efficiency in the traditional model with pumping at forbidden transitions (see Fig. 7.7).

Knowing the cooling power, we can determine the time dependence of the temperature in the model system under study. The equation describing the temperature evolution with time can be written as [29]

$$C_m \rho_0 V_0 \frac{dT}{dt} = P_{tot}(T) + P_{load}(T). \tag{7.24}$$

Here, C_m is the specific heat of the cooled sample, ρ_0 is the mass density, V_0 is the irradiated volume of sample, and P_{load} is the environmental heat load that includes the convective, conductive, and radiative contributions. On the assumption that convective and conductive loads are negligible, the environmental heat load is defined by the radiative load given by the expression

$$P_{load}(T) = \sigma_B A_s \varepsilon_s (T_c^4 - T^4), \tag{7.25}$$

where σ_B is the Stefan–Boltzmann constant; A_s and ε_s are the surface area and emissivity of the sample, respectively; and T_c is the temperature of the chamber surrounding the sample. Equation 7.25 is written for conditions in which the chamber emissivity is equal to unity and the chamber surface area is much larger than the surface area of the sample. Thus, taking into account Eqs. 7.19, 7.20, and 7.25, we can rewrite the differential Eq. 7.24 as

$$C_m \rho_0 V_0 \frac{dT}{dt} = -\eta_{cool}(T) N_1(T) \hbar(\omega_1 - \omega_2)\gamma + \sigma_B A_s \varepsilon_s (T_c^4 - T^4). \tag{7.26}$$

Using Eq. 7.26 we can get the time dependence of the temperature of the irradiated volume of the sample. For that purpose we must define how the cooling efficiency and the number of electrons, which participate in the cooling cycle, depend on the temperature. The numerical calculations of time dependence of temperature we organize in the following order. Starting at room temperature, we find the solutions of the two systems of equations (Eqs. 7.4 and 7.9) for the level populations. After that, we can define the cooling efficiency and the number of electrons for a certain temperature. With the use of Eq. 7.26 we determine the temperature decrease throughout the cooling cycle, the duration of which is equal to γ^{-1}. Then, we repeat the cycle with the new temperature value. The calculations are terminated if the temperature stops changing.

The results of numerical simulation with the initial condition $T(0) = T_c = 300$ K are shown in Figs. 7.6 and 7.7. In the calculations, we use the following parameters: $V_0 = 2.12 \times 10^{-4}$ cm^3, $N_e = 13$, $n_{ion} = 3.7 \times 10^{21}$ cm^{-3} (for the Yb:CaF_2 crystal doped to the 5 mol% Yb content), and $W_{nr} = 5$ s^{-1}. The energies of the SSSs, E_m, are indicated in Fig. 7.3. Other parameters are defined before. The calculations are performed for different values of quantity β. The case of dipole-allowed transitions corresponds to $\beta = 2.0 \times 10^{-5}$, whereas the other three values of β are close to the case of dipole-forbidden transitions with the different background absorption coefficient α_{im}. It is shown in Fig. 7.6 that in the latter case the minimum attainable temperature substantially depends on the concentration of transition metal impurities in the crystal. Reducing the concentration of the impurity ions leads to deeper cooling of the crystal system. For example, if β changes from 6.0×10^{-3} to 4.0×10^{-3}, the minimum attainable temperature decreases from 101 K to 85 K (see insert in Fig. 7.6). This situation is typical for the traditional cooling model with direct pumping at the 4f–4f transitions [6]. The case of $\beta = 2.0 \times 10^{-5}$ simulates the situation with coherent pumping through the 5d level of the RE ion proposed here. In this case, the minimum attainable temperature is 74 K for current purity materials. It should be noted that the specified minimum attainable temperatures are reached within 5 s. The temperature dependencies of the cooling efficiency for the same values of β are shown in Fig. 7.7. It can be seen that the cooling efficiency of the doped system decreases if the temperature decreases. From these dependencies, we can find that the cooling efficiency in the proposed model ($\beta = 2.0 \times 10^{-5}$) is higher by 14% than the cooling efficiency in the traditional model ($\beta = 2.0 \times 10^{-3}$) if the optical absorption coefficient controlled by dipole-allowed transitions is 2 orders of magnitude larger than the absorption coefficient controlled by dipole-forbidden transitions. In other cases, the difference is much larger.

In addition, the cooling power in the proposed model is higher than that in the model with direct pumping at the 4f–4f transitions. As follows from Eqs. 7.19, 7.20, and 7.26, a substantial contribution to the increase in the cooling power is made by the increase in the number of electrons N_1 involved in the cooling process, apart from the increase in the cooling efficiency. In the case of coherent pumping, the population of the SSSs of the ground level, which is transferred to the SSSs of the excited level, is larger than the population of the upper SSS of the ground level of the ion. The ratio of the difference between these populations to the population of the upper SSS of the ground level is defined as

$$\Delta N = \frac{\left\langle \rho_{44}^2(\tau) \right\rangle + \left\langle \rho_{55}^2(\tau) \right\rangle}{\left\langle \rho_{22}^2(0) \right\rangle} - 1 . \tag{7.27}$$

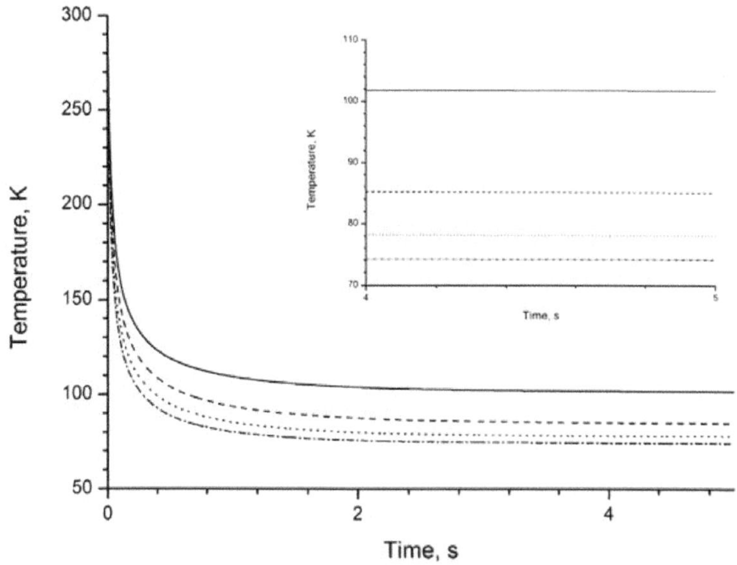

Figure 7.6 Time dependencies of the temperature in the process of laser cooling at different β: (solid line) 6.0×10^{-3}, (dashed line) 4.0×10^{-3}, (dotted line) 2.0×10^{-3}, and (dash-dotted line) 2.0×10^{-5}.

The temperature dependence of ΔN is shown in Fig. 7.8. From this dependence, it can be seen that the number of electrons from lower SSSs of the ground level, participating in the cooling cycle, increases with decreasing temperature. It is shown in Fig. 7.8 that below 120 K, the difference of the populations in the order of magnitude is greater than the population of the upper SSS of the ground level. This effect is due to the presence of coherent pumping and the existence of vibronic interaction. In the case of direct pumping, this population is divided between the ground and excited states in half (in the limiting case of saturation of absorption). Consequently, at the excitation intensities corresponding to saturation of absorption at the 4f–4f transitions, the cooling power in the model of coherent pumping is several times larger than the cooling power in the model of direct pumping of electron levels of the RE ion.

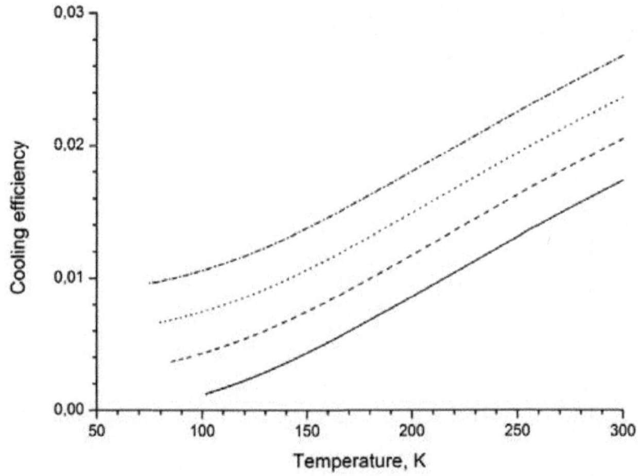

Figure 7.7 Temperature dependencies of the cooling efficiency at different β: (solid line) 6.0×10^{-3}, (dashed line) 4.0×10^{-3}, (dotted line) 2.0×10^{-3}, and (dash-dotted line) 2.0×10^{-5}.

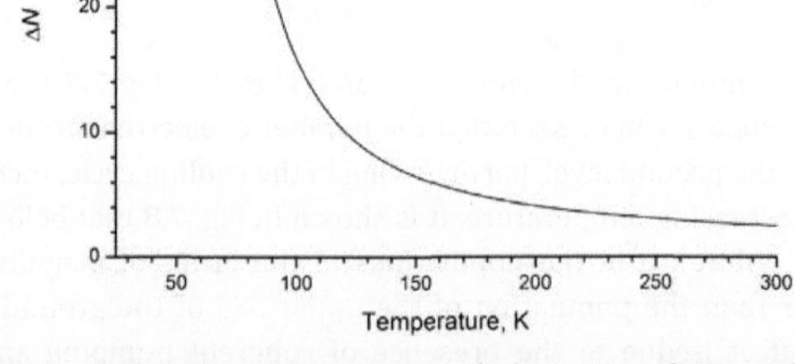

Figure 7.8 Temperature dependence of the ratio of the difference between the population of the SSSs of the ground level, which is transferred to the SSSs of the excited level, to the population of the upper SSS of the ground level (see Eq. 7.27).

7.4 Conclusions

A model of optical cooling of RE-doped crystals on coherent pulse pumping of the SSSs of RE ions is considered in this chapter. It is shown that the use of STIRAP through the dipole-allowed 5d level of the RE ions improves the characteristics of the process of laser cooling of solid systems in comparison to the corresponding characteristics obtained on pumping at the forbidden 4f–4f transitions. The absorption efficiency in the model proposed in the study is 14% higher than the absorption efficiency in the traditional model of direct 4f–4f pumping. In addition, in the proposed model, the power and cooling rate of the system are increased at least by a factor of 2 compared to these parameters in the traditional model. This is due to the fact that the number of electrons involved in the cooling cycle is larger in the case of coherent pumping than in the case of direct 4f–4f pumping. As a result, it becomes possible to implement deeper cooling for Yb^{3+}-doped crystals with purity that can be currently attained. This advantage of the proposed model seems to be of particular importance. Our calculations show that the minimum attainable temperature substantially depends on the concentration of transition metal impurities in the case of the traditional model. In this context, it is worth noting that the key method of improving the cooling efficiency in the case of pumping at forbidden optical transitions is to reduce the content of side impurities in the crystal matrix, which presents a rather difficult technological problem. Moreover, the proposed model is not required to use the nonresonant cavity for maximizing the absorbed laser power. Using coherent pumping, it becomes possible to reach 74.1 K in a time of 5 s. Since the optical cooling occurs at the electron transitions between the SSSs, the limiting minimum temperature of cooling in the above-developed model is the same as in other anti-Stokes fluorescence (ASF) models [6].

This investigation shows importance of the nonadiabatic effects for the problem of the laser cooling of solids doped by RE ions. The vibronic interaction model used in our calculations specifies the mechanism of interaction between the ion electrons and ion vibrations in the crystal matrix. And the density matrix formalism allows to describe coupling of the ion vibrations with the vibrations of host material by introducing a mechanism of heat loss. In the considered case, the loss is attributed to the cooling process. In addition, the establishment of the Boltzmann distribution at SSSs can be easily explained by the coupling between the vibronic subsystem and the phonon reservoir.

As we consider a multilevel system that resonantly interacts with photonic and vibronic subsystems, the density matrix formalism is applied. In the framework of this formalism we use the statistical approximation that implies that the system temperature is constant throughout the laser cooling cycle. However, the number of quanta of pseudolocal vibrational modes decreases in that time interval. Despite the fact that our approach takes into account the temperature dependence of the quantities that determine cooling performance, a more rigorous approach to thermodynamics of quantum-mechanical systems calls for the nonequilibrium statistical operator method [30].

In the above calculations, no account is taken of the temperature dependencies of the parameters involved in Eqs. 7.4 and 7.9 and their solution (the energy separations between levels, the decay rates, and the vibrational frequencies of the host matrix). However, using the temperature dependence of these parameters, obtained theoretical estimations can be only specified but cannot be substantially changed.

Since VUV radiation is efficiently absorbed in the extremely short length, the proposed model may be utilized for laser cooling of small solid-state objects that are used in up-to-date optoelectronics. Also, this model can be helpful in constructing logic elements of quantum computers.

References

1. Epstein, R. I., Buchwald, M. I., Edwards, B. C., Gosnell, T. R., Mungan, C. E. (1995). Observation of laser-induced fluorescent cooling of a solid, *Nature*, **377**, pp. 500–503.

2. Mungan, C. E., Buchwald, M. I., Edwards, B. C., Epstein, R. I., Gosnell, T. R. (1997). Laser cooling of a solid by 16 K starting from room temperature, *Phys. Rev. Lett.*, **78**(6), pp. 1030–1033.

3. Epstein, R. I., Brown, J. J., Edwards, B. C., Gibbs, A. (2001). Measurements of optical refrigeration in ytterbium-doped crystals, *J. Appl. Phys.*, **90**(9), pp. 4815–4819.

4. Patterson, W., Bigotta, S., Sheik-Bahae, M., Parisi, D., Tonelli, M., Epstein, R. (2008). Anti-Stokes luminescence cooling of Tm^{3+} doped BaY_2F_8, *Opt. Express*, **16**(3), pp. 1704–1710.
5. Nemova, G., Kashyap, R. (2010). Laser cooling of Er^{3+}-doped solids, *Opt. Commun.*, **283**, pp. 3736–3739.
6. Seletskiy, D. V., Hehlen, M. P., Epstein, R. I., Sheik-Bahae, M. (2012). Cryogenic optical refrigeration, *Adv. Opt. Photonics*, **4**, pp. 78–107.
7. Fernandez, J., Garcia-Adeva, A. J., Balda, R. (2012). Anti-Stokes laser-induced cooling in rare-earth doped low phonon materials, *Opt. Mater.*, **34**, pp. 579–590.
8. Ghasemkhani, M., Albrecht, A. R., Melgaard, S. D., Seletskiy, D. V., Cedeberg, J. G., Sheik-Bahae, M. (2015). Intracavity-enhanced optical refrigeration of Yb:YLF crystal to cryogenic temperatures, *Proc. SPIE,* **9380**, p. 938003.
9. Fleischhauer, M., Imamoglu, A., Marangos, J. P. (2005). Electromagnetically induced transparency: optics in coherent media, *Rev. Mod. Phys.*, **77**, pp. 633–673.
10. Hemmer, P. R., Turukchin, A. V., Shahriar, M. S., Musser, J. A. (2001). Raman-exited spin coherences in nitrogen-vacancy color centers in diamond, *Opt. Lett.*, **26**(6), pp. 361–363.
11. Zhao, Y., Wu, C., Ham, B. S., Kim, M. K., Awad, E. (1997). Microwave induced transparency in Ruby, *Phys. Rev. Lett.*, **79**(4), pp. 641–644.
12. Kolesov, R. (2005). Coherent population trapping in a crystalline solid at room temperature, *Phys. Rev.*, **72**, p. 051801(R).
13. Rand, S. C. (2013). Raman laser cooling of solids, *J. Lumin.*, **133**, pp. 10–14.
14. Ivanov, A. V., Rozhdestvensky, Y. V. (2015). Laser cooling of doped crystals by methods of coherent pumping, *Proc. SPIE,* **9380**, pp. 93800S (1-10).
15. Ivanov, A., Rozhdestvensky, Y., Perlin, E. (2015). Coherent pumping for fast laser cooling of doped crystals, *J. Opt. Soc. Am. B*, **32**(5), pp. B47–B54.
16. Perlin, Y. E., Kaminskii, A. A. (1985). Nonradiative transitions of the trivalent lanthanides in insulating laser crystals, *Phys. Status Solidi B*, **132**, pp. 11–40.
17. Lo, D., Makhov, V. N., Khaidukov, N. M., Krupa, J. C., Gesland, J. Y. (2004). Upconverted VUV luminescence of Nd^{3+} and Er^{3+} doped into $LiYF_4$ crystals under XeF-laser excitation, *J. Lumin.*, **106**, pp. 15–20.

18. Fischer, G. (1984). *Vibronic Coupling* (Academic, San Diego).
19. Bersuker, I. B., Polinger, V. Z. (1989). *Vibronic Interactions in Molecules and Crystals* (Springer, Berlin).
20. Perlin, Y. E., Wagner, M. (eds.) (1984). *The Dynamical Jahn-Teller Effect in Localized Systems* (North-Holland, Amsterdam).
21. Perlin, Y. E., Tsukerblat, B. S. (1974). *Electron-Vibrational Interaction in Optical Spectra of Impurity Paramagnetic Ions* (Shtintsa, Kishinev) (in Russian).
22. Saiko, A. P., Kuz'min, V. S. (1997). Lasing and Dicke superradiation by vibronic transitions in impurity crystals, *Laser Phys.*, **7**(2), pp. 384–386.
23. Vaikjärv, T., Hizhnyakov, V. (2014). Time-dependent pseudo Jahn-Teller effect: phonon-mediated long-time nonadiabatic relaxation, *J. Chem. Phys.*, **140**, p. 064105.
24. Klauder, J. R., Sudarshan, E. C. G. (1968). *Fundamentals of Quantum Optics* (Benjamin, New York); Glauber, R. J. (2007). *Quantum Theory of Optical Coherence. Selected Papers and Lectures* (Wiley-VCH, Weinheim).
25. Vitanov, N. V., Halfmann, T., Shore, B. W., Bergmann, K. (2001). Laser-induced population transfer by adiabatic passage techniques, *Annu. Rev. Phys. Chem.*, **52**, pp. 763–809.
26. Leniec, G., Kaczmarek, S. M., Boulon, G. (2005). EPR and optical properties of CaF_2:Yb single crystals, *Proc. SPIE,* **5958**, p. 595825.
27. Szczurek, T., Schlesinger, M., Jezowska-Trzebiatowska, B., Legendziewicz, J., Strek, W. (eds.) (1985). *Rare Earth Spectroscopy* (World Scientific, Singapore).
28. Misra, P., Dubinskii, M. A. (eds.) (2002). *Ultraviolet Spectroscopy and UV Lasers* (Marcel Dekker, New York).
29. Edwards, B. C., Anderson, J. E., Epstein, R. I., Mills, G. L., Mord, A. J. (1999). Demonstration of a solid-state optical cooler: an approach to cryogenic refrigeration, *J. Appl. Phys.*, **86**(11), pp. 6489–6493.
30. Andrianov, S. N., Samartsev, V. V. (1998). Laser cooling of impurity crystals, *Laser Phys.*, **8**(1), pp. 14–18.

Chapter 8

Bulk Cooling Efficiency Measurements of Yb-Doped Fluoride Single Crystals and Energy Transfer–Assisted Anti-Stokes Cooling in Co-Doped Fluorides

Azzurra Volpi, Alberto Di Lieto, and Mauro Tonelli
Dipartimento di Fisica-NEST Instituto di Nanoscienze CNR,
Universita`di Pisa, L. B. Pontecorvo 3, I-56127, Pisa, Italy
volpi@df.unipi.it

8.1 Introduction

Optical cooling of solids, based on anti-Stokes fluorescence, has tremendously advanced in the last years. Since its first experimental demonstration in 1995 [5], the temperature drops achieved have dramatically increased from a few degrees up to more than 150. Operations in the cryogenic regime have been reached in ytterbium (Yb)-doped fluoride single crystals [13, 14, 21]. To date, the lowest temperature achieved is 93 K [12] and novel schemes based on energy transfer processes prospect significant enhancements

Laser Cooling: Fundamental Properties and Applications
Edited by Galina Nemova

ISBN 978-981-4745-04-8 (Hardcover), 978-981-4745-05-5 (eBook)
www.panstanford.com

of intrinsic efficiencies, pushing down the minimum achievable temperatures.

Truly vibration-free, cryogenic laser cooling is a much sought after solution for several space-based technologies. The use of mechanical coolers indeed pushes to complex system solutions in order to minimize microphonic vibrations, which limit the detectors' performances. Accessing temperatures well below the thermoelectric coolers' (TECs) cutoff (170 K), the implementation of cryogenic optical coolers would provide an ideal solution for cooling down detectors in the temperature range below 170 K. Moreover, optical coolers would offer an efficient alternative to TECs in the higher temperature range, below 200 K, where TECs exhibit limited wall-plug efficiency. Currently, passive radiators, relying on thermal radiative emission into deep space, provide the main vibration-free solution for cooling down detectors below 170 K, down to 50 K. The use of these devices, however, imposes several orientation constraints and shielding requirements to build up the refrigeration structure. Furthermore large emitting surfaces are necessary to achieve low temperatures and relatively high cooling powers. The radiated power, and hence the available cooling power, is directly proportional to the panel surface as well as to the fourth power of the temperature to reach. Consequently, due to size limitations, the performances rapidly decrease as the temperature to achieve drops, especially below 100 K. The optical cooling technology offers, instead, advantages of extreme compactness and cooling efficiencies almost independent of dimensions, enabling significant size and weight reduction. In addition to that, optical cooling allows for simpler designs, without orientation and shielding constraints, and possesses a much higher potential for intrinsic efficiency enhancement through the development of novel materials. Additional advantages for space applications, offered by the optical cooling technology, are a long lifetime (only limited by the lifetime of the pumping laser), low electromagnetic interference, low sensitivity to magnetic fields, and extremely high reliability.

All these features intensely motivated the research toward device implementation as well as investigation in material development to uplift the intrinsic efficiency of cooling crystals, which are currently of the order of a few percent from room temperature. Various techniques have been exploited and many are under investigation. Laser cavities were introduced to enhance the absorption of incident laser photons and so the available cooling power at low temperatures [12–14, 21]. Suitable configurations were exploited to reduce parasitic heat loads. Novel alternative schemes based on photon localization effects in nanocrystalline powders were recently suggested to promote the input absorption [19]. The superirradiance regime was proposed to increase the cooling process rate [15].

While several methods have been proposed and can be employed to improve the cooling performances, energy transfer–assisted anti-Stokes processes in co-doped systems have been recently demonstrated to provide a route to enhance the intrinsic cooling efficiency of active materials. The first experimental evidence was achieved with Yb–Tm co-doped systems. Efficiency enhancements, over the single Yb doping performances, have been demonstrated via co-doped YLF single crystals [24]. Preliminary results achieved in our laboratory are presented in the following section.

This chapter reports on cooling efficiency measurements performed on various Yb-doped fluoride single crystals grown in our laboratory. The investigation of the cooling process for increasing Yb concentrations in $LiYF_4$ (YLF) single crystals resulted in improved cooling efficiency via significant decrease of detrimental parasitic absorption. Energy transfer processes between Yb ions and low concentrations of rare earth ion impurities were investigated to analyze the effect of detrimental impurities that participate in multiphonon relaxation of excited ions. Studies on Yb–Tm-controlled co-doping led to the first observations of energy transfer–enhanced Yb anti-Stokes efficiency [24]. Preliminary results achieved with Yb–Tm co-doped samples are presented.

The following section introduces the anti-Stokes efficiency model for rare earth–doped materials. Experimental results and a description of experimental apparatus and instruments are reported in Sections 8.3–8.5.

8.2 Optical Cooling Model for Rare Earth–Doped Materials

The idea of using rare earth ions to achieve anti-Stokes cooling in the solid phase was firstly proposed by Kastler in 1950 [9]. The interest for these ions basically relies on their electronic configuration. In the triply ionized state, indeed, the 4f valence electrons are heavily shielded from the surrounding crystal field by the filled 5s and 5p shells, which have a larger spatial extension (Fig. 8.1). When inserted as impurities in a solid, trivalent rare earth ions hence weakly interact with the crystal field of the surrounding. As a result, these ions exhibit atomic-like spectra even when embedded in hosting matrices and the 4f–4f optical transition can be excited in transparent solids. Such transitions amongst 4f states, parity forbidden in the free ions, are weakly allowed in hosting crystals because of induced dipoles due to the interaction of ions with the crystal field, mixing 4f states with different parities.

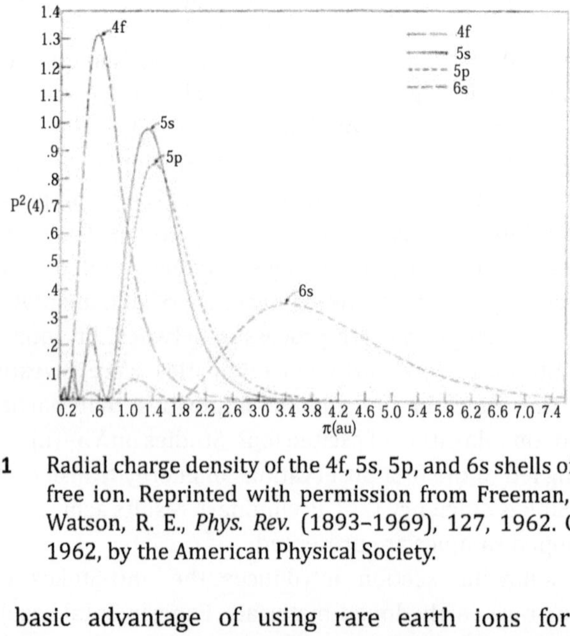

Figure 8.1 Radial charge density of the 4f, 5s, 5p, and 6s shells of the Gd^{3+} free ion. Reprinted with permission from Freeman, A. J. and Watson, R. E., *Phys. Rev.* (1893–1969), 127, 1962. Copyright 1962, by the American Physical Society.

The basic advantage of using rare earth ions for optical refrigeration, however, is related to the extremely high radiative efficiency achievable with 4f–4f transitions. The efficient screening of valence electrons strongly reduces the electron–phonon coupling and suppresses nonradiative decays due to multiphonon emissions from excited 4f states.

Additionally, the screening of the 4f orbital minimizes the Stokes shift of the emission associated with the 4f–4f transitions, allowing states with relatively a large absorption cross section to be utilized for the anti-Stokes cooling process. Such a process is basically enabled by the interaction of active ions with the crystal field, which causes, at the first order of the perturbation theory, a Stark splitting of the free ion energy levels, with a magnitude of the order of 10^2 cm^{-1}. Such splitting leads to energy widths for the 4f manifolds of the order of a few hundreds of cm^{-1}, sufficient for heat removal via anti-Stokes emission. By exciting the lower-energy inter-Stark transition, annihilation of lattice phonons occurs to reequilibrate the Boltzmann distribution of the Stark populations. Fluorescence at higher average energies than pumping photons follows, carrying out thermal energy from the system. Thermalization of the electronic excitations with the phonon field (10^{-12} s) occurs at a much faster timescale than spontaneous emission (10^{-3} s). Bulk cooling of the solid, however, occurs only if the decay is predominantly radiative. Loss channels such as nonradiative decay rates or energy transfer from excited ions thus need be minimized, as well as trapping of anti-Stokes photons.

In the ideal anti-Stokes process each emitted photon removes an average energy equal to the difference between the emitted and the absorbed photons. The cooling efficiency can be defined as the ratio of the energy absorbed from the lattice to the input energy. The efficiency of the ideal anti-Stokes process η_c^{id} can thus be formalized as

$$\eta_c^{\text{id}} = \frac{h\nu_f - h\nu}{h\nu} = \frac{\lambda}{\lambda_f} - 1 \tag{8.1}$$

where $\nu_f(\lambda_f)$ is the mean emission frequency (wavelength) of the electronic transition and $\nu(\lambda)$ the frequency (wavelength) of pumping photons. In real cases, however, in addition to the anti-Stokes process, heat-generating mechanisms, related to multiphonon quenching and impurity-mediated absorption, can occur after laser excitation. Furthermore, fluorescence photons can be trapped due to reabsorption and internal reflection phenomena.

The currently used model for anti-Stokes efficiency in rare earth–doped solids, addressing competing radiative and nonradiative optical processes and loss channels, was developed by Epstein et al. [22]. In this model the probability of a pumping photon to be converted into an anti-Stokes photon escaping the material is quantified by two parameters, the external quantum efficiency (EQE) of the electronic transition and the background absorption coefficient of the material. The EQE parameter defines the radiative efficiency of the electronic transition, addressing the probability of photon trapping. The background absorption coefficient defines the actual absorption efficiency, accounting for loss mechanisms that lower the number of absorbed photons that may initiate the anti-Stokes process. In this model, the cooling efficiency, dependent on temperature T and pumping wavelength λ, is formalized as [20, 22]

$$\eta_c(\lambda,T)=\eta_{ext}\eta_{abs}(\lambda,T)\frac{\lambda}{\lambda_f(T)}-1 \tag{8.2}$$

where η_{ext} represents the EQE of the electronic transition and $\eta_{abs}(\lambda, T)$ the absorption efficiency, defined as

$$\eta_{abs}(\lambda,T)=\frac{\alpha_r(\lambda,T)}{\alpha_r(\lambda,T)+\alpha_b} \tag{8.3}$$

where $a_r(\lambda, T)$ is the resonant absorption of the electronic transition and α_b the background absorption parameter of the material. Net cooling requires a positive value of η_c. Such condition imposes stringent requirements on the value $\eta_{ext}\eta_{abs}$, that is, on the material efficiency. Considering, at temperature T, a detuning of magnitude $k_B T$ between emitted and absorbed photons, net cooling ($\eta_c > 0$) requires

$$\eta_{ext}\eta_{abs} > 1-\frac{k_B T}{h\nu_f} \tag{8.4}$$

Considering a transition gap of 1 eV, such as for the Yb transition, $\eta_{ext}\eta_{abs} > 98\%$ is required to achieve net cooling starting from room temperature. If the temperature is reduced to 100 K, the efficiency requirement increases to $\eta_{ext}\eta_{abs} > 99\%$ for cooling to occur. Both an EQE close to unity and minimized background absorption are thus fundamental conditions to achieve efficient cooling. Stringent requirements on the crystal quality follow.

It is well known that the EQE parameter is basically related to the probability of nonradiative decay and to the efficiency of photon extraction. In the formula $n_{ext} = \eta_e W/\eta_e W_r + W_{nr}$, η_e is the fluorescence extraction efficiency and W_r and W_{nr} the radiative and nonradiative decay rates, respectively, of the electronic transition. Fluorescence photons basically get trapped inside the material due to reabsorption and total internal reflection, while nonradiative decays are basically related to multiphonon relaxations and phonon-assisted energy transfers.

The choices of electronic transition and hosting material are fundamental to minimize the efficiency of nonradiative quenching and photon-trapping phenomena. Optical transitions of trivalent rare earth ions, embedded in fluoride hosts, are preferable as they allow for high EQE values due to the low electron–phonon coupling and low cutoff of optical phonon energy of fluoride crystals. Furthermore the low refractive index of fluoride crystals allows for high-fluorescence escape efficiencies, reducing photon trapping due to total internal reflections. Among trivalent rare earths, however, Yb^{3+} ions possess superior properties that enable, when embedded in fluoride hosts, extremely reduced nonradiative decay rates and hence exceptionally high EQE values. Firstly, the absence of a higher excited state, above the first excited manifold $^2F_{5/2}$, hinders nonresonant Yb–Yb energy transfer processes, which are likely for rare earth ions to cause efficient quenching of excited states via multiphonon release. The absence of higher excited states also enables the use of high concentrations of Yb ions, to enhance absorption efficiency, without foster detrimental energy transfer processes. Secondly the large energy gap of the Yb transition, around 1 μm, inhibits multiphonon relaxation processes, reaching negligible rates in fluoride hosts. Furthermore, high-power pump sources (>10 W CW) emitting around 1 μm are readily available. All these features contributed to the extremely high cooling performances achieved in Yb-doped fluoride materials.

Currently, the most critical parameter limiting the efficiency of such materials is the background absorption coefficient, α_b. This parameter reckons with all the parasitic processes competing with the resonant absorption of the electronic transition. Although its nature and origin are still under investigation, background absorption is basically related to unwanted impurities embedded within materials during the synthesis. Optically active contaminants introduce additional energy levels, which are likely to cause absorption of pumping photons, phonon-assisted energy transfer, and multiphonon quenching of electronic excitations. Spurious laser scattering due to crystal defects also contributes to parasitic absorption, violently heating up the lattice. All these processes significantly degrade the absorption efficiency of the anti-Stokes process, reducing the number of pumping photons that can initiate the anti-Stokes process. Nonradiative impurity-mediated processes also degrade the EQE of the electronic transition. The cooling mechanism can even be compromised if the efficiency of the heat-generating processes is not balanced by the efficiency of the anti-Stokes cooling process [11]. Several studies identify the severe impact of transition metals on the Yb anti-Stokes efficiency [7, 14]. Divalent ions of 3d metals, such as Cu, Fe, Co, and Ni, have strong absorption bands around 1 μm and thus efficiently absorb pump power, lowering the number of pump photons that can initiate the anti-Stokes process. A large background absorption decrease was observed in YLF crystals with the reduction of Fe^{2+} ions in cooling samples, resulting in net efficiency enhancement. Rare earth impurities were also observed to influence the efficiency of the Yb anti-Stokes process, degrading the EQE via nonradiative relaxation and heat-generating energy transfer processes [3]. Any other kind of impurities that participate in nonradiative quenching of excited ions degrades the anti-Stokes efficiency, leading to poor cooling performances. Calculations indicate that further purification of starting materials, leading to a significant background absorption decrease, is the way to improve the current state-of-the-art cooling efficiencies, pushing down minimum achievable temperatures, while increasing the cooling powers available at low temperatures. Due to the chemical inertness of binary fluorides, however, the control of impurities in binary fluorine materials is quite complicate and the purification below the part-per-million level is still a quite challenging process. One approach currently under investigation is provided in Ref. [6].

The background absorption coefficient and EQE parameters are normally used to quantify the cooling performances of investigated materials. Such parameters can be estimated from the fit of experimental data of cooling efficiency with the model curve in Eq. 8.2, where data of the mean emission wavelength and resonant absorption coefficient need to be inserted for each material. Such spectroscopic parameters need to be previously determined from optical investigations of the electronic transition, through absorption and emission measurements.

8.3 Experimental Setup

Before performing cooling efficiency measurements, the crystalline active materials were spectroscopically characterized by absorption and static and dynamic fluorescence measurements. In this section the experimental apparatus and instruments employed for spectroscopic and cooling measurements are described.

8.3.1 Spectroscopic Setup

Absorption measurements were performed by means of a dual-beam design spectrometer Varian Cary 500, operating between 180 nm and 3200 nm. The absorbance range of the machine is 0–10 with resolution 0.001. The nominal wavelength resolution is 0.01 nm in the UV-visible range and 0.04 nm in the IR. Polarized absorbance spectra were acquired by placing two identical Glam–Thompson polarizers on the reference and sample paths in order to linearly polarize the beam incident onto the sample.

To measure the fluorescence spectra, the sample was excited by means of a continuous-wave (cw) laser at a fixed wavelength and the intensity of the emission was registered as a function of the wavelength. In the case of a large overlap between absorption and emission bands, such as for the Yb transition, reabsorption phenomena can cause significant distortions of the intrinsic fluorescence spectra. At shorter wavelengths, where strong absorption is also present, the emissions peaks tend to be decreased as the emitted radiation is partially reabsorbed before it exits the material. Depleted fluorescence is hence observed in this region.

To minimize distortions of the observed fluorescence due to reabsorption-trapping artifacts, the sample was mounted on a pinhole support, which enables one to minimize the overlap between the volume of the sample that is excited and the volume of the sample from which fluorescence is observed. At the same time, the emission signal can be acquired perpendicularly to the excitation beam in order to minimize spurious laser scattering. Polarized pumping can be accurately performed as well as observation of selective fluorescence from each optical axis. The fluorescence from the sample, after laser excitation, was modulated by using a rotating chopper, with 77 Hz frequency, and focused on the entrance slit of a monochromator equipped with interchangeable gratings (300, 600, 1200 grooves/mm) and input and output slits adjustable between 0.01 and 2 mm. The fluorescence signal was then acquired by using a suitable detector placed at the exit of the monochromator. For IR fluorescence the monochromator was equipped with the 600 grooves/mm grating (with dispersion of 5.4 nm/mm) and a 0.1 mm slit aperture; the signal was acquired by means of an indium antimonide (InSb) photodiode (EG&G Judson) cooled to 77 K with liquid nitrogen. The working range of the detector extends from 1 to 5.5 μm and the sensitivity is 2.5 mA/W at 5 μm. For the visible–near-IR (NIR) region the monochromator was equipped with the 1200 grooves/mm grating (with dispersion of 2.77 nm/mm) and a 0.2 mm slit aperture; the signal was detected by means of a photomultiplier tube (Hamamatsu R943-02). Its spectral response ranges from 160 nm to 930 nm, with a sensitivity of 65 mA/W at 82 nm. The output signal from the detector was processed by a lock-in amplifier that used the chopper frequency as reference and stored on a PC. Polarized spectra were measured by placing a Glam–Thompson polarizer in front of the entrance slit of the monochromator. Raw data acquired were calibrated for the optical response of the apparatus, collecting the fluorescence spectra of an incandescent lamp with emission spectra equivalent to a black body at 3000 K. The optical response curve of the apparatus was derived combining acquired fluorescence spectra from the lamp and theoretical black-body emission spectra.

For dynamic fluorescence measurements, the emission temporal decay was investigated after laser excitation of the sample with a pulsed source. For the Yb transition, the pinhole method [10] was used to minimize lengthening of the intrinsic lifetime due to reabsorption phenomena. To implement such a method, the sample was mounted behind a pinhole mask and the temporal decay of the fluorescence was measured as a function of the aperture radius. The intrinsic radiative lifetime was then extrapolated as the value corresponding to a zero hole radius [10]. This method was not used for static fluorescence measurements, because acquisition of polarized spectra was necessary as well as polarized pumping along a single optical axis.

The excitation source for dynamic measurements was a tunable titanium-Al_2O_3 (Ti:Sa) pulsed laser with an emitting wavelength in the range between 750 nm and 870 nm and a maximum pulse energy of 9 mJ. The pulse duration was 30 ns and the repetition rate 10 Hz.

For measurements of the Yb transition lifetime, the Ti:Sa beam was tuned to a maximum in the absorption (~960 nm). The fluorescence from the sample was focused on the entrance slit of the monochromator, equipped with the 600 grooves/mm grating and locked at a fixed wavelength, corresponding to a maximum emission peak. The fluorescence signal was then acquired by means of an InSb photodiode. The signal from the detector was amplified, by fast amplifiers, and then processed by using a digital oscilloscope, triggered at the pulse repetition frequency of the laser excitation source.

8.3.2 Cooling Setup

Cooling efficiency measurements were performed in a single-pass configuration by monitoring the heating/cooling temperature change of the sample after laser excitation at varying wavelengths within the Yb absorption spectrum. A layout of the cooling setup is shown in Fig. 8.2. The sample was mounted inside a steel vacuum chamber ($10 \times 10 \times 10$ cm^3), evacuated at a pressure of 10^{-5} Pa, in order to minimize convective heat load on the sample. Inside the vacuum chamber, the sample was mounted on a stainless steel support suspended on two optical fibers of 120 μm diameter. Such a scheme was designed in order to minimize parasitic conductive heat loads, ensuring thermal isolation of the sample from the surrounding. Four windows on the chamber walls enabled excitation and acquisition of the sample temperature simultaneously.

Four diode laser sources, with central wavelengths between 940 and 1060 nm, were used to excite the sample. All the junctions were mounted to excite the sample along the *c* axis, which allows the highest absorption coefficient. Antireflection-coated BK7 windows were used to pass the laser beams trough the vacuum chamber. To measure the temperature of the sample, contactless thermometry techniques were used. Two methods were employed, a thermal camera with a microbolometer sensor and a spectroscopic setup for acquisition of fluorescence spectra from the sample, which allow one to implement the differential luminescence technique (DLT).

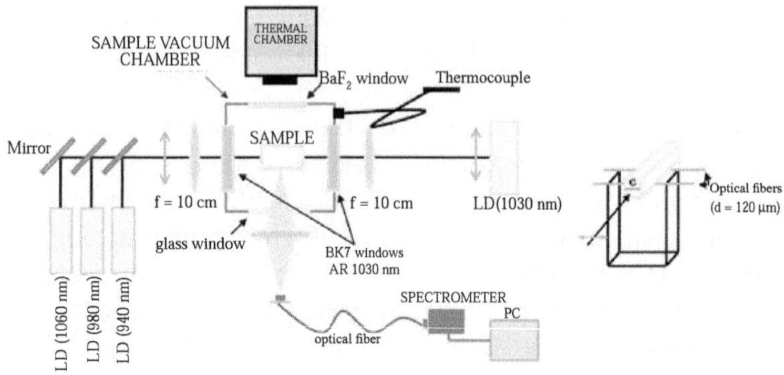

Figure 8.2 Layout of cooling setup for cooling efficiency measurements. On the right is shown a sketch of the sample mounting inside the vacuum chamber.

A long-wavelength infrared (LWIR) thermal camera (Raython 2500 AS) with a photoconductive sensor, made of an uncooled amorphous silicon microbolometer array, was used for the smaller temperature changes, below 5 K. The output of the camera is a gray-scale image file with 8-bit pixels, where dark pixels correspond to hotter areas, while colder areas show up as lighter pixel regions. The nominal resolution of is 0.2 K. A BaF_2 window, which is transparent up to 10 μm, was used to allow the thermal sensor to acquire the fluorescence from the sample. For each pumping wavelength, a thermal video was registered while the sample was excited with the laser beam. The video started with the absence of laser excitation and extended until the thermal equilibrium was reached. From the images acquired, the temperature change of the sample with respect to the surrounding was derived. Thermal images were acquired at a rate of 0.1 Hz. The stability of the chamber temperature was monitored by means of a suitable thermocouple. The thermal camera was previously calibrated in order to determine the proportionality relationship between the pixel values and the temperature of the sample. A linear working range of ±5 K was measured.

For larger temperature changes the DLT was used [21]. Such a technique is based on changes in the shape of the emission spectra as a function of temperature. As the temperature lowers, the decrease of phonon excitations reflects in changes of the electronic excitations distribution of the Stark sublevels. The vibronic broadening of the electronic emission spectra decreases as a function of temperature; consequently the emission lines narrow and become more intense.

The DLT uses the difference between integrated areas of emission to measure the temperature changes.

Emission spectra from the sample, excited with laser beams inside the vacuum chamber, were acquired from a glass window. As for the thermal camera, the acquisition of the spectra started with the absence of laser excitation and extended until thermal equilibrium was reached. A fiber-coupled spectrometer, operating between 200 and 1100 nm with 1 nm resolution, was used. A previous calibration was performed to connect the differential spectrum signal to the temperature value. For the calibration measurements, the cooling setup geometry was closely resembled in order to cause similar distortion effects due to reabsorption. The temperature of the sample was varied by using a Peltier cell and the fluorescence spectra were acquired as a function of the temperature. The estimated resolution for the DLT was of the order of 1 K.

8.4 Experimental Results

This section reports on results of cooling efficiency measurements performed on Yb-doped YLF single crystals grown in our facility. The EQE and background absorption parameters were evaluated from experimental data of cooling efficiency and outcomes of spectroscopic analysis.

The effect of increasing concentrations of Yb ions on the anti-Stokes efficiency was investigated. Larger concentrations of Yb ions resulted in efficiency enhancements due to diminishing background absorption. Different relative concentrations of rare earth impurities were observed to result in depleted EQE values in the case of higher amounts of specific rare earths ions in certain samples. Energy transfer processes between Yb and low concentrations of impurities were investigated. Significant efficiency enhancements, over the single Yb doping performances, were achieved in Yb–Tm co-doped YLF single crystals, providing the first experimental evidence of energy transfer–assisted anti-Stokes cooling. Preliminary results prospect appreciable improvements for low-temperatures operations, besides huge potentialities for intrinsic efficiency enhancements of active materials.

Prior to cooling experiments, spectroscopic measurements were performed to evaluate the resonant absorption coefficient and the mean emission wavelength of the Yb transition. Bulk cooling experiments were performed in a single-pass configuration measuring the heating/cooling temperature change as a function of the pumping wavelength.

8.4.1 Investigation of Anti-Stokes Cooling in YLF Single Crystals with Varying Yb Doping Levels

The anti-Stokes efficiency was systematically studied in YLF crystalline hosts with varying Yb doping levels. Single crystals doped Yb at 5 at%, 7.5 at%, and 10 at% were investigated.

The increase of active ion concentration basically results in enhancement of the absorption coefficient of the material, which is particularly sought after in the excitation region for the anti-Stokes process, at $\lambda > \lambda_f$. This region indeed corresponds to the long-wavelength tail of the absorption spectrum, where resonant absorption is low. Enhancements of Yb concentration, increasing the resonant absorption, enable improvements of the absorption efficiency for the anti-Stokes process: $\eta_{abs}(\lambda,T)=\dfrac{\alpha_r(\lambda,T)}{\alpha_r(\lambda,T)+\alpha_b}$.

Intensification of reabsorption-related trapping phenomena, however, needs to be considered when the active ion concentration is increased.

Prior to cooling experiments, absorption and static and dynamic fluorescence measurements were performed for each doping level in order to calculate the resonant absorption coefficient and the mean emission wavelength and, furthermore, in order to check for anomalies.

Figure 8.3a shows the polarized absorption spectra acquired for the YLF:5%Yb sample at room temperature. The multipeak structure of the spectra is related to the inter-Stark transitions between Stark sublevels of the two manifolds. The highest absorption, over the whole spectrum, is achieved through selected excitation along the *c* axis. The strongest peak in the YLF:5%Yb is 8.36 cm^{-1} at 959.6 nm. Similar behaviors were observed for the higher Yb doping levels. The spectra showed a similar structure but increasing magnitude. In Fig. 8.3b are reported the absorption spectra of the $^2F_{5/2}$ excited manifold acquired, in the same experimental conditions, for the three doping levels (5%, 7.5%, and 10%) in π-polarization. As can be seen from the graph, the magnitude of the resonant absorption increases linearly with the Yb concentration, leading to improvement of about a factor of 2 between 5% and 10% Yb doping in the excitation region for anti-Stokes.

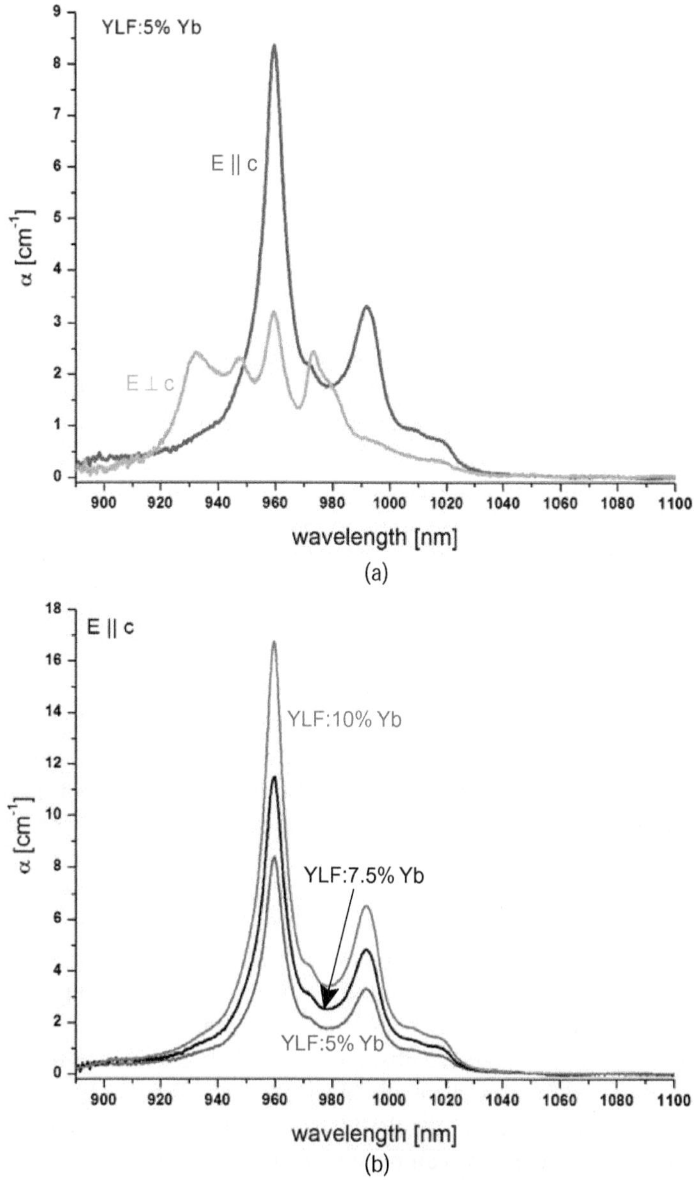

Figure 8.3 (a) Polarized absorption coefficient of YLF:5%Yb acquired. (b) Absorption coefficient of YLF:5%Yb, YLF:7.5%Yb, and YLF:10%Yb for π-polarization.

Following excitation of the $^2F_{5/2}$ manifold, the spontaneous emission yields a broadband fluorescence spectrum $I(\lambda)$ with a mean wavelength of

$$\lambda_f = \frac{\int I(\lambda)\lambda d\lambda}{\int I(\lambda)d\lambda} \tag{8.5}$$

An excitation tuned to wavelengths longer than λ_f is required to achieve anti-Stokes emission.

Polarized fluorescence spectra were acquired for the 5%, 7.5%, and 10% Yb-doped YLF samples after laser excitation of the Yb $^2F_{5/2}$ manifold at 940 nm. Fluorescence spectra registered for the YLF:5%Yb and YLF:10% samples are shown in Fig. 8.4.

Although the structure of emission spectra does not change as the Yb concentration increases, the increase of reabsorption-related effects for larger concentrations of active ions influences the intensity distribution of the emission.

Due to the large overlap between absorption and fluorescence bands, photons that are spontaneously emitted by excited ions tend to be efficiently reabsorbed by other ions in the ground state. Increasing the concentration of Yb ions, radiation-trapping phenomena, due to reabsorption, tend to be more efficient, leading to more depleted emissions at higher energies, as can be observed by comparing spectra in Fig. 8.4.

Fluorescence spectra were used both to calculate the mean emission wavelength parameter and the absorption coefficient via the reciprocity method [17]. This technique is used to improve the accuracy of the absorption coefficient measurement in the long-wavelength tail of the spectrum, that is, the excitation region for anti-Stokes emission. Due to the low value of the absorption in this region, direct measurements normally exhibit large noise. Using reciprocity theory, instead the resonant absorption is derived from fluorescence spectra. The emission spectra exhibit a considerably higher signal-to-noise ratio in this wavelength region and hence improved accuracy for the absorption coefficient evaluation. Figure 8.5 reports, for comparison, the absorption coefficient of YLF:5%Yb for the π-polarization calculated from direct measurements and by using reciprocity. As can be seen from the graph, the reciprocity theory well reproduces the behavior of the absorption coefficient with significantly higher accuracy.

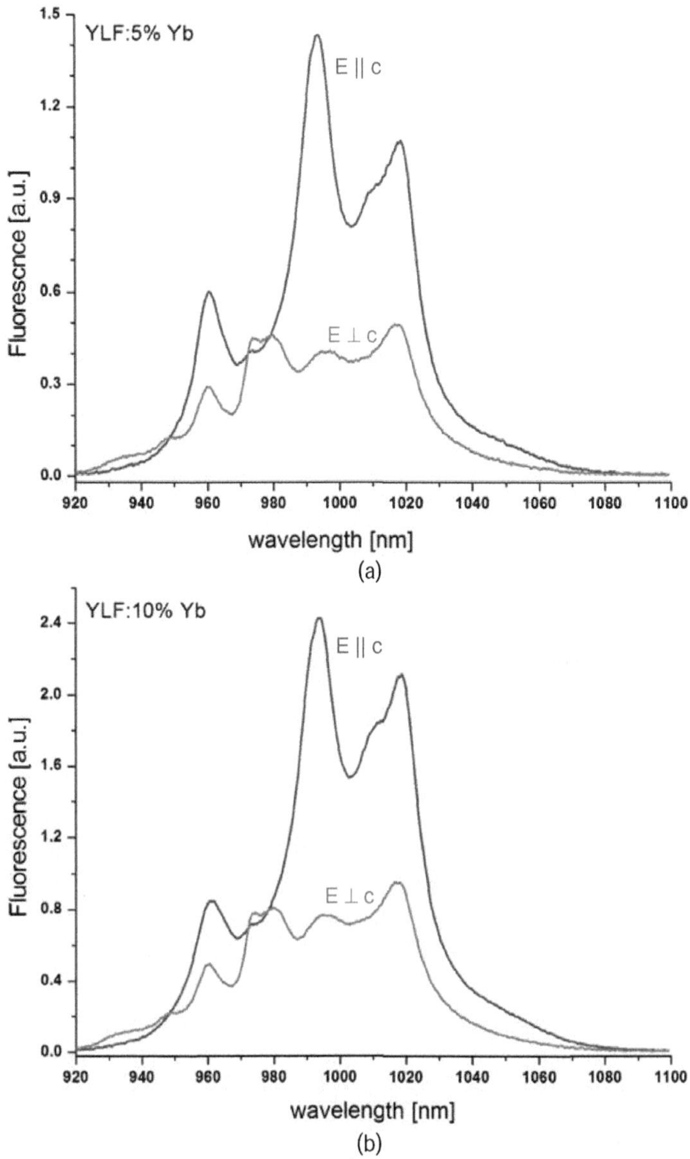

Figure 8.4 Polarized fluorescence spectra of YLF:5%Yb (a) and YLF:10%Yb (b), acquired, between 920 and 1100 nm with a resolution 0.5 nm, after laser excitation of Yb ions at 940 nm along the *c* axis.

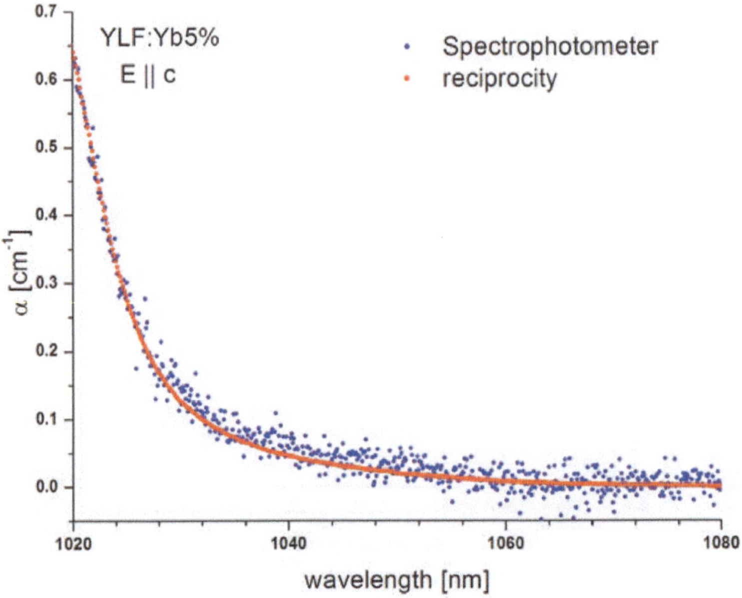

Figure 8.5 Absorption coefficient of YLF:5%Yb calculated from direct measurements (blue data points) and by using reciprocity theory, starting from fluorescence spectra (red data points), for π-polarization.

Dynamic fluorescence measurements, performed by means of the pinhole method (see Section 8.3.1), yielded, for all the three doping levels, nearly the same value for the Yb radiative lifetime, τ = 2.1 ± 0.2 ms, in agreement with literature data [2].

Static and dynamic fluorescence measurements were used to calculate the β-τ emission cross sections required to implement the reciprocity method. In Fig. 8.6 are shown the β-τ emission cross sections calculated for the YLF:5%Yb and the YLF:10%Yb samples.

From the fluorescence spectra acquired, the mean emission wavelength of the Yb transition was calculated by taking the first moment of the fluorescence function (Eq. 8.5). For increased Yb doping levels, a slight red shift of the λ_f value was observed. Such an effect can be ascribed to the increase of reabsorption trapping phenomena, which strengthen with the increase of the active ion concentration. The spectroscopic parameters of the mean emission

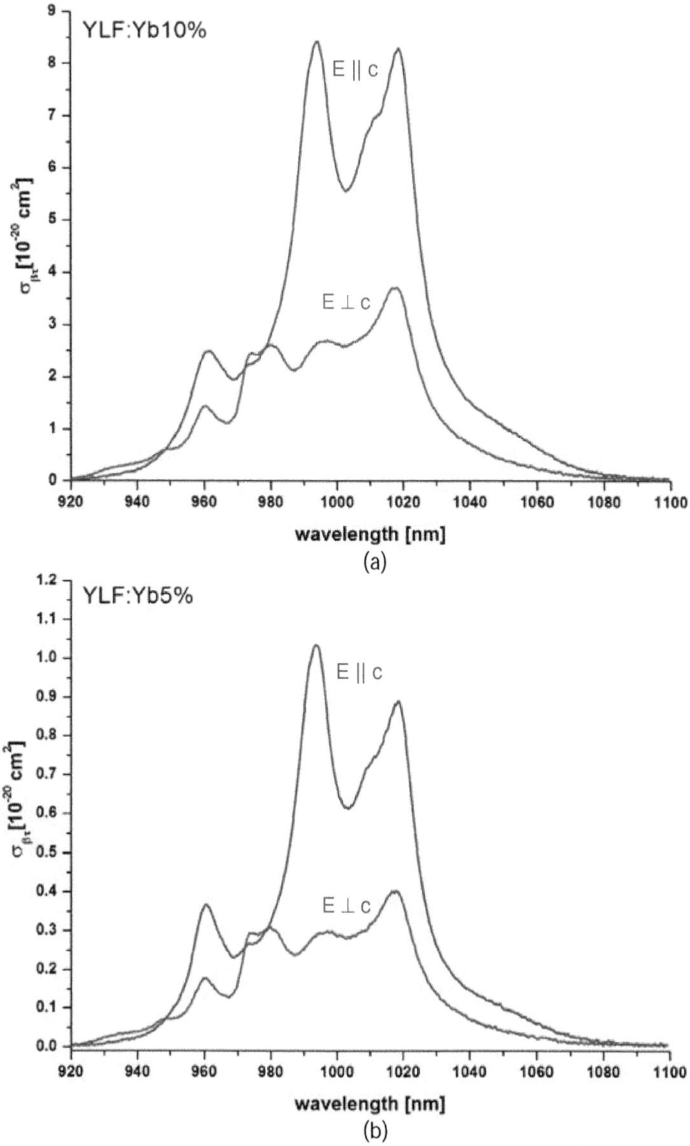

Figure 8.6 Polarized β-τ emission cross sections of YLF:5%Yb (a) and YLF:10%Yb (b).

wavelength and the resonant absorption coefficient obtained from spectroscopic measurements were used to build the model curve of cooling efficiency in Eq. 8.2. Experimental data of cooling efficiency were acquired by measuring the heating/cooling temperature changes of the sample as a function of pump photon energy. The apparatus used for these measurements is detailed in Section 8.3.2. Basically, the sample was placed inside a vacuum chamber, suspended on two optical fibers of 120 μm diameter, in order to minimize conductive heat loads. The chamber was evacuated at low pressure (10^{-5} Pa) in order to minimize convective heat transfer. Four different laser sources, with a central wavelength between 940 nm and 1060 nm, were used to excite the sample along the *c* axis. For each pumping wavelength, the temperature change was measured while the sample was excited with the laser beam, starting from the absence of laser irradiance until thermal equilibrium, under laser pumping, was reached. Figure 8.7 shows the typical temporal evolution of the temperature, under laser excitation in the anti-Stokes region.

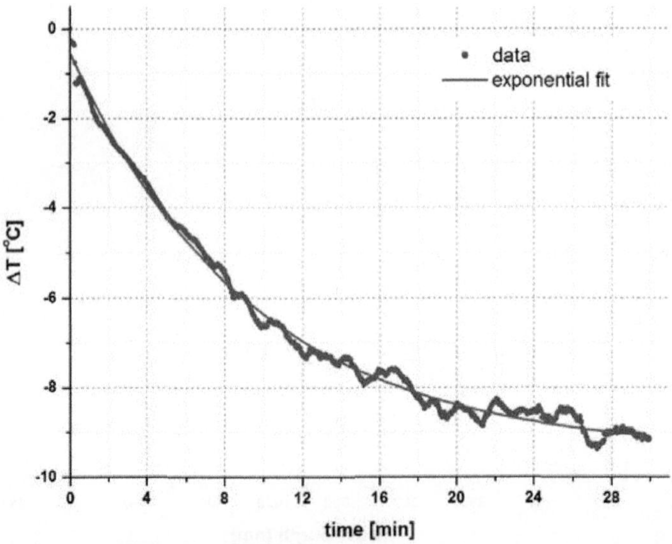

Figure 8.7 Temporal evolution of temperature, and relative exponential theoretical fit, for the YLF:5%Yb sample under laser excitation at 1024 nm with 1.5 W of pump power.

Data analysis confirmed the exponential change of the sample temperature, as expected from thermodynamics.

Assuming black-body radiation as the dominant heat load, the cooling efficiency results were proportional to the ratio of the temperature change to the absorbed power. Experimental data of cooling efficiency were measured for all the available pumping wavelengths. The experimental data of cooling efficiency were fitted with the model curve (Eq. 8.2). From the fit procedure, the EQE and background absorption parameters were evaluated for each sample. Results obtained for the three different Yb doping levels are reported in Table 8.1 [24].

Table 8.1 EQE and background absorption values estimated for YLF single crystals with varying Yb doping levels

	EQE	α_b (10^4 cm^{-1})
YLF:5%Yb	0.992 ± 0.001	4.0 ± 0.4
YLF:7.5%Yb	0.983 ± 0.001	2.7 ± 0.4
YLF:10%Yb	0.982 ± 0.001	2.1 ± 0.5

Large efficiency enhancements were observed for increasing Yb concentrations, due to a significant decrease of background absorption. Figure 8.8 reports experimental data points and a relative model curve for the 5%, 7.5%, and 10% Yb-doped YLF samples.

As can be seen from the graph, the peak efficiency increases with the Yb concentration due to decreased background absorption losses. A slight depletion of the EQE, however, is observed for larger Yb concentrations. Such an effect is ascribable to the enhancement of reabsorption-related phenomena, which red-shift the mean emission wavelength of the electronic transition.

Data collected result in good agreement with those reported in Ref. [4], providing good validation of our experimental setup, besides indication of the high optical quality of our active materials.

Further increases of Yb concentration in YLF crystals are currently under investigation. The balance between detrimental reabsorption effects and virtuous resonant absorption enhancements needs to be previously studied.

Although the total concentration of contaminants inside cooling crystals was tested at the 5N level, different relative concentrations of rare earth impurities at the part-per-million level were observed to cause appreciable variations of cooling efficiency [3]. Two YLF single crystals both doped with 5% Yb but containing different relative concentrations of impurities, due to the use of different raw material batches, exhibited dissimilar cooling performances. Although both samples possessed total amounts of impurities below 10 ppm, a depleted EQE of a few tenth % was measured in one sample, containing higher amounts of Er and Ho impurities. The investigation of energy transfer processes in these samples showed Er and Ho impurities to induce significant heat dissipation, via multiphonon emission in internal processes and direct transfer of excitation from Yb ions [3]. However, while some impurities, such as Er and Ho ions, appear to induce detrimental heat generation in energy transfer from Yb ions, other rare earth impurities can result in virtuous processes, energetically favorable for the Yb anti-Stokes process. Motivated by previous studies on the effect of rare earth impurities on the Yb anti-Stokes efficiency, the Yb–Tm co-doping was investigated. Efficiency enhancements via energy transfer–assisted anti-Stokes cooling were achieved. Results are reported in the following section.

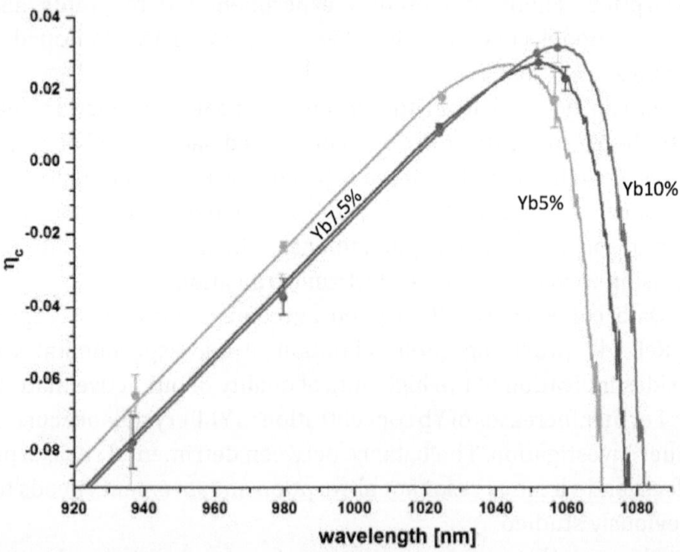

Figure 8.8 Experimental data points and fit model curve of cooling efficiency for Yb-doped YLF samples of variable doping levels.

8.4.2 Yb–Tm Energy Transfer–Enhanced Anti-Stokes Efficiency in Yb-Doped YLF Single Crystals

A novel scheme for anti-Stokes efficiency enhancement, based on energy transfer processes, has been investigated in co-doped YLF crystals [24]. Energy transfer induced by impurities typically results in loss channels, reducing the anti-Stokes efficiency via multiphonon quenching of excited ions, hence providing heat release in the lattice. In our work, we identified a virtuous energy transfer mechanism, based on Yb–Tm co-doping, which resulted in efficiency enhancement of the Yb anti-Stokes process via cooperative sensitization of Yb pairs, followed by radiative emissions at average higher energies than pumping photons, without phonon release.

A theoretical scheme for anti-Stokes efficiency enhancement via energy transfer processes was firstly suggested by Dong et al. [4] in 2013 for a Tm–Ho co-doped system. In such scheme, the energy transfer induced by interaction between Ho ions excited to the $^5I_7$ state and Tm ions in the ground state is employed to enhance the efficiency of the Ho anti-Stokes process around 2 μm. Via Ho de-excitation $^5I_7 \rightarrow {}^5I_8$, few Tm ions get sensitized to the $^3F_4$ state by absorbing the energy transferred from Ho excitations. Radiative emission from the $^3F_4$ state, which occurs at higher energy than Ho anti-Stokes emission, that is, higher than pumping photons, provides an extra release of thermal energy from the material. Such a mechanism, however, strongly suffers from detrimental energy transfer to upper energy levels. Moreover relatively high-power pumping sources around 2 μm are not readily available. As far as we know, these kinds of co-doped systems have been only theoretically investigated.

The Yb–Tm system instead exhibits several advantages. First of all, Yb-doped fluorides currently enable the best laser cooling performances, and efficiency enhancements via Yb–Tm co-doping appear promising to achieve significant improvements of current efficiencies, pushing down the minimum achievable temperature, while uplifting the available cooling powers at low temperatures. Furthermore, due to the respective energy level structure of the two ions, for specific relative concentrations, Yb–Tm endothermic energy transfer can occur via cooperative sensitization of Yb ions, providing an extra release of thermal energy from the lattice.

The effect of Yb–Tm energy transfer on the Yb cooling efficiency was investigated for low Tm doping levels. Net enhancement of anti-Stokes efficiency was achieved via Yb–Tm energy transfer in co-doped YLF single crystals, providing the first experimental evidence that some energy transfer processes can be usefully employed to uplift the efficiency of the Yb anti-Stokes cooling process, supplying an extra release of thermal energy.

The cooling performances of a YLF:5%Yb single crystal, where a controlled Tm doping of 16 ppm (0.0016%) was added, were investigated. Its performances were compared to single Yb-doped YLF samples. A net efficiency increase, over the single 5% Yb doping, was observed with the addition of a controlled Tm doping.

Figure 8.9 reports results of cooling measurements and a fit model curve of the Yb–Tm co-doped YLF sample. Along with data relative to the co-doped system are reported, for comparison, cooling efficiency data relative to the barely Yb-doped 5% and 10% YLF single crystals.

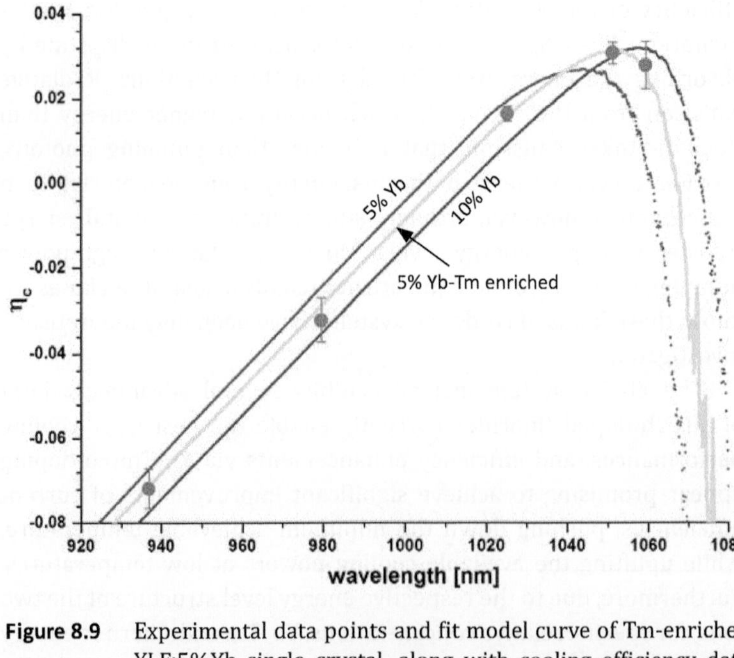

Figure 8.9 Experimental data points and fit model curve of Tm-enriched YLF:5%Yb single crystal, along with cooling efficiency data relative to single Yb doped at 5% and 10% YLF samples.

The fit of experimental data returned the following values for the YLF:5%Yb-0.0016%Tm sample: η_{ext} = 0.988 ± 0.001 and α_b = (1.9 ± 0.4) × 10^{-4} cm^{-1}. For the cooling parameters relative to the Yb-doped YLF single crystals we refer to Table 8.1. The addition of controlled Tm doping in the 5% Yb-doped sample resulted in a dramatic decrease of the background absorption parameter, with respect to the single 5% Yb doping. Parasitic absorption and peak efficiency values comparable to those obtained by increasing the Yb doping level up to 10% were achieved via Yb–Tm co-doping. In addition to that, the Yb–Tm co-doped sample, containing a lower Yb concentration, exhibited a larger EQE with respect to the higher Yb doping levels, which enabled a similar background absorption decrease. Larger Yb concentrations uplift the absorption efficiency but the strengthening of reabsorption phenomena tends to limit the efficiency enhancement due to depleted fluorescence escape rates. Virtuous energy transfer instead can improve the anti-Stokes efficiency, supplying an extra release of thermal energy, but without suffering from EQE reductions due to reabsorption-related effects.

A comparative spectroscopic analysis between single Yb-doped and Yb–Tm co-doped samples was the starting point to investigate the processes involved in the Yb–Tm energy transfer and their contribution to the Yb anti-Stokes efficiency.

At first, the discrepancy in cooling performances was checked to be unrelated to different absorption and emission properties of Yb ions within Tm enriched and single Yb-doped samples. In Fig. 8.10 are shown the polarized absorption and fluorescence spectra, acquired in the same experimental conditions, for the Tm-enriched YLF:5%Yb and the single 5% Yb-doped YLF sample. The spectra acquired did not exhibit any appreciable difference, within the experimental uncertainty, in size or structure. Dynamic fluorescence measurements returned similar values for the Yb transition lifetime as well: τ = 2.1 ± 0.2 ms. The dissimilar cooling behavior could thus be considered unrelated to different absorption and emission cross sections of Yb ions.

However, visible fluorescence observed in laser cooling experiments provided an indication that different energy transfer processes were dominant in case of Yb–Tm co-doping. While single Yb-doped YLF crystals showed a typical green emission when excited in laser cooling experiments, the Yb–Tm co-doped sample exhibited a characteristic blue fluorescence. To investigate processes involved in the energy transfer from excited Yb ions, the emissions outside the Yb region were observed in detail.

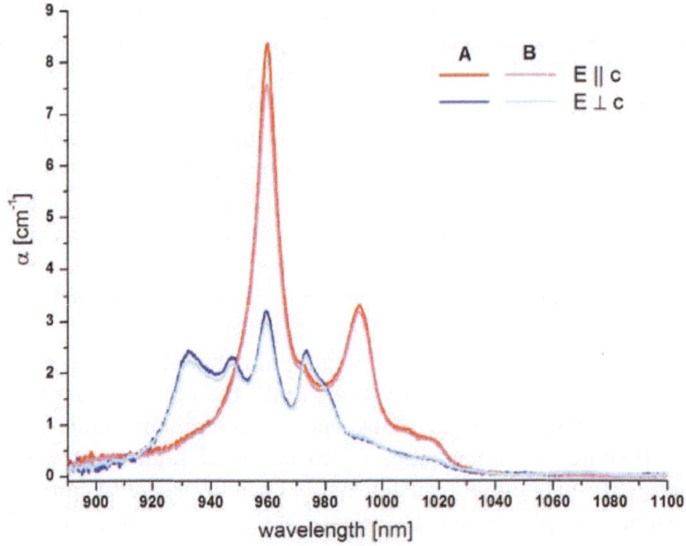

(a)

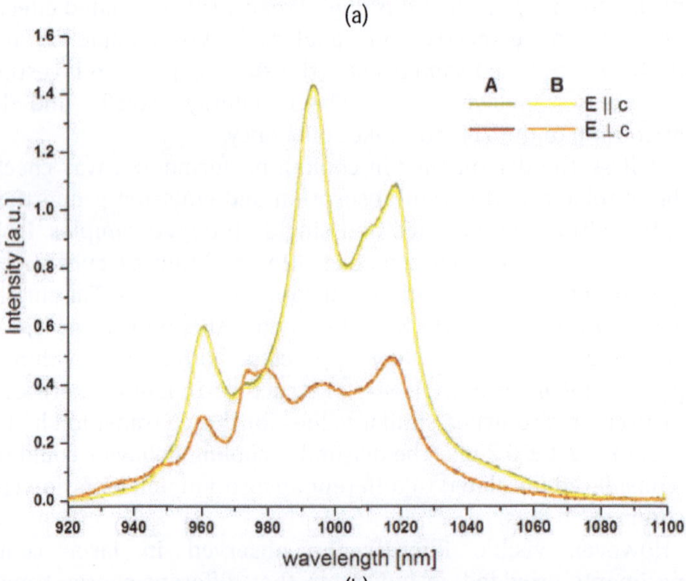

(b)

Figure 8.10 Polarized absorption (a) and fluorescence (a) spectra of the Yb transition acquired, in the same experimental conditions, for the Tm-enriched YLF:5%Yb sample (labeled as A) and the single doped YLF:5%Yb (labeled as B). Fluorescence spectra were acquired by exciting Yb ions at 940 nm along the *c* axis.

Visible and NIR fluorescence spectra were acquired, in the same experimental conditions, for the Tm-enriched sample and the single Yb-doped samples after laser excitation of the Yb ions at 940 nm. Elemental analysis, performed by LA-IPC-MS [23], identified similar amounts of other rare impurities. The discrepancies in emission spectra between Tm-enriched and single Yb-doped samples could thus be attributed exclusively to processes involved in the Yb–Tm energy transfer. Figure 8.11 reports, for simplicity, only emission spectra relative to Yb–Tm co-doped YLF and YLF:10%Yb samples, which exhibit a comparable value of background absorption. All the signals in the spectra are in arbitrary units, but the scale has been respected, expected for the IR signal, because of a different detector.

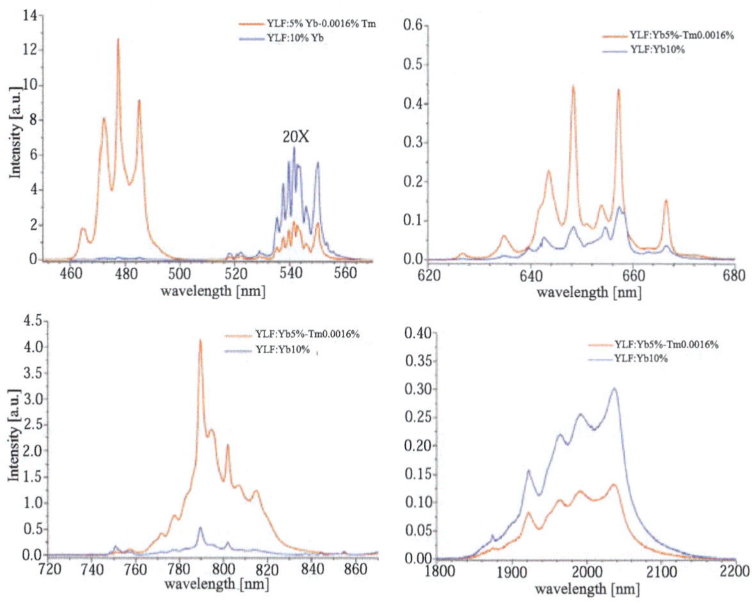

Figure 8.11 Emission spectra from the Tm-enriched YLF:5%Yb sample (red data points) and YLF:10%Yb (blue data points) acquired, in the same experimental conditions, after laser excitation of the Yb transition at 940 nm along the *c* axis.

Although the structure and energy positions of the emission bands were similar for all the samples, the Tm-enriched sample clearly exhibited intense emissions bands at higher average energies than that of excitation photons, which were strongly suppressed in the YLF samples doped only with Yb ions. The blue emission, between 450 and 500 nm, is dominant along with intense NIR fluorescence, between 750 and 850 nm, and red fluorescence between 630 and 680 nm.

The atomic-like structure of the emissions in Fig. 8.14 suggested trivalent rare earth impurities as the sources of such bands. Test samples of YLF single crystals, doped with other rare earth ions, and tabulated energies of Stark sublevels [8] were employed to identify the elemental sources of the observed fluorescence.

The emission bands in the blue, red, and NIR region were identified with the Tm transitions $^1G_4 \rightarrow {}^3H_6$, $^3H_4 \rightarrow {}^3H_6$, and $^1G_4 \rightarrow {}^3F_4$, respectively. The green and 2 μm bands, which exhibited similar intensity in the co-doped and single Yb-doped YLF samples, instead, were identified with combined emissions from Er and Ho ions.

The internal transitions within Tm ions that give rise to blue, red, and NIR fluorescence hence appeared those involved in a virtuous energy transfer process, favorable for the anti-Stokes cooling mechanisms.

In the Yb–Tm system, energy transfer typically results in blue upconversion via nonresonant three-photon processes [1]. These mechanisms involve three sequential steps of nonradiative transfers of excitation from excited Yb ions to Tm, with strong heat release via multiphonon emission. However, when the Yb concentration is dominant over Tm ions, a cooperative sensitization process between Yb ions becomes more likely than nonresonant three-photon upconversion, due to the strengthening of the Yb–Yb interaction. In the cooperative sensitization process a pair of excited Yb ions couple and transfer their energy to one Tm ion, which gets excited to the $^1G_4$ manifold [16, 18]. Such a mechanism is peculiar of the Yb–Tm system because the energy of the acceptor state (Tm, $^1G_4$) is about twice the energy of the donor excited state (Yb, $^2F_{5/2}$) and nonradiative quenching of the acceptor state is prevented by the absence of underlying states close enough in energy.

To identify which mechanism was dominant in the Yb–Tm co-doped sample, the fluorescence intensity at 477 nm was measured as a function of the pump power at 940 nm. The fit of experimental data returned a power law with an exponent value of 2.3 ± 0.2, which indicates that the excitation of the $^1G_4$ manifold was mostly a two-photon process, that is, it essentially occurred via cooperative sensitization.

The virtuous contribution of the Yb–Tm energy transfer to the Yb anti-Stokes cooling mechanism was observed to be basically related to the cooperative sensitization process. Indeed, while the three-photon energy transfer involves strong phonon release and thus detrimental heat generation, the cooperative sensitization process is energetically favorable for heat removal. Via cooperative sensitization of Yb pairs, the Tm $^1G_4$ manifold is excited just above the first Stark sublevel. Annihilation of lattice phonons is thus required for thermalization of the electronic population within the $^1G_4$ manifold. The following radiative emission from the $^1G_4$ manifold thus supplies an extra release of thermal energy from the lattice, which usefully contributes to the anti-Stokes process in thermal energy removal.

In addition to de-excitation of the $^1G_4$ manifold to the ground state, red fluorescence provides an additional channel of radiative emission at higher energy than pumping photons, which does not involve phonon release. It is worth to notice that the energy width of the $^1G_4$ manifold at 10 K is about 600 cm^{-1}, comparable to that of the Yb excited state (400 cm^{-1}), and thus a similar amount of thermal energy removal can be involved in both the Yb anti-Stokes cycle and the Yb–Tm energy transfer.

In addition to that, several virtuous processes can be responsible of the excitation of Tm ions to the $^3H_4$ state. At first, radiative emissions from the $^1G_4$ manifold provide excitation of the $^3H_4$ manifold without heat dissipation. Energy transfer processes between excited Tm ions and Yb ions in the ground state (Tm $^1G_4$, Yb $^2F_{7/2}$)→ (Tm $^3H_6$, Yb $^2F_{5/2}$) can contribute to populate the $^3H_4$ manifold with simultaneous excitation of the Yb transition on the long-wavelength tail of the absorption spectrum. Furthermore, endothermic Tm–Tm upconversion ($^3F_4$, $^3F_4$)→ ($^3H_6$, $^3H_4$) can likely contribute to excitation of the $^3H_4$ manifold, subtracting thermal energy from the lattice. Due to the higher energy gap of the $^3F_4$→$^3H_4$ transition with respect to the $^3F_4$→$^3H_6$ transition, indeed, such a process requires phonons annihilation and thus thermal energy removal from the lattice. The 10 K mismatch of about 1400 cm^{-1} requires absorption of some phonons at room temperature. A relatively long lifetime (of about 2 ms) was measured for the $^3H_4$ manifold, which indicates a negligible rate for the cross-relaxation process ($^3H_6$, $^3H_4$)→ ($^3F_4$, $^3F_4$).

For specific relative concentrations of Yb donors and Tm acceptors, several virtuous processes can be involved in the Yb–Tm energy transfer, providing a positive contribution to the anti-Stokes cooling process in heat removal.

Although further investigations are required in order to quantify the efficiency of energy transfer processes as a function of decreasing temperature, results achieved so far provide the first experimental evidence that selected energy transfer processes can be employed to enhance the efficiency of the anti-Stokes process. The potentialities of this scheme still need to be fully explored. However, preliminary results appear promising for further enhancement of the Yb efficiency in fluoride crystals toward reaching temperatures close to 80 K, improving at the same time the cooling powers available at low temperatures.

So far efficiency enhancements in Yb-doped fluorides have been achieved, increasing the concentration of active ions. Such an approach, however, is strongly limited because of reabsorption phenomena, which become more efficient with the increase of the Yb concentration. Controlled energy transfer processes provide, instead, an efficient alternative that relies on supplying an additional channel of thermal energy removal with efficiency only dependent on the relative concentrations of donors and acceptors ions.

In addition to that, the Yb–Tm energy transfer could be usefully employed to inhibit, a posteriori, energy transfer processes to detrimental impurities, which involve heat generation in internal processes or excitation transfer. Such a possibility reduces the stringent constraints on the purity of starting materials, with huge applicative effects. As mentioned earlier, purification of binary fluorides below a certain grade is a quite difficult process, due to the chemical inertness of these materials. The possibility to inhibit a posteriori energy transfer to detrimental impurities would provide an efficient tool to limit detrimental effects due to impurity-mediated processes, hence imposing much less mandatory purity requirements.

Investigations on the Yb–Tm energy transfer as a function of temperature and relative concentrations of donors and acceptors are currently underway in our laboratory in order to optimize the efficiency of the energy transfer process and study its efficiency as the temperature decreases.

8.5 Concluding Remarks

The efficiency of the optical cooling process has been investigated in several Yb-doped YLF single crystals grown in our facility. The increased Yb doping level resulted in significant efficiency enhancement via diminishing background absorption. A virtuous mechanism, based on energy transfer processes, has been identified, providing the first experimental evidence of energy transfer–assisted anti-Stokes cooling. Preliminary results achieved via Yb–Tm co-doping in YLF crystals show a remarkable increase in cooling performances with respect to the single Yb doping, promising for enhancements of the cryogenic state-of-the-art performances achieved in single Yb-doped fluorides. Such an approach possesses huge potentialities toward intrinsic efficiency enhancements of active materials, providing an extra release of thermal energy via virtuous energy transfer from excited Yb ions. Moreover it participates in inhibiting energy transfer processes to detrimental impurities, reducing the effects of impurity-mediated effects on the Yb anti-Stokes efficiency.

Acknowledgments

The authors would like to acknowledge I. Grassini for her competence and care in preparing the samples. A. Volpi acknowledges support by the European Space Agency under grant no. 4000108074/13/NL/PA, "Cooling Effect on Fluoride Crystals."

References

1. Auzel, F. (2004). Upconversion and anti-Stokes processes with f and d ions in solids, *Chem. Rev.*, **104**, pp. 139–173.
2. De Loach, L. D., Payne, S. A., Chase, L. L., Smith, L. K., Kway, W. L., Krupke, W. F. (1993). Evaluation of absorption and emission properties of Yb^{3+} doped crystals for laser applications, *IEEE J. Quantum Electron.*, **29**, pp. 1179–1191.

3. Di Lieto, A., Sottile, A., Volpi, A., Zhang, Z., Seletskiy, D. V., Tonelli, M. (2014). Influence of other rare earth ions on the cooling efficiency in Yb:YLF crystals, *Opt. Express*, **22**, pp. 28572–28583.
4. Dong, G. Z., Zhang, X. L., Li., L. (2013). Energy transfer–enhanced laser cooling in Ho3+ and Tm3+ co-doped lithium yttrium fluoride, *J. Opt. Soc. Am. B*, **30**, pp. 939–944.
5. Epstein, R. I., Buchwald, M., Edwards, B., Gosnell, T., Mungan, C. (1995). Observation of laser-induced fluorescent cooling of a solid, *Nature*, **377**, pp. 500–502.
6. Hehlen, M. P., Epstein, R. I., Inoue, H. (2007). Model of laser cooling in the Yb3+ -doped fluorozirconate glass ZBLAN, *Phys. Rev. B*, **75**, p. 144302.
7. Hehlen, M. P., Boncher, W. L., Melgaard, S. D., Blair, M. W., Jackson, R. A., Littleford, T. E., Love, S. P. (2014). Preparation of high-purity LiF, YF3 and YbF3 for laser refrigeration, *Proc. SPIE*, **9000**, p. 900004.
8. Kaminski, A. A. (1996). *Crystalline Lasers: Physical Processes and Operating Schemes* (CRC Press).
9. Kastler, A. (1950). Some suggestions concerning the production and detection by optical means of inequalities in the populations of levels of spatial quantization in atoms. Application to the Stern and Gerlach and magnetic resonance experiments, *J. Phys. Radium*, **11**, pp. 255–265.
10. Kuhn, H., Fredrich-Thornton, S. T., Krankel, C., Peters, R., Petermann, K. (2007). Model for calculation of radiation trapping and description of the pin-hole method, *Opt. Lett.*, **32**, pp. 1908–1910.
11. Melgaard, S. D. (2013). *Cryogenic Optical Refrigeration: Laser Cooling of Solids Below 123K*, PhD dissertation, University of New Mexico, Albuquerque, NM.
12. Melgaard, S. D., Albrecht, A., Hehlen, M. P., Seletskiy, D. V., Sheik-Bahae, M. (2014). Optical refrigeration cools below 100K CLEO 2014, *OSA Tech. Dig.*, paper FTh4D.4.
13. Melgaard, S. D., Seletskiy, D. V., Di Lieto, A., Tonelli, M., Sheik-Bahae, M. (2013). Optical refrigeration to 119K, below National Institute of Standards and Technology cryogenic temperature, *Opt. Lett.*, **38**, pp. 1588–1590.
14. Melgaard, S. D., Seletskiy, D. V., Polyak, V., Asmerom, Y., Sheik-Bahae, M. (2014). Identification of parasitic losses in Yb:YLF and prospects for optical refrigeration down to 80K, *Opt. Express*, **22**, pp.7756–7764.

15. Nemova, G., Kashyap, R. (2011). Alternative technique for laser cooling with super irradiance, *Phys. Rev. A,* **83**, p. 013404.

16. Ovsyakin, V. V., Feofilov, P. P. (1966). Cooperative sensitization of luminescence in crystal activated with rare earth ions, *Sov. Phys. JETP Lett.*, **11**, pp. 317–318.

17. Payne, S. A., Chase, L. L., Smith, L. K., Kway, W. L., Krupke W. F. (1992). Infrared cross-section measurements for crystal doped with Er^{3+}, Tm^{3+} and Ho^{3+}, *IEEE J. Quantum Electron.*, **28**, pp. 2619–2630.

18. Pei., X., Hou, Y., Zhao, S., Xu, Z., Teng, F. (2005). Frequency upconversion of Tm3+ and Yb3+ codoped LiYF4 synthesized by hydrothermal method, *Mater. Chem. Phys.*, **90**, pp. 270–274.

19. Ruan, X. L., Kaviany, M. (2006). Enhanced laser cooling of rare earth ion doped nanocrystalline powders, *Phys. Rev. B,* **73**, p. 155422.

20. Seletskiy, D. V., Hehlen, M. P., Epstein, R., Sheik-Bahae, M. (2012). Cryogenic optical refrigeration, *Adv. Opt. Photonics,* **4**, pp. 78–107.

21. Seletskiy, D. V., Melgaard, S. D., Bigotta, S., Di Lieto, A., Tonelli, M., Sheik Bahae, M. (2010). Laser cooling of solids to cryogenic temperatures, *Nat. Photonics,* **4**, pp. 161–164.

22. Sheik-Bahae, M., Epstein, R. I. (2007). Optical refrigeration, *Nat. Photonics,* **1**, pp. 693–699.

23. Tiepolo, B., Bottazzi, P., Palenzona, M., Vannucci, R. (2003). A laser probe coupled with ICP: double focusing sector; field mass spectrometer in situ analysis of geological samples and U-Pb dating of zircon, *Can. Mineral.,* **41**, pp. 259–272.

24. Volpi, A., Di Lieto, A., Tonelli, M. (2015). Novel approach for solid state cryocoolers, *Opt. Express,* **23**, pp. 8216–8226.

Chapter 9

Interferometric Measurement of Laser-Induced Temperature Changes

B. Rami Reddy
Alabama A&M University, Department of Physics, Normal, AL 35762, USA
rami.bommareddi@aamu.edu

9.1 Historical Development of Temperature Sensors

Temperature is the most important parameter in scientific and industrial environments. The development of temperature sensors is continually evolving for more than two centuries. Any material whose property changes quickly with temperature and is reversible, can be used as a thermometer. This property can be pressure, volume, length, refractive index, color, etc. The efforts of Fahrenheit and later Celsius led to the development of the mercury-in-glass thermometer. Modern infrared (IR) thermometers scan the forehead quickly to reveal the body temperature. Some of the popular devices are thermocouples, resistive temperature devices (RTDs), thermistors,

Laser Cooling: Fundamental Properties and Applications
Edited by Galina Nemova

ISBN 978-981-4745-04-8 (Hardcover), 978-981-4745-05-5 (eBook)
www.panstanford.com

thermostats, silicon-based sensors, and IR pyrometers [1]. Each type of device operates in a specific temperature range. They are installed in consumer appliances, process industries, and aircrafts. Thermocouples are cheap and rugged, some operate up to 1800°C, and they are installed in furnaces, combustion engines, chemical reactors, and many others. Silicon sensors operate up to 150°C and are used in electronic circuit boards. RTDs are wire windings of copper or another metal, whose resistance increases linearly with temperature, and they are useful for temperature measurement up to a few hundred degrees. Conventional thermometers are inexpensive, but they have limited applications.

Though several different types of temperature probes are available there is no ideal device that works in all types of environments, for temperature measurement. Optical materials are either heated up or cooled down when a laser beam passes through them [2, 3]. Some materials require longer times to reach thermal equilibrium. Contact thermometers are not suitable for temperature measurement along the beam path. These measurements require noncontact sensing devices, which are preferred in many cases. To address these specific needs optical techniques are investigated for temperature measurement, using rare earth ion–doped solids. These are alternate techniques for temperature measurement when conventional techniques are not suitable.

9.1.1 Optical Sensor Technologies

IR thermometers and imagers are noncontact devices that use IR radiation to infer the temperature of an object. With the advent of lasers a range of optical sensors are developed. Optical measurements involve the detection of light. Any parameter that changes with temperature can be used to estimate the sample temperature, such as changes in emission intensity, lifetime, optical path length, or phase. Optical sensors can be configured as point sensors or distributed temperature sensing (DTS) systems and the latter uses mainly optical fibers [4]. Intensity measurement sensors are based on Raman scattering [5], fluorescence [6], or spectral transmission changes [7]. As the temperature of a material increases, higher vibrational levels of molecules gain population due to Boltzmann distribution. Raman scattering methods took advantage of this phenomenon to estimate the temperature in the combustion studies [5]. Raman scattering is the principle behind the operation of distributed temperature sensor. If a fiber's temperature changes at one or more points along the length of a fiber, thermal effects induce lattice vibrations locally, and a propagating pulsed laser beam undergoes Raman scattering at those points. Time taken by the backscattered Raman signals to arrive at the input end of the fiber is measured by the optical time domain reflection (OTDR) technique to locate the heated spots. The intensity of the anti-Stokes line is temperature dependent and the ratio of anti-Stokes to Stokes line intensities is used to derive the fiber temperature. Commercial devices measure temperatures up to 700°C with a linear resolution of 1 m. A fiber Bragg grating (FBG) is produced by creating a periodic variation in the refractive index in the fiber core, which reflects a specific wavelength. The reflected light wavelength depends on the grating period and the latter depends on the grating temperature. This principle was used to design FBG temperature sensors for several applications [8], including laser cooling studies [9].

9.1.2 Transmission Spectral Measurements for High-Temperature Measurement

The device consists of an optical fiber made of sapphire, Y_2O_3, or another material [7]. One end of the fiber is coated with a rare-earth-ion-doped film that is in contact with the surface under measurement. The light output of the fiber passes through a band-pass filter (or spectrometer) before being detected, whose intensity is related to the sample temperature. This device is useful for temperature measurement from 400°C to 2000°C and operates even in reactive and corrosive environments. It has a fast response time of 1 ms and a temperature resolution of 1°C.

9.1.3 Interferometric Sensors

Interferometric sensors are demonstrated using several different techniques and configurations [10–14]. In the prelaser era interferometers utilized either mercury or sodium light sources whose coherence lengths are very small. So the commercial products were mainly centered on testing optical surfaces or components

[15]. After the advent of lasers the applications of interferometry grew exponentially over time, which was attributed to the long coherence lengths of the laser beams [16]. Two coherent beams can be derived from a laser source by the division of wavefront or division of amplitude, which are suitable for the design of several kinds of interferometers. Interferometers can be classified as two-beam or multiple-beam devices. The fringe intensity varies sinusoidally in a two-beam interferometer. In a multiple-beam interferometer, each pair of beams contributes a Fourier component to the fringe pattern. Interferometers are classified based on the method used to derive the two beams. If the radiation coming from a point source passes through two or more apertures, it is said to be separated into beams by division of wavefront; they are (a) Young's double-slit experiment, (b) Lloyd's mirror, and (c) Rayleigh interferometer [14]. Most of the interferometers derive the two beams by the division of amplitude. Interferometers based on division of amplitude are the Mach–Zehnder interferometer, Michelson interferometer, Sagnac interferometer, and lateral shearing interferometer [10, 14]. For many applications, it is desirable to have an optical arrangement in which the two interfering beams travel along separate paths before they are recombined. This led to the development of a large number of interferometers for specific purposes. A polarization interferometer results when light is separated into ordinary and extraordinary rays (by using a calcite prism, Wollaston prism, Rochon prism, or Nicol prism). The Mach–Zehnder interferometer can be set up in a large space to accommodate big chambers in its path length for in situ monitoring. Alignment of the optics is relatively simpler if the mirrors are replaced by pentaprisms [17]. To facilitate a variable path difference between the two beams, the mirrors are replaced by corner cubes in the Michelson interferometer, and the reflecting mirrors are replaced by mirror pairs in the Mach–Zehnder interferometer. Such a modified version was used in the design of a wavemeter [18, 19]. Straight-line fringes of equal thickness are produced by the Michelson interferometer whenever a wedge-shaped air gap is introduced between one mirror and the image of the other in the former. A Sagnac interferometer is extremely stable because the two beams travel in opposite directions in a closed path.

The beam splitters and mirrors of a Mach–Zehnder interferometer are adjusted so that the optical beams are approximately parallel and the paths travelled by both the beams are equal; the beams form a rectangle or a parallelogram. Thus a collimated input beam generates two wavefronts. If the wavefronts make a small angle with each other a nonlocalized interference pattern is seen, consisting of equispaced straight fringes parallel to the line of intersection of the wavefronts. If there is an angle between the beams after exiting the second beam splitter then localized fringes are formed. On the other hand if there is no angle between the beams then nonlocalized fringes are formed. The Mach–Zehnder interferometer is a much more versatile instrument than the Michelson interferometer because the optical beams are widely separated [20–23]. The beams travel the path only once and the fringes can be localized at any desired point. Because of this, it has been extensively used in studies of gas flow, combustion, plasma density, and diffusion, where changes in refractive index occur, which can be related to changes in pressure, temperature, or the relative concentrations of different components of a mixture [13, 23]. However, adjustment of the interferometer is tedious to get fringes of good visibility, since a displacement of a mirror results in a shift of the plane of localization as well as a change in the optical path difference (OPD). The need for this procedure can be eliminated by using pentaprisms in place of mirrors.

9.2 Rare Earth Luminescence Spectroscopy and Temperature Sensors

Rare earth ion–doped solids are extensively investigated because of their applications for the development of lasers, luminescent devices, optical memories, and optical refrigerators [24–26]. For a clear understanding of laser cooling/heating a brief description of rare earth spectroscopy is given here. In general, rare earths form trivalent compounds. Rare earth ions have an unfilled 4f shell whose active electrons are shielded from the crystal field by the 5s and 5p orbitals. Consequently the crystal field splitting is smaller than the spin–orbit splitting. As a result the center of gravity of each multiplet approximately remains the same as that of the free ion but exhibits slight changes from one host to the other. In the gas phase, transitions among the levels of $4f^n$ configuration are forbidden. However, when the ions are embedded in a solid medium they are weakly allowed due to admixing of $4f^n$ wave functions with those of $4f^{n-1}5d^1$ wave functions at noncentro symmetric sites. Excited levels have relatively long lifetimes and group theory predicts allowed transitions for $\Delta J \leq 6$ [27]. Excited ions in a solid relax radiatively and nonradiatively [28], which are represented by a line arrow and a wavy arrow in Fig. 9.1. The nonradiative relaxation depends on the guest–host combination. For a given material the nonradiative relaxation depends on the energy gap between the excited level and its lower level, the cutoff phonon frequency of the host material, and the phonon density. The cutoff phonon frequency depends on the reduced mass of the molecule and force constants. For example, the cutoff phonon frequencies in LaF_3, $LaCl_3$, and $LaBr_3$ are 350, 260, and 175 cm^{-1}, respectively [29]. When the energy gap ΔE_{32} between levels 3 and 2 is bridged by one or two phonons the upper level relaxes nonradiatively. If the energy gap ΔE_{10} between levels 1 and 0 requires eight or more phonons, nonradiative relaxation does not occur between levels 1 and 0. For intermediate energy gaps as ΔE_{21} between levels 2 and 1 both radiative and nonradiative relaxation phenomena occur simultaneously. For intermediate energy gaps the higher levels relax radiatively and nonradiatively. Therefore, the measured lifetime τ of the excited level i is related to sum of the relaxation rates as [28]

$$\tau^{-1}(i) = \Sigma_{ij}\omega_r + \Sigma_{ij}\omega_{nr} \tag{9.1}$$

where $\Sigma\omega_r$ is the total radiative relaxation rate to all the lower levels and $\Sigma\omega_{nr}$ is the total nonradiative relaxation rate to all the lower levels, represented by j. The nonradiative relaxation rate increases at higher temperatures and hence the lifetime of the excited level decreases at higher temperatures [28]. Based on this phenomenon fluorescence lifetime sensors were designed to measure the sample temperature [30]. However, this technique is not relevant for the laser cooling studies because the nonradiative relaxation rate has to be negligible for laser cooling of solids. However, fluorescence intensity ratio based sensors are useful for temperature measurement if the energy gap between levels 2 and 3 is comparable to thermal energy. In general, emission occurs from level 2 and a weak emission from level 3. As the sample temperature increases, the population in level 3 increases due to thermalization and the intensity of λ_{30} increases at higher temperatures. This is true for Yb^{3+}-doped solids.

The population N_i (i = 0 to 3) of an energy level is dependent on temperature and the population ratio is given by

$$\frac{N_2}{N_0} = \frac{g_2}{g_0} e^{-\Delta E_{20}/kT} \tag{9.2}$$

$$\frac{N_1}{N_0} = \frac{g_1}{g_0} e^{-\Delta E_{10}/kT} \tag{9.3}$$

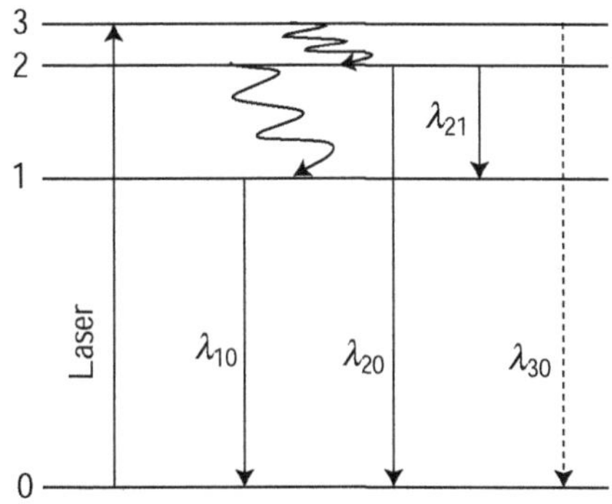

Figure 9.1 Energy levels and relaxation phenomena in a rare earth ion-doped solid. Radiative and nonradiative relaxation phenomena are represented by downward straight-line arrows and wavy arrows. Emission from level 3 (dotted line) occurs if its population is sustained by thermalization.

After taking a ratio of the above equations, it yields

$$\frac{N_2}{N_1} = \frac{g_2}{g_1} e^{-\Delta E_{21}/kT} \tag{9.4}$$

where g_i is the degeneracy of level i and ΔE_{ij} is the energy gap between levels i and j, T is the sample temperature, and k is Boltzmann's constant. The transition intensity, I, between levels 1 and 0 is given by

$$I_{10} \propto N_1 \nu_{10} A_{10} \tag{9.5}$$

where ν_{10} is the frequency corresponding to the wavelength λ_{10} and A is the transition probability. The intensity ratio, R, of the transitions for $2 \rightarrow 0$ and $1 \rightarrow 0$ gives

$$R = \frac{I_{20}}{I_{10}} = \frac{N_2 \nu_{20} A_{20}}{N_1 \nu_{10} A_{10}} = \frac{\nu_{20} A_{20}}{\nu_{10} A_{10}} \frac{g_2}{g_1} e^{-\frac{\Delta E_{21}}{kT}} = (\text{const}) e^{-\frac{\Delta E_{21}}{kT}} \tag{9.6}$$

The above equation can be rearranged as

$$\ln R = a - \frac{\Delta E_{21}}{kT} = a - \frac{b}{T} \tag{9.7}$$

where a and b are constants. A plot of logarithmic R versus T increases linearly with temperature. A plot of ln R versus $1/T$ decreases linearly with temperature. At lower temperatures the ratio R is smaller. However, this method requires the development of a calibration plot by measuring the intensity ratio as a function of sample temperature. An unknown sample temperature can be derived by utilizing the calibration plot and from a measurement of the fluorescence intensity ratio. Such a method was demonstrated for rare earth–doped samples for high-temperature measurements [31]. This technique should work equally well even if the sample temperature decreases. For low-temperature measurement the above equation predicts sample temperature accurately if the peaks' intensities are measured accurately. Because of these reasons, a linear variation was observed in a limited temperature range, up to 140°C in an Er^{3+}-doped silica waveguide [32] and up to 96°C in an Er^{3+}-doped silica fiber [33], though mK sensitivity was reported in the past [34].

9.2.1 Differential Luminescence Measurement

The differential luminescence technique works on similar principles. Populations among different energy levels, for example, levels 2 and 3 of Fig. 9.1, change whenever the sample temperature changes. Consequently spectral features, relative intensities of different peaks, linewidths, and in some cases the fluorescence band shape change significantly. In this case a calibration plot needs to be developed in advance by cooling the sample to different temperatures and acquiring the spectra with a charge-coupled device (CCD) camera. Spectral subtraction reveals whether the material is heating or cooling. The technique reveals very small changes in temperature [35].

9.2.2 Yb^{3+}-Doped YAG Spectra

The transmission spectrum of Yb^{3+}-doped YAG was obtained by exciting the sample with a global lamp source and collecting the sample transmission with a low-resolution spectrometer. The resulting spectrum was mathematically manipulated to depict the absorption spectrum (Fig. 9.2). It reveals a strong absorption at 940 nm and a weaker absorption at 1030 nm. The absorption strength at 915 nm is weaker than that at 940 nm. The laser wavelength 915 nm excites the higher stark levels of the $^2F_{5/2}$ multiplet (Fig. 9.4). Excited-state ions relax nonradiatively (multiphonon relaxation) to the lowest Stark level of the excited state that emits at 1030 nm. The laser-induced heating occurs mainly along the path of the laser beam. In such cases, indirect methods are used to estimate the sample temperature accurately. One such method is Raman scattering, in which the peak intensities depend on the populations of the levels involved, which is a function of the sample temperature. Here we are proposing interferometric methods to measure heating or cooling along the beam path in real time.

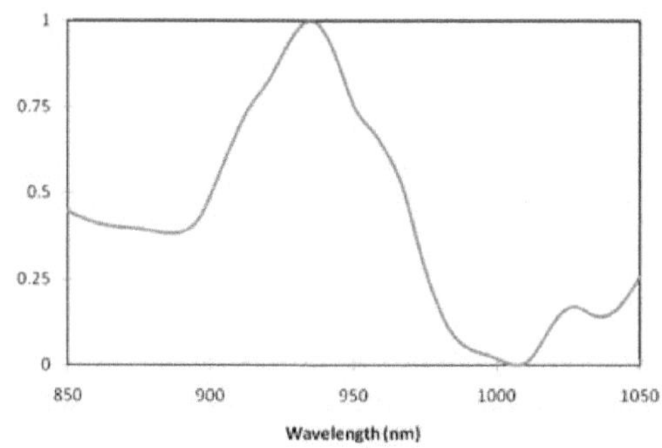

Figure 9.2 Absorption spectrum of YAG:Yb^{3+} [36].

Figure 9.3 Fluorescence observed under 915 nm laser excitation.

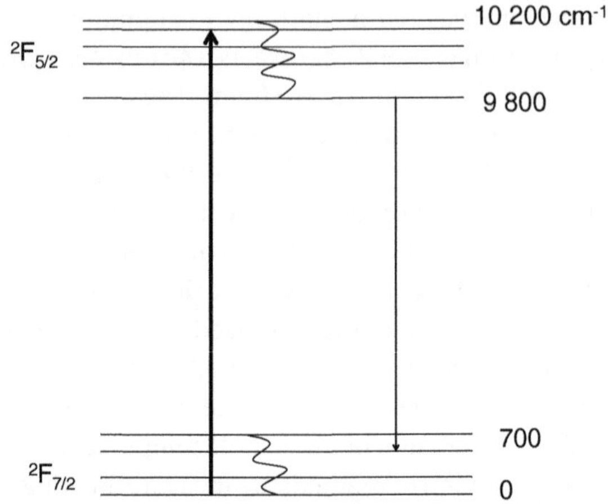

Figure 9.4 Partial energy level diagram of YAG:Yb^{3+}. Upward and downward arrows represent laser excitation and fluorescence. Wavy lines represent nonradiative relaxation.

9.3 Mach–Zehnder Interferometer

A Mach–Zehnder interferometer was set up with a 514.5 nm green laser beam to measure temperature changes in Yb^{3+}-doped YAG induced by another laser [36]. The 514.5 nm green laser beam is separated into two parts by a beam splitter, BS1, and the two beams are arranged to form a rectangle (Fig. 9.5). Beam 1 passes through the sample. Beams 1 and 2 are recombined by the beam splitter BS2. Mirrors M1 and M2 are adjusted so that a wedge forms between the two wavefronts. As a result linear fringes are produced, which are projected on a screen by a lens (Fig. 9.6). A detector having a pinhole opening monitors the intensity of a fringe. A Ti:sapphire laser beam (915 nm) is launched into the sample that propagates in a direction opposite to that of the green laser beam. Both the beams are collinear and counterpropagating. The 915 nm laser excites Yb^{3+} ions from the ground state to the excited state, $^2F_{5/2}$ (Fig. 9.4). Excited-state ions exhibit strong emission at 1030 nm (Fig. 9.3). During radiative relaxation some ions end up in higher Stark components of the ground state, $^2F_{7/2}$ (Fig. 9.4). The nonradiative relaxation processes in the ground and excited states heat the material along the beam path. Surface-mounted contact thermometers may not reveal the sample temperature accurately if the material conductivity is small. The optical path length of the beam is dependent on sample temperature because the refractive index and sample length are dependent on sample temperature. In the interferometric arrangement the fringes appear to move when the sample temperature is continuously changed. As a result intensity of the fringe pattern varied continuously with time, and the detector output followed such a variation (Fig. 9.7). Material heating due to nonradiative relaxation occurs instantaneously when it is exposed to a 915 nm laser beam. Therefore, the period of oscillation is small in the beginning. The period of oscillation increases with time because the crystal temperature is slowly approaching the steady-state value. Therefore, the oscillatory output of the detector exhibits chirping [36]. Once it reaches steady-state value the material temperature does not change anymore and hence the oscillation stops. The detector output is constant if the material does not absorb the probe laser beam. For 300 mW of a 915 nm laser beam, the sample temperature increased 7.1 ± 0.8°C [31]. A similar situation occurs when the heating laser beam was blocked (Fig. 9.8).

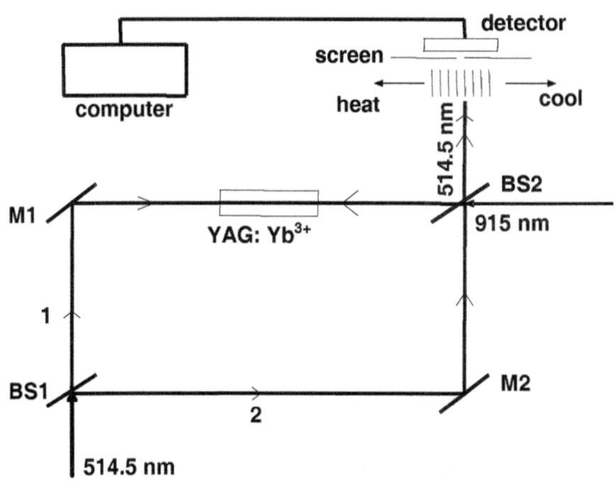

Figure 9.5 A 915 nm laser beam heats the YAG:Yb^{3+} sample. A 514.5 nm beam was used to setup a Mach–Zehnder interferometer to monitor temperature changes [36].

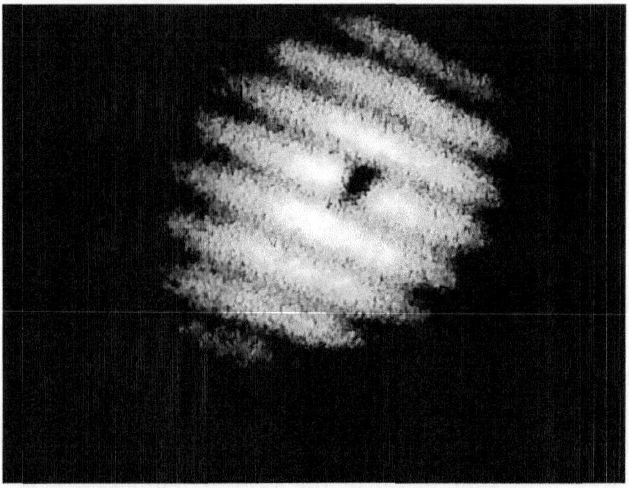

Figure 9.6 Interferometer fringes produced by a 514.5 nm laser. The dark spot is the pinhole in front of the detector [36].

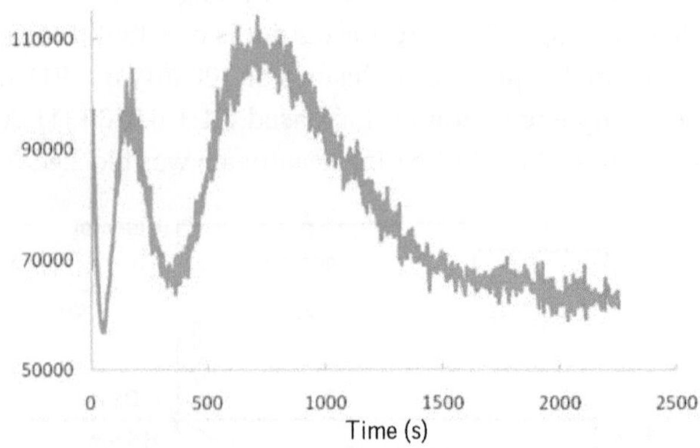

Figure 9.7 Temporal variation of the detector output during laser Ti:sapphire heating.

9.3.1 Theory of Mach–Zehnder Interferometer

The laser beam passing through the sample changes its temperature. As the sample temperature is varied its refractive index changes. So the effective optical path length ηL of beam 1 changes continuously and this change causes the fringes to move. The apparent speed of the fringes depends on the rate of change $d\eta/dt$ and hence on $dT/$

dt. The path lengths of beams 1 and 2 are $\eta_1(L_2 + L_1 - L_3) + \eta_2 L_3$ and $\eta_1(L_1 + L_2)$, where η_1 and η_2 are refractive indices of air ($\eta_1 = 1.0$) and YAG ($\eta_2 = 1.823$), respectively; L_1 and L_2 are the dimensions of the rectangle; and L_3 is the sample length. The OPD between the two beams 1 and 2 is given by OPD = $(\eta_2 - \eta_1)L_3$. Here, η_2 and L_3 are dependent on sample temperature.

The refractive index and sample length are temperature dependent, which are described by

$$\eta_3 = \eta_{0+} \frac{\partial \eta}{\partial T} \Delta T \tag{9.8}$$

$$L_3 = L_{0+} \frac{\partial L}{\partial T} \Delta T \tag{9.9}$$

where η_0 and L_0 are the room-temperature values of refractive index and sample length, respectively. When the Ti:sapphire laser beam passes through the sample its temperature changes from T_1 to T_2. However, the change in OPD is given by

$$\Delta(\text{OPD}) = \Delta(\eta_2 L_3 - \eta_1 L_3) \tag{9.10}$$

Since $\eta_1 = 1.0$ for air,

$$\Delta(\text{OPD}) = \Delta(\eta_2 L_3) - \Delta(\eta_1 L_3) = \Delta\eta_2 L_3 + \eta_2 \Delta L_3 - \Delta L_3 \tag{9.11}$$

$$\Delta(OPD) = L_3 \frac{\partial \eta}{\partial T} \Delta T + \eta_3 \frac{\partial L}{\partial T} \Delta T - \frac{\partial L}{\partial T} \Delta T \tag{9.12}$$

Hereafter $L_3 = L_0 = L$ (sample length) and $\eta_2 = \eta_0 = \eta$ (sample refractive index).

$$\Delta(OPD) = L \frac{\partial \eta}{\partial T} \Delta T + \eta \frac{\partial L}{\partial T} \Delta T - \frac{\partial L}{\partial T} \Delta T \tag{9.13}$$

$$\Delta(OPD) = L\eta \left(\frac{1}{\eta} \frac{\partial \eta}{\partial T} \Delta T + \frac{1}{L} \frac{\partial L}{\partial T} \Delta T - \frac{1}{L\eta} \frac{\partial L}{\partial T} \Delta T \right) \tag{9.14}$$

Equation 9.14 can be written as

$$\Delta(OPD) = L\eta \left(\frac{1}{\eta} \frac{\partial \eta}{\partial T} + \alpha - \frac{\alpha}{\eta} \right) \Delta T \tag{9.15}$$

where α is the linear thermal expansion coefficient of the material, given by

$$\alpha = \frac{1}{L} \frac{\partial L}{\partial T} \tag{9.16}$$

If N fringes translated for a given temperature change ΔT, then $\Delta(\text{OPD}) = \lambda N$. Equation 9.15 reduces to

$$\lambda N = L\eta\left(\gamma - \frac{\alpha}{\eta}\right)\Delta T \tag{9.17}$$

where

$$\gamma = \frac{1}{\eta}\frac{\partial \eta}{\partial T} + \alpha \tag{9.18}$$

The temperature variation of the sample along the beam path is given by

$$\Delta T = \frac{\lambda N}{L\eta\left(\gamma - \frac{\alpha}{\eta}\right)} \tag{9.19}$$

From the above equation it is clear that ΔT is a function of N, L, η, α, and γ. So the accuracy of the ΔT estimate depends on the accuracy with which these parameters were measured. The uncertainty in temperature change ΔT_u is calculated by the method of standard deviations, as follows.

$$\Delta T_u = \sqrt{\left(\frac{\partial T}{\partial N}\right)^2 (\Delta N_u)^2 + \left(\frac{\partial T}{\partial L}\right)^2 (\Delta L_u)^2 + \left(\frac{\partial T}{\partial \eta}\right)^2 (\Delta \eta_u)^2 + \left(\frac{\partial T}{\partial \alpha}\right)^2 (\Delta \alpha_u)^2 + \left(\frac{\partial T}{\partial \gamma}\right)^2 (\Delta \gamma_u)^2} \tag{9.20}$$

where the partial derivatives are obtained by differentiating Eq. 9.19 with respect to each parameter as

$$\frac{\partial T}{\partial N} = -\frac{\lambda}{L\eta}\frac{1}{\left(\gamma - \frac{\alpha}{\eta}\right)} \tag{9.21}$$

$$\frac{\partial T}{\partial L} = -\frac{\lambda}{\eta}\frac{\text{N}}{\left(\gamma - \frac{\alpha}{\eta}\right)}\frac{1}{L^2} \tag{9.22}$$

$$\frac{\partial T}{\partial \eta} = \frac{-\gamma\lambda}{L}\frac{N}{\left(\gamma - \frac{\alpha}{\eta}\right)^2}\frac{1}{\eta^2} \tag{9.23}$$

$$\frac{\partial T}{\partial \alpha}=\frac{\lambda}{L}\frac{N}{\left(\gamma-\frac{\alpha}{\eta}\right)^2}\frac{1}{\eta^2} \tag{9.24}$$

$$\frac{\partial T}{\partial \gamma}=\frac{\lambda}{L}\frac{N}{\left(\gamma-\frac{\alpha}{\eta}\right)^2}\frac{\eta}{\eta^2} \tag{9.25}$$

The errors are estimated by substituting the respective values into Eqs. 9.20 to 9.25.

As soon as the Ti:sapphire laser beam was turned on the fringes appeared to move and the detector output was oscillatory. For a 300 mW laser beam it took about 2000 s to reach the steady-state value and the detector output revealed three fringes in this time interval (Fig. 9.7). By substituting the values for the YAG:Yb^{3+} sample length L = 0.015 m, the number of fringes N = 3, refractive index η = 1.823 [37], coefficient of thermal expansion α = 7.8 × 10^{-6}/K, fractional change in optical path length at 310 K is γ = 1.1 × 10^{-5}/K [38, 39], and λ = 514.5 nm in Eq. 9.19, the change in sample temperature is estimated as 7.1±0.8°C. The reliability of the temperature estimate depends on the accuracy [31] of the other parameters α, β, and γ. The sample absorption at 915 nm is small and hence the temperature change of the material is small. If the Ti:sapphire laser is turned off the crystal should cool down to its steady-state value. One such curve observed is shown in Fig. 9.8. The crystal temperature almost reached the steady-state value in about 900 s. However, the time taken to reach the steady-state value during the crystal cooling is smaller than that observed during crystal heating. The direction of motion of the fringes during crystal cooling should be opposite in the direction to that of heating. To reveal this feature another curve was recorded, which is shown in Fig. 9.9. When the oscillatory detector output was decreasing, the pump laser beam was blocked manually. The detector output starts increasing (phase reversal), which confirms that the material is cooling. From 0 to 1000 s (Fig. 9.9) the crystal was heated by the Ti:sapphire laser beam, which was blocked after 1000 s. The detector output was on the trailing side of the peak value at time, t = 1000 s, and the fringe period was relatively longer between 500 to 1000 s. When the Ti:sapphire laser was blocked the material cools faster and hence the fringe period was smaller from

1000 to 1200 s [36]. As the crystal temperature was reaching the steady-state value the fringe period increased again between 1200 and 2000 s. These results indicate that the crystal temperature does not reach the steady-state value instantaneously. The required time depends on the crystal parameters like thermal conductivity. If there is any residual absorption at 514.5 nm due to the presence of trace amounts of an unknown impurity, that also will contribute to the material heating. At time $t = 0$, whether the signal rises or falls depends on the proximity of the detector pinhole with respect to the fringe and the direction of fringe movement.

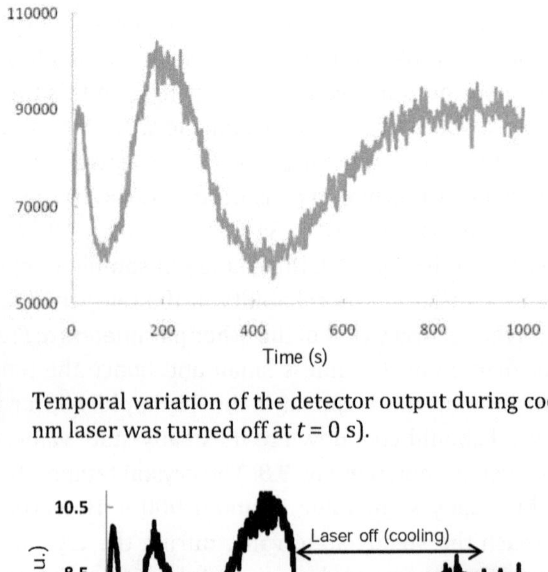

Figure 9.8 Temporal variation of the detector output during cooling (915 nm laser was turned off at $t = 0$ s).

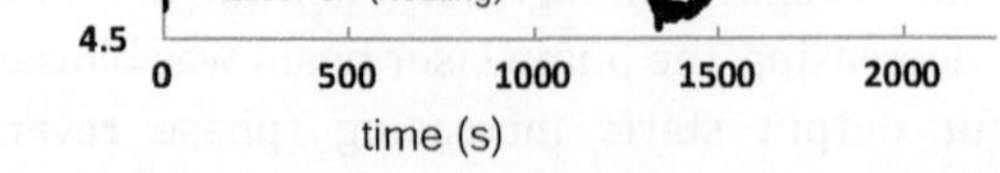

Figure 9.9 Temporal evolution of the detector output during the time of laser heating (0 to 1000 s) and laser off/cooling (1000 to 2000 s).

Though we have demonstrated the application of the Mach–Zehnder interferometer for laser heating the same configuration works equally well to investigate laser cooling of solids. Of course, the difference is in the direction of fringe motion, as depicted in Fig. 9.9. The only stringent requirement is that the heating (or cooling) laser and the probe laser beams overlap in the crystal.

9.4 Optical Heterodyne Technique

In general, a typical interferometer involves two beams, of which one is a reference beam and the other is a measurement beam. The measurement beam either gets reflected or passes through the material under investigation. Interferometry deals with either fringe counting or phase measurement [10, 14–16]. In a fringe-counting experiment an ordinary He–Ne or another low-power laser of long coherence length is required. For phase measurements a stabilized two-frequency Zeeman laser is required [40]. If a Zeeman laser is not available, the required two frequencies can be produced with a modulator [41].

9.4.1 Principle of the Optical Heterodyne Detection Technique

Optical frequencies are of the order of 6×10^{14} Hz or larger. When such a wave falls on a detector, it does not respond to such fast variations. So, the detector output is a DC current whose amplitude is proportional to the intensity of light (Fig. 9.10). The two Zeeman laser frequencies ν_1 and ν_2 are slightly different, and their difference $\Delta\nu = \nu_2 - \nu_1$, is in the radiofrequency (RF) range. When two such optical waves, ν_1 and ν_2, fall on a detector, its output is a DC current that is modulated by the RF current at $\Delta\nu$. The frequency of the RF component equals the difference between the two optical frequencies (1.5 MHz for HP 5501A laser and 250 kHz for OPTRALITE laser). If the two beams travel slightly different path lengths and then fall on a detector, the resultant amplitude is given by [40]

$$E = E_1 + E_2 = Ae^{i(k_1 z_1 - \omega_1 t)} + Ae^{i(k_2 z_2 - \omega_2 t)}, \qquad (9.26)$$

where A, ω_i, k_i, and z_i ($i = 1, 2$) are, respectively, the amplitude, angular frequency, wavenumber, and path length of each wave. Detector output current, I, is proportional to the intensity of the beams,

$$I = EE^* = I_0 + I_{RF} \qquad (9.27)$$

where

$$I_{RF} = I_0 \cos[2\pi\Delta\nu t + \Delta\Theta(z)] \quad (9.28)$$

where $\Delta\Theta(z) = \Theta_1(z_1) - \Theta_2(z_2)$. $\Theta_1(z_1)$ is the phase of the reference beam and $\Theta_2(z_2)$ is the phase of the measurement beam. As the material is cooled (or heated) the optical path length travelled by the measurement beam changes and, hence, its phase, $\Theta_2(z_2)$, also changes. Consequently the phase difference between the two waves, $\Delta\Theta(z)$ changes continuously so long as the material temperature changes. $\Delta\Theta(z)$ is related to the path length change Δz as

$$\Delta\Theta(z) = (2\pi/\lambda)\Delta z \quad (9.29)$$

where λ is the wavelength corresponding to the center frequency of the Zeeman laser.

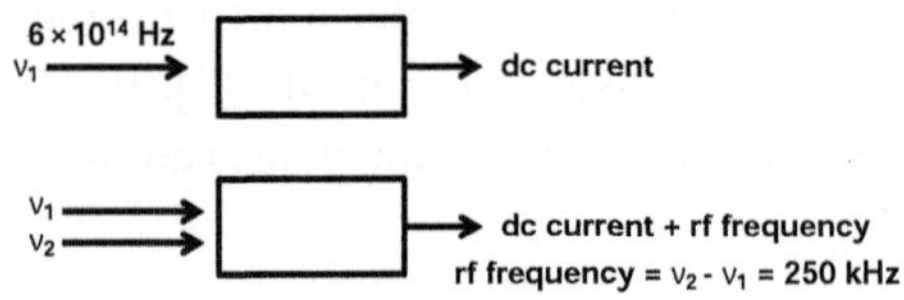

Figure 9.10 Conceptual diagram of optical heterodyne detection. The rectangle represents the detector.

$\Delta\Theta$ is measured by a phase meter. The OPTRALITE laser head provides the 250 kHz RF for reference purposes. For an HP laser the reference beam has to be derived from a second detector output. The reference and measurement RFs (detector outputs of equal frequency) are given to the two inputs of an HP 3575A phase meter, whose output (in DC volts) is proportional to the phase difference between the two RF inputs. The optical heterodyne technique was applied in the past for accurate measurement of surface roughness of coated mirrors [40, 42] and crystal growth rate measurement [43].

9.4.2 Application of Optical Heterodyne Detection Technique to Laser Cooling/Heating of Solids

Figure 9.11 shows the experimental configuration of a Mach–Zehnder interferometer configuration, required to implement optical heterodyne detection scheme for laser cooling or heating of solids.

A milliwatt Zeeman laser produces two collinear and orthogonally polarized beams, whose frequencies differ by 1.5 MHz for an HP laser and 250 kHz for an OPTRALITE laser. For some applications, a setup similar to that of a Mach–Zehnder interferometer is preferred because the measurement beam passes through the sample once. The p- (|) and s- (•) polarized beams are spatially separated by a polarizing beam splitter (PBS). Based on the optical configuration the mathematical model has to be developed accordingly. For the measurement of temperature changes in the experiments of laser cooling or heating of solids, the heterodyne technique works equally well. Whenever a material is heated or cooled its length as well as refractive index change. A measurement of optical path length variation of a beam, which passes through the material, reveals the sample temperature change. So, prior knowledge of the temperature dependence of the material's refractive index and thermal expansion coefficient is needed. For the implementation of the optical heterodyne technique a Mach–Zehnder interferometer configuration (Fig. 9.11) is a straightforward approach whose mathematical description is discussed below. The two orthogonally polarized beams of the Zeeman laser are spatially separated by a PBS, of which one is a reference beam (ν_1) and the other is a measurement beam (ν_2) that passes through the material under investigation. In this case, the material is heated or cooled by another laser beam that gets reflected by the dichroic mirror (DM), which is collinear with the probing beam but passes in a direction opposite to the probe beam (Fig. 9.11). The laser beam either heats up or cools down the material, depending on the sample and beam wavelength. Say ν_1 and ν_2 travel approximately equal distances, and their OPD is within the coherence length of the laser. The refractive indexes of air and the material are represented by η_1 and $\eta 2$, respectively. L_1 and L_2 are the lengths of the interferometer arms and L_3 is the sample length. D_1 and D_2 represent the optical path lengths of the reference and measurement beams, respectively, which are given by [44]

$$D_1 = \eta_1(L_1 + L_2) \tag{9.30}$$

$$D_2 = \eta_1 L_2 + \eta_1(L_1 - L_3) + \eta_2 L_3 \tag{9.31}$$

The OPD is given by

$$\text{OPD} = D_2 - D_1 = \eta_2 L_3 - \eta_1 L_3 \tag{9.32}$$

If we assume that the refractive index of air, $\eta_1 = 1$, then

$$\text{OPD} = \eta_2 L_3 - L_3 \tag{9.33}$$

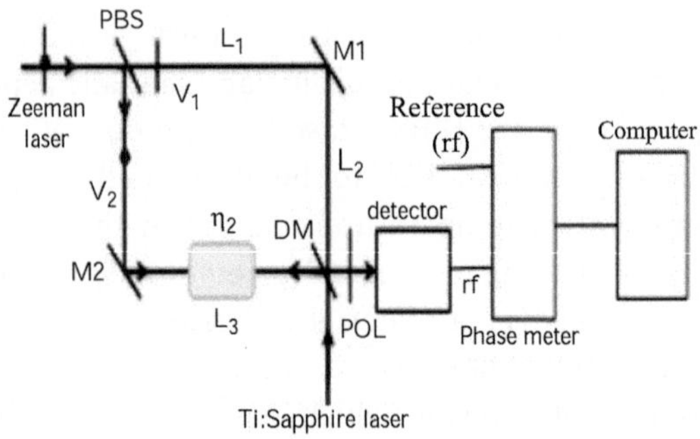

Figure 9.11 Optical configuration of the heterodyne detection scheme.

Hereafter, we drop the subscripts and use η for the refractive index of the material and L for sample length

$$\text{OPD} = \eta L - L \tag{9.34}$$

As the sample temperature changes η and L change accordingly. Therefore a change in the OPD is given by

$$\Delta(\text{OPD}) = \Delta(\eta L) - \Delta L = \Delta\eta L + \eta\Delta L - \Delta L \tag{9.35}$$

The refractive index and sample length are temperature dependent. Equation 9.35 is similar to Eq. 9.11 and further simplification will reduce this equation to

$$\Delta(\text{OPD}) = \frac{\partial\eta}{\partial T}\Delta T L + \eta\frac{\partial L}{\partial T}\Delta T - \frac{\partial L}{\partial T}\Delta T \tag{9.36}$$

$$\Delta(\text{OPD}) = \eta L\left(\frac{1}{\eta}\frac{\partial\eta}{\partial T} + \frac{1}{L}\frac{\partial L}{\partial T} - \frac{1}{\eta L}\frac{\partial L}{\partial T}\right)\Delta T \tag{9.37}$$

where α and γ have their usual meaning. The change in the OPD is simplified as

$$\Delta(\text{OPD}) = \eta L\left(\gamma - \frac{\alpha}{\eta}\right)\Delta T \tag{9.38}$$

The OPD is related to the phase difference between the two beams as

$$\Delta\Theta = \frac{2\pi}{\lambda}\Delta(\text{OPD}) = \frac{2\pi}{\lambda}\eta L\left(\gamma - \frac{\alpha}{\eta}\right)\Delta T \tag{9.39}$$

By rearranging the above equation, the temperature difference is estimated as [44]

$$\Delta T = \frac{\lambda\Delta\Theta}{2\pi\eta L\left(\gamma - \frac{\alpha}{\eta}\right)} \tag{9.40}$$

The probe laser wavelength is 632.8 nm. The values of α and γ are estimated from the sample data [38, 39] using Eqs. 9.14 and 9.15. $\Delta\Theta$ can be measured to an accuracy of 0.1° with a phase meter (resolution of Model HP 3575A phase meter). For YAG material $\alpha = 7.8 \times 10^{-6}$, $\gamma = 11 \times 10^{-6}$, $L = 0.0105$ m, and $\eta = 1.823$. If $\Delta\Theta$ is 0.1°, which is the phase meter resolution, then Eq. 9.40 predicts a temperature change of 1.4 mK (theoretical limit). This accuracy is comparable to that obtained with the fluorescence ratio method [34] and the differential luminescence measurement method [35]. However, the measurement of such a small change in temperature may not be possible because the accuracy is limited by the uncertainty, which is calculated by the propagation of errors [45] as

$$\Delta T_{\text{u}} = \sqrt{\left(\frac{\partial T}{\partial\theta}\right)^2(\Delta\theta_{\text{u}})^2 + \left(\frac{\partial T}{\partial L}\right)^2(\Delta L_{\text{u}})^2 + \left(\frac{\partial T}{\partial\eta}\right)^2(\Delta\eta_{\text{u}})^2 + \left(\frac{\partial T}{\partial\gamma}\right)^2(\Delta\gamma_{\text{u}})^2 + \left(\frac{\partial T}{\partial\lambda}\right)^2(\Delta\lambda_{\text{u}})^2 + \left(\frac{\partial T}{\partial\alpha}\right)^2(\Delta\alpha_{\text{u}})^2} \tag{9.41}$$

where the subscript "u" refers to the uncertainty in the parameter and the partial derivatives are evaluated from Eq. 9.40. The uncertainty in wavelength, $\Delta\lambda_{\text{u}}$ is zero because the probe laser wavelength does not change. Therefore, the accuracy in ΔT estimation depends on the uncertainties in the estimates of η, α, γ, and $\Delta\Theta$ measurement.

$$\frac{\partial \text{T}}{d\theta} = \frac{\lambda}{2\pi}\frac{\Delta\theta_{\text{d}}}{\left(\gamma - \frac{\alpha}{\eta}\right)L\eta} \tag{9.42}$$

$$\frac{\partial T}{dL} = -\frac{\lambda}{2\pi}\frac{\theta_{\text{d}}}{\left(\gamma - \frac{\alpha}{\eta}\right)\eta}\frac{\Delta L}{L^2} \tag{9.43}$$

$$\frac{\partial T}{d\eta} = -\frac{\lambda}{2\pi}\frac{\theta_d}{\left(\gamma - \frac{\alpha}{\eta}\right)}\frac{1}{\eta^2} \tag{9.44}$$

$$\frac{\partial T}{d\alpha} = -\frac{\lambda}{2\pi}\frac{\theta_d}{\left(\gamma - \frac{\alpha}{\eta}\right)^2}\frac{1}{L\eta^2} \tag{9.45}$$

$$\frac{\partial T}{d\gamma} = \frac{\lambda}{2\pi}\frac{\theta_d}{\left(\gamma - \frac{\alpha}{\eta}\right)^2}\frac{1}{L\eta} \tag{9.46}$$

9.5 Michelson Interferometer

The experimental configuration of a Michelson interferometer used for the measurement of laser cooling/heating of solids is shown in Fig. 9.12. Fringes are produced by the superposition of two coherent beams, of which one is a reference beam and the other is a measurement beam. The latter either passes through or gets reflected from the sample and its optical path length changes whenever the sample is heated or cooled. So the fringe intensity, measured by a detector, changes continuously with time because of a change in the optical path length, ηL. Optical path length changes because the sample length, L, and refractive index, η, are temperature dependent. A He–Ne laser is separated into two beams by a beam splitter, BS1, of which one is retroreflected by mirror M1 (reference beam) and the other passes through the sample and is reflected by mirror M2 (measurement beam). The path lengths from BS1 to M1 and M2 are approximately equal. The retroreflected beams are recombined by the same beam splitter and directed toward mirror M3. After reflection from M3 the beams pass through a color filter, CF, and a lens, which projects interference fringes on a screen. A small wedge is created between the reflected wavefronts to produce linear fringes [2]. A Pr^{3+}-doped YAG sample is used for laser heating studies. An argon ion laser beam resonantly excites the Pr^{3+} ion to its excited levels [46]. Excited ions relax radiatively and nonradiatively (Fig. 9.13) and the latter contributes to sample heating. The argon ion laser beam is reflected by beam splitter BS2, which enters the sample collinearly with the He–Ne laser. The color filter blocks the heating beam from reaching the screen. Both the length and the refractive index of a material change whenever it is heated. A measurement of optical path length variation of the beam that passes through the material reveals the sample temperature change. So, prior knowledge of the temperature dependence of the material's refractive index and thermal expansion coefficient is needed [38]. The results obtained for Pr^{3+}-doped YAG are summarized here [2].

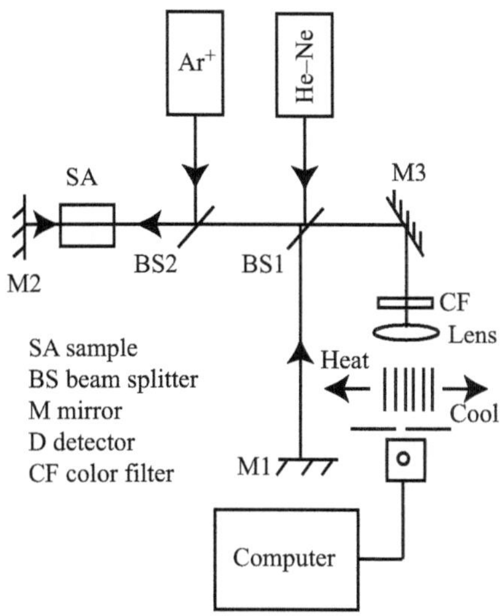

Figure 9.12 Michelson interferometer setup used to measure laser-induced heating of YAG:Pr^{3+}.

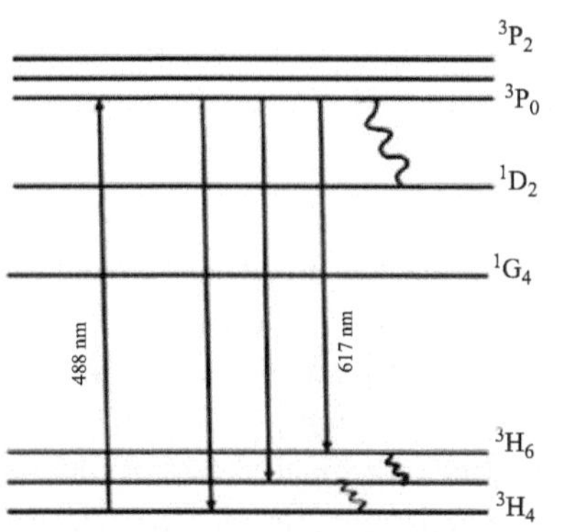

Figure 9.13 Partial energy level diagram of Pr^{3+}-doped YAG depicting radiative and nonradiative relaxations.

9.5.1 Theory of Michelson Interferometer

The OPD of the two beams is given by

$$\begin{aligned} \text{OPD} &= \text{D}_1 - \text{D}_2 = 2\,\eta_1(L_1 - L_2) + 2\eta_2 L_2 - 2\eta_1 L_1 \\ &= 2L_2(\eta_2 - \eta_1) = 2L_2(\eta_2 - 1) = 2L(\eta - 1) \end{aligned} \tag{9.47}$$

where η_1 was set to 1.0 for air. Hereafter the sample length L_2 and index η_2 will be replaced by L and η, respectively. L_1 is the distance from BS1 to the mirror (M1 or M2) and L_2 is the sample length. The temperature dependence of the refractive index and sample length are given by Eqs. 9.8 and 9.9. L_0 and η_0 are the room-temperature values of sample length and refractive index, respectively; hereafter they are represented by L and η without the subscripts. The variation in OPD is given by

$$\Delta(\text{OPD}) = 2[\Delta(\eta L) - \Delta L] = 2[\Delta\eta L + \eta\Delta L - \Delta L] \tag{9.48}$$

After applying the procedure, described by Eqs. 9.8 to 9.12 the above equation reduces to

$$\Delta(\text{OPD}) = 2L\eta\left(\frac{1}{\eta}\frac{\partial\eta}{\partial T} + \alpha - \frac{\alpha}{\eta}\right)\Delta T \tag{9.49}$$

where α and γ have their usual meanings, as explained before. If N fringes translated for a given temperature change, ΔT, then $\Delta(\text{OPD}) = \lambda N$. Equation 9.49 reduces to

$$\lambda N = 2L\eta\left(\gamma - \frac{\alpha}{\eta}\right)\Delta T \tag{9.50}$$

The temperature variation of the sample along the beam path is given by

$$\Delta T = \frac{\lambda N}{2L\eta\left(\gamma - \frac{\alpha}{\eta}\right)} \tag{9.51}$$

From the above equation it is clear that ΔT is a function of N, L, η, α, and γ. So the accuracy of ΔT estimate depends on the accuracy with which these parameters were measured.

The YAG:Pr^{3+} sample length L_0 is 0.01048 m, and its room-temperature refractive index η_0 is 1.823 at 632.8 nm [37], its linear thermal expansion coefficient α is 7.8×10^{-6}/K, the fractional change in the optical path length γ is 11×10^{-6}/K [38, 39], and the probe laser wavelength is 632.8 nm. When 115 mW of a focused Ar^+ beam entered the sample the detector output was oscillatory, as expected, and it approached the steady-state value in 500 s (Fig. 9.14). The fringe period is fast (40 s) as soon as the Ar^+ laser was turned on and eventually it increased to 240 s for the last fringe. This variation in the time period or chirping is due to the nature of sample heating. As soon as the Ar^+ laser is turned on, nonradiative relaxation sets in, which heats the sample fast. Eventually the sample temperature reached a steady-state value in about 500 s and the detector recorded a total of 4.75±0.25 fringes (N) by this time. The number of waves or fringes was measured manually from the plot. After substituting these numbers in Eq. 9.51 the change in sample temperature along the beam path is estimated to be 11.7±1.0 K [2]. The temperature change is small because a small fraction of the incident power is absorbed by the sample, which is a function of sample absorption coefficient at the laser wavelength. The quantum efficiency of the $^3P_0$ level is less than 1; in other words some of the excited ions relax nonradiatively (represented by the wavy line in Fig. 9.13) to the $^1D_2$ level, which contributes to heating. Moreover, the ions in the lower levels $^3H_5$, $^3H_6$, $^3F_2$, and $^3F_3$ levels also undergo nonradiative relaxation, which is represented by wavy lines. All these nonradiative relaxation pathways contribute to a temperature change of 11.7±1.0 K. The uncertainty is large because of the following reasons. Using a micrometer sample length can be measured to an accuracy of 10^{-5} m. The fringe intensity is measured by a detector whose output is sinusoidal. From the signal plot the number of waves (or fringes) can be measured to an accuracy of 0.25. The uncertainties in the refractive index, the coefficient of thermal expansion, and the fractional change in optical path length [38, 39] are 10^{-3}, 0.4×10^{-6}/K, and 0.4×10^{-6}/K, respectively. The uncertainty in the ΔT calculation is similar that given by Eqs. 9.20 to 9.25.

The sample cools down if the argon ion laser beam is blocked. The direction of fringe motion is expected to be opposite to that of heating. This has been demonstrated experimentally and a sample plot is depicted in Fig. 9.15. At time t = 0 s the Ar^+ laser was on and the detector output was varying, as expected. At time t = 135 s the detector output was rising, and the Ar^+ laser beam was blocked. As expected the detector output started decreasing because the material is cooling and the fringes moved in opposite directions. During the cooling process the material slowly reached thermal equilibrium and the detector output reached the steady-state value in about 100 s after the Ar$^+$ laser beam was blocked. The number of fringes (or waves) measured depends on the Ar$^+$ laser power and whether the beam was focused in the material or not. To verify this dependence an unfocused 488 nm laser beam and a He–Ne beam were launched collinearly into the YAG:Pr^{3+} sample. As expected the number of waves or fringes measured increased with the laser power (not shown). Chirping is present in all the measurements. This confirms that the plot shown in Fig. 9.14 is real and not due to random fluctuations in the signal.

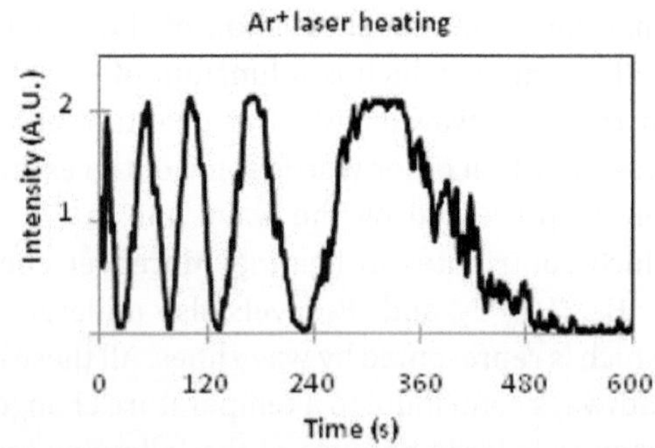

Figure 9.14 Temporal variation of the detector output obtained when 115 mW of an Ar$^+$ laser beam was launched through the sample. The sample was located at the focus. The Ar$^+$ laser was on from $t = 0$ to 600 s [2].

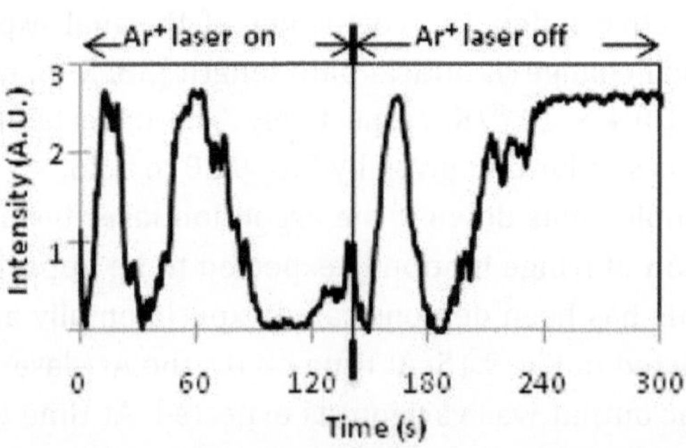

Figure 9.15 Temporal variation of the detector output during heating (Ar$^+$ laser was on from $t = 0$ to 135 s) and cooling (Ar$^+$ laser was off from $t = 135$ to 300 s) periods [2].

The results of interferometric measurements are validated by a thermocouple. For this purpose, the sample was mounted in a metallic block, which was partly immersed in a water beaker (Fig. 9.16). The glass beaker was heated on top of a hotplate. This facilitates the sample heating and cooling in a controlled fashion. A thermocouple was immersed into the water beaker at half height to measure the water temperature. The sample holder assembly was inserted along the path of the measurement beam of the Michelson interferometer. The sample does not absorb the 632.8 nm beam and hence there is no nonradiative relaxation in the sample. The sample temperature was increased by heating the water. As expected the detector output produced an interference pattern and produced five fringes in 800 s (Fig. 9.17a). Thermocouple measurements revealed a 14°C change in sample temperature (increased from 22.5°C to 36.5°C) during this time interval. The detector output is noisy because the optical table was not isolated from floor vibrations and the laser power was not stabilized. The change in sample temperature is estimated using Eq. 9.51 as 12.3 ± 1.0°C, which is in reasonable agreement with that measured by the thermocouple. When the electric power was turned off to the hotplate sample temperature was slowly decreasing, and the detector output revealed three fringes in 2400 s (Fig. 9.17b). This corresponds to a temperature change of 7.4 ± 0.8°C, which is in agreement with that measured by the thermocouple, whose temperature decreased from 35°C to 27.6°C. The agreement was reproducible. During the cooling process the water bath was allowed to cool to room temperature naturally in ambient air, which was a slow process because the heat capacity of water is a large value. Hence, there is perfect agreement between the value given by the thermocouple and that estimated by Eq. 9.51. This experiment confirms that the interferometric measurement of temperature is reliable. The detector output is free from chirping if the heating/cooling rate is constant. However, the agreement is not perfect between the optical and thermocouple measurements during sample heating, because of the procedure used in performing the experiment and human error. The hotplate dial was set at 200°C to heat the water. The temperature of the hotplate increased very fast to reach the set value. However, there was some time lag (couple of minutes) before the thermocouple senses higher temperatures, because of the delay involved in heating the water. When the temperature started to increase the computer was manually triggered for data acquisition. This process has generated some error in timing the events.

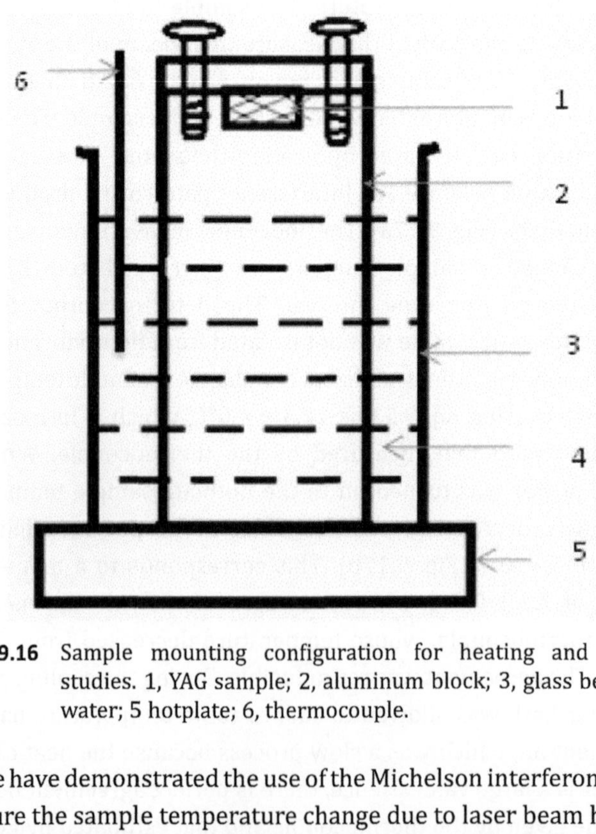

Figure 9.16 Sample mounting configuration for heating and cooling studies. 1, YAG sample; 2, aluminum block; 3, glass beaker; 4, water; 5 hotplate; 6, thermocouple.

We have demonstrated the use of the Michelson interferometer to measure the sample temperature change due to laser beam heating. An analytical model was developed to estimate the temperature change that depends on fringe (or wave) counting. The passage of a 115 mW focused laser beam through the sample raised its temperature by 11.7 ± 1.0 K. Optical interferometric measurement of temperature agreed with that measured by a thermocouple. This method applies equally well to estimate the temperature changes in laser cooling of solids also.

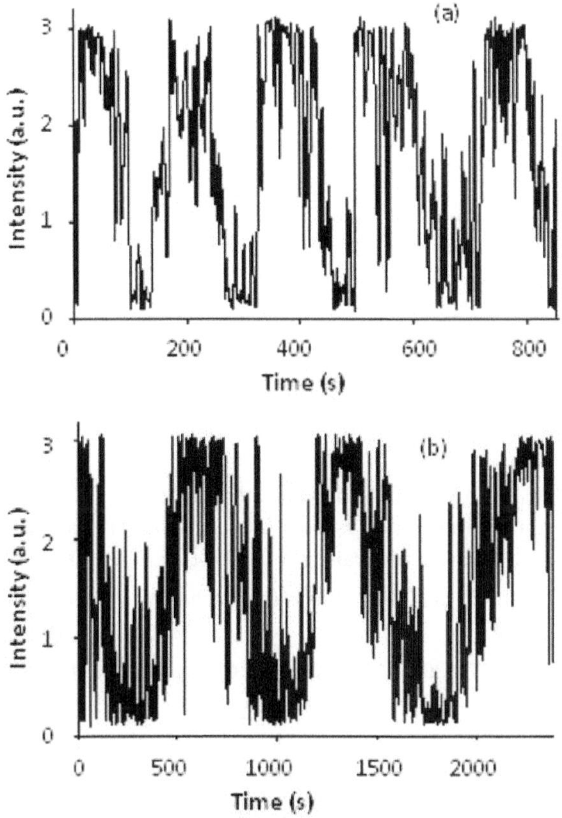

Figure 9.17 Temporal variation of the detector output measured when the sample holder was (a) heated in a water bath from 22.5°C (t = 0 s) to 36.5°C (t = 853 s) and (b) naturally cooled in ambient air from 35°C (t = 0 s) to 27.6°C (t = 2400 s).

References

1. Baker, B. (1998). *Temperature Sensing Technologies* (Microchip Technology), AN679.
2. Farley, C. W., Reddy, B. R. (2011). Interferometric measurement of laser heating in Praseodymium doped YAG crystal, *Appl. Opt.,* **50**, pp. 526–531.
3. Epstein, R. I., Buchwald, M. I., Edwards, B. C., Gosnell, T. R., Mungan, C. E. (1995). Observation of laser induced fluorescent cooling of a solid, *Nature,* **377**, pp. 500–503.

4. Soto, M. A., Nannipierri, T., Signorini, A., Lazzeri, A., Baronti, F., Roncella, R., Bolognini, G., Pasquale, F. D. (2011). Raman-based distributed temperature sensor with 1 m spatial resolution over 26 km SMF using low repetition-rate cyclic pulse coding, *Opt. Lett.*, **36**, pp. 2557–2559.
5. Lapp, M., Hartley, D. L. (1976). Raman scattering studies of combustion, *Combust. Sci. Technol.*, **13**, pp. 199–210.
6. Wade, S. A., Collins, S. F., Baxter, D. W. (2003). Fluorescence intensity ratio technique for optical fiber point temperature sensing, *J. Appl. Phys.*, **94**, pp. 4743–4756.
7. Chubb, D. L., Wolford, D. S. (2000). Rare-earth optical temperature sensor, *Rev. Sci. Instrum.*, **71**, pp. 2233–2237.
8. Zhan, Y., Wu, H., Yang, Q., Xiang, S., He, H. (2008). Fiber Bragg grating sensors for high temperature measurement, *Opt. Lasers Eng.*, **46**, pp. 349–354.
9. de Lima Filho, E. S., Nemova, G., Loranger, S., Kashyap, R. (2014). Direct measurement of laser cooling of Yb:YAG crystal at atmospheric pressure using a fiber Bragg grating, *Proc. SPIE*, **9000**, p. 90000I.
10. Hariharan, P. (2003). *Optical Inteferometry* (Academic Press, San Diego, USA).
11. Yoshizawa, T. (2009). *Handbook of Optical Metrology: Principles and Applications* (CRC Press, Boca Raton, USA).
12. Wilkie, D., Fisher, S. A. (1963). Measurement of temperature by Mach-Zehnder interferometry, *Proc. Inst. Mech. Eng.*, **178**, pp. 461–470.
13. Sandhu, S. S., Weinberg, F. J. (1972). A laser interferometer for combustion, aerodynamics and heat transfer studies, *J. Phys. E: Sci. Instrum.*, **5**, pp. 1018– 1020.
14. Steel, W. H. (1983). *Interferometry* (Cambridge University Press, New York, USA).
15. Malacara, D. (1992). *Optical Shop Testing* (John Wiley & sons, New York, USA).
16. Toal, V. (2012). *Introduction to Holography* (CRC Press, New York, USA).
17. Hariharan, P. (1969). Modified Mach-Zehnder interferometer, *Appl. Opt.*, **8**, pp. 1925–1926.
18. Kowalski, F. V., Teets, R. E., Demtröder, W., Schawlow, A. L. (1978). An improved wavemeter for cw lasers, *J. Opt. Soc. Am.*, **68**, pp. 1611–1613.
19. Lee, L. S., Schawlow, A. L. (1981). Multiple-wedge wavemeter for pulsed lasers, *Opt. Lett.*, **6**, pp. 610–612.

20. Luo, H., Sun, Q., Xu, Z., Liu, D., Zhang, L. (2014). Simultaneous measurement of refractive index and temperature using multimode microfiber-based dual Mach–Zehnder interferometer, *Opt. Lett.*, **39**, pp. 4049–4053.

21. Lu, P., Chen, Q. (2010). Asymmetrical fiber Mach–Zehnder interferometer for simultaneous measurement of axial strain and temperature, *IEEE Photonics J.*, **2**, pp. 942–952.

22. Dullo, F. T., Sokolov, V., Chauvet, C., Lindecrantz, S., Solbø, S. A., Hellesø, O. G. (2015). Application of Mach–Zehnder interferometer based on long period grating structure for salinity measurement in water environment, *Proc. SPIE,* 8988 89881T-1.

23. Ashrafi, Z. N., Ashjaee, M., Askar, M. H. (2015). Two-dimensional temperature field measurement of a premixed methane/air flame using Mach–Zehnder interferometry, *Opt. Commun.*, **345**, pp. 55–63.

24. Koechner, W., Bass, M. (2003). *Solid-State Lasers* (Springer, New York, USA).

25. Moerner, W. E. (1988). *Persistent Spectral Hole Burning: Science and Applications, Topics in Current Physics*, Vol. 44 (Springer, New York, USA).

26. Epstein, R. I., Sheik-Bahae, M. (2009). *Optical Refrigeration: Science and Applications of Laser Cooling Of Solids* (Wiley VCH Verlag, Weinheim).

27. Dieke, G. H. (1968). *Spectra and Energy Levels of Rare-Earth Ions* (John Wiley & Sons, USA).

28. Weber, M. J. (1967). Probabilities for radiative and nonradiative decay of Er^{3+} in LaF_3, *Phys. Rev.*, **157**, pp. 262– 272.

29. Riseberg, L. A., Moos, H. W. (1967). Multiphonon orbit-lattice relaxation in $LaBr_3$, $LaCl_3$ and LaF_3, *Phys. Rev. Lett.*, **19**, pp. 1423–1426.

30. Kamma, I., Kommidi, P., Reddy, B. R. (2008). Design of a high temperature sensing system using luminescent lifetime measurement, *Rev. Sci. Instrum.*, **79**, p. 096104.

31. Reddy, B. R., Kamma, I., Kommidi, P. (2013). Optical sensing techniques for temperature measurement, *Appl. Opt.*, **52**, pp. B33–B39.

32. Messias, D. N., Vermelho, M. V. D., Gouveia-Neto, A. S., Aitchison, J. S. (2002). All optical integrated upconversion fluorescence based point temperature sensing system using Er^{3+} doped silica-on-silicon waveguides, *Rev. Sci. Instrum.*, **73**, pp. 476–479.

33. Paez, G., Strojnik, M. (2003). Erbium-doped optical fiber fluorescence temperature sensor with enhanced sensitivity, a high signal-to-noise

ratio and a power ratio in the 520-530- and 550-560 nm bands, *Appl. Opt.*, **42**, pp. 3251–3258.

34. Paez, G., Strojnik, M. (2003). Experimental results of ratio-based erbium-doped-silica temperature sensor, *Opt. Eng.*, **42**, pp. 1805–1811.
35. Patterson, W., Soto, E., Fleharty, M., Sheik-Bahae, M. (2007). Differential luminescence thermometry in laser cooling of solids, *Proc. SPIE.*, **6461**, p. 64610B.
36. Farley, C. W., Reddy, B. R. (2011). Mach-Zehnder interferometric measurement of laser heating/cooling in Yb^{3+}: YAG, *Proc. SPIE*, **7951**, p. 795101.
37. YAG data sheet, VLOC, FL (www.vloc.com).
38. Fan, T. Y., Daneu, J. L. (1998). Thermal coefficients of the optical path length and refractive index in YAG, *Appl. Opt.*, **37**, pp. 1635–1637.
39. Wynne, R., Daneu, J. L., Fan, T. Y. (1999). Thermal coefficients of the expansion and refractive index in YAG, *Appl. Opt.*, **38**, pp. 3282–3284.
40. Sommergen, G. E. (1981). Optical heterodyne profilometry, *Appl. Opt.*, pp. 610–618.
41. Huang, C. C. (1984). Optical heterodyne profilometer, *Opt. Eng.*, **23**, pp. 365–370.
42. George, M. C., Reddy, B. R., Jagannath, H., Perera, G. M., Venkateswarlu, P. (1991). Measurement of surface roughness using a noncontact profilometer, in *Laser Spectroscopy and Nonlinear Optics of Solids*, Radhakrishna, S., Tan, B. C., eds. (Springer Verlag/Narosa, New Delhi, India), pp. 147–152.
43. Kim, Y. K., Reddy, B. R., George, T. G., Lal, R. B. (1998). Optical heterodyne interferometry technique for solution crystal growth rate measurement, *Opt. Eng.*, **37**, pp. 616–621.
44. Reddy, B. R. (2014). Applications of optical interferometer techniques for precision measurements of changes in temperature, growth and refractive index of materials, *Technologies*, **2**, pp. 54–75.
45. Taylor, J. R. (1997). *An Introduction to Error Analysis: The Study of Uncertainties in Physical Measurements*, 2nd Ed. (University Science Books, Sausalito, CA).
46. Kamma, I., Kommidi, P., Reddy, B. R. (2009). High temperature measurement using luminescence of Pr^{3+} doped YAG and Ho^{3+} doped CaF_2, *Phys. Status Solidi* C, **6**, pp. S187–S190.

Chapter 10

Fluoride Glasses and Fibers

Mohammed Saad
Thorlabs, Inc., 56 Sparta Avenue, Newton, NJ 07860, USA
msaad@thorlabs.com

10.1 Introduction

During the last three decades, optical fibers have experienced an extraordinary and rapid development. They are being used, as passive and active mediums, in various industrial applications, in fields as diverse as telecommunication, medicine, aerospace, defense, spectroscopy, sensing, laser power delivery, fiber lasers, etc.

Optical fibers are now critical parts in many high-tech devices. Of course fiber parameters have to be tailored to each application needs. Unfortunately, there is no single fiber material that can fulfill all types of application needs. Engineers and researchers have to make some compromises when choosing the right fiber material for the right application. There are some important criteria that one has to consider when choosing an optical fiber to be used in a specific application. The first criterion is technical. The glass should have the required properties, such as transmission window, glass

Laser Cooling: Fundamental Properties and Applications
Edited by Galina Nemova

ISBN 978-981-4745-04-8 (Hardcover), 978-981-4745-05-5 (eBook)
www.panstanford.com

characteristic temperatures, and durability. The second criterion is economical. The technology should be mature enough to provide high-quality optical fibers at an effective cost with a reasonable yield.

Silica fiber technology is the most establish technology so far. Billions of dollars have been spent to develop this technology and bring it to the current level. This was mainly motivated by the huge telecommunication market. Standard silica optical fibers, used in the telecommunication field, are produced in thousands of kilometers, with high mechanical strength and ultralow loss. The fiber loss is approaching the theoretical value, which means that the material has ultrahigh purity and the production process is well under control. However, these fibers cannot be used for application above 2 microns since the material is opaque. More recently, silica photonic crystal fibers (PCFs), with some special design, negative curvature fiber [1], have been shown to transmit at few wavelengths over 2 microns. But these fibers are still at the R&D phase. Of course, applying the same design to materials that are already transmitting in the midinfrared region will enlarge the transmission spectrum to longer wavelengths for infrared materials, too.

For applications requesting fibers with either transmission above 2 microns or continuous transmission from the ultraviolet to the midinfrared region, there are few infrared materials that can be considered, such as heavy metal fluorides [2], chalcogenides [3], heavy oxides such as tellurite [4], phosphates [5], single crystals such as sapphire [6], and polycrystalline materials. These fibers are commonly called exotic fibers. Again, one has to choose among these materials according to the application requirements. However, few of these materials can be drawn into high-quality optical fibers that can meet industrial application needs.

Amorphous materials have definitely some advantages when compared to crystalline materials. Fibers made from amorphous materials can have a core cladding structure and have low loss. They are also more robust and very flexible. They can also be drawn into long-length fibers. Furthermore, fiber parameters can be tailored to application needs by adjusting core and cladding compositions.

Among all exotic materials, fluoride glass fiber technology is the more advanced technology so far [7]. In fact, in the last two decades, significant progress has been achieved and the technology is now mature enough to meet many industrial application requirements. Fluoride fibers are now being used in some industrial applications, including the most demanding ones, such as defense and aerospace.

Almost all different types of fibers have been produced with fluoride glasses, including doped and undoped multimode and single-mode fibers. Some exotic shapes such as hexagonal, square, and D-shaped fibers have been reported as well.

Beside the state of the technology, fluoride glasses show also outstanding optical properties that make them the material of choice for many infrared and even multispectral applications. They have a low refractive index and a low and negative *dn/dt*. They can be heavily doped and co-doped with any rare earth ions for active applications, such as fiber lasers and amplifiers. Stable fluoride glass composition can contain up to 100,000 ppm of rare earth ions. This concentration is ranging from only a few hundreds to a few thousands of ppm in silica fibers. In fact, lanthanum fluoride is one of the elements entering some fluoride glass compositions, and it can be substituted by any other rare earth ion without compromising glass quality. Consequently, compact fiber laser can be made using a short length, a few centimeters, of doped fluoride glass fiber.

10.2 Bulk Glass

Glass formation rules established for oxide glasses cannot be used to predict glass formation in exotic glass systems and especially ionic glasses such as halide glasses. So, nobody was expecting that glass formation can be that rich in complex halide systems.

The first fluoride glass has been reported based on beryllium fluoride (BeF_2) glass systems, as reported by Van Uitert [8]. This glass has a low refractive index (1.30) and a glass transition temperature of 250°C. However, these glasses were never used for any practical applications, mainly because their infrared transmission is not that far from silica and also because beryllium fluoride is very toxic.

In the midseventies, a new family of fluoride glasses was accidentally discovered at Rennes University by Michel Poulain [9]. Michel was working on crystalline phases of $BaZrf_6$ and $LnZrF_7$ systems. He was mixing different rare earth fluorides with zirconium fluoride (ZrF_4) and barium fluoride. It turned out that the structure of some crystals showed quite big free sites. After calculation, it turned out that the free sites can contain an anion such as sodium (Na). So Poulain decided to add sodium fluoride to the mixture, to see if the sodium ions can occupy the free sites. He mixed the chemicals in a celled nickel tube under vacuum and placed the celled tube in a furnace at 800°C. The day after he took the nickel tube and cooled it rapidly by doping it into cooled water at room temperature. So the melt was quenched from 800°C to room temperature very quickly. Poulain was surprised to see that the mixture was not anymore a powder, as he was expecting, but a piece of glass. The first heavy metal fluoride glass was born! He obtained in fact a colored glass. After analysis, it turned out that the sample is amorphous and has a wider transmission than silica glasses. The real fluoride glass adventure started!

Since then, tens of research laboratories across the world have participated in the development of fluoride glass and fiber technology. Hundreds of glass systems have been reported. And scientists realized that glass formation in halide systems is more common than expected. More interestingly, the glass-forming area is quite large, so some glass parameters can be tailored for each specific application need. This includes refractive index n, glass transition temperature T_g, the coefficient of thermal expansion (CTE), etc.

Three main glass families have been considered representative of fluoride glasses in general: fluorozirconate, fluoroaluminate, and fluoroindate. They are, respectively, based on zirconium fluoride (ZrF_4), aluminum fluoride (AlF_3), and indium fluoride (InF_3). Generally, the glass family name will use the name of the chemical that has the higher molar percentage in the composition. Therefore, fluorozirconate is used for zirconium fluoride (ZrF_4) base glasses, fluoroaluminate for aluminum fluoride (AlF_3) base glasses, and fluoroindate for indium fluoride (InF_3) base glasses. Indium fluoride glasses have the largest transmission window, and aluminum fluoride glasses have the highest resistance to liquid water. In fact, some aluminum fluoride glasses can be placed in hot water for many days without showing any degradation.

ZBLAN and ZBLA are two standard compositions reported in the fluorozirconate glass family. These two compositions have been widely used for almost all research programs and applications involving fluoride glasses. In fact, they are the most stable compositions in all fluoride glass history. This doesn't mean that they are the only composition that can be drawn into high-quality optical fibers. We have been able to draw into fibers many different compositions in different glass systems.

Like any multicompound glass, the preparation process is very critical. In fact, using the same composition one can end up with bad or good glass, depending on the raw materials' quality and which preparation process was used. There are as many preparation processes as research teams that were involved in fluoride glass research. Therefore, tremendous work has been carried out to optimize the fluoride glass preparation process [10–13]. Of course, and it's not surprising, that the preparation process has to be optimized for each glass family and even in some cases for each glass composition.

More complex systems, using a mix of halide ions such as chlorides and fluorides [14–17] have been also explored and quite stable compositions were reported. However, these glasses are not stable enough to be drawn into high-quality optical fibers, or can't be used in any practical application, because they are hydroscopic. Nevertheless, the study of these systems can provide us with interesting data that help scientists understand glass formation in exotic systems.

Stable compositions, with some interesting properties, have been also reported in aluminum fluoride [18] and indium fluoride glass systems [19]. These compositions contain quite high concentrations of thorium fluoride. Because of the radioactivity of thorium, these glasses have never been used in any application. Some of these glasses have very interesting and unique properties, such as ultraresistance to water attack. The glass can be left in hot water for days without showing any degradation. Aluminum fluoride glasses containing 20 molar percent of thorium fluoride (ThF_4) have a low refractive index, too, around 1.45; a high glass transition temperature around 420°C; and a low CTE, around 140×10^{-7} K^{-1}. Thus, only few research groups have investigated these systems. The stable composition in this system is BATY, containing 20 molar% BaF_2, 28.7% AlF_3, 22.6 % ThF_4, and 28.7% YF_3.

In general, quite stable compositions contain minimum three compounds can be cast into bulk samples with a few millimeters thickness. Most of these compositions cannot be drawn into optical fibers. Glass stability is most often drastically improved by adding few percent of one or two additional fluorides. As an example, the addition of 3% to 4% AlF_3 to two standard zirconium fluoride glasses ZBN (ZrF_4, BaF_2, NaF) and ZBL (ZrF_4, BaF_2, LaF_3) has improved considerably the glass stability [20]. Glass samples with a few centimeters thickness can be obtained. Therefore, the most stable compositions in fluorozirconate glass systems contain 3% to 4% of aluminum fluoride. Table 10.1 shows the composition in molar percent of the two more stable compositions in the fluorozirconate family. These two compositions have been widely used in all studies and applications concerning fluoride glasses.

Table 10.1 Standard fluorozirconate glass, ZBLAN and ZBLA, compositions in molar%

	ZrF_4	BaF_2	LaF_3	AlF_3	NaF
ZBLAN	53	20	4	3	20
ZBLA	57	34	5	4	

10.2.1 Glass Preparation

The modified chemical vapor deposition process (MCVD) has shown to be the process of choice to produce high-quality optical fibers in silica fiber technology. This process can be used when starting materials entering glass composition have quite similar physical and chemical properties. The MCVD process allows us to use ultrahigh-purity materials and also more importantly to keep the materials clean during the process. Thus, the impurities' contribution to the loss has become negligible. Consequently, the commercial-standard silica fibers have attenuation close to the theoretical value.

Unfortunately, the MCVD process cannot be used to prepare most of the multicompound glasses, including fluoride glasses. In fact, stable fluoride glass compositions contain minimum four to five chemicals, with physical and chemical properties quite different. Thus, it is difficult and even impossible to process all starting materials, entering the glass composition, at the same range of temperature and pressure. However, binary thin films have been reported by Jacoboni et al. [21] using chemical vapor deposition.

This process is only limited to binary thin films and can't be used to prepare bulk glasses and preforms.

As far as fluoride glasses are concerned, the preparation process commonly used is melting and casting. In this case ultrahigh-purity starting materials are required. And raw materials' purity is a big concern.

In fact, many chemical molecules and ions, such as transition metals [22], rare earth elements [23], water [24], CO_2 [25], and some organic molecules, have their fundamental absorption in the transmission window of fluoride glasses. And only a very small amount of these impurities in the glass will have a huge absorption peak in the fiber spectrum. Thus, from the beginning the purity of raw materials was a critical issue. At that time no ultrahigh-purity raw materials were available, especially those entering stable glass compositions. As an example, Table 10.2 shows attenuation induced by some transition metals and rare earth elements [22]. As an example, –OH and transition metal ion concentration has to be kept in the ppb level. There are two challenges here: The first one is to be able to produce the material with a low impurity level, and the second is to be able to measure this impurity level with good accuracy. Another challenge is that many raw material suppliers are not really interested in making some chemicals with such ultrahigh-purity materials since the market is too small. So sometimes you have to produce or purify your own raw materials.

Table 10.2 Attenuation (dB/km/ppm) induced by some transition metals and rare earth elements

Elements	2 μm	2.5 μm	3 μm
Transition metals	(dB/km/ppm)	(dB/km/ppm)	(dB/km/ppm)
Fe	90	28	2
Co	130	31	4
Ni	90	6	0.5
Cu	3	0.14	0.01
Rare earth			
Ce	–	–	7.7
Pr	27	1.8	0.4
Tb	25	–	14.3

Various research programs have been dedicated to develop ultrasensitive analysis methods and preparation processes to produce and characterize ultrahigh-purity starting materials, especially those entering stable glass compositions. Since the most stable fluoride glass compositions have been reported in zirconium fluoride glasses, these programs have focused first on ZrF_4, HfF_4, NaF, BaF_2, AlF_3, and rare earth elements LnF_3 [26, 27].

Three preparation processes have been developed to prepare fluoride glasses; only the two first ones have been widely used [28]:

- Direct melting of anhydrous fluoride, in platinum, gold, or glassy carbon crucibles, under a dry atmosphere. The melting temperature can vary from 800°C to 1000°C and the meting duration from one to three hours. At the end of the melting step, the glass is cooled down to around 650°C to 700°C and then poured into a metallic mold preheated at a temperature lower but close to the glass transition temperature. The mold is then cooled down to room temperature in a few hours. Once the mold is at room temperature, the glass is de-molded and polished using a standard polishing process.
- The ammonium bifluoride process, which consists of using fluoride and oxides as starting materials. The oxides are converted into fluoride by ammonium bifluoride (NH_4FHF) during the fluorination step prior to melting [29]. The fluorination process takes place at a temperature ranging from 300°C to 400°C during two to three hours, depending on the composition and glass family. Once the fluorination is completed, he glass is obtained by melting the mixture under a dry and controlled atmosphere. The melting temperature and schedule are similar to the process above. The fluorination reactions of zirconium oxide and indium oxide are given below as an example.

  $$ZrO_2 + 2NH_4FHF \Leftrightarrow ZrF_4 + 2H_2O + 2NH_3$$

  $$In_2O_3 + 3NH_4FHF \Leftrightarrow 2InF_3 + 3H_2O + 3NH_3$$

  Ammonium bifluoride's reaction with oxides produces water. Thus, one has to be sure that water is removed during the glass preparation process. In fact water has its fundamental absorption at 2.87 microns in a fluoride glass matrix. This is too close to the minimum attenuation wavelength, 2.6 microns. G. Fonteneau has reported that 1 ppm of water induces 5000 dB/km [30]. Different reactive atmospheres (RAP) have been also used, during glass processing, to remove water traces in the final glass. HCl, CCl_4, and NF_3 [31] have been widely used.

- The sol gel process has been also reported by Mohammed Saad and Marcel Poulain [32]. This process consists of using organometallic chemicals as starting materials. These chemicals are mostly liquid at room temperature, and thus, they can be easily purified compared to solid raw chemicals. In fact, a liquid can be purified using and combining different purification methods that can't be used for solid chemicals, such as distillation, phase separation, and extraction.

First we mix chemicals of ions entering the glass composition. After hydrolysis we obtain a transparent oxide gel. The gel is then dried to eliminate the excess of organic solvent and water. The dried gel is fluorinated using gaseous HF. The fluoride glass is obtained by melting the fluorinated gel in a platinum crucible and casting it into a brass mold.

10.3 Glass Characterization

Once the glass samples are cast and de-molded, they go through different characterization processes. The glass transition temperature T_g, glass crystallization temperature T_x, and melting temperature T_m are obtained using differential scanning calorimetry (DSC). The thermal expansion coefficient α is obtained by thermomechanical analysis (TMA). Glass transmission spectra are obtained using ultraviolet, visible, and infrared spectrometers. And the refractive index is obtained by the Abbe refractometer as well as other methods using different laser wavelengths.

10.3.1 Differential Scanning Calorimetry

The DSC technique is used to determine glass's characteristic temperatures. It is also used to determine some kinetic and crystallization parameters, such as the glass critical rate. Figure 10.1 shows a typical DSC curve of a glass sample. This curve is obtained by heating, at a given heating rate, 35 to 40 mg of glass and a reference in a microfurnace. Exothermic and endothermic heat flows are plotted as a function of the temperature or time. So, we can determine the glass transmission temperature T_g, the glass crystallization temperature T_x, and the glass melting temperature T_m.

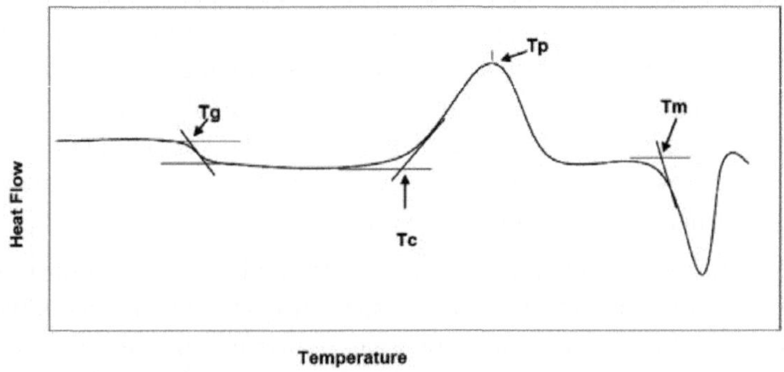

Figure 10.1 Typical DSC scan of a glass sample.

DSC is also used to estimate the stability of glass and crystallization parameters. Some stability criteria have been proposed to estimate glass stability. These criteria use characteristic temperatures: glass transition T_g, glass crystallization T_x, crystallization peak T_p, and glass melting T_m. Ruby was the first one who proposed a criterion to estimate glass stability using DSC measurement data [33]. This criterion is described by the equation below:

$$\text{Hr} = (T_x - T_g)/(T_m - T_x)$$

A stable glass will have a small $(T_m - T_x)$ and a large $(T_x - T_g)$. That means a crystallization peak close to melting temperature and far from glass transition temperature.

The Hr criterion has been simplified to Hr′; in fact most of the time the melting temperature is not well defined in multicompound systems.

$$\text{Hr}' = (T_x - T_g)/T_g$$

Saad and Poulain proposed a new stability criterion S in 1985 [34]. The new criterion is defined as below:

$$S = (T_x - T_g)(T_p - T_x)/T_g$$

The new criterion takes into account the crystallization peak shape. A stable glass will have its crystallization temperature far from its glass transition temperature (large (T_x – T_g)), and a wide crystallization peak (large (T_p – T_x)). The *S* criterion is more sensitive to small composition changes than Hr is. Consequently, it is more useful for composition optimizations. In fact, a small composition change will not affect significantly the characteristic temperatures but can significantly affect the stability of the glass. Figure 10.2 shows a comparison of the three stability criteria with a composition change in 69 ZrF_4; (23 – *x*) BaF_2; (6 + *x*) SrF2; 2AlF_3.

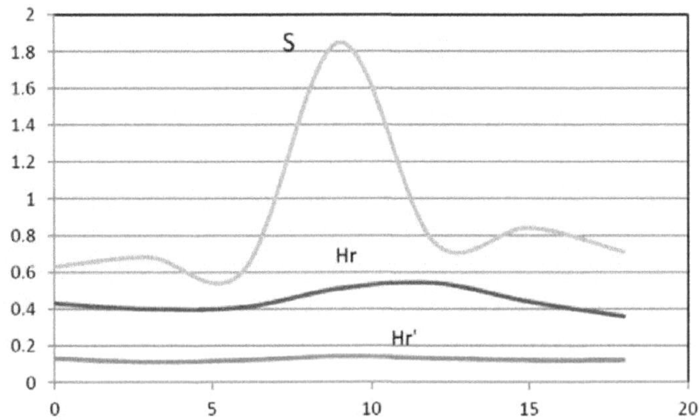

Figure 10.2 Comparison of H′, Hr, and S glass stability criteria.

Of course the ultimate way to determine glass stability is to determine the thermodynamic and kinetic parameters of each composition. We can also determine the critical cooling rate (CCR); this is the slowest cooling rate at which one can cool down a glass from the liquid state without crystallization.

The CCR can be calculated from time-temperature-transformation (TTT) curves [35]. This method is time consuming and quite complicated to use. Indeed, many measurements are required to plot TTT curves. Furthermore, TTT curves can only be obtained for quite stable compositions. Generally, glass stability criteria are used when exploring the glass formation region in a new system. The DSC duration is about 20 to 25 minutes. Furthermore, we can use disposable aluminum pans to perform the measurement, rather than small platinum crucibles, which are difficult to clean. When the glass formation region and the stable composition region have been defined using the stability criterion, one can use the CCR to determine the most stable composition of the system among all stable compositions. Table 10.3 shows the CCR values for the two standard fluorozirconate glasses ZBLAN and ZBLA. The ZBLAN composition has the lowest CCR among all fluoride glass compositions reported so far [36]. In fact, after melting, the ZBALN composition glass can be cooled slowly in the crucible and no crystallization occurs.

Table 10.3 Critical cooling rates for ZBLAN and ZBLA

Glass	Rc [°C/s]
ZBLAN	0.13
ZBLA	0.28

Table 10.4 shows some properties of some fluoride glass families. The glass transition temperature varies between 250°C and 450°C, the CTE varies between 140 and 200 × 10^{-7} K^{-1}, and the refractive index in the visible varies between 1.45 and 1.60. Consequently, no antireflective coating is required for many applications. The Fresnel reflections are low in the 3% to 4% range per end face in the visible and even less in the infrared.

Table 10.4 Some fluoride glass properties

Glass Family	Glass Transition T_g (°C)	Refractive index at 589.3 nm	Coefficient of Thermal expansion $10^{-7}K^{-1}$
ZrF_4	230–310	1.48–153	150–200
InF_3	290–320	1.48–1.53	170–190
AlF_3	360–420	1.42–1.46	150–180
ZnF_2	280–320	1.50–1.54	170–200
CdF_2	300–360	1.48–1.50	140–160

10.3.2 Glass Transmission

Fluoride glasses have a large transmission window. They are transparent from the ultraviolet to the midinfrared region without any absorption peak. The transmission window depends on the glass family and composition. The transmission is limited in the ultraviolet by electronic transition and in the infrared by the multiphonon. Figure 10.3 shows a comparison between zirconium and indium fluoride glass transmissions.

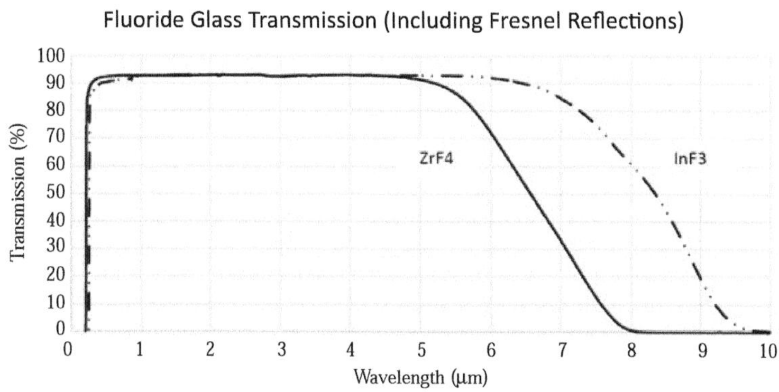

Figure 10.3 ZrF_4 and InF_3 fluoride glass transmissions (9 mm thickness).

Indium fluoride glasses have a larger transmission window than zirconium fluoride glasses. This is due to the phonon energy of indium fluoride glasses, which is lower than that of zirconium fluoride glasses.

Both glass families offer a large choice of compositions. So glass parameters can be adjusted to meet different application needs.

Fluoride glasses offer many advantages to be used for bulk optics, such as lenses, filters, and windows. We demonstrated that aspheric lenses can be obtained by molding and diamond point turning techniques. Figure 10.4 shows a photo of large lenses of ZBLAN obtained by the diamond point turning technique. The surface roughness is 2 nm.

10.3.3 Refractive Index and Dispersion

The refractive index is an important parameter for the optical designer to carefully design optical systems and optical waveguides. In fluoride glass systems, this parameter can be linearly adjusted to each application need by adjusting the composition. In fact most of the glass systems have a quite wide glass formation area. Furthermore, large substitution of any ion by another from the same column of the periodic table of the element is possible. As an example, ZrF_4 can be totally replaced by hafnium fluoride (HfF_4). We can go from the ZBLAN to the HfBLAN composition, where 53% of ZrF_4 is totally replaced by HfF_4.

Figure 10.4 ZBLAN convex and concave lenses made by the single-point diamond turning technique.

The refractive index of fluoride glasses is ranging from 1.45 to 1.60 at the sodium D wavelength 589 nm.

Lipeng Zhang et al. have studied relationships between the refractive index and dispersion of fluoride glasses and their compositions [37]. They found a good agreement between the calculated and measured values. The equation used to calculate the refractive index at different wavelengths is given below, where λ_i and λ_j are, respectively, the ultraviolet and infrared absorption wavelengths of the glass, f_i and f_j are the constants related to oscillator strength.

$$n_{\lambda}^{1} - 1 = \sum_{i=1}^{k} f_i \, \lambda^2(\lambda^2 - \lambda_i^2) + \sum_{j=1}^{m} f_j \lambda^2(\lambda^2 - \lambda_j^2) \tag{10.1}$$

One has to take into consideration glass processing, especially when glass compositions contain chemicals that can sublimate during glass preparation.

Fluoride glasses show low dispersion, lower than silica glasses. The Abbe number is about 75. The Abbe number is given by the equation below:

$$V_D = (n_D.1)/(n_F - n_c) \tag{10.2}$$

where n_D is the refractive index at the D sodium line 589.3 nm, n_F is the refractive index at the hydrogen blue line 486.1 nm, and n_C is the refractive index at the hydrogen red line 656.3 nm.

The refractive index at different wavelengths can be obtained also using the Sellmeier empirical equation, where n is the refractive index and μ_i are the Sellmeier coefficients, which are determined experimentally.

$$n(\lambda)=\sqrt{\frac{\mu_0\lambda^2}{\lambda^2-\mu_3^2}+\frac{\mu_1\lambda^2}{\lambda^2-\mu_4^2}+\frac{\mu_2\lambda^2}{\lambda^2-\mu_5^2}+A} \tag{10.3}$$

Table 10.5 and Table 10.6 show, respectively, the refractive indexes of ZBLAN and indium fluoride glasses measured at different wavelengths.

Table 10.5 ZBLAN glass measured refractive index at different wavelengths

ZrF_4 Glass	
Wavelength λ (nm)	**Refractive index $n\lambda$**
405	1.506
486.1	1.504
587.6	1.4995
632.8	1.49807
656.3	1.4974
1060	1.49198
1550	1.48883
3500	1.47447
4500	1.46312

Table 10.6 InF_3 glass measured refractive index at different wavelengths

InF_3 Glass	
Wavelength λ	**Refractive index $n\lambda$**
4500	1.45886
2000	1.47384
1064	1.47828
587.6	1.48545
485	1.49005
404.7	1.49645

Using the Sellmeier equation one can calculate the refractive index at any wavelength. Figure 10.5 shows the refractive index of ZBLAN glass. The Sellmeier coefficients are given in Table 10.7.

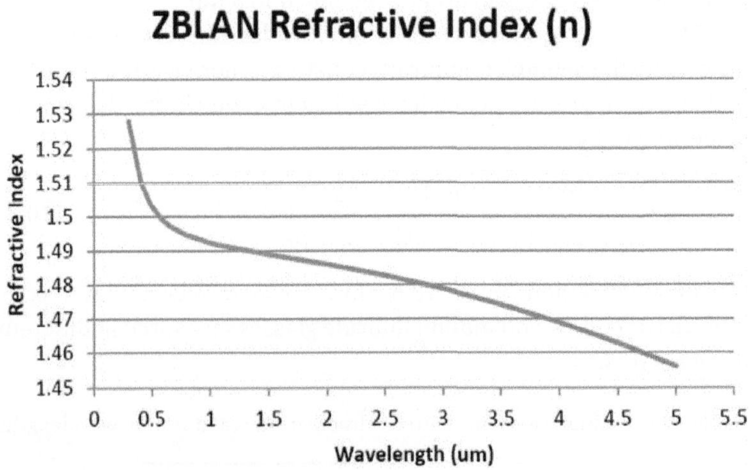

Figure 10.5 Dispersion of ZBLAN glass.

Table 10.7 Sellmeier coefficients for ZBLAN glass

Coefficient	Cladding
μ_0	0.705674
μ_1	0.515736
μ_2	2.204519
μ_3	0.087503
μ_4	0.087505
μ_5	23.80739
A	1

Figure 10.6 displays the refractive index of indium fluoride (InF_3) glass at different wavelengths. The Sellmeier coefficients are given in Table 10.8.

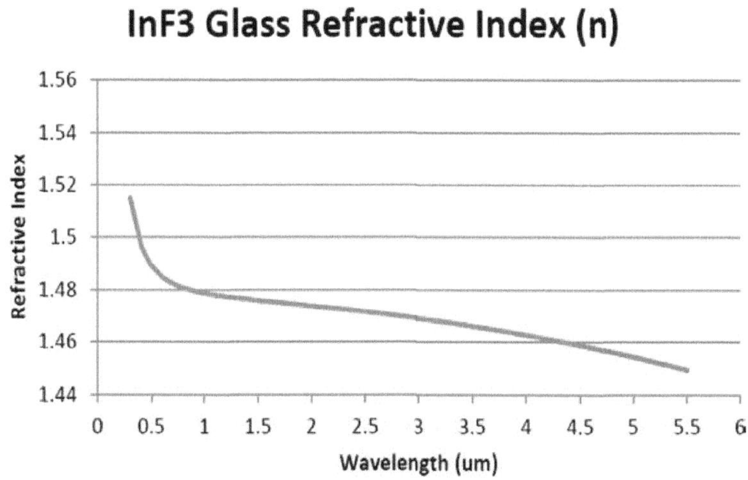

Figure 10.6 Dispersion of indium fluoride glass.

Table 10.8 Sellmeier coefficients for indium fluoride glass

Coefficient	Cladding
μ_0	0.68462594
μ_1	0.4952746
μ_2	1.4841315
μ_3	0.0680833
μ_4	0.11054856
μ_5	24.4391868
A	1

10.4 Optical Fiber

Fluoride glasses are, in fact, the only material that transmits light from the ultraviolet up to the midinfrared and can be drawn into high-quality optical fibers. Furthermore, as they are transparent in the visible region, one can use visible light and standard equipment for troubleshooting and optical alignment.

10.4.1 Fiber Drawing

Fluoride glass fibers are produced using the same drawing technique as the one used for silica fibers. They are drawn from solid preforms. This process is very effective and allows good control of fiber parameters such as fiber diameter and core cladding concentricity. Furthermore, fibers with exotic design can also be produced using this process, such as hexagonal, D-shaped, and square fibers, and PCFs [38].

There are few soft glass drawing towers in the world, but these are all different. In fact, they are all homemade. Generally, they are similar to the silica one but shorter. This is due to the fact that glass-drawing temperatures for soft glasses are much lower than those of silica. Consequently, the distance between the drawing furnace and the coating cup can be kept much shorter than with silica.

The drawing furnace is the more important piece of equipment, since it is the equipment especially designed for fluoride or soft glass fiber drawing. All other equipment pieces, fiber diameter measurement, coating cup, UV lamp, etc., are exactly the same as those used on the silica drawing tower. However, some settings have to be adjusted according to soft glass drawing parameters, which are different from those used for silica. The drawing feed, drawing temperature, drawing speed, and coating pressure ultraviolet lamp power are all different. Fluoride glass preforms are also much smaller than silica ones. The preform diameter can vary from 8 to 16 mm and 100 to 150 mm in length.

The double-crucible technique, which has been developed to draw low–melting temperature oxide glasses, has been also used to draw fluoride fibers [39]. A few hundreds of meters of single-mode and multimode fibers have been obtained. This technique is effective to draw long-length fibers but suffers from poor core cladding concentricity and crystal and bubble formation at the core cladding interface.

Figure 10.7 shows a typical drawing tower for soft glass. It is constituted with a translation table with micron accuracy to feed the preform into the drawing furnace, a drawing furnace, fiber diameter measurement with submicron accuracy, a coating cup, concentricity measurement, and an ultraviolet lamp. All these pieces of equipment are controlled by the computer.

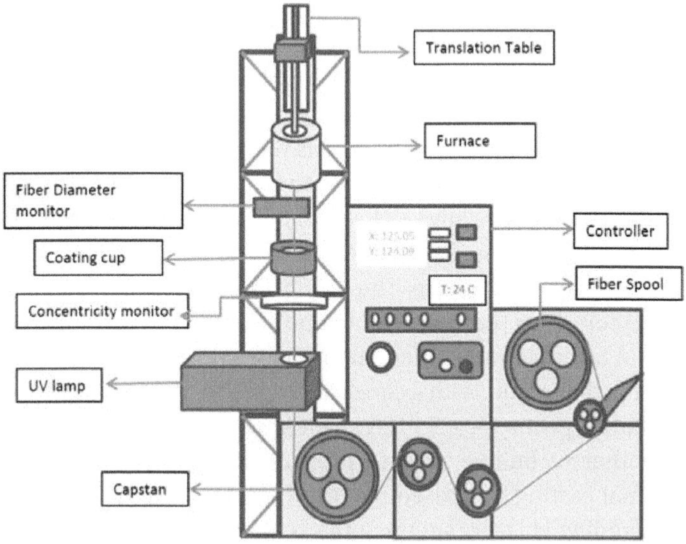

Figure 10.7 Soft glass drawing tower.

Almost all types of fibers have been produced with fluoride glasses. Single-mode and multimode fibers are prepared with good geometrical, optical, and mechanical properties. For example, 125 micron fibers are obtained with a diameter variation of ±1 micron. Figure 10.8 shows fiber diameter variation during the drawing process of a 125 micron indium fluoride fiber. The diameter is automatically controlled.

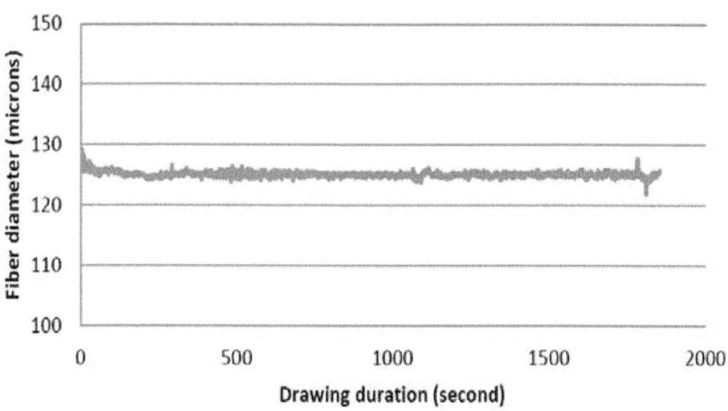

Figure 10.8 A 125 um InF_3 fiber's diameter variation during the drawing process.

Different methods are used to make fluoride glass preforms. The easiest one is built-in casting [40]. This technique consists of pouring clad glass into a preheated vertical mold. A few seconds later the bottom of the brass mold is opened, and the unsolidified glass flows out. We obtain a clad tube. The core is then poured into the cladding tube. This technique is quite simple and yields fiber with quite good attenuation and very low –OH content. The main advantage of this technique is that the core cladding interface is never exposed to the laboratory atmosphere. The only drawback is we end up with a tapered core. This technique cannot be used to produce single-mode fibers. In fact, it is very difficult to prepare a preform using the built-in casting technique, with a core cladding ratio of 0.072, which is the core cladding ratio for a 9 /125 single-mode fiber.

Another technique used to prepare fluoride glass preforms is rotational casting, proposed by Tran [41]. First we produce a tube with cladding glass by spinning the cladding glass at 3500 to 5000 rpm in a rotational jig. Then the core is poured into the tube. In this case also the core cladding interface is never exposed to the laboratory atmosphere. The core tapering issue is eliminated or drastically reduced. The mechanical design of the jig is very critical. One has to avoid any mold vibration during the spinning step. In fact, any mold vibration will result in as off-center tube.

The third technique to prepare performs is the rod-in-tube technique. This technique is similar to rotational casting, but the core is not poured into the cladding tube. The core is a solid rod that is inserted into the tube prior to drawing. In this case, the core cladding interface is exposed to the laboratory environment; thus care has to be taken to avoid scratches and contamination, especially with water. The insertion step is critical, too. One has to avoid scratching the inner surface during this step.

The rotational casting technique is also used to produce cladding tubes with a small inner diameter, required to produce single-mode fibers.

For some specific fibers, one has to combine some or all these techniques to be able to produce some specific fibers.

Figure 10.9 shows a 9 micron core single-mode fiber and a 450 micron core multimode fiber. The multimode fiber is made using the built-in casting method, and the single-mode fiber is made using the rod-in-tube process. Multiple drawings are necessary to draw the final single-mode fiber. The number of drawings depends on the fiber core dimensions. The two types of fibers have been cleaved using a standard cleaver with adjusted parameters. The two cleaves have been obtained using a Vytran cleaver. They look very good, meaning that fibers are stress free. Both cleaves' quality is similar to that of silica fibers.

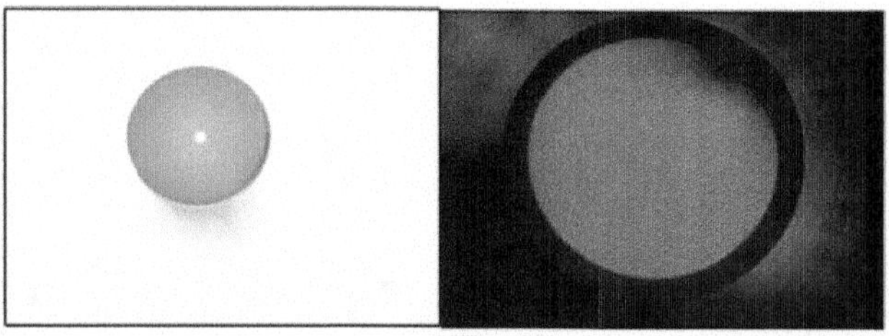

Figure 10.9 A 9 μm core single-mode and a 450 μm core multimode zirconium fluoride glass fiber.

10.4.2 Fiber Attenuation

Attenuation is without doubt the property of fluoride glass that more attracts the attention of scientists and industrial communities. In fact, when we compare attenuation between silica and fluoride fibers, as shown in Fig. 10.10, we understand why reducing the loss was the first technical race that took place between different research groups. The theoretical attenuation is 2 orders of magnitude lower for fluoride than for silica [42]. Tens of university and industrial research groups across the world have participated to develop this technology, including Centre National d'Etudes des Telecommunications (CNET) and Alcatel in France, Corning and Lucent and the Naval Research Laboratory (NRL) in USA, Nippon Telegraph and Telephone (NTT) in Japan, and British Telecom in the United Kingdom.

Unfortunately, after two decades of intense development, and few tens of millions of dollars in funding, we are still far from the goal. However, quite low loss has been reported by some groups. Kanamori et al. from the NTT have reported 0.7 dB/km [43] and Tran et al. from the NRL have reported 0.9 dB/km [44]. This loss is a considerable achievement, considering the state of the technology at that time, 1986, only eight years after the discovery of the first fluoride glass. These values have been obtained on short lengths of fibers and still 1 to 2 orders of magnitude higher than theoretical values and even higher than already achieved attenuation for silica fiber.

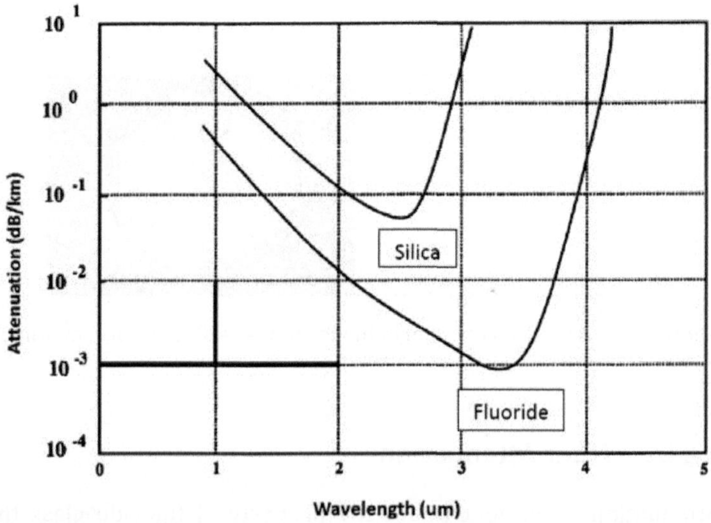

Figure 10.10 Comparison of theoretical loss of fluoride and silica fibers.

Total attenuation is due to intrinsic and extrinsic attenuations, as shown by the equation below:

$$\alpha_{\text{Total}} = \alpha_{\text{Intrinsic}} + \alpha_{\text{Extrinsic}} \tag{10.4}$$

Extrinsic loss can be either absorption or scattering, due to impurities, such as transition metals, rare earth elements, water, microbubbles, crystals, complex anions such as sulfate and phosphate [45], dissolved gases, and core cladding interface defects.

Intrinsic loss is due to the material matrix absorption and scattering, including ultraviolet absorption, Rayleigh scattering, and multiphonon absorption. Rayleigh scattering is due to density and refractive index variation. Intrinsic attenuation can be written as follows:

$$\alpha_{\text{Intrinsic}} = \alpha_{\text{UV}} + \alpha_{\text{Rayleigh}} + \alpha_{\text{Multiphonon}}$$

$$\alpha_{\text{Intrinsic}} = A_1 \exp\left(\frac{A_2}{\lambda}\right) + \frac{B_1}{\lambda^4} + C_1 \exp\left(\frac{-C_2}{\lambda}\right) \tag{10.5}$$

Above 2 microns the ultraviolet absorption is negligible and the loss is mainly due to Rayleigh scattering and multiphonon absorption. And above 3 microns, the Rayleigh scattering becomes negligible and the loss in mainly dominated by multiphonon absorption.

For more details, there are two interesting books dedicated to fluoride glass and fiber attenuation measurement [46, 47].

The cut-back method is commonly used to measure fiber attenuation. Using new Fourier transform infrared spectroscopy (FTIR) from ABB we were able to measure attenuation for both single-mode and multimode fibers. Of course the length of the fiber to be measured is limited, less than 40 m. For longer fibers one need a supercontinuum source that covers the transmission window of the fiber. The alternative solution is to use, especially for single-mode fibers, different lasers.

Current commercial fibers, including single-mode and multimode fibers, have attenuation ranging from 50 to 100 dB/km. This is more than enough for short- and medium-length applications. These fibers are produced with very good yield, to be offered at a reasonable cost.

Figure 10.11 shows the attenuation curve of zirconium multimode fiber from the ultraviolet up to 4.5 microns in the midinfrared, without any absorption peaks. The small absorption peaks at a short wavelength are due to some transition metal impurities remaining in the starting materials. No –OH peak is visible on the spectrum around 2.9 microns, which means that the –OH concentration is extremely low. In fact, the glass is prepared in a very controlled environment. The –OH concentration inside the glove boxes is maintained lower than 0.1 ppm.

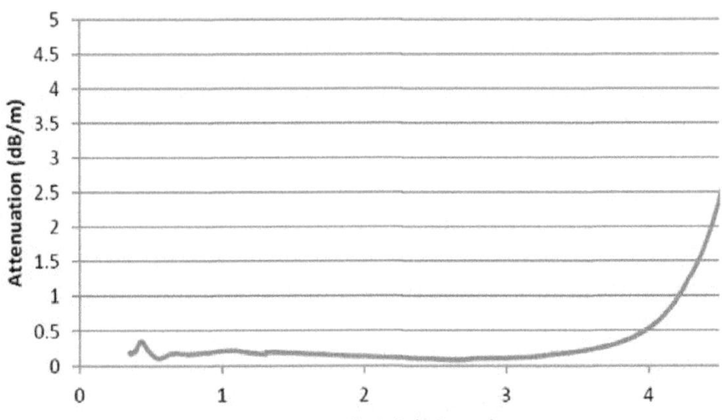

Figure 10.11 ZrF_4 multimode fiber attenuation.

Indium fluoride glass fibers have a transmission window larger than that of zirconium fluoride glass fibers. This is due to the fact that indium fluoride has a lower phonon energy than zirconium fluoride. Figure 10.12 shows a comparison of transmission curves of indium and zirconium multimode fibers. Indium fluoride glass fibers extend the application field of fluoride glasses and fibers.

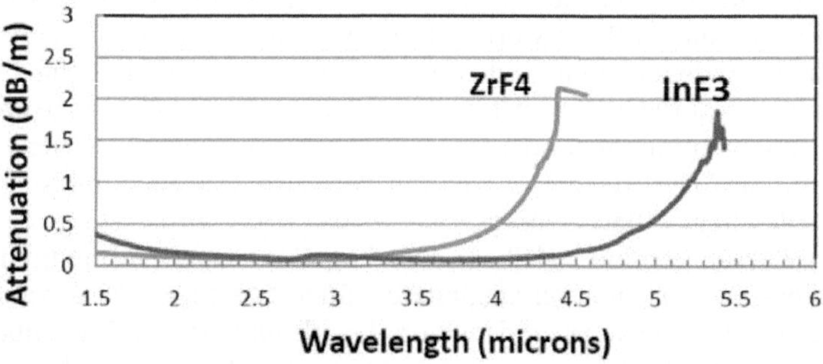

Figure 10.12 ZrF_4 and InF_3 multimode fiber attenuation (comparison).

10.4.3 Mechanical Properties

Unlike optical properties of fluoride glass fibers, only limited work has been reported on mechanical properties of exotic fibers. The reasons are the cost of the fiber and the poor mechanical quality of the fiber at the beginning of the development. One can't draw any good conclusion when using fiber that had many extrinsic defects, and the preparation process is far from perfect. In fact the mechanical quality of a fiber will depend on the mechanical properties of the material itself and also on its preparation process, including glass preparation, preform preparation, and the fiber-drawing process. So, for mechanical studies, one has to take into account where the fiber comes from, which process has been used to make the glass, the preform preparation process, and the fiber-drawing conditions. In fact, there is no standard process to produce exotic optical fibers. There are as many processes as groups that were involved in the development of these fibers.

The mechanical strength of all exotic optical fibers is without doubt a big concern. Their practical use is ultimately linked to their mechanical properties. In fact, most exotic fibers have low strength and are far from being used in practice without problems.

Considering short- and medium-length applications, the solidity of the fibers is more important than their optical loss. A fiber with good strength and even only 95% transmission will be more useful than a fiber with ultralow loss 99.9% transmission and that is mechanically very weak.

Of course, there is no exotic fiber that can have a mechanical strength comparable to silica fiber strength. This is due to the nature of the material and its structure and the strength of interatomic bonds.

As far as fluoride glass fibers are concerned, the theoretical strength is about 50% to 60% lower than silica fiber strength. The theoretical strength of fluoride glass fibers has been estimated to be around 5 to 6 GP; this is between 700 and 800 kpsi [48, 49]. Theoretically, in practice one can expect obtaining fluoride glass fibers with 40% to 50% the strength of current silica fibers, which would be extraordinary. Unfortunately, we are still very far from this reality. In fact, the preparation process of exotic glasses is still far from perfect. However, significant progress has been made in this area during the last two decades, and quite high-tensile-strength fibers have been reported. Irphotonics, now Thorlabs, has reported fibers with strength ranging from 75 to 130 kpsi and a bending strength around 200 kpsi. This is the highest strength ever reported for long-length midinfrared fibers. Figure 10.13 shows the Weibull plot for tensile strength for a 200 micron InF_3 multimode fiber. The Weibull plot has been obtained using a 50 cm gauge, as described in the FOTP 28 telecommunication standards. Figure 10.14 shows the Weibull plot of two-point bending strength. Bending strength is always higher than tensile strength because only a short length of a fiber is submitted to stress during the test. There is still room for improvement.

High mechanical strength makes the fibers easy to use and work with. In fact fluoride glass fibers can be striped, cleaved, and spliced using standard equipment. However, the equipment's parameters have to be adjusted.

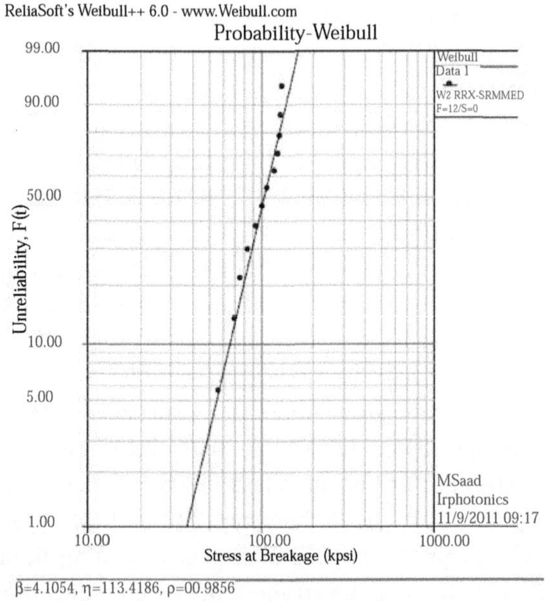

Figure 10.13 Weibull plot for InF_3 multimode fiber tensile strength.

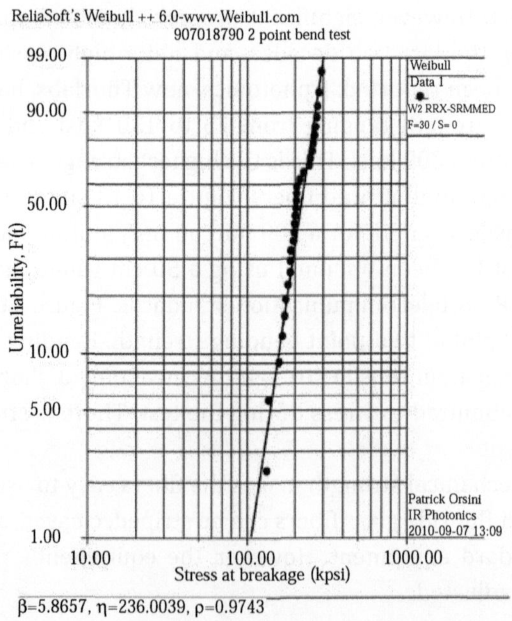

Figure 10.14 Two-point bending strength for InF_3 multimode fiber.

10.4.4 Rare Earth–Doped Fibers

An optical fiber can be either a passive or an active medium. A passive optical fiber will only transmit light from point A to point B without changing its nature or its wavelength. An active fiber will change the nature of light going through it under certain conditions. The fiber has to be doped with active ions, such as rare earth elements. Fluoride glasses are interesting and excellent hosts for rare earth elements. In fact, rare earth ions can be part of the glass compositions. Stable compositions contain more than 100,000 ppm of rare earth ions without compromising glass stability. For comparison, the rare earth concentration in silica glass fiber is only a few hundreds to a few thousands of ppm according to the glass composition.

Active applications include fiber lasers, amplifiers, and laser cooling. Tremendous work has been reported in these areas. Over 35 laser lines from the ultraviolet to the midinfrared region have been reported using fluoride glass fibers doped or co-doped with different rare earth ions. A very interesting and complete review has been reported by Xiushan Zhu and N. Peyghambarian from the College of Optical Science at Arizona University [50]. When this review was written, the maximum output power reported for erbium-doped fluoride fiber at 2.94 microns was 10 w, which was reported by the Laval University team in Quebec City in Canada. Since then this output has been tripled and a 30 W erbium laser at 2.9 microns has been reported recently by the same team [51].

More recently, Quimby and Saad reported, according to lifetime and spectroscopy measurement and theoretical modeling, that efficient lasing by a 4.5 um fiber laser is possible using dysprosium-doped indium fluoride glass fibers using a cascade lasing scheme [52].

Laser cooling, which is the subject of this book, and will be deeply addressed by experts, has been demonstrated using rare earth–doped fluoride glasses and fibers. Most of the laser cooling work has been done using ytterbium fluoride–doped fluoride glasses and fibers [53–55]. However, thulium fluoride–doped fluoride glasses have been also reported [56].

References

1. Yu, F., Wadsworth, W. J., Knight, J. C. (2012). Low loss silica hollow core fibers for 3–4 μm spectral region, *Opt. Express*, **20**(10), pp. 11153–11158.
2. Poulain, M. (1995). New trends in halide glass compositions, *J. Non-Cryst. Solids*, **184**, pp. 103–108.
3. Conseil, C., Bastien, J.-C., Boussard-Plédel, C., Zhang, X.-H., Lucas, P., Dai, S., Lucas, J., Bureau, B. (2012). Te-based chalcohalide glasses for far-infrared optical fiber, *Opt. Mater. Express*, **2**(11), pp. 1470–1477.
4. Wang, J. S., Vogel, M. E., Snitzer, E. (1994). Tellurite glass: a new candidate for fiber devices, *Opt. Mat*er., **3**, pp. 177–203.
5. Guorong, C., Wen, L., Jijian, C. (2000). Improved properties for precision molding phosphate glasses, *Am. Ceram. Bull.*, pp. 44–46.
6. Feigelson, R. S., Kway, W. L., Route, R. K. (1984). Single crystal fiber by the laser-heated pedestal growth method, *Proc. SPIE,* **0484** (Infrared Optical Materials and Fibers III), p. 133.
7. Saad, M. (2009). Fluoride glass fiber: state of the art, *Proc. SPIE*, **7316**, p. 73160N.
8. Van Uitert, L. G., Wemple, S. H. (1978). ZnCl2 glass: a potential ultralow-loss optical fiber materials, *Appl. Phys. Lett.*, **33**, pp. 57–58.
9. Poulain, M., Poulain, M., Lucas, J., Brun, P. (1975). Verres flores au tetrafluorure de zirconium proprietes optiques d'un verre dope au Nd^{3+}, *Mater. Res. Bull.*, **10**, pp. 243–246.
10. Tregoat, D., Fonteneau, G., Lucas, J. (1985). Verres fluores a base de metauxlours. Traitement des impuretes oxydes et hydroxides sous atmosphere controlee, *Mater. Res. Bull.*, **20**, pp. 179–185.
11. Robinson, M. (1985). Preparation and purification of fluoride glass starting materials, *Mater. Sci. Forum*, **5–6**, pp. 19–33.
12. Tran, D. C., Sigel, G. H., Bendow, B. (1984). Heavy metal fluoride glasses and fiber, a review, *J. Lightwave Technol.*, **LT2**(5), pp. 566–586.
13. Nakai, T., Mimura, Y., Hokiwa, H., Shinbori, O. (1986). Dehydration of fluoride by NF3 processing, *J. Lightwave Technol.*, **LT4**, pp. 87–89.
14. Adan, J. L., Ricordel, C., Lucas, J. (1997). New composition of low phonon energy fluoride and chloro-fluoride glasses, *J. Non-Cryst. Solids*, **213–214**, pp. 30–35.
15. Matecki, M., Poulain, Mi., Poulain, Ma. (1983). Cadmium halide glasses, *J. Non-Cryst. Solids,* **56**(1–3), pp 81–86.

16. Poulain, Mi., Matecki, J. M., Mouric, L., Poulain, Ma. (1983). Cadmium halide glasses II. Chloride glass, *Mater. Res. Bull.*, **18**(5), pp. 631–636.

17. Matecki, M., Poulain, Ma. (1992). Composition adjustments in fluorochloride glasses, *J. Non-Cryst. Solids,* **140**, pp. 82–86.

18. Poulain, Mi., Poulain, Ma. Matecki, M. (1986). Rare-earths in fluoride glasses: LnF3, AlF3, ThaF4, BaF2 systems and related glasses, *J. Non-Cryst. Solids*, **51**(2), pp. 201–215.

19. Bouaggad, A. (1986). PhD thesis, Rennes University, France.

20. Lecoq, A., Poulain, Ma. (1980). Etude phenomenologique du role stabilisateur de l'aluminium dans les verres au tetrafluorure de zirconium, *Verres Refct.*, **34**(3), pp. 333–342.

21. Jacoboni, C., Boulard, B., Baniel, P., Poignant, H. (1987). Vapor deposition of fluoride glasses, *Mater. Sci. Forum*, **19–20**, pp. 253–258.

22. Ohishi, Y., Mitachi, S., Kanamori T., Manabe, T. (1983). Optical absorption of 3d transition metal and rare-earth elements in zirconium fluoride glasses, *Phys. Chem. Glasses*, **24**, pp. 135–140.

23. France, P. W. (1987). PhD thesis, Chapter 6, Sheffield University.

24. France, P. W., Carter, S. F., Williams, J. R., Beales, K. J., Parker, J. M. (1984). OH-absorption in fluoride glass infrared fibers, *Electron. Lett.*, **20**(14), pp. 607–608.

25. Drexhage, M. G. (1985). Preparation and properties of high quality bulk fluoride glasses, *Mat. Sci. Forum*, **5**, pp. 1–18.

26. Fujiura, K., Ohishi, Y., Sakaguchi, S., Ternuma, Y. (1989). Synthesis of high-purity zirconium fluoride for fluoride glass fibers by chemical vapor deposition, *J. Am. Ceram. Soc.*, **72**(8), pp. 1368–1372.

27. Withers, H. P. Jr., Monk, V. A., Cooper, G. A. (1989). High purity zirconium tetra-fluoride for fluoride glass applications, *Proc. SPIE*, **1048**, pp. 72–77.

28. Saad, M. (2009). High purity fluoride glass synthesis: a review, *Proc. SPIE*, **7228**, p. 72280G.

29. Poulain, M. (1991). Chapter 1 Fluoride glass compositions and processing, in *Fluoride Glass Fiber Optics*, Aggarwal, I., Lu, G., eds. (Academic Press, Boston), pp. 1–35.

30. Fonteneau, G., Tregoat D., Lucas, J. (1985). Determination du Coefficient d'Extinction de OH- dans les verres fluores a base de metaux lourds, *Mater. Res. Bull.*, **20**(9), pp. 1047–1051.

31. Nakai, T., Mimura, Y., Tokiwa, H., Shinmori, O. (1986). Dehydration of fluoride glasses by NF3 processing, *J. Lightwave Technol.*, **LT-4**, 1.

32. Saad, M., Poulain, Ma. (1995). Sol-gel synthesis of fluorozirconate glasses, *Am. Ceram. Bull.*, **74**(8), pp. 66–71.

33. Hruby, A. (1972). Evaluation of glass-forming tendency by means of DTA, *Czech J. Phys.*, **B22**, pp. 1187–1193.

34. Saad, M., Poulain, M. (1987). Glass forming ability criterion, *Mater. Sci. Forum*, **19/20**, pp. 11–18.

35. Uhlmann, D. R. (1983). Glass formation, a contemporary view, *J. Am. Ceram. Soc.*, **66**, pp. 95–100.

36. Busse, L. E., Lu, G., Tran, D. C., Sigel, G. H. (1985). A combined DSC/optical microscopy study of crystallization in fluorozirconate glasses upon cooling from melt, *Mater. Sci. Forum*, **6**, pp. 219–228.

37. Lipeng, Z., Fuxi, G., Wang, P. (1994). Evaluation of refractive-index and material dispersion in fluoride glasses, *Appl. Opt.*, **33**(1), pp. 50–56.

38. Jiang, X., Joly, N. Y., Finger, M. A., Wong, G. K., Babic, F., Saad, M., Russell, P. S. (2013). Close to three-octave-spanning supercontinuum generated in ZBLAN photonic crystal fiber, *Advance Solid-State Lasers Congress*, JTh5A, p. 6.

39. Tokiwa, H., Mimura, Y., Nakai, T., Shinbori, O. (1985). Fabrication of long single mode and multimode fluoride-glass fibers by double crucible technique, *Electron. Lett.*, **21**(24), pp. 1131–1132.

40. Mitachi, S., Miyashita, T., Kanamori, T. (1981). Fluoride glass cladded optical fibers for mid-infrared rat transmission, *Electron. Lett.*, **17**, pp. 591–592.

41. Tran, D. C., Fisher, C. F., Sigel, G. H. (1982). Fluoride glass preforms prepared by rotational casting process, *Electron. Lett.*, **18**(15), pp. 657–658.

42. Shibata, S., Horiguchi, M., Jinguchi, K., Mitachi, S., Kanamori, T., Manabe, T. (1981). Prediction of loss minima in infrared optical fibers, *Electron. Lett.*, **17**, pp. 775–777.

43. Kanamori, T., Sakaguchi, S. (1986). Preparation of elevated NA fluoride optical fibers, *Jpn. J. Appl. Phys.*, **25**, pp. 468–470.

44. Tran, D. C., Levin, K. H., Burk, M. J., Fisher, C. F., Brower, D. (1986). Preparation and properties of high optical quality IR transmitting glasses and fibers based on metal fluorides, *Proc. SPIE*, **618**, pp. 48–50.

45. Poulain, Ma., Saad, M. (1984). Absorption loss due to complex anions in fluorozirconate glasses, *J. Lightwave Technol.*, **LT-2**(5), pp. 599–602.

46. France, O. W., Drexhage, M. G., Parker, J. M., Moore, M. W., Carter, S. F., Wright, J. V. (1990). *Fluoride Glass Optical Fibers* (CRC Press).

47. Aggarwal, I. D., Lu, G. (1991). *Fluoride Glass Fiber Optics* (Academic Press, Harcourt Brace Jovanovich, San Diego, CA).

48. Carter, S. F. (1990). Mechanical properties, in *Fluoride Glass Optical Fibers,* P. W. France, ed., Blackie USA and Canada (CRC Press), p. 219.

49. Poulain, Ma., Gougeon, N., Maze, G. (2003). Mechanical properties of fluoride glass fibers, *Proc. SPIE,* **4940**, pp. 30–46.

50. Xiushan, Z., Peyghambarian, N. (2010). High-power ZBLAN glass fiber lasers: review and prospect, *Adv. OptoElectron.,* **2010**, Article ID. 501956, 23 pages.

51. Fortin, V., Bernier, M., Bah, S. T., Vallee, R. (2015). 30W fluoride glass all-fiber laser at 2.94 μm, *Opt. Lett.,* **40**(12), pp. 2882–2885.

52. Quimby, R., Saad, M. (2013). Dy-fluoroindate fiber laser at 4.5 μm with cascade lasing, *Adv. Solid State Lasers,* paper AM2A.7.

53. Epstein, R. I., Buchwald, M. I., Edwards, B. C., Gosnell, T. R., Mungan, C. E. (1995). Observation of laser-induced fluorescent cooling of a solid, *Nature,* **377**, pp. 500–503.

54. Rayner, A., Friese, M. E. J., Truscott, A. G., Heckenberg, N. R., Rubinsztein-Dunlop, H. (2001). Laser cooling of a solid from ambient temperature, *J. Mod. Opt.,* **48**(1), pp. 103–114.

55. Murtagh, M. T., Sigel Jr., G. H., Fajardo, J. C., Edwards, B. C., Epstein, R. I. (1999). Laser-induced fluorescent cooling of rare earth–doped fluoride glasses, *J. Non-Cryst. Solids,* **253**, pp. 50–57.

56. Hoyt, C. W., Hasselbeck, M. P., Sheik-Bahae, M. (2003). Advances in laser cooling of thulium-doped glass, *J. Opt. Soc. Am. B,* **20**(5), pp. 1066–1074.

Chapter 11

Crystal Growth of Fluoride Single Crystals for Optical Refrigeration

Azzurra Volpi, Alberto Di Lieto, and Mauro Tonelli
Dipartimento di Fisica-NEST Instituto di nanoscienze CNR,
Universita'di Pisa, L. B. Pontecorvo 3, I-56127, Pisa, Italy
volpi@df.unipi.it

This chapter reports on the crystal growth of Yb-doped fluoride single crystals for solid-state laser cooling. In the last years, the advances enabled by Yb-doped fluorides have tremendously increased the interest in this technology. Cryogenic operations have been achieved as well as competitive steady-state cooling powers in the temperature range below 250 K. Minimum achievable temperatures are approaching 80 K. Truly vibration-free and extremely compact, the optical cooling technology provides an ideal solution for cooling down detectors, in space applications below the thermoelectric cooler (TEC) barrier (170 K), where Peltier coolers cannot operate and other techniques lead to complex solutions. The current state-of-the-art results also prospect efficiency improvements in the higher-temperature range, between 170 and 220 K, where the TEC exhibits limited performances.

Laser Cooling: Fundamental Properties and Applications
Edited by Galina Nemova

ISBN 978-981-4745-04-8 (Hardcover), 978-981-4745-05-5 (eBook)
www.panstanford.com

Efficient cooling performances, however, are firstly related to the optical quality of active materials. Stringent requirements on the purity and structural quality need to be satisfied in order to achieve net bulk cooling. Even though the purity of starting materials plays a crucial role, the optical quality of finished crystals essentially relies on the quality of the growth process. Low impurity concentrations or isolated structural defects can be easily introduced during the synthesis with tremendous effects on the efficiency of the anti-Stokes process, which removes heat from the material. Careful techniques, ultracontrolled conditions, and high-purity environments are basic requirements to ensure the sufficient optical quality grade.

In our facility several Yb-doped fluoride crystals have been grown for laser cooling experiments. Cryogenic operations were achieved in 5 at% Yb-doped $LiYF_4$ (YLF) single crystals in cooling experiments carried out at the University of New Mexico [12, 17]. More recently, varying Yb doping levels have been investigated and controlled co-doped systems have resulted in efficiency enhancements via energy-transfer-assisted anti-Stokes processes. The efficient performances measured in cooling experiments reflect the high optical quality of the grown samples. In the following sections, crystal growth techniques and methods for assessment of crystal quality are presented.

11.1 Introduction

In the solid phase thermal energy is largely contained in the vibrational modes of the lattice. Upon interaction with laser light, heat can be removed from the system through annihilation of lattice phonons.

Optical cooling in solids is based on the principle of anti-Stokes luminescence, where coherent excitation below the mean emission energy of the electronic transition induces absorption of energy from the lattice, and hence annihilation of lattice phonons, via ultrafast thermalization of the electronic excitations (Fig. 11.1). Absorption of photons at one frequency is followed by spontaneous re-emission of blue-shifted photons. The extra energy is removed from the lattice, resulting in bulk net cooling.

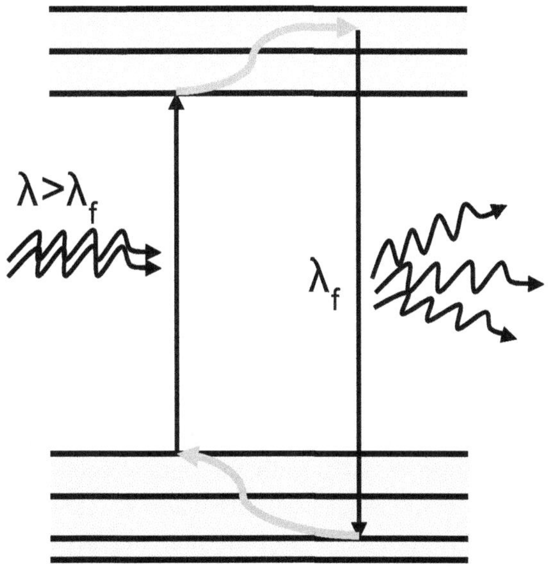

Figure 11.1 Schematic of anti-Stokes cooling. Following coherent excitation in the low-energy tail of the absorption spectrum ($\lambda > \lambda_f$), spontaneous emission of blue-shifted photons occurs via annihilation of lattice phonons. Ultrafast thermalization of the electronic excitations leads to absorption of energy from the lattice through inelastic scattering with the vibrational modes. The extra energy is removed from the lattice through fluorescence of blue-shifted photons.

The idea of using optical radiation to lower thermal energy from vibrational modes, through fluorescence upconversion, was firstly proposed by Prighsheim in 1929 [16]. First observations, however, required the invention of lasers. Anti-Stokes cooling was firstly achieved in gaseous systems [6] and in 1995 the first experimental demonstration in solid materials was accomplished. At the Los Alamos National Laboratory, Epstein et al. observed net cooling of a Yb^{3+}-doped ZBLANP glass by optically pumping the 1 μm transition [7]. The prior invention of purification techniques for fluorozirconate glasses was pivotal for such achievement, enabling the synthesis of high-purity active materials. Subsequent optimization of experimental conditions led to optically cool a highly purified Yb-doped ZBLAN glass to 208 K from room temperature [18].

Dramatic improvements of cooling performances, however, have been achieved only recently with high-quality fluoride single crystals. Conversely to glassy matrices, in crystalline hosts the long-range order leads to sharp homogeneous splitting of rare earth energy, levels resulting in dramatic enhancement of resonant absorption. Fluoride crystals also possess a low phonon energy, which strongly reduces nonradiative decays rates, leading to high radiative efficiencies. Additionally, crystalline hosts stoichiometrically allows for a high doping level and thus the resonant absorption of the electronic transition can be largely enhanced by just increasing the concentration of active ions.

High-purity Yb-doped fluoride single crystals have definitely led to cooling operations in the cryogenic regime. The first breakthrough result was achieved in 2010 with the optical cooling of a 5 at% Yb-doped $LiYF_4$ (YLF) single crystals down to 155 K from room temperature [17]. Optimization of the pumping scheme led to the first operation below the NIST-defined cryogenic temperature (123 K) [12]. The increase of Yb doping level up to 10 at% in ultrapurified YLF single crystals enabled the first optical cooling below 100 K [14]. Current operations, however, are limited by impurity-mediated background absorption. Calculations indicate that further purification of starting materials would enable operations close to the liquid nitrogen temperature (77 K) [13]. Due to the chemical inertness of binary fluoride, however, purification of raw materials below the part-per-million level is still a quite challenging process. One approach currently under investigation is detailed in Ref. [10].

Cryogenic achievements have dramatically increased the technological interest in this field. Current operations and cooling powers available in the temperature range of 90–250 K definitely allow one to consider in earnest device implementation. Truly vibration free, such a solution is much sought after for space applications. The inherent absence of vibration and stunning compactness, indeed, would permit significant improvements for several high-precision space-based technologies, including focal plane infrared (IR) imaging detectors, ultrastable cavity lasers for atomic clocks and gravitation wave detection, and X-ray and gamma ray sensors. In the temperature range of 80–150 K solid-state optical coolers would also enable the use of superconductor devices.

With the advances achieved in cooling experiments, the investigation in the synthesis of more and more pure and defect-free fluoride crystals has attracted enormous interest as well as the investigation of purification techniques for raw materials. Efficient cooling performances are primarily related to the quality of active materials, and both device implementation and fundamental research require high-quality materials for efficient operations and further improvements of the current state of the art, pushing down the minimum achievable temperature close to the thermodynamic limit.

High optical purity and structural quality are basic conditions for efficient bulk cooling. Impurity-mediated processes and crystallographic defects introduce severe background absorption, thereby lowering the anti-Stokes efficiency. Spurious electronic and vibrational bands cause trapping of pump photons as well as nonradiative quenching of the electronic excitations via energy transfer and multiphonon relaxation. Besides degrading the absorption efficiency, such processes lower the external quantum efficiency of the electronic transition and generate heat dissipation via multiphonon emission. Maximizing the cooling efficiency firstly requires minimizing impurity concentrations. High structural quality is mandatory as well. Crystal defects, such as microbubbles, dislocations, and cracks, cause strong losses of pump power through scattering processes of pumping photons, violently heating up the lattice.

In addition to degrade the anti-Stokes efficiency, all these processes generate detrimental heating. Such heat dissipation can readily compromise the cooling process if the efficiency of heat-generating processes is not surpassed by the efficiency of the anti-Stokes process, which removes heat from the lattice. As shown in several experiments, even sparse distribution of defects and very low concentration of impurities heavily affect, and even compromise, the cooling performances [5, 13]. Stringent requirements on optical purity and structural quality of active materials come after.

Although a minimum concentration of impurities is imposed by the purity grade of starting materials, the absence of structural defects and polycrystalline nuclei as well as the inclusion of additional impurities are exclusively related to the quality of the synthesis procedure. Growth is a very critical process for the optical quality of finished materials. Impurities can be easily incorporated during the synthesis and defects can readily form due slightly unsuitable thermal and kinetic growth conditions. As shown in the following, ultra-accurate procedures, controlled conditions and highly purified environments are basic requirements to achieve the sufficient optical quality grade.

Fluoride single crystals possess exceptional physical properties as hosting materials for anti-Stokes cooling. The growth of high-purity samples, however, requires specific conditions due to the strong reactivity of fluorine. The main advantage of these materials, over other crystalline hosts, is the low cutoff energy (300–500 cm^{-1}) of optical phonons that effectively suppresses the efficiency of phonon-mediated relaxation processes, leading to high radiative quantum efficiencies. The lower phonon energy of bromides, chlorides, and iodides does not offer any appreciable advantage if compared to the complexity associated with the extreme hygroscopicity typical of heavy halides. Fluoride crystals are instead not hygroscopic. Additionally, they have the advantage of low refractive index and good thermal and mechanical properties. The low refractive index reduces photon trapping due to internal reflections at the surfaces, while good thermal conductivity enables efficient heat removal, and the relatively high mechanical hardness lends benefits to sample processing and surface polishing. Largely used for years as gain media for efficient solid-state lasers even in commercial devices, rare earth–doped fluorides are currently under investigation for a variety of novel applications, such as upconversion efficiency enhancement in solar cells [8, 9], quantum computation, and particle detection.

Due to the strong reactivity of fluorine compounds with moisture, however, the growth process must be carried out under ultracontrolled conditions in order to avoid detrimental contaminations. The most difficult impurities to avoid when growing fluorides are oxygen complexes, such as OH^- hydroxyl radicals, which efficiently substitute for F^- ions in the crystalline structure. Such complexes, when incorporated inside crystals during the growth, strongly degrade the overall optical performances due to spurious absorption bands in the ultraviolet (UV) and near-infrared (NIR), which act as efficient traps for electronic excitations in energy transfer and multiphonon relaxation processes. Especially in the case of Yb^{3+} ions, inclusion of OH^- ions makes nonradiative decay the dominant relaxation process from the excited state, dramatically lowering the radiative efficiency of the electronic excitation.

The growth of fluorides doped with Yb^{3+} ions imposes further conditions, increasing the complexity of the synthesis process. Due to the tendency of Yb^{3+} ions to reduce to the 2+ state, a suitable oxidant atmosphere need to be used in order to prevent any inclusion of Yb^{2+} impurities inside growing crystals, which form from reduction of Yb^{3+} ions. In addition to suppress the efficiency of cooling process, reducing the number of Yb^{3+} active ions, the inclusion of Yb^{2+} ions inside growing crystals significantly deteriorates the optical quality. Such ions indeed possess large absorption bands in the visible region, which are likely to participate in nonradiative transfer of excitations from Yb^{3+} ions, preventing anti-Stokes emission. Fluorinating agents, however, can lead to other kind of contaminations and hence require proper use.

In the following section, methods and growth details are presented for different Yb-doped samples grown in our facility for laser cooling applications. Methods used for optical quality assessment after the growth are also discussed.

11.2 The Czochralski Growth Method

The Czochralski technique is nowadays most commonly used for the synthesis of bulk semiconductors, oxide and fluoride single crystals and synthetic gemstones for photonic applications. Originally invented in 1918 by Czochralski [4], this process is capable of fabricate defect-free single crystals of high optical quality, with a large size, at a relatively fast growth rate [2, 15].

Crystal growth is involved with the control of a phase transition. Growth methods from melt, such as the Czochralski process, are based on the control of the solid–liquid-phase transition to induce the crystallization. The liquid is slightly undercooled below its freezing point and the nucleation of a single-crystal domain is induced by using a crystal seed with the same crystal structure and composition of the material to be grown. A principle diagram is shown in Fig. 11.2.

Raw materials are normally stored in powder form. The mixture of starting materials is first melted inside a crucible and then brought in an undercooled phase, lowering the temperature of the liquid. In the undercooled phase, the liquid tends to crystallize. The crystallization is induced at the solid–liquid interface by using a crystal seed, dipped into the melt. Once the nucleation occurs, the growth starts. The crystal is pulled from the melt slowly moving upward the crystal seed, while it is rotated, at the same time, in order to ensure thermal symmetry, hence preventing the growing crystal to assume an irregular cross-sectional shape related to the temperature distribution in the melt.

Fundamental condition for the Czochralski growth is that the temperature of the solid–liquid interface must be the lowest in the liquid. If it is not, spurious nucleation can occur.

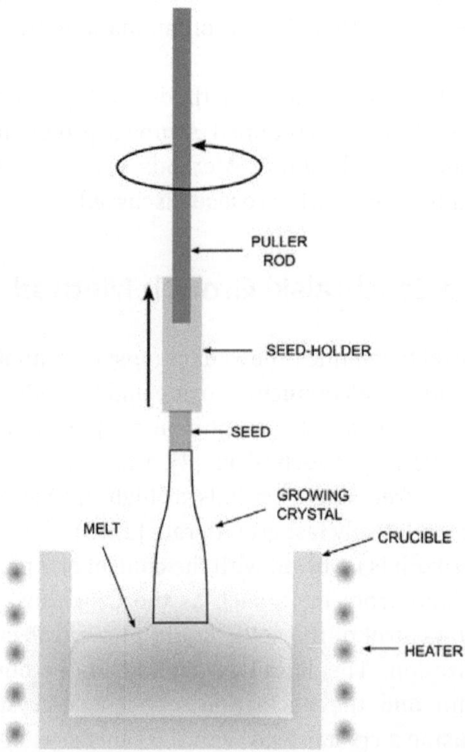

Figure 11.2 Schematic of Czochralski growth.

The basic requirement to grow defect-free single crystals is the use of a slow pulling rate. Fast growth speeds commonly give rise to inclusion of dislocations, nucleation of polycrystalline domains, and lack of transparency. For fluorides typical pulling rates vary between some mm/h and tens of mm/h. The optimal growth rate depends on the physical and thermal properties of the material. A good compromise between growing as fast as possible and the introduction of defects that this produces is normally chosen. Rotation rates of a few revolutions/min are normally sufficient to achieve defect-free samples.

One of the main advantages of the Czochralski method is the possibility to control the diameter of the crystal while it is growing. The crystal diameter at the solid–liquid interface depends on both the temperature of the melt and the crystal pull rate and can thus be controlled by changing both these parameters. Decreasing the temperature of the melt or the pulling rate leads to diameter enlargements, while the opposite is needed for shrinking the diameter. To minimize formation of defects, however, the diameter control is commonly run by changing the melt temperature, while maintaining fixed the pulling rate.

Ultra-accurate control of the melt temperature is fundamental condition for the growth process. Irregular temperature changes reflect in sudden diameter variations. In proximity of such diameter changes, clouds of defects typically form due to fast growth rates in certain directions. Temperature instabilities also lead to spurious nucleation in the melt. Controlled diameter changes and uniform growth at sufficient slow rates require fine-tuning and high stability of the melt temperature. Suitable heating systems thus need to be employed.

A key point for the Czochralski growth process is the quality of the crystal seed, which defines the crystalline structure and the orientation of the unit cell in the nucleation process. An ideal seed would be a defect-free oriented single crystal of the same composition of the material to be grown. However, the basic requirements to be satisfied are the same crystal structure and similar lattice parameters. Especially for the growth of novel materials, if crystal seeds are not available, a wire or a tube, properly shaped, can be used to induce the nucleation of a polycrystalline domain, from which single crystals can be obtained, alternatively increasing and decreasing the diameter.

A further advantage of the Czochralski technique is the possibility to carry out a necking operation which both enables prevention of the propagation of nonaxial defects from the seed to the growing crystal and favors the propagation of a single-crystal domain in the growing crystal.

Such a procedure simply consists of making the crystal diameter as small as possible by increasing the melt temperature. Shrinking the crystal diameter either favors the propagation of a single-crystal domain, reducing the number of crystallites propagated into the crystal, or allows one to interrupt the propagation of nonaxial dislocations within the growing crystals (see Fig. 11.3). When a sufficient length of narrow crystal has been grown, the diameter can be enlarged to the desired value. The necking process can be repeated several times, if necessary.

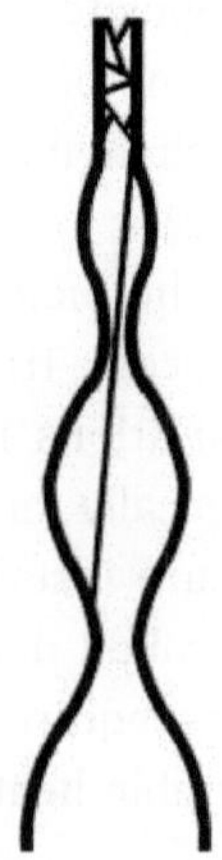

Figure 11.3 Schematic of necking technique.

Although a vacuum chamber would be the purest environment as regard contaminations, the Czochralski process is commonly carried out in controlled atmosphere as the molten material cannot be stable at low pressures. To minimize instabilities which typically lead to evaporation of material from the melt surface, the growth chamber is usually filled, at proper pressure, with a purified inert gas. The kind of gases constituting the growth atmosphere depends on the materials to be grown. Argon is normally chosen for the growth of fluoride crystals, while tetrafluoromethane (CF_4) is typically used to build up an oxidant atmosphere.

The basic requirement to avoid inclusion of contaminants during the growth process is to keep the melt surface completely free from foreign particles. Impurities deposited on the melt surface can readily get incorporated inside growing crystals, compromising the optical quality of samples. In addition to introducing loss channels through spurious absorption bands, foreign particles at the growth interface can originate irregular anisotropic growth rates, leading to dislocations. Spurious nucleations on the melt surface are another typical result of impurities deposited on the melt surface. Careful cleaning operations must be performed before start of the growth.

The maximum sizes that can be produced by the Czochralski technique are only related to the dimensions of the furnace. Basically, the volume of the crucible and the distance that the seed holder can move determine the maximum crystal dimensions. Typical sizes of crystal boules used for optical applications are of the centimeter scale.

Basic components of a Czochralski furnace are a growth chamber and a pulling system. The pulling of the crystal from the melt takes place inside the growth chamber, where are placed the heater element and the crucible containing the raw materials. The pulling system is designed to control the rototranslation movements of the seed holder shaft, required either to bring the crystal seed in contact with the melt or to pull the growing crystal from the melt surface. A detailed description of the furnace, with all its components, designed at the NMLA Laboratory for the growth of fluorides single crystals is outlined in Section 11.4.

11.3 Crystal Growth of Fluorides

Due to the strong electronegativity of F^- ions, for the growth of fluorine-based crystalline compounds, specific requirements need to be met in order to achieve high optical quality. Oxygen complexes, which form when oxygen and water molecules are present during the growth process, are the most detrimental impurities to avoid when growing fluorides. Among these complexes, hydroxyl radicals, OH^-, are the most favorite to form as they efficiently substitute for F^- ions in fluoride crystalline structures. These ions exhibit strong UV electronic absorption and IR vibrational bands [3], which cause either trapping of pumping radiation or luminescence quenching of electronic excitations. Spurious IR bands, however, are the most detrimental for the efficiency of the cooling process preventing anti-Stokes fluorescence via multiphonon relaxation, that is, heat generation. In particular, IR absorption traps due to hydroxyl radicals (around 3400 cm^{-1}) cause significant multiphonon quenching even for large-gap (>4000 cm^{-1}) electronic transition, which possess a negligible nonradiative decay rate in pure fluoride crystals. This is the case of Yb-doped materials, where the presence of OH^- contaminants dramatically reduces the radiative efficiency of the electronic transition. In such cases anti-Stokes fluorescence is hindered and heat is generated via multiphonon emission. Depending on the impurities' concentration, the cooling process can be even compromised due to a faulty balance between heating and virtuous cooling processes.

The basic condition to prevent formation of OH^- radicals in the melt is to avoid interaction of raw materials with oxygen and water at high temperatures. Specific conditions on the growth conditions and raw powders' storage and handling follow.

An essential requirement for furnaces dedicated to the growth of fluorides is a high-vacuum system which enables the evacuation of the growth chamber at low pressure, removing any presence of oxygen and water molecules. Pressures of the order of 10^{-5} Pa are normally sufficient for purification of the growth environments. Storage of raw powders and preparation of the starting mixture require proper procedures as well. Before loading into the furnace, the fluorine raw materials must be stored inside sealed boxes. Weighting and mixing of primary components inside the crucible must be carried out inside a moisture-controlled environment.

To prevent instabilities of the molten material at high temperatures, the growth process is typically carried out under inert atmosphere. After high-vacuum cleaning, high-purity (5N) argon is pumped in overpressure inside the growth chamber. For the growth of fluorides doped with rare earth ions that tend to reduce from the triply ionized state to the 2+ state, such as Yb^{3+}, high-purity (5N) tetrafluoromethane (CF_4) is typically used as fluorinating agent. A proper percentage of CF_4 is pumped inside the furnace together with argon. The dissociation of CF_4 at high temperatures releases F^- ions, which create an oxidant environment, preventing triply charged Yb ions to reduce to double-charged states.

The composition of the growth atmosphere is a critical parameter for the optical quality of Yb-doped crystals and requires a previous dedicated investigation. A lack of CF_4 concentration in the growth atmosphere leads to the formation of detrimental Yb^{2+} impurities in the melt. An excess of CF_4 concentration has multiple drawbacks as well. Firstly, it likely leads to carbon contamination inside the melt. Additionally, a CF_4-enriched atmosphere is rather corrosive for furnace components, especially in the case of extended use. The deposition of a dark scum inside the furnace inside the furnace is the most common result of an excess of CF_4 in the atmosphere composition.

Melting temperatures of fluoride materials are normally below 1000°C. Resistive heating systems are the most commonly used for furnaces dedicated to the growth of these materials. Maximum temperatures achievable are of the order of 1500°C. In resistive systems, heat is generated via the Joule effect by the passage of a current through a conductor surrounding the crucible, which contains the raw materials. High stability and fine-tuning of the melt temperature can be achieved. A sufficiently low vertical thermal gradient can be attained, thanks to the high uniformity of the thermal field generated by the resistive element.

For the furnace components, materials resistant to the corrosive atmosphere at high temperatures must be used. Usual materials employed are high-purity graphite and platinum. The former is suitable to manufacture the heater element. Graphite combines conductive behavior and excellent electrical resistivity for Joule heating at high temperatures. It can operate up to 2500°C in an inert atmosphere. Moreover, compared to most of heating element materials, it can be readily machined in self-supporting tubular shapes, suitable to surround the crucible and thus providing uniform heating inside.

Regarding crucibles, the basic requirement is to avoid any contaminations of raw materials at high temperatures. The crucible thus must be made of high-purity materials which possess high stability under the temperature and pressure conditions of growth and hence do not interact with the material to be grown even at high temperatures. Low-porosity materials are preferable. Among materials chemically and thermally compatible to the fluorides' growth conditions, platinum, graphite, and glassy carbon are more suitable, as they additionally are not wetted by the fluoride melting. Platinum is, however, the best choice as regard to purity. Relatively chemically inactive, it is highly pure and has advantages of excellent resistance to corrosion and stability at high temperatures. Its melting temperature is about 1770°C. Being malleable and ductile, it can be readily hammered into thin sheets or drawn into thin wires. Such properties make high-purity platinum the best choice not only for crucibles but for all the components that directly contact the melt, such as seed holders, wires to fasten the crystal seed, and cleaning tools.

While the melting point is reached, foreign particles tend to condense above the melt surface, producing a dark, thin film. Especially in the case of a CF_4-enriched atmosphere, the extended use of a corrosive atmosphere tends to increase the porosity of graphite components, and the foreign particles, which are released when the furnace is heated, tend to get absorbed more by the pores. Suitable cleaning techniques must be employed before start the growth. Clean surfaces, free from foreign particles, appear completely clear.

A typical technique used for the growth of fluorides under a CF_4-enriched atmosphere consists in freezing the melt surface by using a crystal seed, which is lowered in contact to melt so as to induce the formation of a large and thin crystal, collecting the impurities deposited above the surface.

After the growth, the crystal boules need to be slowly cooled to room temperature before be extracted from the furnace. Slow cooling rates need to be used in order to prevent thermal shock which can cause internal cracks or phase changes. Typical cooling rates are of the order of tens of Celsius degrees per hour.

11.4 The Growth Facility at the NMLA Laboratory of Pisa University

The growth facility for fluoride single crystals at the NMLA Laboratory of Pisa University consists of two home-made Czochralski furnaces, properly conceived for the growth of this kind of materials. The growth of bulk cooling materials is carried out in a dedicated furnace, specifically designed for the growth of fluorides under CF_4-enriched atmosphere. The furnace is equipped with a high-vacuum system, resistive heating and computer-controlled optical diameter control. A schematic of the apparatus is shown in Fig. 11.4, along with a photograph of its exterior.

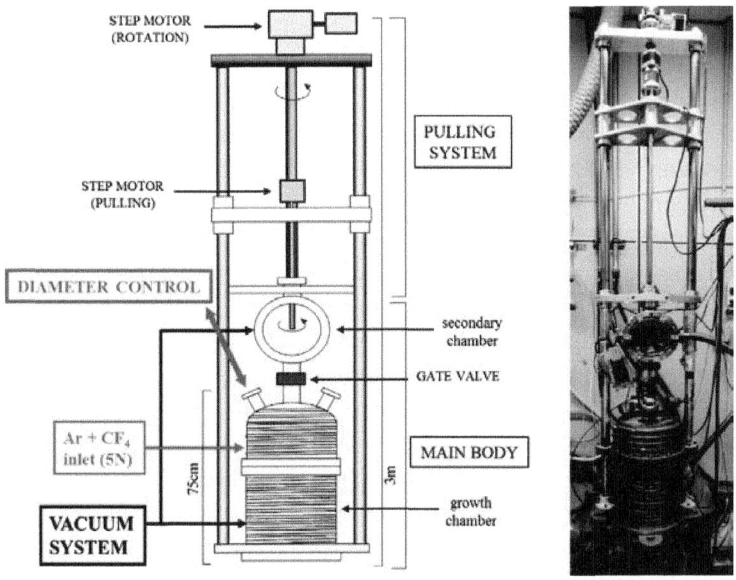

Figure 11.4 Schematic (a) and picture (b) of the Czochralski furnace at the NMLA Laboratory of Pisa University.

The main body of the furnace is composed of a primary growth chamber connected by a gate valve to a secondary vacuum cell placed above. Inside the growth chamber are produced the high temperatures and pressure conditions necessary for the crystal growth.

The secondary cell has been designed to enable access to the bottom end of the puller shaft without altering the temperature and pressure conditions of the growth chamber. This possibility is crucial for the cleaning operation. Once the gate valve is closed, the access to the bottom end of the puller allows to repeatedly lower the cleaning tools into the melt, while maintaining completely unchanged temperature and pressure conditions of the melt surface between subsequent plunges.

Cooling to room temperature the molten material and degradation of the growth atmosphere would be otherwise necessary to replace the cleaning tools after being plunged into the melt, with consequent poor efficiency of the cleaning operation. Additionally, such a design is crucial to start the growth process by lowering the seed in contact with the melt, without altering the conditions of the melt surface, once the cleaning operation is completed.

The growth chamber below is constituted of a stainless steel cylindrical body with a removable spherical cap. By removing the cap, the crucible can be loaded inside the kiln. On the top of the cap, three circular windows enable monitoring of the interior of the chamber and hence the growth process. The entire exterior of the chamber is water-chilled through a tight coil of copper pipes, soldered on the external surface, within which cold water flows. Inside the chamber are placed the heater element and the crucible containing the raw materials to be melt. The heater element consists of a graphite resistance with cylindrical geometry, which accommodates the crucible in the middle. Such geometry has been properly conceived to achieve uniform heating of the crucible and hence a uniform temperature in the melt. By changing the vertical position of the crucible within the height of the resistor, the proper thermal gradient at the interface can be selected. Below the resistance are placed the two electrodes. A schematic of the mounting is shown in Fig. 11.5.

A central hole on the bottom of the crucible holder enables the connection with a thermocouple, by which is measured the temperature inside the growth chamber. Outside the furnace, the thermocouple is connected to a digital multimeter, which returns the temperature value measured on the bottom of the crucible. Beneath the crucible holder, a flange, by means of a feedthrough, enables moving the crucible within the structure of the resistor. Two thermal shields are positioned around the resistor to limit heat leakage outward.

The temperature control inside the furnace is operated by means of a PID (proportional–integral–derivative) control, which automatically adjusts the output current of the power supply accordingly to the discrepancy between measured and set values in the feedback loop. Fine temperature changes of the order of few tenths of Celsius degree can be achieved. The temperature stability in growth conditions is less than tenth of a Celsius degree.

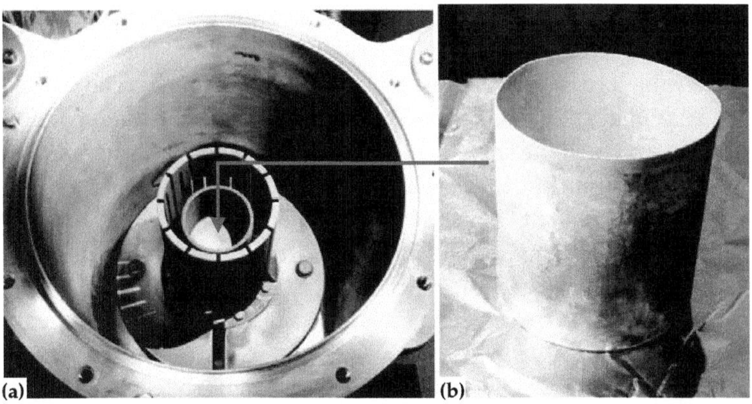

Figure 11.5 (a) Interior of the growth chamber. The heater element, with cylindrical geometry, accommodates the crucible holder in the middle. Beneath the resistance are placed the two electrodes. (b) Platinum crucible used for the growth of Yb-doped YLF single crystal.

A high-vacuum system enables evacuation at low pressure both the growth chamber and the secondary cell. The apparatus is comprised of two sequential vacuum pumps, connected separately to the two chambers. A prevacuum is operated through a mechanical rotary pump, reaching a pressure of 10^{-1} Pa. A turbomolecular pump is sequentially employed to operate high vacuum, up to a level of 10^{-6} Pa.

A suitable system is installed outside the furnace to pump inside the growth chamber controlled concentrations of high-purity (5N) Ar and CF_4 gases. The apparatus is designed to separately fill in Ar and CF_4. The gas amount pumped inside the chamber is controlled by a pressure gauge installed at the entrance valve with 10^2 Pa sensitivity.

The pulling system, which produces and controls the rototranslation movement of the seed holder shaft, is installed above the main body of the furnace (see Fig. 11.4). Basically, the apparatus consists of a stainless steel puller shaft, controlled by two stepper electrical motors, one for vertical translation and one for revolution. The mechanical design ensure steady, fine and smooth movements of the seed holder shaft, essential to avoid thermal shock and irregularities of the growth rates with consequent degradation of the structural quality of growing crystals. The arrangement ensures the bottom end of the rod to move back and forth from the secondary chamber to the bottom of the crucible. Minimum displacements achievable are of the order of the tenth of a millimeter. The slowest translational speed is 0.1 mm/h. The stepper motor that produces the revolution of the shaft enables rotation rates of the order of a few revolutions/min, normally required to attain thermal symmetry.

The puller shaft is terminated by a dedicated extension (seed holder), which consists of a cylindrical tube, made of high-purity platinum (99.95%). On the bottom end of the seed holder extension, crystal seeds and the cleaning tools are fastened by using high-purity platinum wires (99.95%).

The diameter of the growing crystal is controlled by using an optical system computer controlled (see Fig. 11.6).

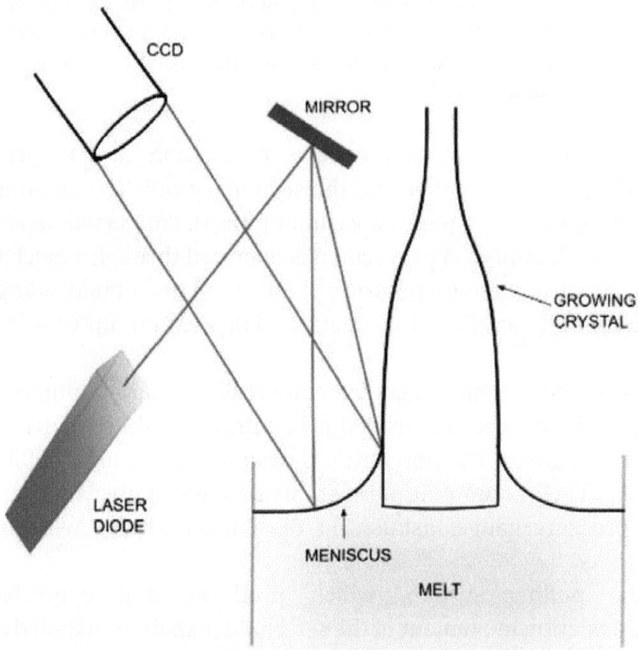

Figure 11.6 Schematic of the optical diameter control system.

The apparatus constitutes a red diode laser, which shines the solid–liquid interface meniscus, and a charge-coupled device (CCD) camera, which acquires the bright point due to the reflection of the laser beam by the meniscus. The laser beam is aligned to produce an asymmetric spot at the interface so that the laser beam can shine the meniscus while it is moving toward the crucible edge when the crystal diameter increases. The reflection position acquired in a revolution period is converted in a diameter value by means of a suitable algorithm. The laser and the CCD camera are arranged inside a compact support placed outside the growth chamber, in proximity of one of the top window of the main body, through which the incident beam is passed and the reflection from the meniscus is acquired. The diameter control is operated by means of a second PID feedback loop, which automatically varies the current supplied to the resistance, and hence the melt temperature, according to the difference between measured and set values of the crystal diameter. Diameter fluctuations attainable in growth conditions are of the order of 1%.

11.5 Crystal Growth of Cooling Materials at the NMLA Laboratory: Recent Results

Several Yb-doped fluorides have been grown in our facility for laser cooling experiments. So far, major results in bulk cooling have been achieved with the growth of high-quality YLF single crystals. Cryogenic operations were achieved in 5 at% Yb-doped YLF samples in cooling experiments carried out at the University of New Mexico [1, 2]. More recently, YLF single crystals with increasing Yb doping levels, up to 10 at%, have been grown, resulting in efficiency increase via significant background absorption diminishing. A novel scheme for Yb efficiency enhancement, based on energy transfer processes, has been demonstrated with the growth of Yb–Tm co-doped YLF single crystals. Remarkable cooling efficiencies were measured in a single-pass configuration with both single Yb-doped and Yb–Tm co-doped samples [21].

The high optical quality achieved in the YLF samples, and verified in laser cooling experiments, is the outcome of a thorough investigation carried out to identify the optimal growth parameters while assessing the crystal quality after the growth. In addition to the purity grade of the starting materials, specific growth parameters, that is, composition of the growth atmosphere, thermal gradients, choice of growth axis, melt composition, and pulling rate, critically affect the crystal quality and hence need to be properly investigated to achieve single crystals with a sufficient quality grade. After the growth, the optical and structural quality of finished samples needs to be assessed. Inclusion of internal defects, polycrystalline nuclei, and major contaminants needs to be investigated with suitable techniques.

This section focuses on these topics. Growth details and crystal quality assessments are discussed for the Yb-doped YLF single crystals grown in our facility.

11.5.1 The YLF Crystal

The YLF crystal is a birefringent insulating material largely used as a laser-active medium, even in commercial devices. The crystal possesses a band gap of about 7 eV and is optically transparent from the UV (180 nm) to the IR (6.7 μm), enabling its use as a crystalline host for almost all the trivalent rare earth ions. It is particularly appreciated for its low indices of refraction and negative thermal lensing. Additionally, among fluorides, it possesses a relative high thermal conductivity and mechanical hardness. The crystal structure is tetragonal with lattice parameter a = 5.197 Å and c = 10.735 Å and the unit cell contains four molecules inside. A scheme of the primitive cell of the YLF crystal is reported in Fig. 11.7. The crystallographic space symmetry group is $I4_1/a$ (Hermann–Mauguin notation). The lattice type is body centered (I) and the crystal structure possess a screw axis (4_1) and a glide plane ($/a$) symmetries.

The unit cell is thus invariant under a rotation of 90° (360°/4) followed by a translation of half the lattice vector (4_1 screw axis symmetry) and for a combination of a translation of half the lattice vector along an a axis and a reflection in that plane ($/a$ glide plane symmetry). Crystals belonging to the tetragonal class are optically uniaxial, which means they possess a unique direction along which the two allowed orthogonal polarizations undergo the same phase velocity. Such direction corresponds to the crystallographic c axis in YLF crystals and defines the optical axis z of the crystal. The other two optical axes x and y possess the same refractive index ($n_x = n_y \neq n_z$) and coincide with the two equivalent crystallographic axes a of the unit cell. Only light traveling along the c axis is not subjected to birefringence inside the crystal. Trivalent RE ions substitutionally enter the Y^{3+} sites with point symmetry S_4 and coordination number

8. Charge compensation is not required and the replacement of Y^{3+} with RE^{3+} ions does not significantly affect the crystal structure due to the similarity of ionic radii sizes. Slight discrepancies in ionic radii, however, can cause segregation phenomena. In the case of Yb doping, the segregation coefficient is expected to be equal to 1 as the ionic radii of Y^{3+} and Yb^{3+} for coordination number 8 in fluorides are 1.16 Å and 1.12 Å, respectively [11]. A detailed list of optical and physical properties of the YLF crystal is reported in Table 11.1.

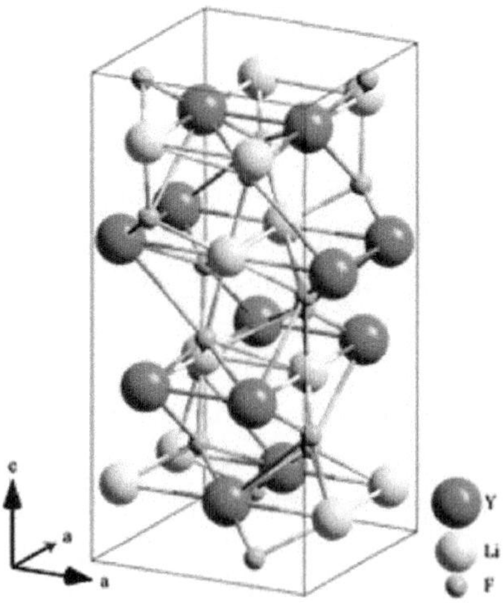

Figure 11.7 Scheme of the unit cell of YLF crystal.

Table 11.1 Main physical and optical properties of the YLF crystal

Thermal conductivity @ 298 K (W/mK)	5.2 (*a* axis) 7.2 (*c* axis)
Transparency range (μm)	0.18–6.7
Refractive index (@ 1 μm)	$n_o = 1.45$ $n_e = 1.47$
Phonon energy (cm^{-1})	450
Density (g/cm^3)	3.99
Specific heat (J/gK)	0.79
Coefficient of thermal expansion (10^{-6} K)$^{-1}$	8.3 (*a* axis) 13.3 (*c* axis)
Hardness (Knoop)	300

11.5.2 The Growth of Yb-Doped YLF Single Crystals

11.5.2.1 The phase diagram of the LiF–YF_3 system

Detailed knowledge of phase diagram is the starting point for any growth process. The phase diagram provides a temperature composition map which represents the thermodynamic phases of the system as a function of temperature and relative concentration of components. Temperature and composition conditions suitable to achieve the growth along the phase transition between a single solid phase and homogeneous liquid need to be identified as well as single equilibrium phases and temperature-dependent phase transitions relative to the compound composition.

The phase diagram relative to the YLF compound is shown in Fig. 11.8 [19]. Such diagram represents all the equilibrium phases and transition lines of the binary LiF-YF_3 system as a function of temperature and relative concentrations of lithium fluoride (LiF) and yttrium fluoride (YF_3), which are the constituents of the YLF crystal. On the left and right sides of the diagram are the behaviors of the pure LiF and YF_3, respectively. For the YLF crystal, the phase diagram indicates a strong incongruent melting behavior, which means that the YLF crystal does not melt uniformly through formation of liquid LiF and YF_3 but, instead, increasing temperature, LiF fuses firstly, while YF_3 stays solid (YLF→ LiF + solid YF_3). A displacement from the stoichiometric composition is thus required in order to intersect the transition line between the YLF solid phase and the homogeneous liquid phase, along which the growth occurs. The peritectic single point occurs at 49 mol% YF_3, the eutectic point occur at 19 mol% YF_3. An ideal phase for the undercooling state is that with the coexisting YLF crystal and homogenous liquid (B), which extend from 19 mol% to 49 mol% YF_3. In principle this whole concentration range would be suitable for growth. However, to maximize the temperature extension for the growth process and to keep a ratio between the components close as possible to 1, a high concentration of YF_3 is typically chosen.

To grow YLF single crystals with a given doping level of rare earth^{3+} ions, a partial amount of Y^{3+} ions need to be replaced by RE^{3+} ions. A suitable concentration of YF_3 hence need to be removed from the starting mixture and replaced with the dopant compound REF_3.

In case of co-doping with two different species of RE^{3+} ions, the YF_3 compound needs to be replaced with appropriate amounts of both the RE compounds. The dopant RE^{3+} ions, present in the melt, will be incorporated in the crystal while it is growing, replacing the Y^{3+} host ions in the unit cells.

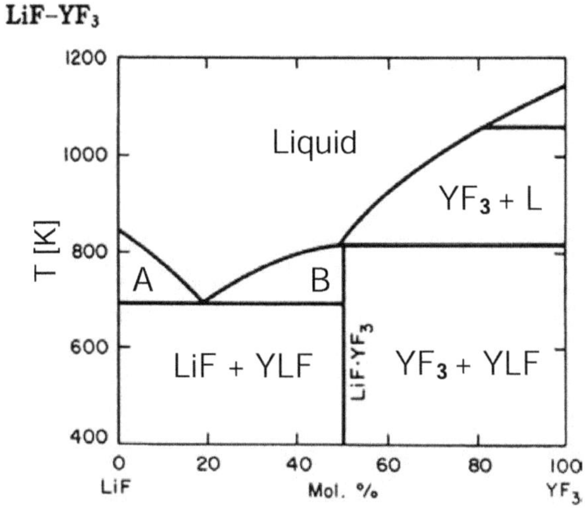

Figure 11.8 Phase diagram of the LiF-YF_3 system. The label A indicates a phase with solid LiF mixed with liquid. The label B indicates a phase with solid YLF and liquid [19].

The final composition in the case of single doping can be expressed as: $Li(Y_{1-x}RE_x)F_4$, where x indicates the molar percentage of RE^{3+} ions. In case of co-doping with two species of rare earth ions $RE1^{3+}$ $RE2^{3+}$ the final composition can be expressed as: $Li(Y_{1-x-y}RE1_xRE2_y)F_4$ where x and y indicates the molar percentage of $RE1^{3+}$ and $RE2^{3+}$ ions, respectively.

11.5.2.2 The growth of YLF single crystals with varying Yb doping levels

The use of high-purity raw materials is the starting point for the growth of efficient cooling crystals. Yb-doped YLF single crystals for laser cooling were grown starting from fluorine raw powders of LiF, YF_3, and YbF_3 with guaranteed 5N purity (99.999%). Such a purity grade ensures that the sum of impurities contained in the binary compounds is less than 10 ppm. Very few companies in the world can supply such high-purity fluorine powders. Raw materials used in this work where provided by AC Materials (Tarpon Spring, FL, USA), a world-class company in this field, specialized in preparation of high-purity binary fluorine for crystal growth, via prefluorination of oxide compounds with hydroxyl acid.

YLF single crystals doped with Yb ions at three different doping levels (5 at%, 7.5 at%, and 10 at%) were grown. For each growth, the mixture of starting materials was prepared inside a moisture-controlled environment, storing the components, in the proper relative concentrations, inside a dedicated high-purity platinum crucible (99.9%). The relative concentrations of host components used were 52.5 mol% LiF and 47.5 mol% YF_3. A suitable concentration of YF_3 was replaced by YbF_3 in order to achieve the desired doping level (5 at%, 7.5 at%, and 10 at%). Total amounts of raw materials between 80 g and 100 g were used for all the growths.

Prior to load the starting materials inside the growth chamber, the furnace was carefully precleaned. Suitable solvents were firstly used. Subsequently, the furnace was evacuated at low pressure and then heated up to 1000°C for several hours while drawing with the vacuum pumps. An ultimate vacuum level of 10^{-6} Pa was reached. After high-vacuum cleaning, the crucible, containing the mixture of starting materials, was loaded inside the furnace. The growth atmosphere was built up pumping inside the furnace argon and CF_4 gases in the proper concentrations. The furnace was then heated up until the melting point of the starting materials was reached. The melt temperature slightly changed as a function of the Yb doping level, between 880°C and 890°C. Impurities emerged on the melt surface were removed by using high-purity (99.9%) Pt tubes or crystal seeds, repeatedly plunged into the melt at different temperatures. When the melt surface appeared completely free from foreign particles, the growth was started by lowering a crystal seed in contact with the melt. The crystal seed was slowly lowered toward the melt surface in order to prevent thermal shock and sudden uncontrollable crystallizations of the surface. For all the growths, a YLF undoped seed, oriented along the *a* axis, was used (see Sec. 11.5.2.1). The molten was slightly undercooled by a few Celsius degrees, below the melting point, when the nucleation started. In the early stage of the growth, fine temperature changes were operated to shrink the crystal diameter down to 1–2 mm. After the growth of a narrow neck, a few millimeters long, the crystal diameter was smoothly enlarged up to 1–2 cm. Subsequently, the growth was carried at a uniform diameter for 5–10 cm. For all the samples the pulling rate was 0.5 mm/h, leading to growth times of several days for each boule. The rotation rate was kept at 5 rpm.

In Fig. 11.9 are shown typical temperature and diameter data acquired during the growth process. Such graphs show typical temperature changes and diameter variation rates for a good growth process. As can be seen from graphs, the melt temperature changed within a few Celsius degrees during the growth process and diameter fluctuations achieved in growth conditions with the PID control are of the order of 1%. The origin for both graphs corresponds to the nucleation. The temporal evolution graph for crystal diameter shows the diameter decrease, operated in the early stage of the growth, for the necking operation, by changing the melt temperature. After the growth of a narrow neck for a few millimeters, the diameter of the growing crystal was enlarged at very slow rates by using the PID control. The red line in the graph represents the set point of the optical diameter control. Blue data points correspond to the experimental measurements of the diameter. A very slow rate was applied to enlarge the diameter of the growing crystal in order to prevent formation of defects due to irregular anisotropic growth rates. The largest diameter value was set around 1 cm. The growth at uniform diameter was carried on for several centimeters. Along with temperature and diameter graphs, in Fig. 11.9 is shown a picture of the finished crystal.

After growth, before being extracted from the furnace, the crystal boules were slowly cooled down to room temperature. Typical cooling rates used to prevent thermal shock were 15°C/h.

As soon as finished crystals are removed from the furnace, some defects can be visually detected. Good-quality Yb-doped fluorides appear transparent, as the Yb^{3+} ions do not have any visible absorption bands and the host is transparent from the UV (180 nm) up to several microns. The regular growth along the *a* axis results in elliptic cross sections with two well-defined lateral edges. Irregular diameter changes are typical indications of internal defects. Polycrystalline domains appear unclear and cracked. In Fig. 11.10 are reported some pictures of grown boules of YLF:5%Yb, YLF:7.5%Yb, and YLF:10%Yb. All the crystal boules appear clear and transparent. In Fig. 11.10b it can be seen the crystal seed with the Pt wire, used to fasten the crystal seed to the seed holder extension, still attached to the crystal boule. In both Fig. 11.10c and Fig. 11.10d is visible the narrow crystal neck operated in the early stage of growth process by changing the melt temperature. Narrow diameters of 1–3 mm (Fig. 11.10c) and 2–3 mm (Fig. 11.10d) were grown for up to 10 mm (Fig. 11.10c). The lateral crystals edges and the elliptic cross section can be observed in Fig. 11.10a,c,d. In Fig. 11.10e the circular cross section is indication of a *c* axis component along the growth axis. Irregular diameter changes as those visible in Fig. 11.10a,b,d did not reflect in the inclusion of defects. An example of excellent diameter stability is provided in Fig. 11.10c.

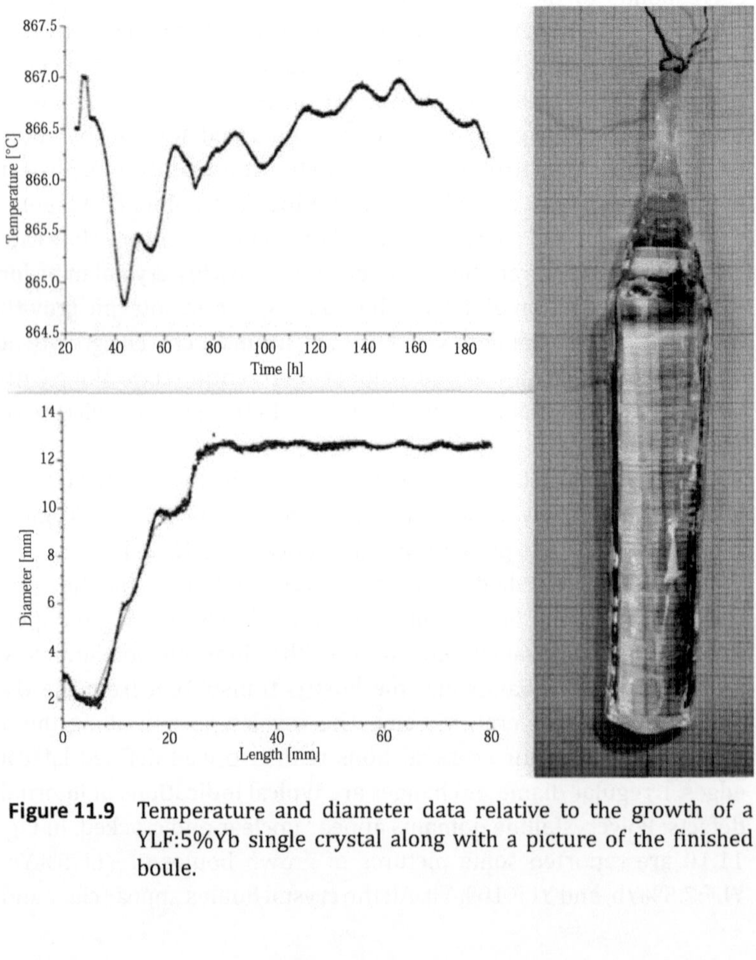

Figure 11.9 Temperature and diameter data relative to the growth of a YLF:5%Yb single crystal along with a picture of the finished boule.

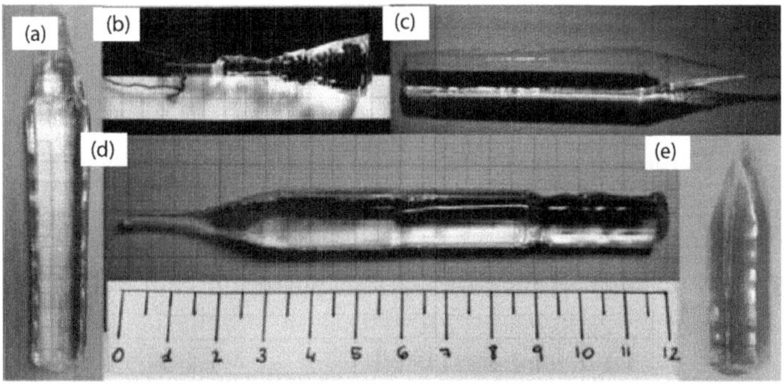

Figure 11.10 As-grown crystals of YLF:5%Yb (a and b), YLF:7.5% (d), and YLF:10%Yb (c and e).

In Fig. 11.11 is highlighted the effect of a faulty growth atmosphere composition on the optical quality of Yb-doped single crystals. In Fig. 11.11a is shown the formation of Yb^{2+} impurities inside Yb-doped fluorides due to a CF_4-depleted growth atmosphere. Crystals grown under this kind of atmosphere incorporate Yb^{2+} ions, which are responsible of the red coloration and hence of the poor optical properties. Such samples are completely useless for optical cooling experiments. On the other side, in Fig. 11.11b is shown a carbon scum deposition both on the crystal surface and on the interior walls of the furnace, due to an excess of CF_4 concentration in the growth atmosphere. Such growth conditions are rather detrimental for the growth process as carbon impurities can be incorporated inside the growing crystals. Moreover, the carbon scum which deposits on the interior of the furnace typically accumulates also on the surfaces of the top windows, making difficult the optical control of the growing crystal diameter. Irregular diameter variations typically occur.

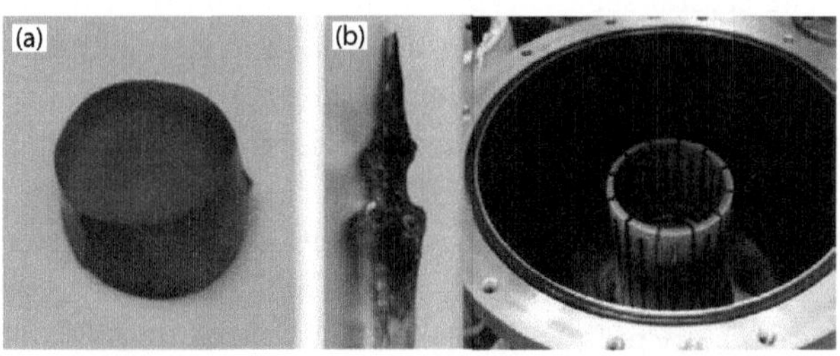

Figure 11.11 (a) Yb^{2+} contamination of Yb-doped YLF single crystals grown under a CF_4-depleted atmosphere. (b) Carbon deposition on crystal surface and furnace interior due to excessive concentration of CF_4 in the growth atmosphere.

11.5.2.3 Preparation of crystal seeds for the growth of Yb-doped YLF single crystals

Prior to the growth of the Yb-doped samples, an undoped YLF single crystal was grown in order to fabricate crystal seeds for the growth of the cooling materials. For the growth of the undoped YLF, the mixture of starting materials was composed of LiF and YF_3 raw powders mixed with the relative concentration 52.5 mol% LiF and 47.5 mol% YF_3. A high-purity (99.9%) platinum crucible dedicated to undoped materials was used. The growth atmosphere was composed 100% of high-purity (5N) argon. Crystal seeds obtained from previous growths of undoped YLF single crystals were not used in order to prevent any inclusion of impurities, not contained in the LiF and YF_3 raw powders. The growth was hence started, inducing self-nucleation around a high-purity (99.99%) Pt wire wounded up in a tubular shape with a narrow diameter. A slightly higher melt temperature are typically necessary to induce the crystallization starting from a Pt wire instead of a crystal seed. During the growth, the crystal was pulled at a rate of 1 mm/h while rotated at a speed of 5 rpm. A picture of the finished boule is shown in Fig. 11.12a. The circular cross section of the crystal boule is indication of growth axis mostly aligned along the *c* axis of the unit cell. In Fig. 11.12b are shown some crystal seeds fabricated from the grown crystal boule. Such samples are typically used in parallelopiped shapes with edges oriented along the optical axes and tip-cut ends. Typical dimensions are 3 × 3 (*c* × *a*) mm^2 cross section and 5–10 mm (*a*) length. The longer axis of the parallelopiped was cut parallel to the *a* axis. This configuration defines a crystal seed oriented along the *a* axis. The crystal seed is typically mounted on the seed holder in order to contact the melt with the longer axis perpendicular to the melt surface. A crystal seed oriented along the *a* axis hence induce oriented nucleation with the growth axis of the growing crystal aligned with the *a* axis of the unit cell.

Although the self-nucleation preferentially tends to occur along the *c* axis, the choice of oriented growth along the *a* axis is related to the optical quality of finished crystals.

The precise orientation of the crystallographic axes of crystal boule were previously determined by X-ray diffraction and the absence of major impurities and scattering centers was checked through spectroscopic and scattering analysis, as detailed in Section 11.3.

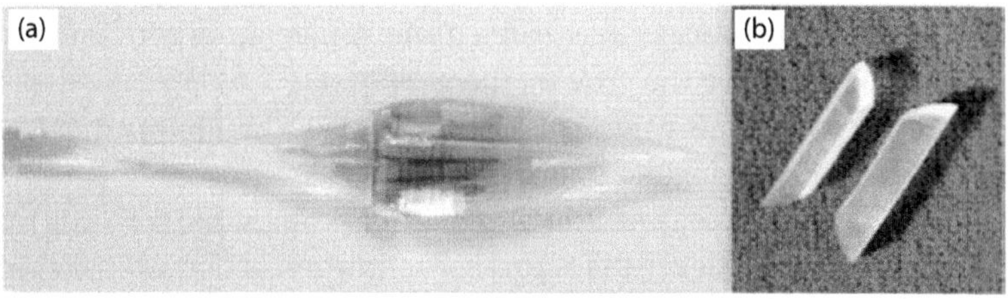

Figure 11.12 (a) As-grown crystal boule of YLF undoped crystal. (b) Seeds of YLF undoped crystal oriented along the *a* axis.

11.5.2.4 The growth of Yb-doped YLF single crystals with controlled Tm doping

A novel scheme for laser cooling, relying on virtuous energy transfer processes, has been investigated in Yb–Tm co-doped YLF single crystals, providing the first experimental evidence of energy transfer–assisted anti-Stokes cooling [21]. Significant efficiency enhancements, over the single Yb doping performances, were achieved via Yb–Tm energy transfer in the co-doped single crystal. The virtuous energy transfer process between Yb and Tm ions provides an additional channel of heat removal via radiative emission at higher average energies than absorbed photons. The potentialities of this approach for intrinsic efficiency enhancements are huge.

A YLF single crystal Yb doped at 5 at% with a controlled Tm doping of 16 ppm was grown. A low concentration of Tm ions was chosen as the starting point for the investigation of the effect of Yb–Tm energy transfer on the Yb anti-Stokes process.

The co-doped sample was grown under similar conditions as those used for the single Yb-doped samples. For the mixture of starting material, the usual concentrations of host components were used: 52.5 mol% LiF and 47.5 mol% YF_3. However, to achieve the Yb–Tm co-doping, the YF_3 compound was partially replaced with appropriate concentrations of both YbF_3 and TmF_3 compounds. The relative concentrations were calculated in order to replace 5% of Y^{3+} host ions with Yb^{3+} ions and 0.0016% of Y^{3+} host ions with Tm^{3+} ions.

Due to the similarity of Yb^{3+} and Tm^{3+} ionic radii sizes, a segregation coefficient equal to 1 is expected, even for the substitution of Y^{3+}/ Tm^{3+} substitution [11].

The growth process was carried out under an argon and CF_4 atmosphere. During the growth, the pulling rate was 0.5 mm/h and the rotation speed 5 rpm. As a crystal seed, a YLF undoped single crystal oriented along the *a* axis was used.

Due to the low concentrations of the TmF_3 compound required for the desired doping level, the actual concentration of Tm was checked after the growth with a trace element analysis performed by means of laser ablation–inductively coupled plasma–mass spectrometry (LA-IPC-MS) [20]. A Tm concentration of 16 ppm was detected. No internal defects related to the Yb–Tm co-doping were detected.

11.5.3 Investigation of Sample Quality: Structural Analysis, Testing of Internal Defects, and Major Contaminants

After growth, the crystal boules were investigated with different techniques in order to assess the optical and structural quality. The local single-crystalline structure and the orientation of the unit cell were investigated through X-ray backscattering diffraction in a Laue chamber. The inclusion of structural defects was assessed observing the propagation of visible laser beams through the volume of samples. Absorption measurements were performed to check the absence of major contaminants.

11.5.3.1 Structural analysis: X-ray backscattering diffraction

Due to the optical anisotropy of the YLF crystal, its spectroscopic properties depend on the crystal orientation and hence need to be selectively investigated for all the independent polarizations. The orientation of the crystallographic axes, and hence of the optical axes, needs to be accurately identified so that samples oriented along the optical axes can be fabricated maximizing absorption and emission cross sections of the materials.

After growth, all the crystal boules were firstly analyzed by X-ray backscattering diffraction in a Laue chamber. Such analysis enables to identify the orientation of the crystallographic axes of the unit cell with high accuracy. Moreover, diffraction patterns acquired in a different region of the samples allow the checking of the single-crystalline structure of the sample.

For the Laue analysis, the sample, fixed on a goniometer holder, is shined with a nonmonochromatic X-ray beam. The precision goniometer, which allows rotations around the three axes, is used to fix the orientation of the crystal with respect of the X-ray direction. The backscatter radiation is collected on a sensor plate, placed between the X-ray source and the crystal, which record the diffraction pattern.

After acquisition of the diffraction pattern in the Laue chamber, suitable software is used to simulate the backscattering of the crystal for a given orientation of the crystallographic axes with respect to the direction of the X-ray beam. Combined rotations around the three axes are applied in order to reproduce the experimental pattern. The correspondence between the acquired pattern and the simulation allows one to identify the orientation of the crystallographic axes, with an accuracy of 1°.

The absence of a polycrystalline domain can be locally checked from the diffraction pattern. The presence of large or double spots instead of single dots in the backscatter pattern is due to diffraction from a staggered lattice and hence indicates a polycrystalline structure.

Some examples of a diffraction pattern relative to the grown crystals are shown in the following section.

In Fig. 11.13a is shown the diffraction pattern acquired in backreflection for the YLF:7.5%Yb boule shown in Fig. 11.10d, by shining the X-ray beam parallel the direction of the two lateral edges. Along with the measured diffraction pattern is shown the theoretical pattern, simulated for the *c* axis parallel to the X-ray direction. The correspondence between the two patterns confirms the growth axis aligned with the crystallographic *a* axis and returned the orientation of the crystallographic axes with accuracy of 1°. Single dots in the diffraction patterns acquired in different areas of the boule ascertain the single-crystalline structure.

In Fig. 11.13b is shown the diffraction pattern acquired for the YLF:10%Yb crystal shown in Fig. 11.10e, by shining the X-ray beam in proximity of the single edge. The X-ray simulation returned a *c* axis component along the growth axis, as suggested by the circular cross section of the sample. The growth axis resulted in the plane *c-a*, tilted about 30° with respect to the *a* axis. In Fig. 11.14 is reported the diffraction pattern acquired for the as-grown crystal boule of YLF undoped, by shining the X-ray beam perpendicularly to the growth axis.

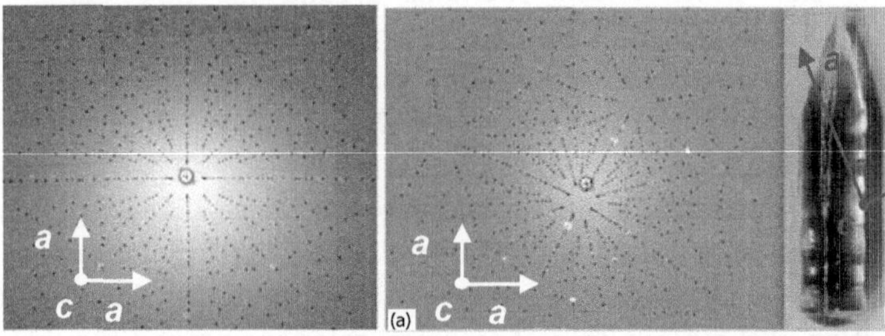

Figure 11.13 (a) X-ray backscatter diffraction pattern of YLF:7.5%Yb (sample shown in Fig. 11.10d), acquired by shining the X-ray beam along the direction of the two lateral edges, and relative simulation. (b) Experimental pattern, acquired by shining the X-ray beam in proximity of the lateral edge, and relative simulation for the YLF:10%Yb sample shown in Fig. 11.10e.

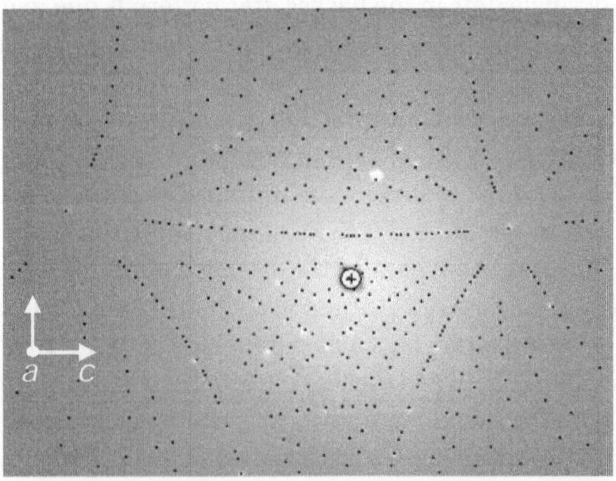

Figure 11.14 Acquired backscatter diffraction pattern and relative simulation of as-grown YLF undoped crystal. The X-ray beam was shone perpendicularly to the growth axis.

The correspondence with the theoretical pattern simulated for the *c* axis along the growth axis indicated that the favorable growth axis for the self-nucleation in YLF crystal is the *c* axis. The measurement of the crystallographic axes orientation returned the *c* axis tilted just a few degrees with respect to the growth axis.

11.5.3.2 Scattering analysis for crystal defect detection

Internal defects, such as cracks, microbubbles, and dislocations can be detected observing the propagation of a laser beam through the volume of the sample [1]. Crystal defects give rise to scattering centers, hence scattering phenomena of laser beams are indication of internal defects. In the case of good crystal quality, scattering is absent. Laser wavelengths not absorbed by the material under investigation should be chosen. For Yb-doped fluorides all the visible wavelengths are suitable for the scattering analysis.

A very sensitive test for internal strains in uniaxial crystals can be performed observing the extinction ratio of the sample, placed between two crossed polarizers with the *c* axis along the direction of propagation of light. In the case of strain, a significant fall of the extinction ratio is detectable.

As-grown boules and finished samples were tested, observing the propagation of visible laser beams by means of a Zeiss microscope. Two different wavelengths were used: 633 nm and 405 nm. In Fig. 11.15 are shown some tests of crystal quality performed with the 633 nm laser beam.

In Fig. 11.15a and Fig. 11.15b is shown the propagation of the 633 nm laser beam inside a cooling sample that possesses all the surfaces polished at a high optical grade. In Fig. 11.15a a single scattering center is detected, and reflections at the entrance and exit surfaces are visible. In Fig. 11.15b is shown a defect-free area of the same sample. Figure 11.15c shows a dense aggregate of scattering center detected in a crystalline boule grown under an unsuitable thermal gradient. On the bottom of Fig. 11.15d–f is shown the axial distribution of scattering center detected in a defected crystalline boule. In Fig. 11.15d is shown the propagation of the laser beam through the neck: a high density of scattering center is visible. In Fig. 11.15e is shown the laser propagation in the middle of the boule: scattering defects are less numerous in this region.

In Fig. 11.15e is shown the bottom of the boule: scattering is almost absent. The 405 nm wavelength was used to check the absence of scattering center in finished cooling samples, where the polished surfaces strongly reduce spurious reflections. Moreover, the shorter laser wavelength increases the scattering cross section, enabling a more refined analysis of the propagation of the laser beam.

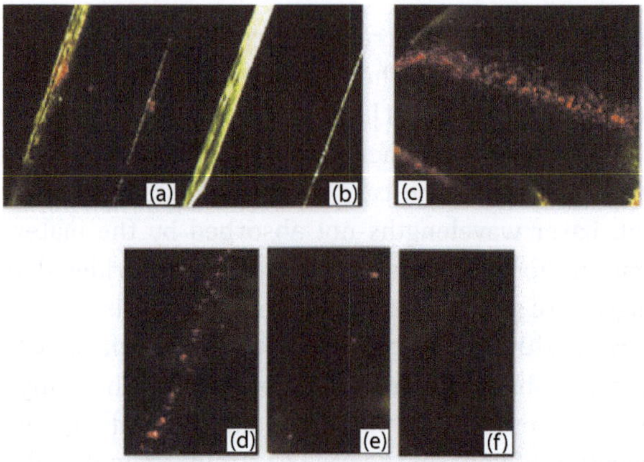

Figure 11.15 Test of crystal quality with 633 laser beam. (a) Single scattering center detected inside a cooling sample; (b) scattering-free area of the same sample; (c) high-density aggregate of scattering centers; (d) propagation of 633 nm laser beam in the neck of a crystalline boule: high density of scattering defects; (e) middle region: less numerous scattering center; and (f) bottom: scattering almost absent.

Internal stress in as-grown crystal boules were investigated by means of crossed polarizers. No significant variation of the extinction ratio were detected, indicating good crystal growth conditions.

11.5.3.3 UV-NIR absorption measurement for detection of major contaminants

Inclusion of major contaminants inside grown crystals can be examinated through absorption measurements. Spurious bands in the absorption spectrum of the sample are the fingerprints of optically active impurities.

Among rare earth ions, the Yb^{3+} ion possesses a peculiar energy level structure, which exhibits a single excited manifold $^2F_{5/2}$ at energies between 1.2 and 1.3 eV with respect to the ground-state $^2F_{7/2.}$ In the absence of major contaminants only the absorption band relative to the Yb transition, which extends from 890 nm to

1100 nm in fluorides, should be detectable. Any additional line in the absorption spectrum can be ascribed to undesired impurities.

As detailed earlier, the most difficult issue when growing Yb-doped fluorides is to avoid the presence of oxygen complexes and Yb^{2+} ions in the melt. When hydroxyl radicals, OH^-, substitute F^- ions in fluoride crystals they give rise to UV electronic bands between 180 and 230 nm and vibrational absorption in the range of 3000–4000 cm^{-1} [3]. The inclusion of OH^- impurities in fluoride crystals, however, is typically detectable from the analysis of the UV absorption spectrum as in this region OH^- complexes exhibit the strongest absorption. The vibrational absorption in the IR is much weaker than UV electronic bands and typically can be detected only at low temperatures. Low-temperature measurements, however, normally are performed when UV absorption bands are detectable in order to provide further support of OH^- contaminations. Yb^{2+} ions in fluoride crystals instead possess large absorption bands between 200 and 600 nm due to $4f$–$5d$ transitions and hence can be detected from the absorption spectrum in this range.

In addition to oxygen complexes and Yb^{2+} ions, impurities of other rare earth and transition metals can be also incorporated inside growing crystals due to small inaccuracies in the cleaning procedures.

To check the absence of major contaminants, for all the grown samples, a broad absorption spectrum from the UV (180 nm) to the NIR (2000 nm) was acquired. Figure 11.16 reports the absorption spectra acquired for the Yb-doped YLF samples. All the spectra exhibit a single absorption band identifiable with the Yb $^2F_{7/2} \rightarrow {}^2F_{5/2}$ transition.

No additional bands can be detected, indicating that the inclusion of major impurities can be ruled out. In particular, the UV spectrum indicates the absence of OH^- contaminants in relevant concentrations, that is, higher than tens of ppm. The absence of a spurious band in the whole visible range proves the efficiency of the growth atmosphere in preventing formation of Yb^{2+} ions in the melt. The inclusion of other kind of contaminants in relevant concentrations can be ruled out as well.

The absorption spectrum of the Yb–Tm co-doped sample showed a similar behavior as that of the single Yb-doped YLF single crystals.

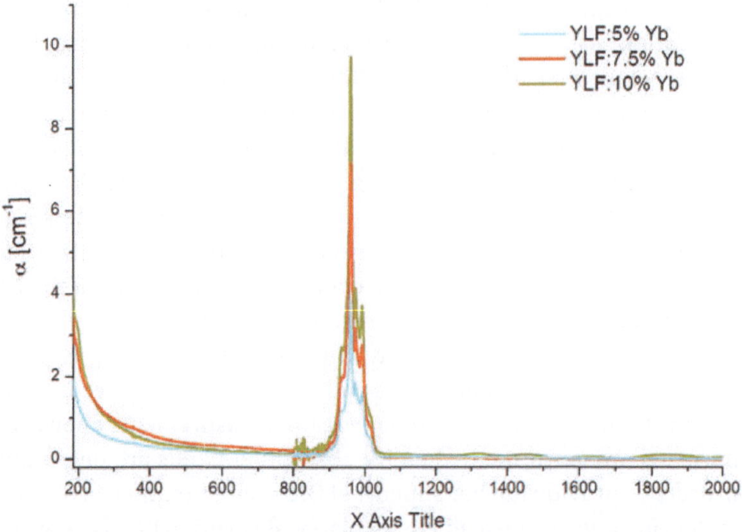

Figure 11.16 Absorption spectra ranging from 180 nm up to 2 μm of YLF:5%Yb, YLF:7.5%Yb, and YLF:105Yb samples. The wavelength resolution is 0.4 nm between 180 and 800 nm and 0.6 nm between 800 and 2000 nm, while the absorption resolution is 0.002 cm^{-1}.

The multipeak structure of the Yb^{3+} absorption band is related to the interaction of Yb^{3+} ions with the electric field of the surrounding crystal, which induces a Stark splitting of the free ions' energy levels of 10^2 cm^{-1} magnitude. For the Yb^{3+} ions occupying the S_4 sites in the YLF host the excited manifold $^2F_{5/2}$ is divided in three sublevels and the ground-state $^2F_{7/2}$ in four sublevels. Phonon-assisted inter-Stark transitions between these sublevels are the origin of the multiple lines visible in the Yb absorption band at room temperature.

The increase of the background absorption between 300 nm and 180 nm is not due to the host absorption, which is transparent in this region, but mainly to Rayleigh scattering from the surfaces. Even if the entrance and exit facets of the samples were polished before absorption measurements, surface roughness and imperfections are still present and cause efficient scattering at short wavelength, increasing the background absorption in such region. The automatic detector change around 800 nm is the source of the noise in this region.

11.5.4 Sample Fabrication

After identification of the crystallographic axes orientation and assessment of the absence of polycrystalline domains, internal defects, and major contaminants, samples for cooling experiments were prepared. Precision cutter machines, equipped with a diamond wire saw, were used to cut samples of the desired orientation and dimensions from the crystals boules. Right angle–cut and Brewster-cut samples were fabricated.

Right angle samples were cut to a parallelopiped shape, with the longer edge parallel to the *a* axis in order to allow the excitation along the *c* axis, which maximizes the absorption in YLF crystals. A layout is shown in Fig. 11.17a. Typical dimensions were 3 × 3 mm^2 (*c* × *a*) cross section and length (*a*) between 5 and 15 mm.

Brewster-cut cooling samples were fabricated to minimize reflection losses. The orientation was designed to eliminate reflections for incident light polarized along the *c* axis of the crystal. Pictures of finished right angle–cut and Brewster-cut samples are shown in Fig. 11.17 and Fig. 11.18.

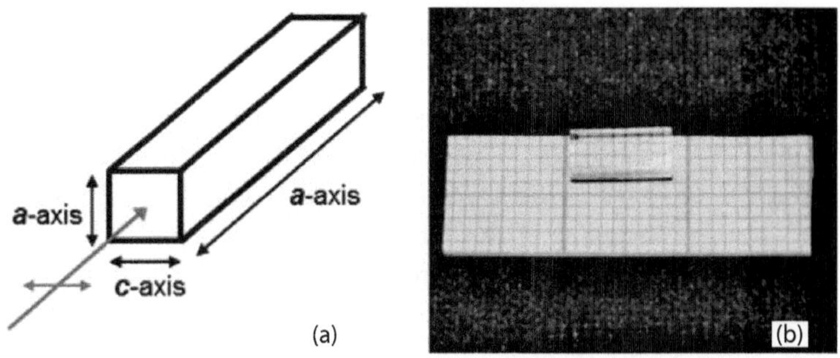

Figure 11.17 (a) Layout of right angle–cut sample prepared for cooling experiments in single-pass configuration. (b) Picture of finished cooling sample of YLF:10%Yb.

All the surfaces of cooling samples were polished to a high optical grade using fine diamond powders. The quality of the crystal surfaces significantly affects the fluorescence escape efficiency; hence high–optical grade polishing is of fundamental importance.

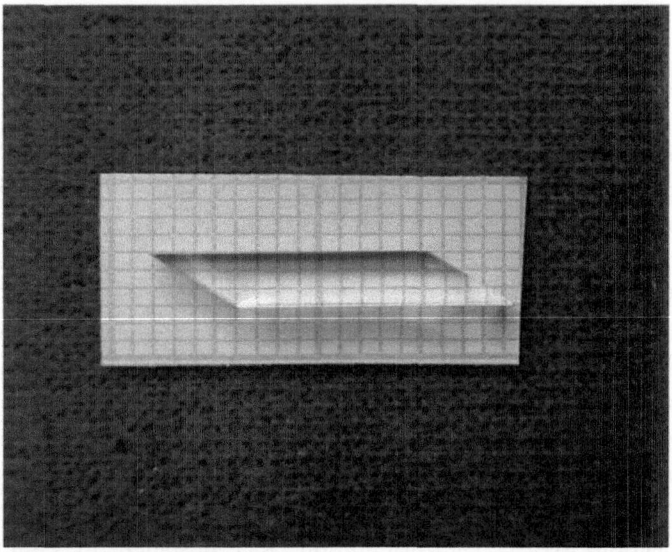

Figure 11.18 Picture of finished Brewster-cut sample of YLF:5%Yb for laser cooling experiments.

11.6 Concluding Remarks

Several Yb-doped fluoride single crystals have been grown in our facility for laser cooling experiments. Growth parameters have been optimized for this kind of materials achieving high optical and structural quality. The remarkable performances achieved in bulk cooling experiments confirmed the high crystal quality. Crystal growth techniques and methods for optical and structural quality assessments were reviewed.

Acknowledgments

The authors would like to acknowledge I. Grassini for her competence and care in preparing the samples and D. Parisi and S. Veronesi for useful discussions and their expertise in crystal growth. A. Volpi acknowledges support by the European Space Agency under grant no. 4000108074/13/NL/PA, "Cooling Effect on Fluoride Crystals."

References

1. Belt, R. F., Uhrin, R. (1991). Top seeded solution growth of Cr3+:LiCaAlF6 in HF atmosphere, *J. Cryst. Growth*, **109**, pp. 334–339.
2. Brice, J. C. (1973). *The Growth of Crystals from Liquids (Series of Monographs on Selected Topics in Solid State Physics)* (North Holland, Amsterdam).
3. Cornacchia, F., Palatella, L., Toncelli, A., Tonelli, M., Baraldi, A., Capelletti, R., Cavalli, E., Shimamura, K., Fukuda, T. (2002). Temperature dependence of impurity quenched luminescence in Tm3+:LiLuF4, *J. Phys. Chem. Solids*, **63**, pp. 197–202.
4. Czochralski, J. (1918). A new method for the measurement of the crystallization rate of metals, *Z. Phys. Chem.*, **92**, pp. 219–221.
5. Di Lieto, A., Sottile, A., Volpi, A., Zhang, Z., Seletskiy, D. V., Tonelli, M. (2014). Influence of other rare earth ions on the cooling efficiency in Yb:YLF crystals, *Opt. Express*, **22**, pp. 28572–28583.
6. Djeu, N., Whitney, W. T. (1981). Laser cooling by spontaneous anti-Stokes scattering, *Phys. Rev. Lett.*, **46**, pp. 236–239.
7. Epstein, R. I., Buchwald, M., Edwards, B., Gosnell, T., Mungan, C. (1995). Observation of laser-induced fluorescent cooling of a solid, *Nature*, **377**, pp. 500–502.
8. Fisher, S., Ivaturi, A., Frolich, B, Rudiger, M., Richter, Kramer, K. W., Richards, B. S., Goldschmidt, J. C. (2014). Upconverter silicon solar cell devices for efficient utilization of sub band gap photons under concentrated solar radiation, *IEEE J. Photovolt.*, **4**, pp. 183–189.
9. Fisher, S., Favilla, E., Tonelli, M., Goldschmidt, J. C. (2015). Record efficient upconverter solar cell devices with optimized bifacial silicon solar cells and monocrystalline BaY2F8:30%Er upconverter, *Sol. Energ. Mater. Sol. Cells*, **136**, pp. 127–136.
10. Hehlen, M. P., Boncher, W. L., Melgaard, S. D., Blair, M. W., Jackson, R. A., Littleford, T. E., Love, S. P. (2014). Preparation of high-purity LiF, YF3 and YbF3 for laser refrigeration, *Proc. SPIE*, **9000**, p. 900004.
11. Kaminski, A. A. (1996). *Crystalline Lasers: Physical Processes and Operating Schemes* (CRC Press, Boca Raton, FL).
12. Melgaard, S. D., Seletskiy, D. V., Di Lieto, A., Tonelli, M., Sheik-Bahae, M. (2013). Optical refrigeration to 119 K, below National Institute of Standards and Technology cryogenic temperature, *Opt. Lett.*, **38**, pp. 1588–1590.

13. Melgaard, S. D. (2013). *Cryogenic Optical Refrigeration: Laser Cooling of Solids Below 123K*, PhD dissertation, University of New Mexico, Albuquerque, NM.
14. Melgaard, S. D., Albrecht, A., Hehlen, M. P., Seletskiy, D. V., Sheik-Bahae, M. (2014). Optical refrigeration cools below 100 K CLEO 2014, *OSA Technical Digest*, paper FTh4D.4.
15. Pamplin, B. (1975). *Crystal Growth* (Pergamon Press, Oxford, UK).
16. Pringsheim, P. (1929). Zwei Bemerkungen über den Unterschied von Lumineszenz-und Temperaturstrahlung, *Z. Phys.*, **57**, pp. 739–746.
17. Seletskiy, D. V., Melgaard, S. D., Bigotta, S., Di Lieto, A., Tonelli, M., Sheik Bahae, M. (2010). Laser cooling of solids to cryogenic temperatures, *Nat. Photonics*, **4**, pp. 161–164.
18. Thiede, J., Distel J., Greenfield, S. R., Epstein, R. I. (2005). Cooling to 208 K by optical refrigeration, *Appl. Phys. Lett.*, **86**, p. 154107.
19. Thoma, R. E., Weaver C. F., Friedman, H. A., Insley, H., Harris, L. A., Yakel, H. A. (1961). Phase equilibria in the system LiF-YF3, *J. Phys. Chem.*, **65**, pp. 1096–1099.
20. Tiepolo, B., Bottazzi, P., Palenzona, M., Vannucci, R. (2003). A laser probe coupled with IPC - double focusing sector: field mass spectrometer in situ analysis of geological samples and U-Pb dating of zircon. *Can. Mineral*, **41**, pp. 259–272.
21. Volpi, A., Di Lieto, A., Tonelli, M. (2015). Novel approach for solid state cryocoolers, *Opt. Express*, **23**, pp. 8216–8226.

Chapter 12

Microscopic Theory of Optical Refrigeration of Semiconductors

Rolf Binder and Nai-Hang Kwong
College of Optical Sciences, University of Arizona, Tucson, AZ 85721, USA
binder@optics.arizona.edu

This chapter is intended to give an overview of some theoretical efforts to analyze optical refrigeration of inorganic semiconductor crystals. A fermionic theory describing the conduction and valence bands, combined with Coulomb interaction between the excited charge carriers, yields the absorption and luminescence spectra that include excitonic effects valid for arbitrary excitonic ionization ratios. The microscopic theory used as an input to the Sheik-Bahae and Epstein semiconductor cooling model yields a cooling analysis that has been evaluated for GaAs. Extensions of the basic cooling model for bulk crystals include light propagation and luminescence re-absorption effects and the effects of passivation layers in doped spatially inhomogeneous structures. While the numerical results presented here are restricted to GaAs, the general theoretical formulation is valid for any direct-gap crystalline semiconductor.

Laser Cooling: Fundamental Properties and Applications
Edited by Galina Nemova

ISBN 978-981-4745-04-8 (Hardcover), 978-981-4745-05-5 (eBook)
www.panstanford.com

12.1 Introduction

The conceptual foundation of optical refrigeration of solids was laid by Pringsheim [1]. Soon after the first major experimental break-through by Epstein et al. [2], who achieved net cooling of an Yb-doped ZBLANP glass, the study of the physics and applications of optical refrigeration of solids became a major worldwide research area, as documented by review articles, including those written by Sheik-Bahae and Epstein [3] and Nemova and Kashyap [4]. Works following the 1995 break-through also included research on the cooling of inorganic semiconductors, which are generally believed to hold promise for cooling to temperatures of about 10 K. The semiconductor research includes various theoretical aspects, for example those discussed by Oraevsky [5], Rivlin and Zadernovsky [6], Sheik-Bahae and Epstein [7], Apostolova [8], and Khurgin [9], as well as experimental studies, including, for example, the ones presented by Gauck et al. [10] and Imangholi et al. [11]. Important recent milestones are the observation of cooling to cryogenic temperatures of Yb-doped YLF crystals reported by Seletskiy et al. [12] and the cooling of nanostructured CdS crystals reported by Zhang et al. [13].

Inorganic semiconductors, such as GaAs or CdS crystals, are commonly characterized by their electronic band structure. In the vicinity of the fundamental bandgap, the band structure can be approximated by two bands, the conduction and the valence band (Fig. 12.1). However, optical transitions involved in absorption and photoluminescence processes are generally not well described by band-to-band transitions, owing to the strong Coulomb interaction between electrons in the conduction band and holes in the valence bands. The Coulomb interaction can give rise to sharp optical resonances called excitonic resonances, which correspond to the bound electron–hole pairs (excitons). Optical refrigeration of semiconductors is dominated by excitonic resonances, both in the absorption (pumping) and the upconverted (anti-Stokes) luminescence process. While at low excitation densities (low pumping intensities) the excitons can be described with simple Bosonic models, such descriptions are not appropriate for optical refrigeration, as the strong pumping gives rise to relatively large excitation densities and hence to the partial ionization of excitons.

A comprehensive and predictive theory must therefore be built upon a Fermionic description of electrons and holes and must be nonperturbative in the Coulomb interaction to allow for the modeling of excitons or, more precisely, a partially ionized exciton gas. We have developed such a microscopic theory and applied to the analysis of several questions important for the understanding of realistic experimental setups.

In this chapter, we will first review the microscopic theory and its basic results, which is the dominant role of excitons played in the cooling at low (few tens of Kelvin) temperatures. We then apply the theory to the description of realistic structures in which the propagation of light (in particular luminescence light) gives rise to effects such as luminescence re-absorption and finite luminescence extraction from the system. Furthermore, in many experiments the pump beam does not cover the entire cross section of the semiconductor crystal, and we study effects of finite lateral beam size on the cooling. Finally, we study more complex systems in which the semiconductor crystal is doped and which additionally may be sandwiched between doped passivation layers, which are intended to reduce nonradiative surface recombination.

12.2 Theoretical Foundation and the Importance of Excitonic Effects

We begin by summarizing our theoretical model for the optical absorption and photoluminescence spectra and their use in the cooling model of Sheik-Bahae and Epstein [7]. Details on the derivation of our theory can be found in Kwong et al. [14], as well as Rupper et al. [15, 16]. As mentioned above, an important question that our theory can answer is this: Given the fact that excitonic resonances dominate semiconductor absorption and luminescence spectra (especially at low temperatures), is the same true for the cooling characteristics? In other words, do we need excitonic effects to achieve cooling at low temperatures?

Starting from an electronic band structure model in which we have (possibly degenerate) valence (v) and conduction (c) bands, which are parabolic and isotropic, we use a microscopic Green's function theory that is nonperturbative in the Coulomb interaction between the charge carriers (electrons in the conduction band and holes in the valence band), most importantly between the electrons and holes, and hence carries full information about the excitonic resonances (bound states of electrons and holes) without treating the exciton as a separate quasi-particle (as is often done in Bosonic models for the excitons).

In this section we start out by analyzing a spatially homogeneous semiconductor structure, where in the presence of a partially ionized exciton gas the frequency-dependent absorption coefficient $\alpha(\omega)$ is given in terms the retarded interband susceptibility $\chi^{R}(q, \omega)$, evaluated in the long-wavelength approximation at approximately zero wavevector **q**,

$$\alpha(\omega) = \frac{4\pi\omega}{n_{b}c}\mathrm{Im}\left(\chi^{R}(\mathbf{q}=0,\omega)\right) \tag{12.1}$$

where n_b is the background dielectric index and c the speed of light in vacuum. Once the absorption coefficient is known, the Kubo–Martin–Schwinger (KMS) relation, Zimmermann [17], can be used to relate absorption to luminescence. In the present case of a bulk semiconductor the the KMS relation reads

$$R(\omega) = \left(\frac{\omega n_{b}}{\pi c}\right)^{2} g(\omega)\alpha(\omega) \tag{12.2}$$

This relation is valid if the system is in quasi-thermal equilibrium at temperature T. Here $R(\omega)$ is the luminescence spectral density, $g(\omega)$ is the Bose function,

$$g(\omega) = \frac{1}{e^{(\hbar\omega - \mu_{e} - \mu_{h})/k_{B}T} - 1} \tag{12.3}$$

where μ_a is the chemical potentials of electrons (a = e) and holes (a = h), respectively, and μ_B is the Boltzmann constant. If, instead of a bulk semiconductor, we have a semiconductor quantum well, the KMS relation reads [18]

$$R(\omega) = \frac{2}{3}\left(\frac{\omega n_{b}}{\pi c}\right)^{2} g(\omega)\mathrm{Im}\sigma^{R}(\omega) \tag{12.4}$$

where the retarded response function $\mathrm{Im}\sigma^{R}(\omega)$ is approximately given by the absorbance $A_{OW}(\omega)$ of the quantum wells, which is defined in terms of reflectance and transmittance as

$$A_{QW}(\omega) \equiv 1 - \frac{I_r}{I_0} - \frac{I_t}{I_0} \approx Im\sigma^R(\omega) \tag{12.5}$$

where I_0, I_r, and I_t are the incident, reflected, and transmitted intensities, respectively.

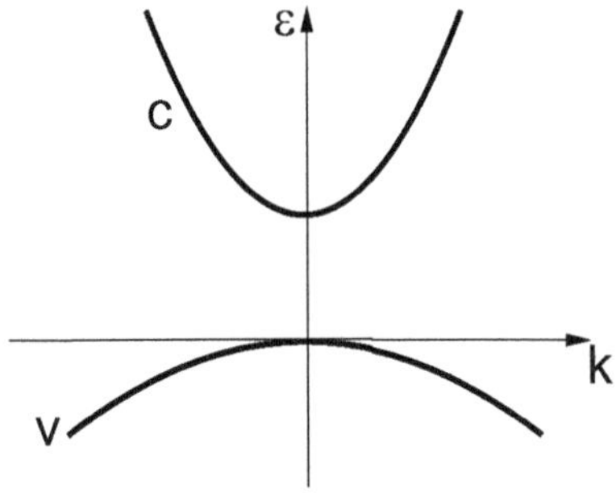

Figure 12.1 Sketch of parabolic band structure (energy versus wavevector) for valence (v) and conduction (c) bands.

The retarded interband susceptibility entering Eq. 12.1 follows from an integral equation that, in our approximation, corresponds to the so-called T matrix approximation, which is a nonperturbative infinite summation of ladder diagrams. It reads

$$\chi^R(\mathbf{q},\omega) = \chi^{0R}(\mathbf{q},\omega) - \sum_{sj\vec{k}\vec{k}'} d_{js}(\mathbf{k}) g_{sj}^{0R}(\mathbf{k},\mathbf{q},\omega) T_{sj}^R(\mathbf{k},\mathbf{k}',\mathbf{q},\omega) g_{sj}^{0R}(\mathbf{k}',\mathbf{q},\omega) d_{js}(\mathbf{k}') \tag{12.6}$$

where the summation is of the band indices labeling the spin-degenerate conduction bands (index s) and spin-degenerate valence bands (index j), and over the wavevectors $\mathbf{k}$ and $\mathbf{k}'$. The susceptibility Eq. 12.6 describes the propagation of electron–hole pairs excited by photon absorption. In the "free" part $\chi^{0R}(q, \omega)$, the electron and hole propagate independently of each other, and the T matrix T_{sj}^R represents interactions between the two. Explicitly,

$$\chi^{0R}(\mathbf{q},\omega) = -\sum_{sj\vec{k}} d_{js}(\mathbf{k}) g_{sj}^{0R}(\mathbf{k},\mathbf{q},\omega) d_{js}(\mathbf{k}) \tag{12.7}$$

where

$$d_{js}(k) \tag{12.8}$$

denotes the interband dipole matrix element (for details of this and related matrix elements see Gu et al. [19]). The imaginary and real parts of the free two-particle Green's function are given by

$$\mathrm{Im}\left\{g_{sj}^{0\mathrm{R}}(\mathbf{k},\mathbf{q},\omega)\right\} = -\frac{1}{2\hbar}\int\frac{d\omega'}{2\pi}\left\{1 - f_s(\omega-\omega') - f_j(\omega')\right\} \times A_s(\mathbf{k}_s,\omega-\omega')A_j(\mathbf{k}_j,\omega') \tag{12.9}$$

and

$$\mathrm{Re}\left\{g_{sj}^{0\mathrm{R}}(\mathbf{k},\mathbf{q},\omega)\right\} = P\int\frac{d\omega'}{\pi}\frac{\mathrm{Im}\left\{g_{sj}^{0\mathrm{R}}(\mathbf{k},\mathbf{q},\omega')\right\}}{\omega'-\omega} \tag{12.10}$$

where

$$f_\mathrm{a}(\omega) = \frac{1}{e^{(\hbar\omega-\mu_\mathrm{a})/k_\mathrm{B}T}+1} \tag{12.11}$$

are the Fermi functions for electrons (a = e) and holes (a = h), and the spectral functions are given by

$$A_\mathrm{a}(\mathbf{k},\omega) = -2\mathrm{Im}\left\{G_\mathrm{a}^\mathrm{R}(\mathbf{k},\omega)\right\} \tag{12.12}$$

In its present stage, our theory does not include a microscopic description of phonons and electron–phonon coupling. However, for optical refrigeration it is important to account for excitonic line-broadening effects since the optical pump beam is usually chosen to be in resonance with the low-frequency tail of the lowest exciton resonance. We therefore include a phenomenological amendment to the spectral function that accounts for such line-broadening effects. The temperature dependence of the line broadening due to coupling of electrons to longitudinal acoustic (LA) and longitudinal optic (LO) phonons is approximately given by

$$\gamma_\mathrm{ph}^0(T) = \gamma_\mathrm{LA}T + \frac{\gamma_\mathrm{LO}}{e^{\Omega_\mathrm{LO}/k_\mathrm{B}T}-1} \tag{12.13}$$

This yields the following phonon-broadened spectral functions for electrons and holes:

$$A_\mathrm{a}^\mathrm{ph}(\mathbf{k},\omega) = \int\frac{d\omega'}{2\pi}A_\mathrm{a}(\mathbf{k},\omega-\omega')\frac{2}{\pi\gamma_\mathrm{ph}^0\cosh(2\hbar\omega'/\gamma_\mathrm{ph}^0)} \tag{12.14}$$

The retarded (superscript R) single-particle Green's function is given by

$$G_{\mathrm{a}}^{\mathrm{R}}(\mathbf{k},\omega)=\frac{\hbar}{\hbar\omega-\varepsilon_{\mathrm{a}}(\mathbf{k})-\Sigma_{\mathrm{a}}^{\mathrm{R,T}}(\mathbf{k},\omega)-\Sigma_{\mathrm{a}}^{\mathrm{R,exc}}(\mathbf{k})-\Sigma_{\mathrm{a}}^{\mathrm{R,CH}}(\mathbf{k})} \tag{12.15}$$

where the unrenormalized single-particle energies (i.e., the parabolic band structure, parametrized in terms of effective masses m_{a}) is given by

$$\varepsilon_{\mathrm{a}}(\mathbf{k}_{\mathrm{a}})=\frac{\hbar^2|\mathbf{k}_{\mathrm{a}}|^2}{2m_{\mathrm{a}}} \tag{12.16}$$

and the real and imaginary parts of the T matrix (superscript T) self-energies (see Kwong et al. [14] and also Schmielau et al. [20] and Piermarocchi and Tassone [21]) are

$$\mathrm{Re}\left\{\Sigma_a^{\mathrm{R},T}(\mathbf{k},\omega)\right\}=P\int\frac{d\omega'}{\pi}\frac{\mathrm{Im}\left\{\Sigma_a^{\mathrm{R},T}(\mathbf{k},\omega')\right\}}{\omega'-\omega} \tag{12.17}$$

and

$$\begin{aligned}\mathrm{Im}\left\{\Sigma_a^{\mathrm{R},T}(\mathbf{k},\omega)\right\}=\sum_{a'\vec{k}'}\int\frac{d\omega'}{2\pi}\left\{g(\omega+\omega')+f_{a'}(\omega')\right\}\\ \times\mathrm{Im}\left\{T_{aa'}^{\prime\prime\mathrm{R}}(\mathbf{p},\mathbf{p},\mathbf{q},\omega+\omega')\right\}A_{a'}(\mathbf{k}',\omega')\end{aligned} \tag{12.18}$$

where $\mathbf{q}=\mathbf{k}+\mathbf{k}'$ and

$$\mathbf{p}=\frac{m_{a'}\mathbf{k}-m_a\mathbf{k}'}{m_a+m_{a'}} \tag{12.19}$$

Furthermore, the real-valued exchange (superscript "exc") self-energy is given by

$$\Sigma_a^{\mathrm{R,exc}}(\mathbf{k})=-\sum_{\mathbf{k}'}W_{aa}(\mathbf{k}-\mathbf{k}')\int\frac{d\omega}{2\pi}f_a(\omega)A_a(\mathbf{k},\omega) \tag{12.20}$$

and the density of the charge carriers follows from

$$n_a=\sum_{\mathbf{k}}\int\frac{d\omega}{2\pi}f_a(\omega)A_a(\mathbf{k},\omega) \tag{12.21}$$

which is an implicit equation for the chemical potential μ_a. The Coulomb hole (superscript CH) self-energy is

$$\Sigma_{\mathrm{a}}^{\mathrm{R,CH}}=-\sum_{\mathbf{k}}\left(W_{\mathrm{aa}}(\mathbf{k})-V(\mathbf{k})\right) \tag{12.22}$$

The "ladder" equation for the T matrix reads

$$T^{\mathrm{R}}_{aa'}(\mathbf{k},\mathbf{k}',\mathbf{q},\omega)=W_{aa'}(\mathbf{k}-\mathbf{k}')+\sum_{\mathbf{k}''}W_{aa'}(\mathbf{k}-\mathbf{k}'',\omega)g^{0\mathrm{R}}_{a}(\mathbf{k}'',\mathbf{q},\omega)T^{\mathrm{R}}_{aa'}(\mathbf{k}'',\mathbf{k}',\mathbf{q},\omega) \quad (12.23)$$

and the quasi-statically screened potential is (with e being the charge of the electron in vacuum)

$$W_{\mathrm{aa}'}(\mathbf{q})=z_{\mathrm{a}}z_{\mathrm{a}'}\frac{4\pi e^2}{\varepsilon_{\mathrm{b}}(q^2+\kappa(q)^2)} \quad (12.24)$$

where $\varepsilon_{\mathrm{b}}=n_{\mathrm{b}}^2$, $\kappa(q)$ is the inverse screening length, and z_{a} the sign for the charge (−1 for electrons and +1 for holes). The unscreened potential is $V(q) = 4\pi e^2/(\varepsilon_{\mathrm{b}}q^2)$. Details on how the effective potential can be treated in a numerical solution of the equation set can be found in Kwong et al. [14]. Note that the above set of equations constitutes a self-consistency problem, as the self-energies depend on the spectral function, but the spectral functions in turn depend on the self-energies.

Once the luminescence spectral density is known, we can use the cooling model developed by Sheik-Bahae and Epstein [7]. We define the radiative recombination coefficient B (for $n_{\mathrm{e}} = n_{\mathrm{h}} \equiv n$) via

$$L(n)=B(n)n^2=\int\frac{d\omega}{2\pi}R(\omega) \quad (12.25)$$

and use the rate equation for the excitation density evaluated under stationary conditions,

$$\frac{dn}{dt}=\frac{\alpha(\omega_a,n)}{\hbar\omega_a}I_{\mathrm{pump}}-An-\eta_{\mathrm{e}}L(n)-Cn^3=0 \quad (12.26)$$

where I_{pump} is the pump intensity (more precisely irradiance), ω_a the frequency of the pump field, A the nonradiative recombination coefficient, η_{e} the extraction efficiency of the luminescence (i.e., the fraction of the luminescence that leaves the crystal), and C the Auger coefficient. Equation 12.26, together with the power balance equation for the net power extracted from the system,

$$P_{\mathrm{net}}=I_{\mathrm{pump}}\left[\alpha(\omega_{\mathrm{a}},n)+\alpha_{\mathrm{b}}\right]-\hbar\omega_{\mathrm{l}}\eta_{\mathrm{e}}L(n) \quad (12.27)$$

where α_{b} is the parasitic background absorption coefficient and

$$\omega_\ell = \frac{\int d\omega\, \omega\, R(\omega)}{\int d\omega\, R(\omega)} \tag{12.28}$$

the mean luminescence frequency, constitutes the basic framework of the Sheik-Bahae–Epstein model.

The break-even nonradiative recombination coefficient A_b, which marks the boundary where cooling becomes possible, is defined by $P_{net}(A_b) = 0$. Fixing all but two parameters, leaving only the experimentally easily adjustable parameters ω_a and n_e as variables, we optimize the break-even condition in the two-dimensional space of ω_a and n_e, resulting in the optimized break-even nonradiative coefficient

$$\bar{A} \equiv \max\left\{ A(\omega_a, n_e)\big|_{P_{net}=0} \right\} \tag{12.29}$$

Figure 12.2 shows the results for the optimized breakeven non-radiative lifetime (defined as the inverse of *A) as a function of temperature. The solid and dotted lines show, respectively, the results of the full calculation and the second-order approximation using the free single-particle Green's function G^0 in the T matrix self-energies. The overall temperature dependence is similar for these two curves (the luminescence spectra in the full T matrix and second-order approximations are quite different, but the resulting differences in the values of the optimal density and radiative recombination coefficient largely compensate each other, leading to similar values for *τ_b), suggesting that many-body correlation effects beyond the T matrix approximation would probably not change the results significantly. In the so-called plasma theory model, the T matrix self-energy is completely neglected, but the excitonic effects in the susceptibility are still taken into account at a nonperturbative level. In other words, in that model the exciton resonance is accounted for in the susceptibility, but the many-particle processes that renormalize the exciton resonance are reduced to pure electron–hole plasma effects. We see that, qualitatively, plasma theory yields results similar to the full theory. Only the model where all Coulomb interactions are neglected, called "free" model, yields qualitatively different results. We see from Fig. 12.2 that the existence of the exciton resonance is the main factor that can make cooling at low temperatures possible. More details can be found in Refs. [14–16].

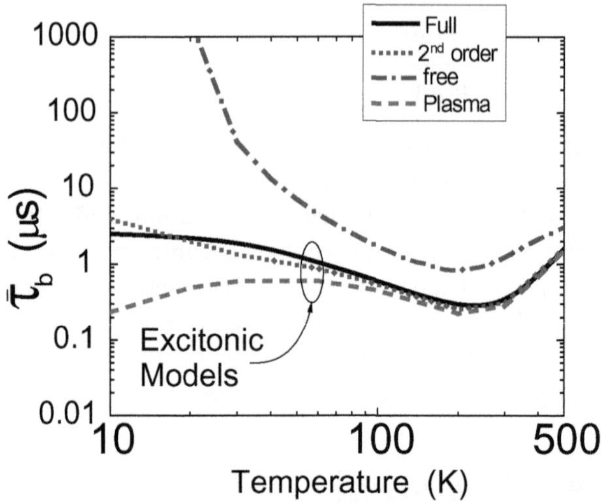

Figure 12.2 Optimized break-even nonradiative lifetime versus temperature for several different theories. For each model, the predicted cooling region is above its curve. Reprinted from Ref. [16], Copyright 2007, by the American Physical Society.

12.3 Effect of Luminescence Propagation and Re-Absorption

In the previous section, we assumed the system to be a perfect infinite crystal. Furthermore, we did not account for luminescence re-absorption and imperfect luminescence extraction in the microscopic theory. Rather, we included the luminescence extraction efficiency η_e as a phenomenological parameter in the cooling model. In the following, we generalize the theory to include light energy transport, luminescence re-absorption, and surface or interface reflection. A typical configuration including the semiconductor sample and a transparent dome, intended to maximize the extraction efficiency (see, for example, Sheik-Bahae and Epstein [7]) is shown in Fig. 12.3.

Light energy transport theory is on the same footing as the microscopic absorption theory. In addition to the electron Green's function, we now have a photon Green's function D, which is renormalized by the presence of the interband polarization.

Evaluated at equal times, the photon Green's function yields the photon Wigner function,

$$n_{\mathbf{q}\sigma}(\mathbf{r},t)=\sum_{\mathbf{k}}\langle c^{\dagger}_{\mathbf{q}-\mathbf{k}/2,\sigma}\,c_{\mathbf{q}+\mathbf{k}/2,\sigma}\rangle e^{i\mathbf{k}\cdot\mathbf{r}} \tag{12.30}$$

where σ denotes the polarization state of the photon density, and $c(c^{*})$ $c(c^{\dagger})$ are the photon annihilation (creation) operators. The photon Wigner function determines the Poynting vector,

$$\begin{aligned}\mathbf{S}(\mathbf{r},t)&=\frac{c}{4\pi}(\mathbf{E}\times\mathbf{H})=\frac{1}{V}\sum_{\mathbf{q}\sigma}\frac{c}{n_b}\,\hat{q}\,\hbar\omega_q\,\mathrm{Re}(n_{\mathbf{q}\sigma}(\mathbf{r},t))\\&=\sum_{\mathbf{q}\sigma}\int d\omega\,\mathbf{J}_{\mathbf{q}\sigma\omega}(\mathbf{r},t)\end{aligned} \tag{12.31}$$

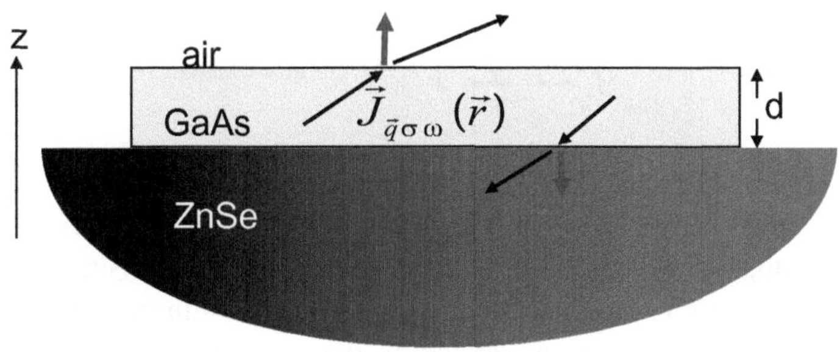

Figure 12.3 Schematic of the geometry analyzed with the spectral energy flux density. In the calculations, we assumed the lateral extension to be infinite and as a consequence the net energy flow to be restricted to positive and negative z directions (thick blue vertical arrows). The extension in the z direction was chosen to be 1 μm, and the partial or total reflection at the GaAs/air and GaAs/ZnSe interfaces were taken into account according to Fresnel's equations.

In the last step, we defined the spectral density of the energy flux $J_{\mathbf{q}\sigma\omega}(r,t)$, which obeys the following transport equation:

$$\left\{\frac{\partial}{\partial t}+\mathrm{v}_{\mathbf{q}\,\omega}\,\hat{q}\cdot\nabla-\mathrm{Im}\left(\Pi^{R}_{\mathbf{q}\,\omega}\right)\right\}\mathbf{J}_{\mathbf{q}\sigma\omega}(\mathbf{r},t)=-\mathbf{q}\,\mathrm{Im}\left(D^{\mathrm{R}}_{\mathbf{q}\,\omega}\right)i\Pi^{<}_{\mathbf{q}\,\omega} \tag{12.32}$$

Here, the second term in the parenthesis on the left-hand side accounts for spatial propagation in the $\hat{q}$ direction with generalized group velocity $v_{\mathbf{q}\omega}$, the third term accounts for luminescence reabsorption, and the right-hand side represents the luminescence source. The photon spectral function $D^{\mathrm{R}}_{\mathbf{q}\omega}$ is usually taken to be a delta function peaked at the free-photon dispersion, $D^{\mathrm{R}}_{\mathbf{q}\omega} \approx \delta(\omega - \omega_{\mathbf{q}})$, with $\omega_{\mathbf{q}} = c\mathbf{q}/n_{\mathrm{b}}$. We will come back to a more general photon spectral function, which includes effects of excitons–polaritons, below.

The numerical solution of the photon transport equation allows us to determine the escaped luminescence rate entering the rate equation of the excitation density, Eq. 12.26,

$$\eta_{\mathrm{e}}\, B(n)\, n^2 = \frac{1}{d} \sum_{z_{\mathrm{surface}}} \sum_{\mathbf{q}\sigma} \int d\omega \frac{1}{\hbar\omega_{\mathbf{q}}}\, \hat{z}_{\mathrm{S}} \cdot \mathbf{J}_{\mathbf{q}\sigma\omega}(z_{\mathrm{surface}})\, t_{\mathbf{q}\sigma\omega}(z_{\mathrm{surface}}) \tag{12.33}$$

and the escaped luminescence energy entering the corresponding rate equation for the net power, Eq. 12.27,

$$\hbar\omega_{\mathrm{l}}\, \eta_{\mathrm{e}}\, B(n)\, n^2 = \frac{1}{d} \sum_{z_{\mathrm{surface}}} \sum_{\mathbf{q}\sigma} \int d\omega\, \hat{z}_{\mathrm{S}} \cdot \mathbf{J}_{\mathbf{q}\sigma\omega}(z_{\mathrm{surface}})\, t_{\mathbf{q}\sigma\omega}(z_{\mathrm{surface}}) \tag{12.34}$$

Here, $t_{\mathbf{q}\sigma\omega}(Z_{\mathrm{surface}})$ is the transmission coefficient according to Fresnel's equations. With the solution of the photon transport equation, we can now predict the value of the extraction efficiency.

In addition to the important effects of luminescence re-absorption and interface reflection, we have also addressed the issue of exciton–polariton propagation. It is frequently assumed that exciton–polariton effects play a negligible role in the cooling theory. In the following we investigate this issue within the limits of photon Wigner function theory. Within this theory, polariton effects enter locally; at each point in space the retarded photon Green's function $D^{\mathrm{R}}_{\mathbf{q}\omega}(\mathbf{r})$ can exhibit anti-crossing between exciton states and photon states. An example is shown in Fig. 12.4 for negligible excitation density and a very low temperature of $T = 5$ K. We chose a temperature this low in order to demonstrate the principle existence of polariton effects (anticrossing). At higher temperatures (we usually investigate cooling between 10 K and 500 K) and higher densities, this effects is practically absent, because the exciton linewidth exceeds the longitudinal-transverse splitting (the relevant energy scale for the anticrossing).

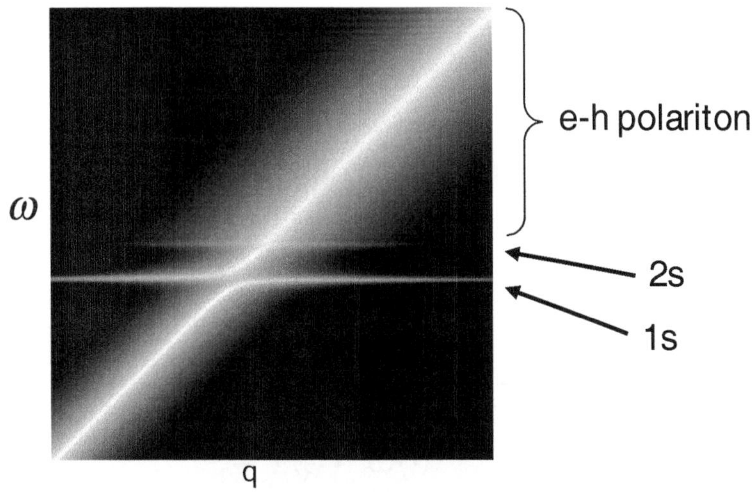

Figure 12.4 Exciton–polariton dispersion at a temperature of 5 K, as seen in the peaks of the logarithm of the photon spectral function $[-2\mathrm{Im}\,D^{\mathrm{R}}_{\mathbf{q}\omega}]$. The anticrossing between the 1s exciton and the photon dispersion is clearly visible, and to a lesser extent even that of the 2s exciton and the photons. In the e–h continuum, the polariton dispersion is dominated by that of free photons. The (quadratic) spatial dispersion of the 1s and 2s excitons cannot be seen in this figure because the chosen $\mathbf{q}$ range is very small.

In our photon transport equation model, the boundary conditions determining the interface reflection and transmission are solved separately for each $(\omega, \mathbf{q})$ value (which is an approximation to the proper boundary conditions). One can assume that luminescence from the lower polariton branch is suppressed at large wavevectors. As far as escaped luminescence is concerned, exactly this happens in our model, as can be seen by comparing Figs. 12.5 and 12.6. We see clearly that the luminescence observed outside the semiconductor is insensitive to polariton effects, even at temperatures as low as 5 K and at negligible excitation density. An effect much more important is the luminescence re-absorption, which is also shown for the example of $T = 5$ K in Fig. 12.6.

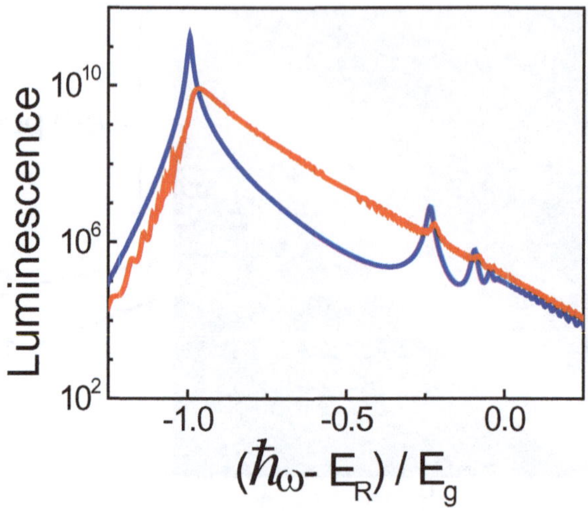

Figure 12.5 Photoluminescence spectra at $T = 5$ K as hypothetically seen inside the semiconductor with (red curve) and without (blue curve) polariton effects. The polariton luminescence contains a large component from the lower polariton branch (at large wavevectors), which however are not transmitted to the outside (see Fig. 12.6).

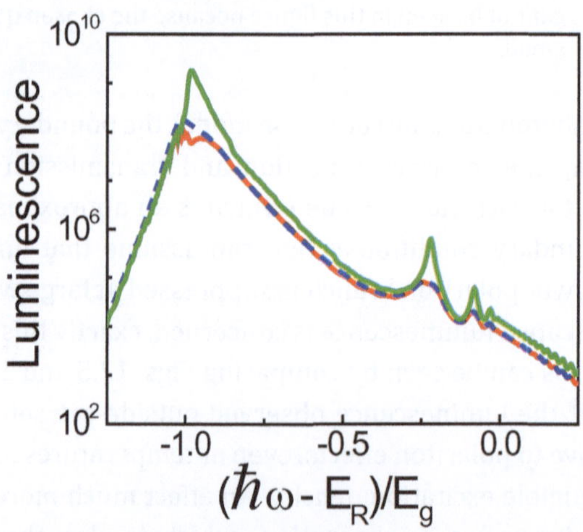

Figure 12.6 Photoluminescence spectra at $T = 5$ K as seen outside the semiconductor with (red solid curve) and without (blue dashed curve) polariton effects (these two curves are practically indistinguishable). The green curve shows results neglecting luminescence re-absorption.

To further extend our analysis, we combine now our theory of luminescence from p-doped semiconductors, see Rupper et al. [22, 23], with the light transport theory outlined above. We formulate the theory of luminescence from a partially ionized exciton gas with the additional luminescence channel between acceptors and conduction band electrons. Including this channel, the spectrally integrated escaped luminescence can be written as

$$L_{\text{outside}} = \eta_e B(n_e) n_e (n_e + N_a) \tag{12.35}$$

and the corresponding spectrally integrated escaped luminescence energy is

$$\hbar\omega_l \eta_e B(n_e) n_e (n_e + N_a) \tag{12.36}$$

Here, N_a is the acceptor density and n_e the optically excited electron density. Using the numerical solution for the spectral energy flux density, we can compute the extraction efficiency and the minority carrier lifetime, the latter being defined as $\tau^{-1} \approx L_{\text{outside}}/n_e$.

Figure 12.7 shows the calculated extraction efficiency. The decrease of η_e with decreasing temperature is expected and can be attributed to the spectral sharpening of the exciton resonance, which in turn leads to an increase of luminescence re-absorption. The sharp rise of η_e at temperatures below 30 K may seem surprising. It can be explained by the onset of electron-to-acceptor luminescence (acceptor freeze-out), because this luminescence is spectrally below the exciton resonance and thus suffers less from re-absorption.

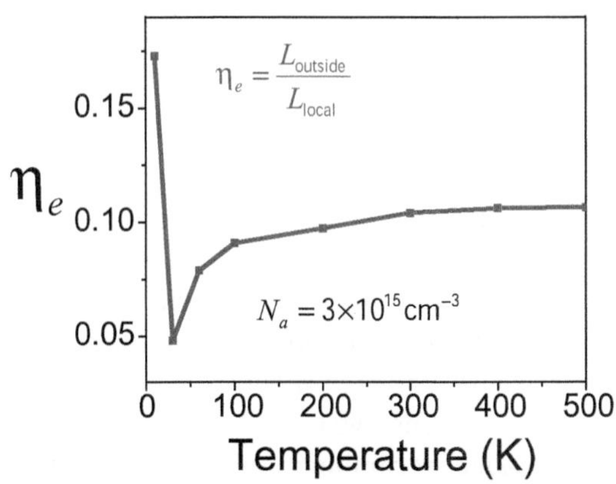

Figure 12.7 Calculated luminescence extraction efficiency for a p-doped structure.

Figure 12.8 shows calculated results for the minority carrier lifetime. These results are in good agreement with the experimental findings from the University of New Mexico team (see Imangholi et al. [11]). Their measurements of the lifetime τ yield the same general trends as a function of temperature (see Fig. 12.8). For example at T = 500 K, the measured value is 8.3 μs, which is very close to our value calculated for a similar structure.

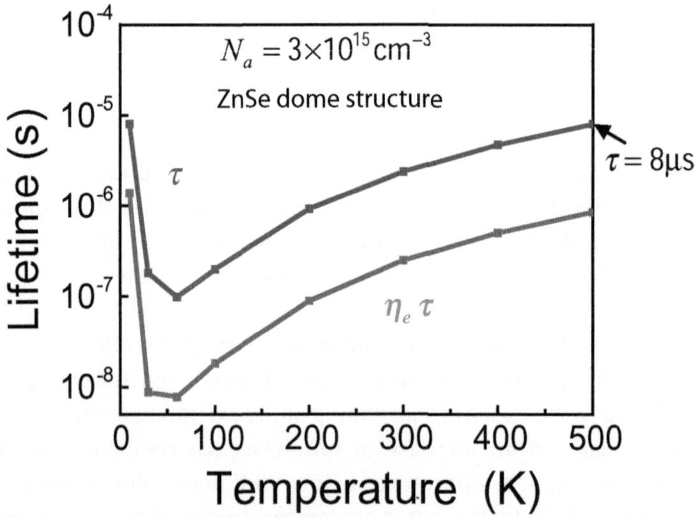

Figure 12.8 Calculated minority carrier lifetime for a p-doped GaAs structure with a ZnSe dome.

12.4 Finite Spatial Beam Profiles

In the following we study the effect of radial spatial photon transport (see also Rupper et al. [24]). In typical experiments, the beam profile is smaller than the extension of the semiconductor sample. For example, the beam profile can be Gaussian (with azimuthal symmetry) and the sample can be disk-like. Typical numbers for the beam waist are 100 μm with a disk diameter of 1000 μm or larger (see Fig.12.9). The cooling process is sensitive to the carrier density. If the density is too large, then Auger recombination and/or band filling prevents cooling. If the density is too low, then nonradiative recombination prevents cooling. As such, the spatially inhomogeneous optical excitation can be expected to have negative consequences for the cooling process. Since the excitation power drops away from the beam center, the cooling threshold can be expected to be reached only in a small fraction within the beam. Hence it would be possible that the sample cools in that narrow region, while it heats in the beam's radial wing. This effect is expected to be exacerbated by the re-absorption of the luminescence propagating (laterally) inside the sample. As is schematically shown Fig. 12.10, a fraction η_e (the extraction efficiency) of the total luminescence generated at a given position inside the sample is able to escape from the sample. The remainder, that is, the fraction $1 - \eta_e$ of the luminescence, propagates laterally inside the sample. If we assume the sample's lateral extension to be much larger than the absorption length (=inverse absorption coefficient) of the emitted light, the laterally propagating luminescence light will be completely re-absorbed. Depending on the specific spatial profile of the re-absorption, this process can also be expected to hinder cooling. Also, lateral charge carrier diffusion can affect the spatial density and power profile, and thus influence the cooling process, but we have not yet addressed this issue in our studies.

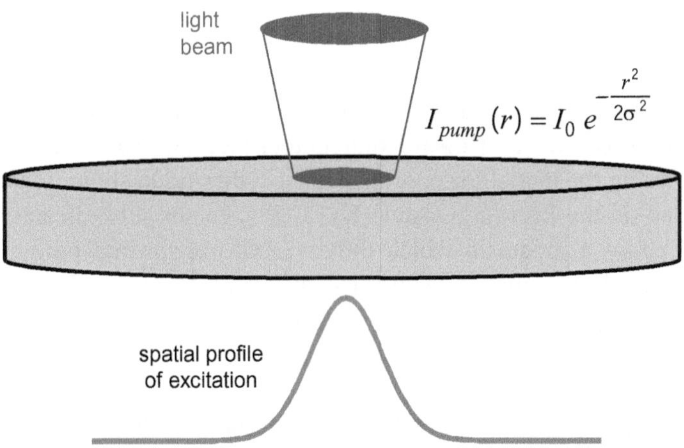

Figure 12.9 Sketch of excitation geometry with a disk-like GaAs sample and a Gaussian beam. Reprinted with permission from Ref. [24], Copyright 2009, Society of Photo Optical Instrumentation Engineers.

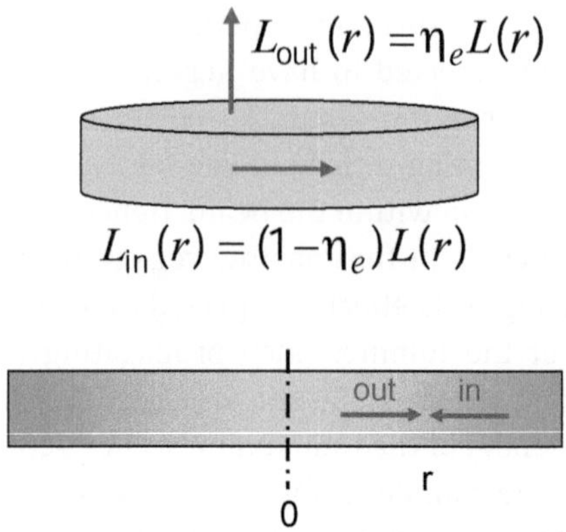

Figure 12.10 (Top sketch) Sketch of luminescence propagation. A fraction η_e of the luminescence produced at a given position r in the sample escapes from the sample, whereas the rest $(1 - \eta_e)$ propagates laterally in the sample and is re-absorbed for sufficiently laterally extended samples. (Bottom sketch) In our model, the re-absorption at position r has two contributions from lateral photon propagation, one from luminescence propagation laterally outward and the other laterally inward. Reprinted with permission from Ref. [24], Copyright 2009, Society of Photo Optical Instrumentation Engineers.

We assume that at each point in the structure, the fraction $1 - \eta_e$ of the locally generated luminescence propagates inside the medium. For simplicity, we assume the propagation direction to be restricted to the radial direction, either outwards (away from the center) or inwards (toward the center) (see Fig. 10, bottom). We are using the model for photon transport discussed above, which is based on the Poynting vector (Eq. 12.31). To simplify calculations, we adopt a model in which photons are transported only along the radial direction (in cylindrical coordinates), that is, $\mathbf{J}_{\mathbf{q},\sigma,\omega}(\boldsymbol{r})$ is nonzero only when $\mathbf{q} \parallel \hat{\mathbf{e}}_r$ (outward flow) or $\vec{q} \parallel -\hat{\mathbf{e}}_r$ (inward flow), where $\hat{\mathbf{e}}_r$ is the radial unit vector. We write

$$\mathbf{J}_{\mathbf{q},\sigma,\omega}(\mathbf{r}) = \delta(q_z)\left[J^{\mathrm{out}}_{\mathbf{q},\sigma,\omega}(r)\delta(\phi_{\mathbf{q}}-\phi) - J^{\mathrm{in}}_{\mathbf{q},\sigma,\omega}(r)\delta(\phi_{\mathbf{q}}-\phi-\pi)\right]\hat{\mathbf{e}}_r \tag{12.37}$$

(where $\phi_{\mathbf{q}}$ is the azimuthal angle of $\mathbf{q}$). The transport equation, including the luminescence source term and the luminescence re-absorption term, with $\alpha(r,\omega) \equiv \alpha(n(r),\omega)$, reads

$$\left\{\pm\frac{1}{r}\frac{\partial}{\partial r}r+\alpha(r,\omega)\right\}J^{\text{out/in}}_{\mathbf{q},\sigma,\omega}(r)=s_{\mathbf{q},\omega,\sigma}(r) \tag{12.38}$$

where the upper sign is for J^{out} and the lower sign for J^{in}, respectively.

Using the boundary conditions $J^{\text{in}}_{\mathbf{q},\sigma,\omega}(\infty)=0$ and $J^{\text{out}}_{\mathbf{q},\sigma,\omega}(0)=J^{\text{in}}_{\mathbf{q},\sigma,\omega}(0)$ (which accounts for the fact that inward flow passes through the origin and then contributes to outward flow), we find the outgoing flux to be

$$J^{\text{out}}_{\mathbf{q}\sigma\omega}(r)=\frac{1}{r}\int_0^{\infty}dr'r's(r',q,\omega)\left(\theta(r-r')e^{-\int_{r'}^{r}dr''\alpha(r'',\omega)}+e^{-\int_{-r'}^{r}dr''\alpha(|r''|,\omega)}\right) \tag{12.39}$$

and the incoming flux is

$$J^{in}_{\mathbf{q}\sigma\omega}(r)=\frac{1}{r}\int_r^{\infty}dr'r's(r',q,\omega)e^{-\int_r^{r'}dr''\alpha(r'',\omega)} \tag{12.40}$$

The luminescence source $s_{\mathbf{q},\omega,\sigma}(r)$ is normalized such that the total in-plane (lateral) luminescence (i.e., $s_{\mathbf{q},\omega,\sigma}(r)$ summed/integrated over all wavevectors, frequencies, and polarization states) yields the fraction 1 – η_e of the globally generated luminescence (integrated over all space). We have modified the rate equations entering the Sheik-Bahae–Epstein model by including the re-absorption calculated with the lateral photon transport model. The modified rate equation for the density profile reads

$$\frac{\alpha(\omega_a,n(r))}{\hbar\omega_a}I_{\text{pump}}(r)-An(r)-B(n(r))n(r)^2-Cn(r)^3+Q_n(r)$$
$$=\frac{dn(r)}{dt}=0 \tag{12.41}$$

Note that the luminescence term (Bn^2) is not multiplied by the extraction efficiency. Instead, we include a source term that accounts for the re-absorption of carriers due to lateral luminescence propagation, which is given in terms of the photon flux and the absorption coefficient,

$$Q_n(r)=\sum_{\mathbf{q},\sigma,\omega}\alpha(r,\omega)|\mathbf{J}_{\mathbf{q},\sigma,\omega}(\mathbf{r})| \tag{12.42}$$

Similarly, we have for the power balance equation

$$P_{\text{net}}(r) = I_{\text{pump}}(r)\left[\alpha(\omega_a, n(r)) + \alpha_b\right] - \hbar\omega_\ell(r)B(n(r))n(r)^2 + Q_e(r) \tag{12.43}$$

with the source term due to luminescence re-absorption

$$Q_e(r) = \sum_{\mathbf{q},\sigma,\omega} \alpha(r,\omega)\hbar\omega \,|\, \mathbf{J}_{\mathbf{q},\sigma,\omega}(\mathbf{r}) \,| \tag{12.44}$$

Figures 12.11 and 12.12 show results with and without the lateral photon transport included. We see that, apart from slight numerical noise, the results for the power is very similar in both cases, implying that lateral photon transport under typical experimental conditions should not be expected to prohibit cooling. As expected, the density profile in the case with lateral photon transport is smeared out beyond the region of optical excitation by the external light beam. However, the off-axis density produced by luminescence re-absorption is much smaller than the density excited directly by the external beam. This explains why the power profiles are very similar in the two cases (with and without lateral photon transport).

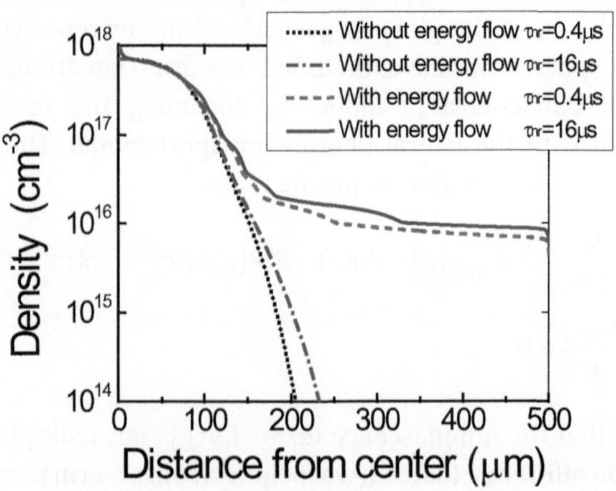

Figure 12.11 Spatial density profile for excitation with a Gaussian beam (σ = 100 μm) at temperature T = 300 K with spatial photon transport. Reprinted with permission from Ref. [24], Copyright 2009, Society of Photo Optical Instrumentation Engineers.

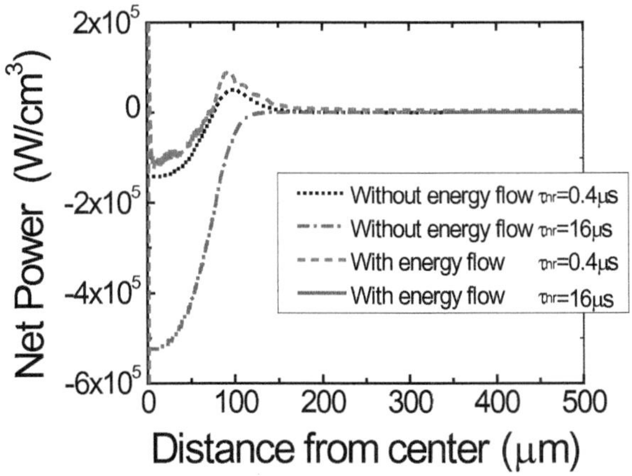

Figure 12.12 Spatial profile of the net power for excitation with a Gaussian beam (σ = 100 μm) at temperature T = 300 K with spatial photon transport. Reprinted with permission from Ref. [24], Copyright 2009, Society of Photo Optical Instrumentation Engineers.

12.5 Theory of Passivation Layers

So far, our theoretical descriptions treated only spatially homogeneous semiconductors, albeit under possibly inhomogeneous excitation conditions. In typical experiments, however, semiconductors are often heterostructures designed to reduce nonradiative recombination at the sample's surface. In particular, GaAs/GaInP structures have been used. In these structures, the GaAs layers are usually unintentionally p-doped, while the surface layers of GaInP are n-doped.

The spatial inhomogeneties that we address in the following section are those related to the doping profile of the semiconductor in the growth direction. We call these longitudinal inhomogeneties in order to distinguish them from the lateral spatial inhomogeneties discussed in the previous section. The structures used in the University of New Mexico experiments are doped structures, and many of them are of the form GaInP/GaAs/GaInP (see Fig. 12.13). The role of the GaInP layers is to minimize nonradiative surface recombination and hence act as a passivation layer. A critical question related to this structure concerns the spatial separation of electrons and holes. It had been known for some time that the band profile of this structure is of type II character, which means that the conduction band has a maximum in the GaAs region. In that case, the electrons tend to accumulate in the energetically more favorable GaInP layers, away from the holes that remain in the GaAs layer. Such a spatial charge separation would hinder photo-luminescence and hence could be detrimental for optical refrigeration. To address this issue, we first determine the spatial band profile and electron and hole density profiles across the n-p-n structures. With that knowledge, we use our microscopic photoluminescence and absorption theory to determine the spatial variation in the luminescence efficiency and the absorption spectra.

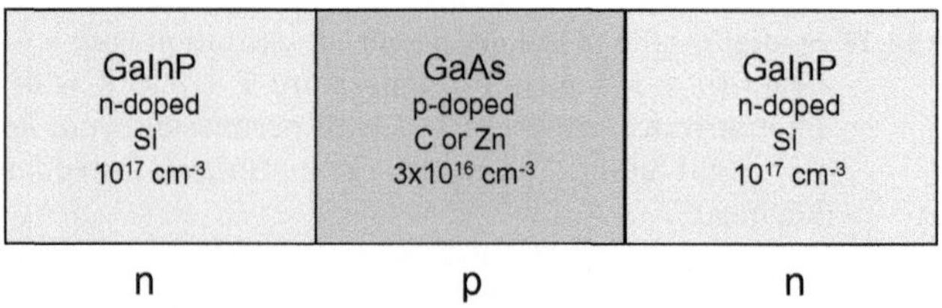

Figure 12.13 Schematic of the doping profile of the structure, which was used in the University of New Mexico experiments, see Imangholi et al. [11], with typical doping concentrations.

The theory for the band and density is an extension of the conventional textbook treatment of heterogeneous and doped semiconductor structures via self-consistent solutions for the Poisson equation and the carrier density profiles. While the conventional textbook treatment is restricted to structures in thermal equilibrium, in which charge carries can only originate from the doping, our generalization allows us to include optically excited carriers (Rupper et al. [22, 23, 25]). The fundamental set of equations consists of the space-dependent acceptor and donor occupation functions

$$f_{\mathrm{a}}(z)=\frac{D_{\mathrm{a}}}{D_{\mathrm{a}}+e^{\beta\left(\varepsilon_{\mathrm{a}}+eV(z)-\mu_{\mathrm{h}}-\varepsilon_{\mathrm{v}}^{0}(z)\right)}} \tag{12.45}$$

and

$$f_{\mathrm{d}}(z)=\frac{D_{\mathrm{d}}}{D_{\mathrm{d}}+e^{\beta(\varepsilon_{\mathrm{d}}-eV(z)-\mu_{\mathrm{e}}-\varepsilon_{\mathrm{c}}^{0}(z))}} \tag{12.46}$$

where $V(z)$ is the self-consistently computed electric potential, $\beta = 1/k_B T$ the inverse temperature, ε_a the acceptor and ε_d the donor energy, $D_{a/d}$ the degeneracies of the acceptor and donor levels, respectively, and $\varepsilon^0_{v/c}(z)$ the energy band profiles of the separated layers. The free electron density in the conduction band is given by the Fermi functions,

$$n_e(z) = 2\sum_{\vec{k}} \frac{1}{1+e^{\beta\left(\varepsilon_e(k)-eV(z)-\mu_e-\varepsilon_c^0(z)\right)}} \tag{12.47}$$

and similarly for the valence band holes,

$$n_h(z) = 2\sum_{\vec{k}} \frac{1}{1+e^{\beta\left(\varepsilon_h(k)+eV(z)-\mu_h-\varepsilon_v^0(z)\right)}} \tag{12.48}$$

The electric field in the GaAs layer is

$$E(z) = \frac{4\pi e}{\varepsilon_{\text{GaAs}}}\left(\int_{z_j}^{z} dz'\, n_{\text{charge}}(z') + \int_{z_e}^{z_j} dz'\, n_{\text{charge}}(z')\right) \tag{12.49}$$

and in the GaInP layer it is

$$E(z) = \frac{4\pi e}{\varepsilon_{\text{GaInP}}}\int_{Z_e}^{z} dz'\, n_{\text{charge}}(z') \tag{12.50}$$

This is determined by the charge density

$$n_{\text{charge}}(z) = n_h(z) + N_d^+(z) - n_e(z) - N_a^-(z) \tag{12.51}$$

with

$$\int dz\, n_{\text{charge}}(z) = 0 \tag{12.52}$$

where the concentration of ionized donors $N_d^+(z)$ follows from the donor occupation function and similarly $N_a^-(z)$ from the acceptor occupation function. Finally, the electric potential is given by

$$V(z) = -\int_0^z dz'\, E(z') \tag{12.53}$$

Figure 12.14 shows a flowchart of the numerical solution algorithm.

Figure 12.15 shows the band profile without the electric potential. It shows that the structure could be called a weak type I structure, in the sense that the potential wells for electron and holes are lined up (they are both in the middle layer), and thus no charge separation would be expected. However, the potential for the electrons is rather shallow, in that sense the type I characteristic could be called weak. Including the effects of the self-consistent electric potential, we see in Fig. 12.16 that a structure in thermal equilibrium is clearly a type II structure, in the sense the electrons and holes reside at different spatial locations. In this case, the electrons experience a potential barrier in the middle layer, and are thus pushed to the side layers, while the holes see a well in the middle layer, thus residing there. The situation changes when we consider an optically excited structure, such as occurs under optical refrigeration conditions. As we see from Fig. 12.16, optical excitation tends to flatten the bands, in this case especially the conduction band, which means that spatial charge separation due to a type II characteristic is not expected to prevail under optical excitation conditions.

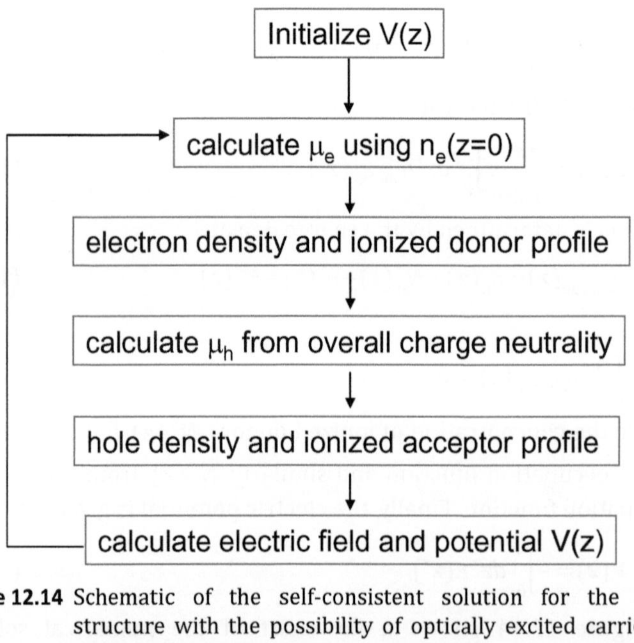

Figure 12.14 Schematic of the self-consistent solution for the n-p-n structure with the possibility of optically excited carrier for given temperature and center electron density $n_e(z = 0)$.

One of the main objective in using an structure with n-doped layers on each side is the reduction of non-radiative recombination. This type of recombination, which inhibits optical refrigeration, is often dominated by surface and interface recombination. The presence of the n-doped passivation layers almost removes the effect of nonradiative recombination at the surface (in this case the outside surface of the n-doped GaInP layers. There remains recombination taking place at the GaAs–GaInP interface, and in the following we discuss the properties of those recombination processes and possible control thereof. The control of nonradiative interface recombination is based on the general idea of keeping the minority carrier concentration in the vicinity of the interface as low as possible. Figure 12.17 shows a sketch of the electron and hole concentration and clarifies that, trivially, holes (electrons) are minority carriers in the n-doped (p-doped) segments. What is not immediately clear from this sketch is whether electrons or holes are minority carriers at the interface.

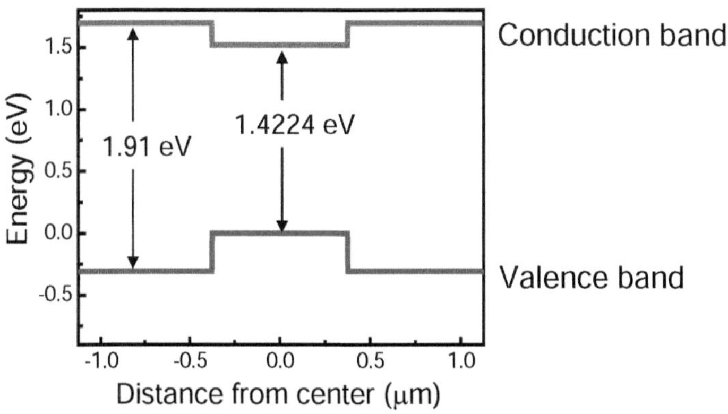

Figure 12.15 Energy band profile. The lower curve is for the valence band (hole density) and the upper curve for the conduction band (electron density). Reprinted with permission from Ref. [23], Copyright 2010, AIP Publishing LLC.

A simple model for nonradiative interface recombination is based on Shockley–Read–Hall processes:

$$\left.\frac{\partial n_\mathrm{e}(z)}{\partial t}\right|_\mathrm{nr} = \left.\frac{\partial n_\mathrm{h}(z)}{\partial t}\right|_\mathrm{nr} = -\sum_{j=1,2} \delta(z-z_j)\, S\, \frac{n_\mathrm{e}(z_j)n_\mathrm{h}(z_j)}{n_\mathrm{e}(z_j)+n_\mathrm{h}(z_j)}$$
$$\sim -\sum_{j=1,2} \delta(z-z_j)\, S\, n_\mathrm{min}(z_j) \qquad (12.54)$$

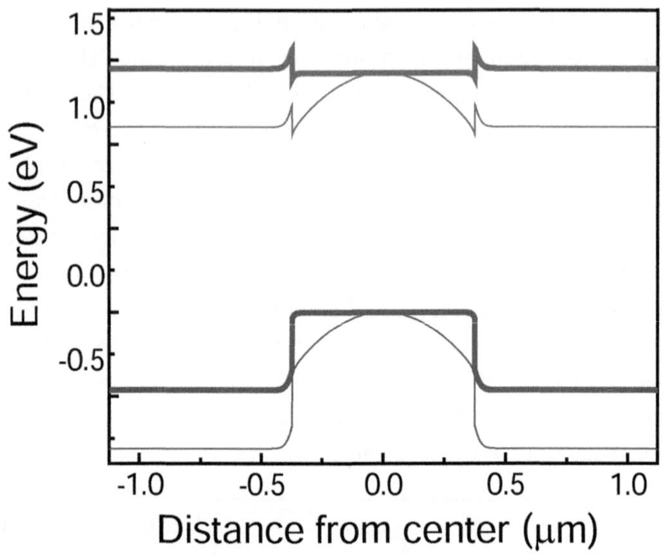

Figure 12.16 Band profile for conduction (upper green curves) and valence (lower red curves) bands. Thermal equilibrium at T = 300 K is shown as thin lines. The type II character (electrons in conduction band pushed away from GaAs region and holes in valence band localized in GaAs region) is apparent. The thick curves show an optically excited structure with an electron density at the center of n_e (z = 0) = 10^{17} cm^{-3}. The optical excitation has led to flat-band conditions for the conduction band.

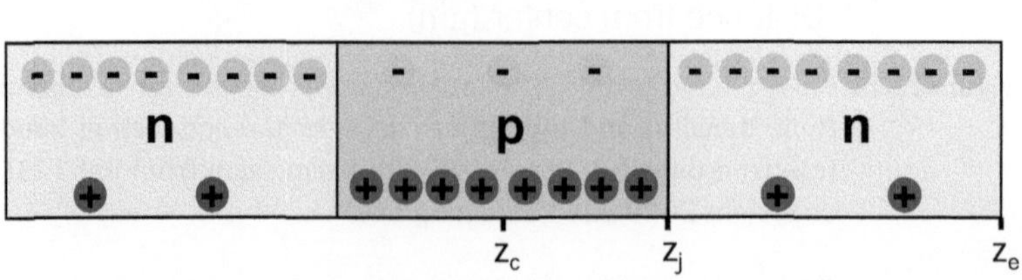

Figure 12.17 Sketch of n-p-n structure with electrons and holes schematically indicated.

where S denotes the surface recombination velocity and

$$n_{\min}(z_j) = \min\{n_e(z_j), n_h(z_j)\} \tag{12.55}$$

is the minority carrier density at interface. Assuming that the optically excited carrier density $n_a^{\text{opt}}(z)$ is much larger than the intrinsic carrier density, we have

$$\left.\frac{\partial n_{\mathrm{e}}^{opt}(z)}{\partial t}\right|_{\mathrm{nr}} = \left.\frac{\partial n_{\mathrm{h}}^{opt}(z)}{\partial t}\right|_{\mathrm{nr}} = -\sum_{j=1,2} \delta(z-z_j)\, S\, n_{\min}(z_j) \quad (12.56)$$

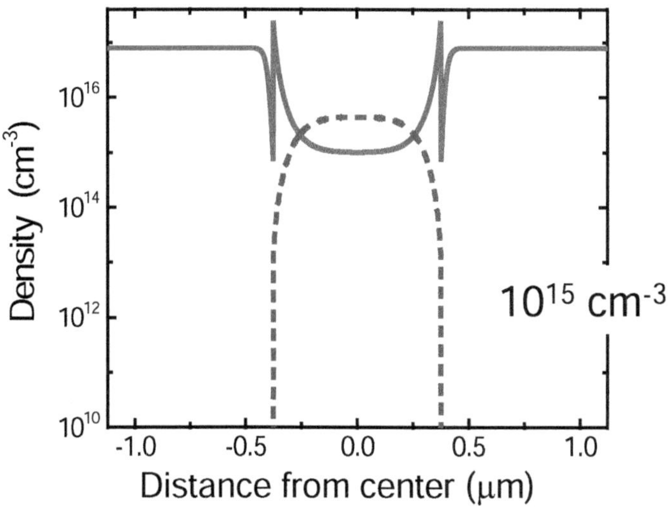

Figure 12.18 Electron (solid green line) and hole (dashed red line) density profile at T = 300 K and with an acceptor density of N_{a} = 3.5 × 10^{15} cm^{-3}. The optical excitation is chosen to yield a carrier density at the center of 10^{15} cm^{-3}. Holes are majority carriers at the center but minority carriers at the interface.

We introduce the average carrier density (averaged over the GaAs layer of thickness d)

$$\bar{n}_{\mathrm{a}} = \frac{1}{d}\int dz\, n_{\mathrm{a}}^{\mathrm{opt}}(z) \quad (12.57)$$

and assume the layer thickness to be sufficiently large so that the average density is the same as the density at the center of the layer, $\bar{n}_{\mathrm{a}} \approx n_{\mathrm{a}}^{\mathrm{opt}}(z_c)$ for a = e or h, we have

$$\begin{aligned}\left.\frac{\partial n_{\mathrm{e}}^{\mathrm{opt}}(z_c)}{\partial t}\right|_{\mathrm{nr}} = \left.\frac{\partial n_{\mathrm{h}}^{\mathrm{opt}}(z_c)}{\partial t}\right|_{\mathrm{nr}} &= -\frac{2S}{d} n_{\min}(z_j) \\ &= -\frac{2S}{d}\frac{n_{\min}(z_j)}{n_{\min}(z_c)} n_{\min}(z_c) \\ &= -\mathrm{A}\, n_{\min}(z_c)\end{aligned} \quad (12.58)$$

where we introduced an effective nonradiative recombination coefficient A, whose inverse is the effective nonradiative decay time

$$\tau_{\text{nr}} = \frac{1}{A} = \frac{d}{2S}\frac{n_{\min}(z_c)}{n_{\min}(z_j)} \tag{12.59}$$

This shows that, as mentioned above, improving (lengthening) the nonradiative lifetime can be achieved by minimizing the minority carrier density at the interface, $n_{\min}(z_j)$.

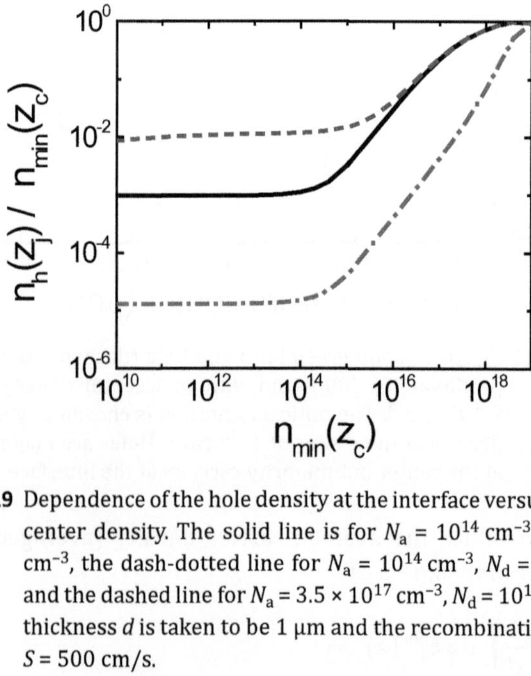

Figure 12.19 Dependence of the hole density at the interface versus minority center density. The solid line is for $N_a = 10^{14}$ cm^{-3}, $N_d = 10^{17}$ cm^{-3}, the dash-dotted line for $N_a = 10^{14}$ cm^{-3}, $N_d = 10^{19}$ cm^{-3}, and the dashed line for $N_a = 3.5 \times 10^{17}$ cm^{-3}, $N_d = 10^{17}$ cm^{-3}. The thickness d is taken to be 1 μm and the recombination velocity $S = 500$ cm/s.

Figure 12.19 shows the dependence of the interface hole density as a function of the optical excitation, here parametrized by the optically excited density at the center, or minority center density $n_{\min}(z_c)$. With increasing $n_{\min}(z_c)$, the interface hole density increases. Since the holes are the minority carriers at the interface, this in turn means that the effective nonradiative decay time decreases, as is shown in Fig. 12.20. However, an increased n-doping level, in our example of $N_d = 10^{19}$ cm^{-3}, is found to decrease the interface hole density (Fig. 12.19) and hence improve (increase) the effective nonradiative lifetime (Fig. 12.20). As shown in Fig. 12.20, such an increased n-doping concentration can, for the idealized model of and infinitely thin interface region, yield lifetimes comparable to the optimized break-even lifetimes discussed above. This, however, assumes that a high n-doping concentration does not open up new nonradiative transitions that would again decrease the effective nonradiative lifetime.

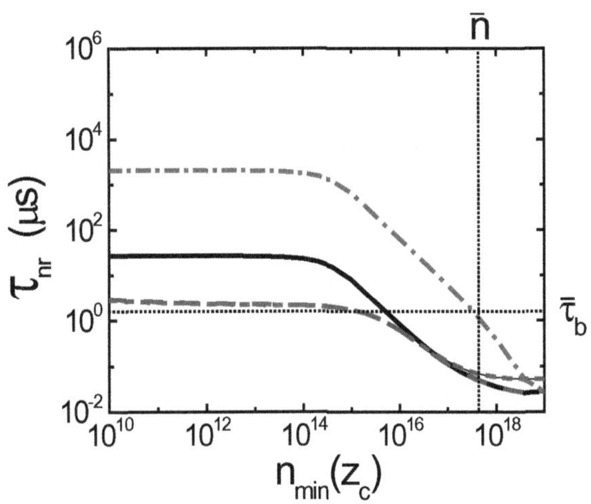

Figure 12.20 Effective nonradiative recombination time as a function of center density. The three different thick line styles correspond to the same doping levels as in Fig. 12.19. The vertical and horizontal dotted lines indicate the optimized cooling density and optimized break-even lifetime, respectively, obtained from the model of bulk semiconductors discussed above in Section 12.2.

Finally, we show in Fig. 12.21 the cooling efficiency as a function of minority center density. We see that positive cooling efficiency can be achieved for moderate minority center densities (in other words, optical pumping that yields densities small enough so that Auger recombination does not become dominant), and high n-doping concentrations in the passivation layers. Further details can be found in Rupper et al. [23].

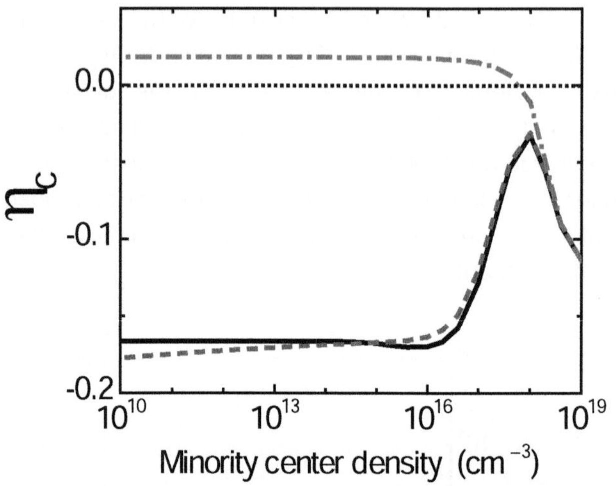

Figure 12.21 Cooling efficiency versus minority center density. The three different line styles correspond to the same doping levels as in Fig. 12.19. The dash-dotted green line shows that cooling should be possible, $\eta_c > 0$, for low to moderate minority center densities (i.e., moderate optical pumping powers) and large n-doping concentration in the passivation layers. Reprinted with permission from Ref. [23], Copyright 2010, AIP Publishing LLC.

Acknowledgments

We thank M. Sheik-Bahae, B. Imangholi, M. Hasselbeck, R. Epstein, and J. Khurgin for helpful discussions. We acknowledge financial support from AFOSR MURI grant no. A9550-04-1-0356 and additional support from JSOP and TRIF Photonics.

References

1. Pringsheim, P. (1929). Zwei Bemerkungenueber über den Unterschied von Lumineszenz- und Temperaturstrahlung, *Z. Phys.*, p. 739.
2. Epstein, R. I., Buchwald, M. L., Edwards, B. C., Gosnell, T. R., Mungan, C. E. (1995). Observation of laser-induced fluorescent cooling of a solid, *Nature,* **377**, pp. 500–503.
3. Sheik-Bahae, M., Epstein, R. I. (2007). Optical refrigeration, *Nat. Photonics,* **1**, pp. 693–699.

4. Nemova, G., Kashyap, R. (2010). Laser cooling of solids, *Rep. Prog. Phys.*, **73**, p. 086501.
5. Oraevsky, A. (1996). Cooling of semiconductors by laser radiation, *J. Russ. Laser Res.*, pp. 471–479.
6. Rivlinand, L., Zadernovsky, A. (1997). Laser cooling of semiconductors, *Opt. Commun.*, **139**, pp. 219–222.
7. Sheik-Bahae, M., Epstein, R. I. (2004). Can laser light cool semiconductors?, *Phys. Rev. Lett.*, **92**, p. 247403.
8. Apostolova, T., Huang, D., Alsing, P. M., Cardimona, D. A. (2005). Comparison of laser cooling of the lattice of wide-band-gap semiconductors using nonlinear or linear optical excitations, *Phys. Rev. A*, **71**, p. 013810.
9. Khurgin, J. (2007). Surface plasmon-assisted laser cooling of solids, *Phys. Rev. Lett.*, **98**, p. 177401.
10. Gauck, H., Gfroerer, T. H., Renn, M. J., Cornell, E. A., Bertness, K. A. (1997). External radiative quantum efficiency of 96 percent from a GaAs/GaInP heterostructure, *Appl. Phys. A*, **64**, pp. 143–147.
11. Imangholi, B., Hasselbeck, M. P., Sheik-Bahae, M., Epstein, R. I., Kurtz, S. (2005). Effects of epitaxial liftoff on interface recombination and laser cooling in GaInP/GaAs heterostructures, *Appl. Phys. Lett.*, **86**, p. 081104.
12. Seletskiy, D. V., Melgaard, S. D., Bigotta, S., Di Lieto, A., Tonelli, M., Sheik-Bahae, M. (2010). Laser cooling of solids to cryogenic temperatures, *Nat. Photonics*, **4**, pp. 161–164.
13. Zhang, J., Li, D., Chen, R., Xiong, Q. (2013). Laser cooling of a semiconductor by 40 kelvin, *Nature*, **493**, pp. 504–508.
14. Kwong, N. H., Rupper, G., Binder, R. (2009). Self-consistent T-matrix theory of semiconductor light-absorption and luminescence, *Phys. Rev. B*, **79**, p. 155205.
15. Rupper, G., Kwong, N. H., Binder, R. (2006). Large excitonic enhancement of optical refrigeration in semiconductors, *Phys. Rev. Lett.*, **97**, p. 117401.
16. Rupper, G., Kwong, N. H., Binder, R. (2007). Optical refrigeration of GaAs: theoretical study, *Phys. Rev. B*, **76**, p. 245203.
17. Zimmermann, R. (1987). *Many-Particle Theory of Highly Excited Semiconductors* (Teubner, Germany).
18. Kwong, N. H., Rupper, G., Gu, B., Binder, R. (2007). The relation between light absorption and luminescence in laser cooling of two-dimensional semiconductor systems, *Proc. SPIE*, **6461**, p. 64610I.

19. Gu, B., Kwong, N. H., Binder, R. (2013). Relation between the interband dipole and momentum matrix elements in semiconductors, *Phys. Rev. B*, **87**, p. 125301.
20. Schmielau, T., Manske, G., Tamme, D., Henneberger, K. (2000). T-matrix approach to the linear optical response of highly excited semiconductors, *Phys. Status Solidi B*, **221**, pp. 215–219.
21. Piermarocchi, C., Tassone, F. (2001). Role of bound pairs in the optical properties of highly excited semiconductors: a self-consistent ladder approximation approach, *Phys. Rev. B*, **63**, p. 245308.
22. Rupper, G., Kwong, N. H., Binder, R. (2008). Theory of luminescence and optical refrigeration in p-doped semiconductors, *Proc. SPIE*, **6907**, p. 690705.
23. Rupper, G., Kwong, N. H., Binder, R., Li, Ch.-Y., Sheik-Bahae, M. (2010). Effect of n-p-n heterostructures on interface recombination and semiconductor laser cooling, *J. Appl. Phys.*, **108**, p. 113118.
24. Rupper, G., Kwong, N. H., Binder, R. (2009). The role of finite spatial beam profiles on photo-luminescence and laser cooling in GaAs structures, *Proc. SPIE*, **7228**, p. 722805-1.
25. Rupper, G., Kwong, N. H., Binder, R. (2010). Theory of time-resolved photo-luminescence and carrier lifetime measurements in GaAs/GaInP heterostructures, *Proc. SPIE*, **7614**, p. 76140D-1.

Chapter 13

Coulomb-Assisted Laser Cooling of Piezoelectric Semiconductors

Iman Hassani Nia and Hooman Mohseni
EECS Department, Northwestern University, 2145 Sheridan Rd., M250, Tech Building Evanston, IL 60208, USA
hmohseni@northwestern.edu

13.1 Introduction

Cooling materials with light is a topic that has attracted tremendous interest in recent years. Researchers usually find the theory counterintuitive, and, in fact, it was a highly disputed idea when Pringsheim first proposed it in 1929 [35]. However, only 66 years later, the first observation of laser cooling of solids was made by Epstein in 1995 in trivalent ytterbium-doped heavy metal fluoride glass [12]. In rare earth ions, such as ytterbium, the electronic 4*f* levels are shielded from their surroundings by the filled 5s and 5p shells, leading to suppression of multiphonon relaxation and a high radiative recombination efficiency favorable for optical refrigeration [36]. Since the first discovery of laser cooling, a methodical

Laser Cooling: Fundamental Properties and Applications
Edited by Galina Nemova

ISBN 978-981-4745-04-8 (Hardcover), 978-981-4745-05-5 (eBook)
www.panstanford.com

approach by many scientist led to a significant improvement of optical refrigeration starting from 0.3° in 1995 by Epstein to more than 193° in 2014 [25]. Such significant progress was the result of optimized purification and doping concentrations of the rare earth ions in the host material. Laser cooling in rare earth–doped systems are considered in Chapters 2–4 of this book.

It should be noted that the practical laser cooling method of solids is mainly based on the anti-Stokes mechanism [29–31], which is schematically depicted in Fig. 13.1. The absorption of a laser in the material creates electron–hole pairs, and the photogenerated carriers go through different scattering and recombination processes before reaching thermal equilibrium. Now suppose that the carriers are excited with energies close to the band edge and, after reaching thermal equilibrium, they absorb a net energy from the lattice vibrations (which are called phonons). As a result, the thermalized carriers have more energy than the absorbed photons' energy. The energy by which the carriers exceed the photon's energy is roughly equal to kT (where k is the Boltzmann constant and T is the temperature). Consequently, the photoluminescence (PL) of the thermalized carriers has a higher energy than the absorbed laser. It is therefore expected that a full extraction of the PL would be equivalent to extraction of the thermal energy from the material to the outside environment by means of radiation and the material cools down. This is quite important because operating the optoelectronic devices at lower temperatures usually reduces their electrical noise level [13, 14].

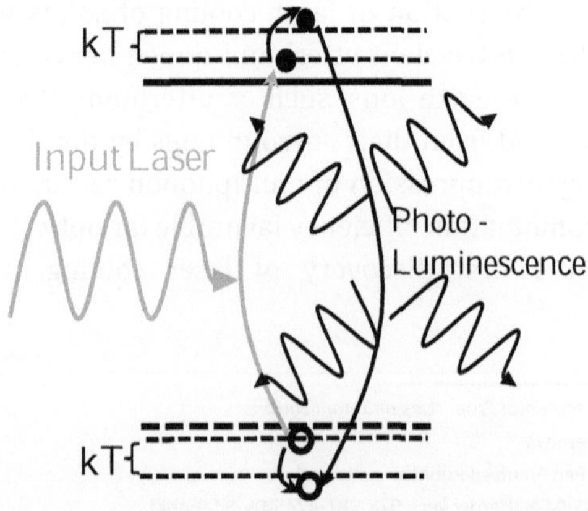

Figure 13.1 The process of anti-Stokes laser cooling of semiconductors.

In practice, there are challenges that can forbid laser cooling of solids by the above simple picture. One important issue is the existence of nonradiative recombination pathways that turn all of the absorbed energy into heat. For semiconductors, the nonradiative recombinations can be divided into two main categories: (1) the Shockley–Reed–Hall (SRH) recombination and (2) the Auger recombination process. The SRH process is mainly due to the interaction of the carriers with crystal nonidealities such as surface states and defects. On the other hand, the Auger recombination arises from the Coulomb interaction and scattering of three carriers; the interaction between two carriers gives energy to the third carrier and scatters it to a state with a higher energy. The third carrier eventually relaxes to a lower energy, losing its excess energy to the phonon bath. To realize laser cooling, both of the aforementioned processes should be suppressed. While the SRH process can be suppressed by increasing the material quality, surface passivation and preventing the carriers from reaching the defective area, the Auger recombination can be suppressed by appropriate band structure engineering [27] and operating at lower laser powers to avoid overly high photogenerated carrier densities. It should be noted that the ratio of the radiative to the nonradiative recombination rates increases as the operating temperature decreases. But unfortunately, at the same time, the thermal energy gained by the carriers (approximately equal to kT) decreases.

Another important challenge is to extract the PL out of the material. For the case of rare earth–doped glass, the refractive index of the host crystal is close to air (for glass it is approximately 1.5). However, the difference between the index of the semiconductors (around 3) and air is high. As a result, the issue of PL trapping arises in semiconductors. The PL trapping results in photon recycling and eventually wastes some portion of the luminescent energy to nonradiative and heat-producing mechanisms. For this reason, effective luminescence extraction methods need to be developed for materials with a higher index of refraction.

To have a better understanding of the above explanations, we note that the cooling efficiency (η_c) can be calculated as [41]:

$$\eta_c = \eta_{abs}\eta_{ex}\frac{E_{pl}}{E_i} - 1 \tag{13.1}$$

where E_{pl} is the average energy of the PL, E_i is the average energy of the incident photons, and η_{ex} and η_{abs} are the external quantum efficiency and the absorption efficiency, respectively. The external quantum efficiency (η_{ex}) can be calculated as [42]:

$$\eta_{ex} = \frac{\eta_e Bn^2}{An + \eta_e Bn^2 + Cn^3} \tag{13.2}$$

where η_e is the extraction efficiency [28], n is the photogenerated carrier density, and A, B, and C are the SRH, radiative, and Auger recombination coefficients, respectively. This parameter is a measure of the fraction of luminescence power extracted out of the material before getting reabsorbed. The absorption efficiency is equal to:

$$\eta_{abs} = \frac{\alpha}{\alpha + \alpha_b} \tag{13.3}$$

where α is the absorption in the active region and α_b is the background absorption due to free carriers, impurities, and the defects of the material. It is straightforward to conclude from Eqs. 13.2 and 13.3 that in order to maximize the net cooling efficiency, both the external quantum efficiency and the absorption efficiency should be maximized. The external quantum efficiency is dependent on the material quality, the extraction efficiency, and the carrier density. The extraction efficiency can be increased to almost 100% using novel techniques that suppress both the total internal reflection and the Fresnel reflection inside a high-index material. However, the material quality sets a limit on maximum attainable external quantum efficiency and absorption efficiency. The materials of choice for semiconductor laser cooling up to now are GaAs [42] and CdS [51]. Considering the bandgaps of these materials, the maximum net cooling efficiency is approximately 2% for gallium arsenide and 1% for cadmium sulfide at room temperature. These values are quite low, rendering the laser cooling of bulk GaAs impossible, and for CdS, only a particular nanoribbon structure has been cooled using a laser. This is due to the insufficient external and absorption efficiencies of these materials. These challenges suggest that mechanisms other than anti-Stokes should be developed and employed in order to have practical laser cryocoolers in the future. While different methods of laser cooling for gaseous systems (such as Doppler cooling, Sisyphus, resolved-sideband cooling, and sympathetic cooling) have been implemented, the laser cooling of solids has remained a vast area to explore new ideas. In the following sections, we start by describing different types of Coulomb interactions in semiconductor quantum wells (QWs), which have the potential for optical refrigeration. The approach that we are going to explain is based on the Coulomb interaction between the photogenerated electron and holes in piezoelectric QW structures. For this reason, the next section is devoted to the topic of piezoelectricity in semiconductors and its previous application. We also discuss the effects of Coulomb interactions in semiconductors in Section 13.4. This provides the essential basis for understanding laser cooling of piezoelectric semiconductors, which is explained in Section 13.2. We finish the chapter with our concluding remarks about limitations and the possible improvement of the described systems in Section 13.8.

13.2 Piezoelectricity in Semiconductors

The piezoelectric effect refers to the creation of polarized charge by applying stress in a material that contains more than one type of atom without geometrical inversion symmetry. To begin with, we consider zinc-blende semiconductors, which are the simplest semiconductor crystals for the investigation of piezoelectricity. The induced polarization (P) can be related to stress (S) using the piezoelectric tensor (e) [2]:

$$P = eS \tag{13.4}$$

In most solids, the stress can be characterized using only six components $(S_{xx}, S_{yy}, S_{zz}, S_{xy}, S_{yz}, S_{xz})$, which constitute the elements of the stress vector S. The first three components ($S_{ii} i = x,y,z$) are called biaxial tensor components, whereas the other three components ($S_{ij} i,j = x,y,z,\ i \neq j$) are the shear stress components. The applied stress results in deformation of the crystal, and the atoms deviate from their equilibrium positions. The stress (S) and the strain (ε) of a crystal are connected by Hooke's law and through the elasticity tensor (C) [5, 44]:

$$S = C\varepsilon \tag{13.5}$$

Similar to the stress, the strain of the material is also comprised of six components. In Voigt notation they are defined as

$$\epsilon_{\alpha,\beta} = \frac{\partial u_\alpha}{\partial x_\beta},\ u_\alpha = u_x\ u_y, u_z x_\beta = x, y, z \tag{13.6}$$

where u_α is the lattice displacement along the direction specified by x_α. The strain tensor is symmetric and the following relation holds [50]:

$$\epsilon_{\alpha\beta} = \epsilon_{\beta\alpha} = \frac{\partial u_\alpha}{\partial x_\beta} + \frac{\partial u_\beta}{\partial x_\alpha} \tag{13.7}$$

Equations 13.4–13.7 provide a complete description of piezoelectric field as a result of different types of stress and strain in the material. It is important to note that because of crystal symmetry, many elements of the elastic constants and the piezoelectric tensor are essentially zero. For example, for a cubic crystal, the values of the uniaxial elastic constants along the three axis are the same because of spatial invariance (C_{ii}=C_{jj}, $i \neq j$). A similar relation holds for shear components. Furthermore, shear strain cannot produce a normal stress (e.g., $C_{41} = 0$) , and as a result the elasticity tensor is simplified to [32]

$$C_{\text{cubic}} = \begin{pmatrix} C_{11} & C_{12} & C_{12} & 0 & 0 & 0 \\ C_{12} & C_{11} & C_{12} & 0 & 0 & 0 \\ C_{12} & C_{12} & C_{11} & 0 & 0 & 0 \\ 0 & 0 & 0 & C_{44} & 0 & 0 \\ 0 & 0 & 0 & 0 & C_{44} & 0 \\ 0 & 0 & 0 & 0 & 0 & C_{44} \end{pmatrix} \tag{13.8}$$

However, for crystals that belong to the hexagonal crystal symmetry group (such as wurtzite GaN compositions), the elasticity constant is described by four components [8]:

$$C_{\text{hcp}} = \begin{pmatrix} C_{11} & C_{12} & C_{13} & 0 & 0 & 0 \\ C_{12} & C_{11} & C_{13} & 0 & 0 & 0 \\ C_{13} & C_{13} & C_{11} & 0 & 0 & 0 \\ 0 & 0 & 0 & C_{44} & 0 & 0 \\ 0 & 0 & 0 & 0 & C_{44} & 0 \\ 0 & 0 & 0 & 0 & 0 & (C_{11} - C_{12})/2 \end{pmatrix} \tag{13.9}$$

The above matrices are used to find the stress in materials grown along (0 0 1) and (0 0 0 1) for the case of cubic and hexagonally symmetric crystals. For all other growth directions, the appropriate Euler rotation matrix should be applied to find the correct elasticity tensor. We focused only on these two crystal symmetries because the highly luminescent semiconductors, as suitable candidates for laser cooling, fall within these two categories.

It should be noted that the converse piezoelectric effect is the generation of strain by applying an electric field across a piezoelectric material. The piezoelectric strain tensor (d) is used to connect these two physical observables:

$$d = C^{-1}e \tag{13.10}$$

where e is the piezoelectric tensor and is described by only one nonvanishing element for III–V semiconductors [1]:

$$e = \begin{pmatrix} 0 & 0 & 0 & e_{14} & 0 & 0 \\ 0 & 0 & 0 & 0 & e_{14} & 0 \\ 0 & 0 & 0 & 0 & 0 & e_{14} \end{pmatrix} \tag{13.11}$$

whereas for wurtzite crystals, it has the form [43]:

$$e = \begin{pmatrix} 0 & 0 & 0 & 0 & e_{15} & 0 \\ 0 & 0 & 0 & e_{15} & 0 & 0 \\ e_{31} & e_{31} & e_{33} & 0 & 0 & 0 \end{pmatrix} \tag{13.12}$$

The above equations form a complete set for finding the displacement with respect to the applied voltage to a piezoelectric material and vice versa. In the next sections, we describe how the separameters can be used to coherently control the lattice vibrations (damping or amplification) in a piezoelectric QW structure. In this regard, we note that there are many reports about the generation of the acoustic phonons in piezoelectric heterostructures by optical excitation. For instance, lattice-mismatched GaN/InGaN QW structures possess a giant piezoelectric field. Upon absorbing short optical pulses by this material, electron–hole pairs are generated, which travel across the multiple–quantum well (MQW) region due to the built-in piezoelectric field. The motion of the photogenerated carriers screen the piezoelectric field and induce a mechanical strain wave and acoustic phonons [6, 7, 24, 39]. The generation of the acoustic phonons alters the refractive index of the material, which manifests itself in the transmission modulation of the probe beam, as shown in Fig. 13.2.

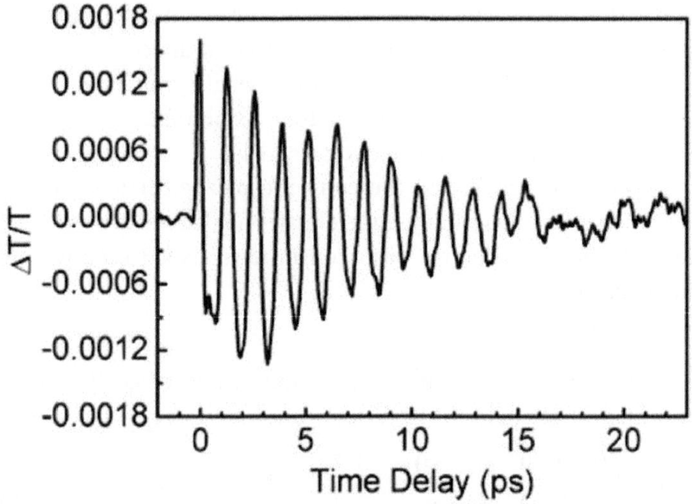

Figure 13.2 Transmission modulation versus time for a 2.2 nm/7 nm $In_{0.1}Ga_{0.9}N/GaN$ MQW due to the presence of acoustic phonons. Reprinted with permission from Ref. [7]. Copyright 2004, AIP Publishing LLC.

We will investigate the opposite of the described process to coherently absorb the acoustic phonons and achieve microscopic laser cooling of piezoelectric heterostructures. This coherent process is based on optomechanical backaction cooling concepts which are the subject of the next section.

13.3 Basics of Optomechanical Cooling and Amplification

The interaction of the light and mater can result in an efficient energy transfer between the mechanical and the optical fields. There are many mechanisms that can be utilized to achieve an efficient energy transfer. In this section, we only focus on backaction optomechanics and we will use the same concept to demonstrate the cooling of large number of phonon modes in a bulk solid. Backaction dynamical interaction can be realized if there is an optical and mechanical resonator and both are coupled to each other. During the recent years many miniature nano- and micro-optomechanical resonators have been fabricated that made the observation of this effect possible. The potential applications include ultrasensitive motion and force sensors used in gravitational wave detectors [3], photonic clocks [18], high-precision accelerometers, atomic force microscopy, and parametric amplification.

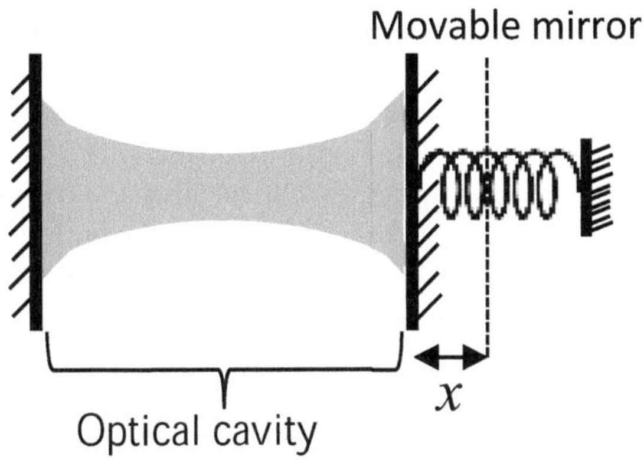

Figure 13.3 Schematics of the system suitable for optomechanical backaction interaction. It consists of a movable mirror (the displacement of the mirror is shown by x) and resonating optical field that is coupled to the mechanical state of the mirror through the optical cavity.

Consider an optical cavity with one movable mirror, as shown in Fig. 13.3. The coupled mechanical and optical equations of motion can be written as follows [22]:

$$\frac{da}{dt} = i\Delta(x)a - \frac{1}{2}\left(\frac{1}{\tau_0} + \frac{1}{\tau_{ex}}\right)a + i\sqrt{\frac{1}{\tau_{ex}}}s \tag{13.13a}$$

$$\frac{d^2x}{dt^2} + \frac{\Omega_m}{2Q_m}\left(\frac{dx}{dt}\right) + \Omega_m^2 x = \left(\frac{1}{m_{eff}}\right)(F_{rp} + F_L) \tag{13.13b}$$

where x is the position of the movable mirror with respect to equilibrium, Δ is the detuning of the laser with respect to the cavity resonance, τ_0 is the intrinsic photon decay rate of the cavity, τ_{ext} is the extrinsic photon decay rate of the cavity (mainly due to the mirror loss), Ω_m is the mechanical resonance frequency, Q_m is the mechanical quality factor of the mirror, and m_{eff} is the effective mass of the mirror. $|a|^2$ is the optical energy stored in the cavity and

$|s|^2$ is the optical power incident on the cavity. F_{ep} is the force due to radiation and F_L is the Langevin force, which is responsible for thermal vibrations of the mirror. According to the above equations, one can figure out that the movement of the mirror changes the detuning (i.e., $\Delta(x) = \Delta + (\omega_0/L)x$, where L is the length of the Fabry–Perot cavity and ω_0 is the resonance radial frequency of the cavity), which in turn changes the optical power inside the optical resonator. On the other hand the change in the optical energy of the resonator results in a coherent change in the optical force exerted on the moving mirror. As we shall see in the following, with a correct phase difference between the dynamical optical force and the mechanical motion, an efficient energy transfer between the optical mode and the mechanical mode can be achieved. To start this analysis, we note that the optical force exerted on the mirror wall is equal to $2|a|^2/(cT_{tr})$ in which T_{tr} is the round-trip transit time ($\sim 2L/c$) and c is the speed of light in the cavity. We expand $|a|^2$ in the weak retardation regime, that is, $1/t >> \Omega_m$ according to Eq. 13.13a to obtain

$$F(t) = \left(\frac{\tau}{cT_{rt}}\frac{2\tau/\tau_{ex}}{4\Delta^2\tau^2+1}\right)\left[1+\left(\frac{8\Delta\tau^2}{4\tau^2\Delta^2+1}\right)\frac{\omega_0}{R}x-\tau\left(\frac{8\Delta\tau^2}{4\tau^2\Delta^2+1}\right)\frac{\omega_0}{R}\frac{dx}{dt}\right]|s|^2 \tag{13.14}$$

where $\tau^{-1} = \tau_0^{-1} + \tau_{ex}^{-1}$ and the first two terms in this series are the adiabatic response of the force to the changes in position of the moving mirror. As the mirror moves, the instantaneous detuning changes which results in the change of optical power and force as described by these two terms. The second term corresponds to the mechanical displacement and hence results in an induced optical stiffness. The power transfer between the optical mode and the mechanical mode can be found by evaluating $F.dx/dt$ and averaging it over one period of mechanical oscillation. By doing so, it turns out that only the third term in the above equation is responsible for a nonzero cooling/amplification rate because it is in phase with the velocity. It is also inferred from the above equation that the sign of the detuning determines the direction of the energy transfer between the optical mode and the mechanical mode. A red-detuned laser leads to cooling of the mechanical vibration that would change the balance in favor of anti-Stokes radiation, as schematically shown in Fig. 13.4, and the reverse is true for a blue-detuned laser. In general in any system based on the backaction mechanism, to have an effective cooling or amplification an in-phase force component with the velocity is required. We note that the force applied by the photons inside the cavity does not need to be the radiation pressure. If photons initiate any kind of mechanism that exert force on the mirror, they can lead to damping/amplification of the mechanical motion and should be considered in the theory. To have a better understanding we expand the light-induced forces (F_{opt}) into its components:

$$F_{\text{opt}} = \sum_n \int_0^t \frac{dF_n(\tilde{t})}{dt} g_n(t-\tilde{t})d\tilde{t} \tag{13.15}$$

where F_n denotes the force exerted by a particular mechanism activated by the photons and g_n is equal to $1 - \exp(-t/\tau_n)$. τ_n is the average time for F_n to reach equilibrium after a sudden change in the position of the moving mirror. By solving the above equation in the limit of very small displacements (which is the case for thermal vibrations), one can find the following relation for the position of the mirror in the Fourier domain:

$$x_\omega = \frac{\sum_n F_{n,\omega}/(1+i\omega\tau_n)+F_{L,\omega}}{K} \frac{\omega_0^2}{\omega_{\text{eff}}^2-\omega^2+i\Gamma_{\text{eff}}\omega} \tag{13.16}$$

where $F_{n,\omega}$ and $F_{\text{L},\omega}$ are the amplitudes of the light-induced force and the Langevin force at the frequency of ω. τ_n is the characteristic delay time of $F_{n,\omega}$. K is the mechanical rigidity of the mirror (which is equal to $m_{\text{eff}}\Omega_{\text{m}}^2$). ω_{eff} and Γ_{eff} are the effective resonance frequency and the damping rate of the moving mirror in the presence of the optical field inside the cavity and can be expressed as follows [26]:

$$\Gamma_{\text{eff}} = \left(\frac{\Omega_{\text{m}}}{Q_m}\right)\left[1-Q_{\text{m}}\sum_n \frac{\omega_0\tau_n}{\omega^2\tau_n^2+1}\frac{K_n}{K}\right] \tag{13.17a}$$

$$\omega_{\text{eff}}^2 = \omega_0^2\left[1+\sum_n \frac{1}{\omega^2\tau_n^2+1}\frac{K_n}{K}\right] \tag{13.17b}$$

where K_n is the light-induced rigidity and is equal to $-\partial F_{n,0}/\partial x$, where $F_{n,0}$ is the static light-induced force at equilibrium. Considering the mechanical dispersion of the moving mirror and spectral properties of the thermal driving force, one can obtain the following formula for amplitude of vibration between $\omega - \delta\omega/2$ and $\omega + \delta\omega/2$:

$$<x_\omega>^2 = \frac{2k_\mathrm{B}T}{\pi K}\frac{\omega_0^2\Gamma}{(\omega_\mathrm{eff}^2-\omega^2)^2+(\Gamma_\mathrm{eff}\omega)^2}\delta\omega \tag{13.18}$$

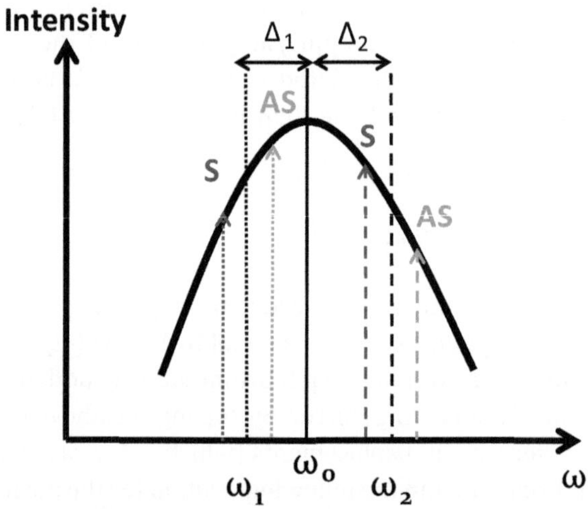

Figure 13.4 The resonator dispersion and stokes (S) and anti-Stokes (AS) components of the photons in the cavity. Two different conditions with negative laser detuning (ω_1) and positive laser detuning (ω_2) are shown. In the first scenario the anti-Stokes component is larger than the Stokes component, leading to the cooling of the mechanical mode. The reverse is true for positive detuning, which results in the amplification of the mechanical motion.

By defining the temperature (T_eff) according to equipartition theorem for a 1D harmonic oscillator $\left(\frac{1}{2}K<x^2>=k_\mathrm{B}T_\mathrm{eff}\right.$, where k_B is the Boltzmann constant) and finding the spectral average of the square of the displacement $<x^2>$) using Eq. 13.18, one can show that

$$T_\mathrm{eff} = T\frac{\omega_0^2}{\omega_\mathrm{eff}^2}\frac{\Gamma}{\Gamma_\mathrm{eff}} \tag{13.19}$$

The above relation shows that both the effective resonance frequency (modified by the induced optical rigidity) and the effective damping rates are responsible for the change in temperature. According to Eq. 13.17 it can be shown that for $K_n = -K$ and $\tau_n = \sqrt{2}/\omega_0$, maximum cooling with minimum temperature of $T_{\mathrm{eff}} = 4\sqrt{2}T / Q_{\mathrm{m}}$ can be achieved. The delay (τ_n) is necessary to provide the phase shift required to transfer the energy between the mechanical and the optical modes, whereas the stiffness increases with the magnitude of the optical force exerted on the mirror as it moves out of its equilibrium condition.

From the above discussion, it can be inferred that to have an effective optomechanical cooling not only a backaction force with correct phase and large magnitude is required, but also the mechanical quality factor should be high. Extension of the above concepts to phonon modes of a bulk material is possible and forms the basis of the Coulomb-assisted piezoelectric cooling, as described in Section 13.6. In a bulk material, each phonon mode can be considered as a mechanical mode with a low quality factor that leads to a low cooling efficiency. However, once the cooling mechanism is activated on all phonon modes of the material, a microscopic drop of temperature is expected. Similar to optomechanical cooling systems, where the minimum achievable temperature is set by the photo-induced heating, in bulk semiconductors parasitic absorption of pump and PL limit the cooling process.

13.4 Coulomb Interaction in Semiconductors and Its Application for Laser Cooling

Coulomb interaction of carriers here refers to the electrostatic attraction between negative and positive charges in a host medium that results in the change of the optoelectronic properties of the material. We will show that these effects are more pronounced in confined 2D (QWs) and 3D (quantum dot systems) because of shorter distances between the electrons and holes. In this regard, the excitonic effect has always been the focus of research and has been one of the most important implications of the Coulomb interaction in semiconductors. The excitons are the photoexcited electrons and holes bound by the Coulomb attractive force. They are similar to excited states of a hydrogen atom, as opposed to free electrons and holes that resemble ionized hydrogen atoms. The excitonic effect is most apparent at low temperatures where the scattering rate with high energy phonons is not high enough to exceed the orbiting rate of the exciton and dissociate it effectively. On the other hand, at high temperatures, the optical phonon broadening is higher which prevent the excitons from resolving in the absorption spectrum of the material. Nonetheless, in the last three decades, the advancement of epitaxial semiconductor growth has led to ultrafine MQWs, which are suitable for harvesting the benefits of excitons, even at room temperature. The reason is that in such confined systems, the radius of the exciton is much smaller along the growth directions, resulting in the reduction of the exciton radius in the lateral directions. This is due to the fact that the energy of the excitons is minimized when the exciton has a spherical shape. When the dimension of an exciton reduces, its binding energy and its orbiting rate increase, and hence the exciton dissociation rate by scattering with high energy phonons decreases. As a result, a more favorable situation for the presence of excitons in lower-dimensional systems is provided.

Previous theories have shown that the excitonic effect is quite beneficial for laser cooling of solids. This is due to the enhanced excitonic absorption of the pump laser below the bandgap energy, and also it is proven beneficial by the amplified radiative decay of the excitons. Rupper et al. [37, 38] has conducted a comprehensive theoretical investigation to find the emission and absorption spectra of bulk GaAs with the presence of excitonic effects. According to that study, the enhancement in excitonic absorption leads to enhancement of radiative decay at higher energies. Furthermore, the Coulomb-induced steepening of the absorption tail helps to have significant absorption of the photons at low energies in order to overcome the adverse effects of the background absorption of the input laser.

Confined systems with delocalized electron and hole wave functions are of particular interest for implementing the Coulomb effects. In such systems, increasing the density of electrons and holes leads to an effective electric field between the particles that either tilts the energy bands or screens the external electric field. Due to the quantum-confined Stark effect (QCSE), the red shift or blue shift of bandgap follows respectively. This dynamical effect leads to an effective nonlinearity that can be utilized for the purpose of laser cooling of solids. One particular scenario is shown schematically in Fig. 13.5. The population of electrons and holes screens the external electric field and blue shifts the transition energy. Therefore, after the carriers reach thermal equilibrium, not only do they gain the thermal energy of kT but also they experience an inherent blue shift due to the Coulomb interaction. As a result, the difference between the energy of the spontaneously emitted photons and the absorbed photons increases, which could be an advantage, according to Eqs. 13.1–13.3, for increasing the laser cooling efficiency. Similar behavior has been demonstrated in type II QW structures, where the excitation of free carriers leads to a blue shift of radiation by the specific type II band alignment [10]. This type of band alignment plays the same role as a built-in electric field and separates the excited electron and hole pairs. This separation directly affects the Auger and radiative recombination rates. In the next section, the experimental and theoretical investigations to clarify these effects are outlined.

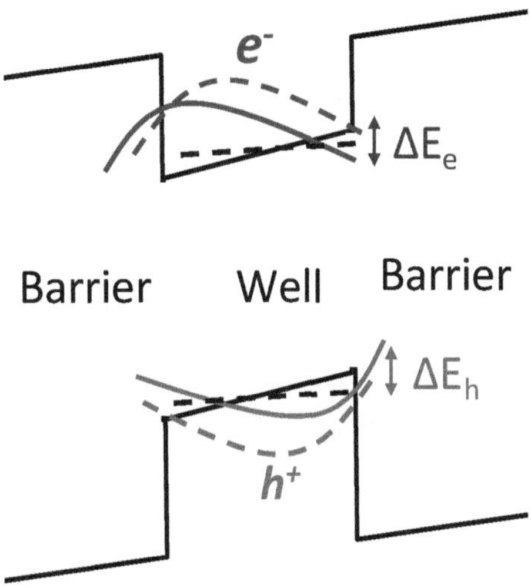

Figure 13.5 Schematics of a quantum well structure with an external applied bias. The electron and hole pairs are delocalized. The population of these delocalized states results in screening of the external electric field. Therefore at equilibrium the tilt in the energy band decreases and electrons and holes get closer to each other, as shown by the dashed lines. The reduction of the effective electric field within the quantum well is associated with the blue shift of electron and hole energy levels, as denoted by ΔE_e and ΔE_h. The blue shift adds to the thermal blue shift of the radiation kT.

13.5 Effect of the Separated Electron and Hole Wave Functions on the Recombination Rates

As explained earlier, confined structures with delocalized electron–hole wave functions can exhibit a blue shift of radiation by means of the Coulomb interaction between the photogenerated carriers. In this section, we briefly discuss the radiative and Auger recombination in such systems. These two parameters play a major role in the external quantum efficiency and the laser cooling efficiency (see Eqs. 13.1–13.3) and should be investigated in detail.

The radiative recombination rate at thermal equilibrium without any photoexcitation is basically equal to the black-body radiation rate, and is also equal to the absorption rate of the material. This ensures that the temporal variations of the carrier densities and the number of photons within the material are zero. For a material with a bandgap of E_g, we can write [20]

$$Bn_i^2 = \frac{8\pi c}{(hc)^3}\int_{E_g}^{\infty}\alpha(h\nu)(n_0(h\nu))^2\frac{(h\nu)^2}{\exp\left(\frac{h\nu}{k_B T}\right)-1}d(h\nu) \tag{13.20}$$

where B is the radiative recombination coefficient, n_i is the intrinsic carrier density, $\alpha(h\nu)$ and $n_0(h\nu)$ are the absorption and the refractive index, respectively, of the material at a photon frequency of ν, and h is Planck's constant. By simplifying the above equation, we obtain

$$B = \frac{8\pi c}{n_i^2(hc)^3}\alpha_g n_g^2(E_g + k_B T)^2 k_B T\ \exp(-E_g/k_B T) \tag{13.21}$$

where α_g and n_g are the absorption and the refractive index at the average PL wavelength. The above equation shows that at thermal equilibrium, the absorption and radiative recombination are proportional to each other. For a QW structure, the interband absorption coefficient, neglecting the excitonic effects, has been found to be

$$\alpha(h\nu) = C\sum_{n,m}|I_{nm}|^2\frac{2}{V}\sum_{k_{||}}|\hat{e}.p_{cv}|^2\,\delta(E_{nm}(k_{||}) - h\nu)(f_v^m - f_c^n) \tag{13.22}$$

where C is a constant of proportionality. The internal summation is taken over all in-plane wavevectors ($\mathbf{k}_{||}$) and the external summation is taken over all electron and hole energy levels (E_{nm}). V is the volume of the crystal, $\hat{e}$ is the unit vector characterizing the polarization of the incident electromagnetic field, and p_{cv} is the momentum matrix element of the corresponding electron and hole states. f denotes the Fermi distribution of the carriers. One can notice that the absorption is proportional to the square of the overlap integral of the electron and hole wave functions (I_{nm}). Since the radiative recombination coefficient is directly proportional to the absorption, the radiative recombination rate also increases with the square of the overlap integral. It is quite evident that the separation of the electron and holes results in the decrease the absorption and the radiative recombination of the material. Now let's see the consequence of electron–hole pair separation on the Auger recombination rate of the QW structures. In general, the behavior of Auger recombination is substantially different in bulk and heterostructures. The latter shows nonthreshold Auger properties that have power–law temperature dependence [19, 20] as opposed to bulk materials with Auger recombination rates that exponentially increase with the temperature [21, 22].

As mentioned earlier in Section 13.1, Auger recombination involves Coulomb scattering between three carriers. In MQW structures, the CHCC Auger recombination is dominant [20]. In this process, one electron and one hole recombine and excite another electron in the conduction band to a higher energy level. The third electron finally thermalizes with the phonon bath, that is, it relaxes to a lower energy level converting its energy to heat. In heterostructures, the momentum conservation is violated along the direction of growth leading to the nonthreshold behavior of the Auger recombination. The suppressed nonthreshold Auger recombination has been observed in type II structures with separated electron and hole wave functions [27]. Here we also provide experimental and theoretical evidences for type I QW systems to show that the nonthreshold Auger is in fact the circumstance for any MQW system with delocalized electron and hole wave functions.

One period of the grown MQW structure is shown in Fig. 13.6. It consist of a 3 nm InGaAs layer for confining the lowest energy state of the heavy holes and a 18 nm InGaAsP layer with the bandgap of 1.43 μm for confining the electrons. Due to the presence of a built-in electric field equal to 45 kV/cm, the energy bands are tilted, and the electron and hole wave functions are separated with respect to each other. There is also a thin (1 nm) InAlAs layer in between the two layers in order to further separate the electrons and holes. The electron barrier is the InAlAs layer which is adjacent to the InGaAsP, and the hole barrier is InP adjacent to InGaAs layer. Both of them have a thickness of 17 nm. All of the described layers are lattice-matched to the InP substrate. To simulate these types of structures both the band lineup and the electron and hole effective masses are required. The energy band lineup for any composition can be obtained using the model solid theory [48]. The effective masses close to the band minima are found using the Luttinger parameters obtained by applying the *k.p* method. These data are necessary to solve the Schrodinger equation self consistently with the Poisson equation. After finding the energy levels and the wave functions, the carrier distribution in different energy levels can be obtained using Fermi–Dirac statistics and the PL spectra can be evaluated. The results have been plotted and compared to the experimental data in Fig. 13.7. The close agreement between the two confirms the good quality of growth and validity of the theoretical calculations.

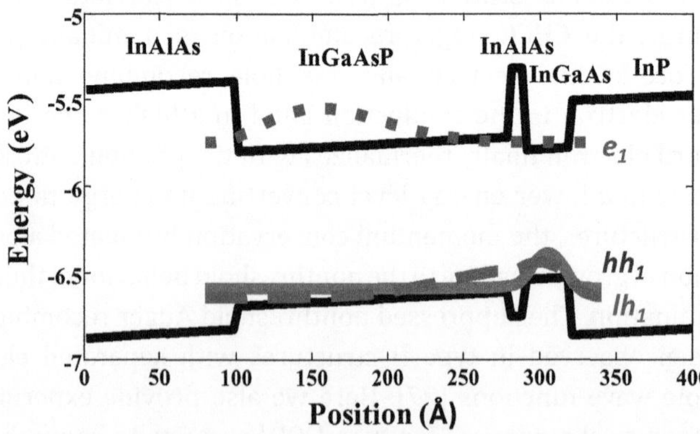

Figure 13.6 The band structure and the wave function of the grown structure used to investigate the Auger recombination process. The electron wave function is shown by the dotted lines, whereas the first heavy-hole and light-hole wave functions are shown by the solid and dashed lines.

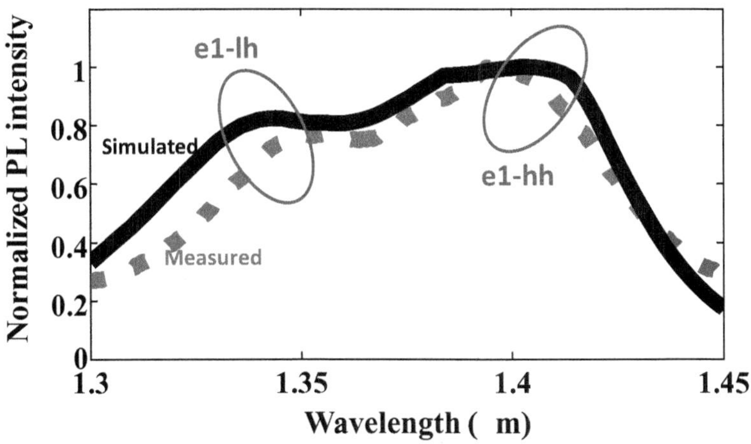

Figure 13.7 The simulated and measured PL spectra. Two peaks corresponding to the electron–light hole and electron–heavy hole recombinations were identified and indicated. The simulated curve is a solid line and the measured curve is shown by the dotted line.

To find the Auger recombination rate experimentally, one way is to perform the excitation-dependent PL measurements. In these experiments, the intensity of the excitation laser is varied and the detected power of the PL is used to extract the recombination coefficient [23]. Figure 13.8 shows the measured PL curves. There are also lines that indicate the slope of the Auger and radiative recombination decay for typical PL–pump intensity curves of such materials. According to the slope of the measured curves, it is quite evident that the intensity of the pump laser was substantially high to make the Auger mechanism dominant in all of the measurements. Therefore these data can be used to obtain the Auger coefficient versus temperature relationship, as shown in Fig. 13.9. The behavior of the Auger recombination can be explained by considering two ranges of the temperatures. For low temperatures, the nonthreshold Auger recombination coefficient is found to be [11]

$$C = \frac{32\sqrt{2}}{3\pi}\omega_{\mathrm{B}} I_{11}^{2} \frac{1}{k_{\mathrm{g}}^{4}} \left[\frac{E_0}{E_{\mathrm{g}}}\right]^{3.5} \frac{U}{k_{\mathrm{B}}T} \frac{m_{\mathrm{e}}}{m_{\mathrm{hh}}} \tag{13.23}$$

where E_{g} is the bandgap, ω_{B} is the electron Bohr frequency, E_0 is equal to the electron energy in the first state, and U is the conduction band offset. The effective electron and heavy-hole masses are denoted by m_e and m_{hh}, respectively, and $k_g = \sqrt{2m_e E_g} / \hbar$. I_{11} is the overlap integral between the first electron and hole levels (n = m = 1). By evaluation of all of the parameters , a very good fit to the experimental data can be found as shown in Fig. 13.9. At high temperatures, however, the carriers occupy higher-energy states and the confinement probability decreases. As a result, the behavior of the Auger recombination becomes closer to the bulk Auger described by [45]

$$C = C_0 \exp\left(-\frac{\mu}{1+\mu}\frac{E_g}{k_B T}\right) \tag{13.24}$$

where μ is ratio of the of the electron to heavy-hole effective mass and C_0 is the constant of proportionality. A good fit using this formula to the Auger coefficient obtained at high temperatures can be found as shown in Fig. 13.9. Due to the reduced overlap integral between the electron and holes, the Auger recombination decreases significantly as compared to other bulk and QW systems made with similar material composition (see Table 13.1).

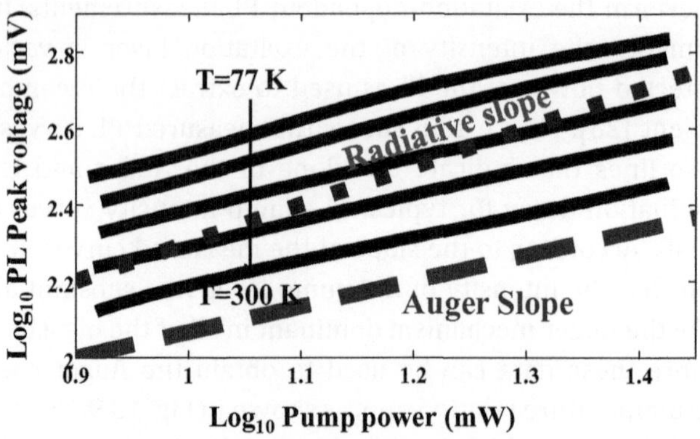

Figure 13.8 The PL peak voltage versus pump power for different temperatures ranging from 77 K up to 300 K. The Auger slope is shown by the dashed line and the radiative slope by the dotted line. The slopes of the measured curves are close to Auger process, which shows that the carrier density is high enough for the Auger process to dominate in all of the measurements.

We also note that the best carrier concentration for the purpose of laser cooling is equal to $\sqrt{A/C}$, where A is the SRH recombination rate [42]. Therefore, the decrease in the Auger recombination coefficient necessitates the use of higher laser powers to achieve maximum laser cooling efficiency.

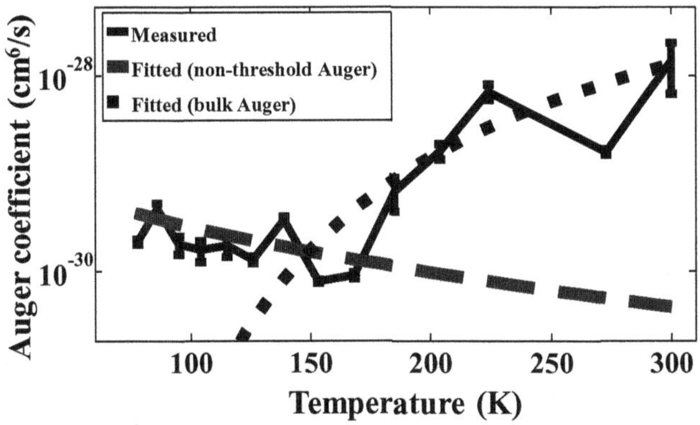

Figure 13.9 The extracted Auger recombination coefficient from the curves in Fig. 13.8. These data are plotted versus the temperature and fitted by nonthreshold and bulk-type Auger, as described in the text.

Table 13.1 Comparison of Auger recombination coefficient at T = 200 K of the introduced structure to the quantum well structures without electron and hole wave function separation and also the bulk material with similar composition

Structure	(cm^6/s)
1.1 Auger coefficient	
1.2 The structure-separated electron 1.3 hole wave function	6.5×10^{-30}
1.4 Type I $In_{0.53}Ga_{0.47}As$ QW (lattice-matched)[a]	1.3×10^{-28}
1.5 Type I $In_{0.46}Ga_{0.54}As$ QW (tensilely strained layer)[b]	6×10^{-29}
1.6 Type I $In_{0.66}Ga_{0.34}As$ QW (compressively strained layer)[c]	7×10^{-29}
1.7 Bulk InGaAsP (Q1.3)[c]	2.5×10^{29}

[a]Ref. [17]
[b]Ref. [16]
[c]Ref. [40]

13.6 Formalism of Coulomb-Assisted Laser Cooling of Semiconductors

To facilitate the laser refrigeration of semiconductors, the QCSE can be used to increase the bandgap after the generation of electron–hole pairs, as we explained in Section 13.4. This way, the emitted photons are guaranteed to have a higher energy than the absorbed ones. The increase of energy level difference of electrons and holes due to the QCSE happens due to the screening of the internal field by the photogenerated carriers. The band deformation caused by photogenerated carriers in asymmetric type II QWs has the same origin. Although we provide a theoretical basis to show that the Coulomb interaction can lead to cooling in piezoelectric materials, the physical interoperation of the energy transfer in this process in nonpiezoelectric materials is still under investigation. In this section, we shall present the most relevant phenomena that may reveal the physics behind this phenomenon in near the future.

The first question that comes to mind is the location of the electrons and holes immediately after creation. According to Refs. [47, 46] in which the dynamic solution of self-consistent Poisson and Schrödinger equations is presented, the electron–hole pairs immediately occupy the empty wave functions (wave functions calculated without the presence of photogenerated carriers); therefore the screening immediately happens but it decreases over time until the self-consistent state is reached. The bandgap increases immediately upon creation of electron–hole pairs and the origin of this energy shift is a question that should be addressed in the future.

Anti-Stokes cooling requires a high enough photon extraction efficiency as well as quantum efficiency. Therefore the implementation of this technique is extremely difficult for bulk semiconductors. The Coulomb interaction between photogenerated electron–hole pairs in strong piezoelectric materials such as GaN/InGaN QWs can assist anti-Stokes laser cooling. By comparing to cavity backaction mechanism [26], we explain how cooling or heating depends upon laser detuning with respect to the bandgap. Therefore, this method can also be also utilized for coherent phonon generation. The process is illustrated in Fig. 13.10a and can be explained as follows: The propagation of phonons with polarization along the growth direction induces piezoelectric field, which coherently changes the bandgap by the QCSE. The change of bandgap results in the change of photogenerated carrier density. On the other hand, the electric field between the electron–hole pairs due to spatial charge separation is dependent upon carrier density. Therefore, the change of carrier density leads to the change of the electric field between the electron–hole pairs and the associated piezostress on the atoms. In this cycle, a backaction mechanism exists: the strain caused by the phonons results in a coherent piezostress due to the Coulomb potential of electron–hole pairs. Therefore, there will be a nonzero power transfer between the phonons and the carriers. This process could lead to net cooling if the energy gained by the carriers is extracted efficiently from the material by radiative recombination, as shown in Fig. 13.10b.

The strain induced by the phonons, generates a piezoelectric field, which through the QCSE changes the bandgap:

$$E_g(t) = C_{\text{Stark}}\left(\frac{d_{31}}{\epsilon}\left(C_{11} + C_{12} - \left(\frac{2C_{13}^2}{C_{33}}\right)\right)\frac{C_{33}}{C_{13}}\left(\frac{u_z(0,t) - u_z(L_{\text{QW}},t)}{L_{\text{QW}}}\right)\right) \tag{13.25}$$

where u_z is the out-of-plane displacement of atoms within the width of the QW structure (L_{QW}) and the Cs are elastic stiffness coefficients. To obtain the above formula, first the in-plane strain component (ε_{xx}) was obtained by multiplying the factor $-\frac{C_{33}}{2C_{13}}$ to the average out of plane strain $\left(\frac{u_z(0,t) - u_z(L_{\text{QW}},t)}{L_{\text{QW}}}\right)$ [52]. The in-plane strain component corresponds to a piezoelectric field with a magnitude equal to $\left(\frac{2d_{31}}{\epsilon}\right)\left(C_{11} + C_{12} - \frac{2C_{13}^2}{C_{33}}\right)\epsilon_{xx}$, where d_{31} is the piezoelectric constant and ε is the electric permittivity of the medium [52]. The piezoelectric field changes the bandgap through the QCSE and this change is calculated by including the Stark coefficient (C_{Stark}). It should be noted that for the QW structure described in here the change of bandgap by this process is much larger than that of deformation potential ($\Delta E_{\text{g,d}}$), that is,

$\frac{\Delta E_{\mathrm{g}}}{\Delta E_{\mathrm{g,d}}} = \left(\frac{2C_{\mathrm{Stark}}d_{31}}{\epsilon(D2+D4)}\right)\left(C_{11}+C_{12}-\frac{2C_{13}^{2}}{C_{33}}\right) >\sim 10$ where D_1 and D_2 are deformation potential parameters.

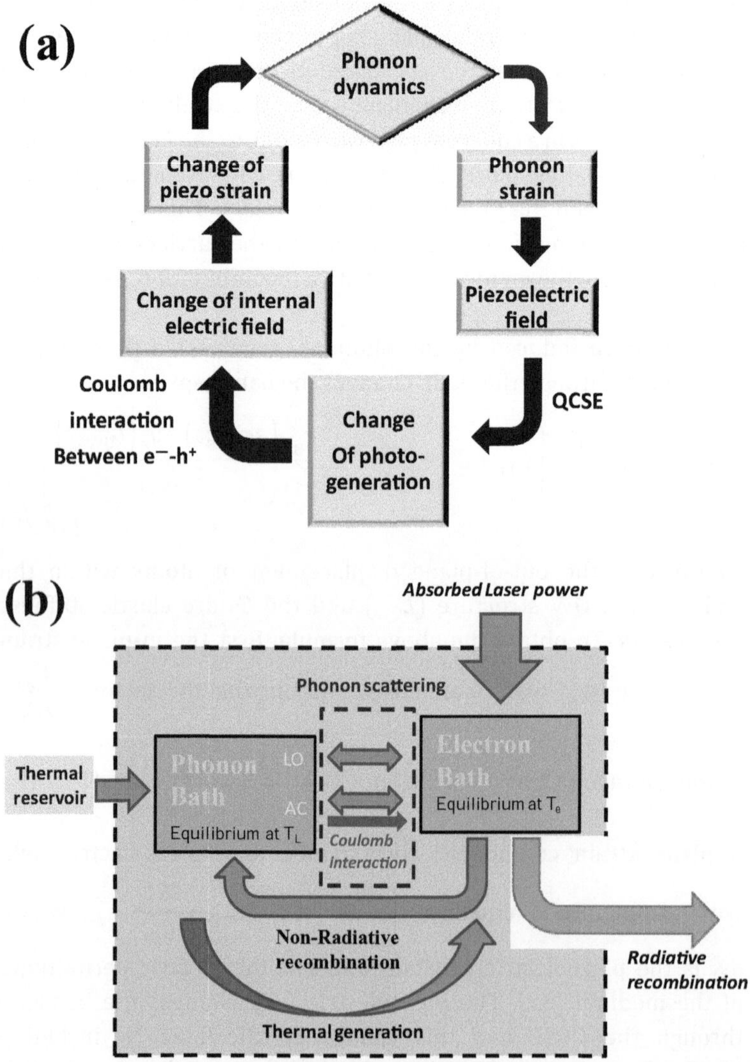

Figure 13.10 (a) The diagram showing the processes that lead to the Coulomb backaction mechanism and (b) the energy flow schematics. The Coulomb cooling increases the acoustic phonon absorption, as shown by the red flash.

In a piezoelectric material, the electrostatic field between the spatially separated electrons and holes leads to an effective piezoelectric stress. It is known that the piezoelectric stress can be related to the electrostatic field using the piezoelectric coefficient [7, 33]: $e_{33}\lambda_{21}n$, in which e_{33} is the piezoelectric coefficient and λ_{21} is the Coulomb strength parameter, which is the ratio of the induced electric field to the carrier density (n). Therefore the change in bandgap (as predicted by Eq. 13.25) leads to the change of carrier density and the piezoelectric stress (σ_{piezo}). To find the change in carrier density we use the rate equation

$$\frac{dn}{dt} = G - (An + Bn^2 + Cn^3) \tag{13.26}$$

where A, B, and C are the SRH, radiative, and Auger recombination rates, respectively. G is the generation rate and is calculated as follows:

$$G(\omega) = \frac{I_{\text{abs}}}{L_{\text{QW}}\hbar\omega} = \frac{I(1 - \exp(-\alpha(\hbar\omega - E_{\text{g}})L_{\text{QW}}) + \dfrac{\alpha_{\text{2PA}}L_{\text{QW}}I}{1 + \alpha_{\text{2PA}}L_{\text{QW}}I}\Big)}{L_{\text{QW}}\hbar\omega} \tag{13.27}$$

where, I_{abs} is the intensity of the absorbed photons, L_{QW} is the length of the QW where the carriers are confined, α is the absorption coefficient that depends on detuning ($\hbar\omega - E_{\text{g}}$), I is the laser intensity, E_{g} is the bandgap, and α_{2PA} is the two-photon absorption coefficient. Assuming small changes in carrier density we can write the small signal form of Eq. 13.27 as follows:

$$\frac{d\Delta n}{dt} = \Delta G - R\Delta n \tag{13.28}$$

where Δ means the change in the quantity of interest and $R = A + 2Bn + 3Cn^2$. By substituting $\Delta n = Re[\Delta N(\omega_q)e^{iw_q t}]$ (ω_q being the radial frequency of the phonon), we find the following equation for $\Delta N(\omega_q)$:

$$N(\omega_q) = \left(\frac{1}{1 + \left(\dfrac{\omega_q}{R}\right)^2} - \left(\frac{i\omega_q R}{R^2 + \omega_q^2}\right)\right)\left(\frac{\Delta G(\omega_q)}{R}\right) \tag{13.29}$$

and consequently

$$\Delta n(t)=\frac{1}{1+\left(\frac{\omega_q}{R}\right)^2}\Delta n_0-\frac{R}{R^2+\omega_q^2}d\Delta n_0/dt \tag{13.30}$$

where Δn_0 is the solution of the rate equation when $\frac{d\Delta n}{dt}\ll\Delta G$. By using $\Delta n_0=\frac{dn}{dE_g}\Delta E_{g,piezo}$, we can write

$$\Delta n(t)=\frac{1}{1+\left(\frac{\omega_q}{R}\right)^2}\left(\frac{dn}{dE_g}\right)\Delta E_{g,piezo}-\frac{R}{R^2+\omega_q^2}\left(\frac{dn}{dE_g}\right)\left(\frac{d\Delta E_{g,piezo}}{dt}\right) \tag{13.31}$$

Therefore the component of the piezostress, which is out of phase with atomic displacement due to phonons, will be equal to

$$\sigma_{piezo,o}=e_{33}\lambda_{21}\frac{1}{1+\left(\frac{\omega_q}{R}\right)^2}\left(\frac{dn}{dE_g}\right)\Delta E_g \tag{13.32}$$

It is then possible to calculate the cooling intensity from the coherent piezoelectric stress:

$$I_{cooling,q}=\sum<\sigma_{piezo}\left(\frac{du_z}{dt}\right)>=\left(\frac{1}{a}\right)\int_0^{L_{e\text{-}h}}<\sigma_{piezo}\left(\frac{du_z}{dt}\right)>dx \tag{13.33}$$

in which the cooling intensity for each atomic layer is calculated by taking an average in the time domain (as denoted by <> signs) and the results from all atomic layers within the average electron-hole separation length ($L_{e\text{-}h}$) has been summed up. The summation was replaced by the integration assuming that the electric field due to charge separation is uniform within the active region and is negligible elsewhere. Using Eqs. 13.25, 13.32, and 13.33 and assuming that $u_z = u_{zm}\cos(k_q.r - \omega_q t)$, where u_{zm} is the amplitude of vibration perpendicular to QW layers and k_q is the wavevector of the phonon, we can obtain the following equation for cooling per unit area of mode q:

$$I_{\text{cooling,q}} = 8\left(\frac{L_{\text{e-h}}}{a\rho V}\right)$$

$$C_{\text{Coulomb}} \frac{1}{1+\left(\frac{\omega_q}{R}\right)^2}\left(\frac{1}{v_s}\right)\left(\frac{1}{L_{\text{e-h}}^2 k_{qz}^2}\right)\left(\frac{k_{qz}}{k_q}\right)(1-\cos(k_{qz}L_{\text{e-h}}))\left(n_q+\frac{1}{2}\right)\hbar\omega_q \tag{13.34}$$

where ρ is the mass density, a is the lattice constant, $L_{\text{e-h}}$ is the average separation of electron and hole wave functions, V is the volume of the crystal, v_s is the group velocity of the phonon mode, k_q and k_{qz} are the amplitudes of the phonon wavevector along the direction of the propagation and along the crystal growth direction, respectively, and n_q is the phonon occupation number of mode q. C_{Coulomb} is an effective elasticity and is equal to the ratio of backaction piezoelectric stress to the average phonon's strain along the direction of epitaxial growth and is expressed as

$$C_{\text{Coulomb}} = \left(\frac{1}{2}\right)e_{33}\lambda_{21}C_{\text{Stark}}\left(\frac{d_{31}}{\epsilon}\left(C_{11}+C_{12}-\left(\frac{2C_{13}^2}{C_{33}}\right)\right)\frac{C_{33}}{C_{13}}\left(\frac{dn}{dE_{\text{g}}}\right)\right) \tag{13.35}$$

The obtained formula for cooling per unit area (Eq. 13.34) shows that the mode with an out-of-plane wavelength of $2.33L_{\text{e-h}}$ experiences maximum cooling. In Fig. 13.11 the phenomena that lead to Coulomb-assisted cooling are shown as a function of time for this particular mode. The change of bandgap and the carrier density due to out-of-plane strain of the phonon leads to a coherent change of piezostress, as shown in Fig. 13.11a–d. The component of the piezostress, which is 180° out of phase with carrier velocity, results in the net cooling power (Fig. 13.11d–f). The total cooling per unit area extracted from the phonons by means of Coulomb interaction is obtained by summing up the contribution of all phonon modes. The cooling per unit area of phonons modes with a specific energy of $\hbar v_s(k_q)k_q$ is equal to $I_{d^3k} = I_{\text{cooling},q} \times \frac{2\pi k_{||q}dk_{||q}k_{zq}}{\left(\frac{2\pi}{L}\right)^3}$, where L is the length of the crystal and $k_{||q}$ is the magnitude of the in-plane wavevector. Thus summing up I_{d^3k} over the whole Brillion zone

$\left(k_{||q} \in \left(0, \frac{\sqrt{2}\pi}{a}\right)\right.$ and $k_{zq} \in \left(-\frac{\pi}{a}, \frac{\pi}{a}\right)$ [34] gives the net cooling per unit area:

$$I_{\text{cooling}} = \iint_{k_{zq}=-\frac{\pi}{a}, k_{||q}=0}^{k_{zq}=\frac{\pi}{a}, k_{||q}=\left(\frac{\sqrt{2}\pi}{a}\right)} \left(\frac{2}{a\pi^2\rho}\right)$$

$$C_{\text{Coulomb}} \frac{1}{1+\left(\frac{\omega_q}{R}\right)^2} \left(\frac{1}{L_{\text{e-h}} k_{qz}}\right) (1-\cos(k_{qz} L_{\text{e-h}})) \left(n_q + \frac{1}{2}\right) \hbar k_{||q} dk_{||q} dk_{zq} \tag{13.36}$$

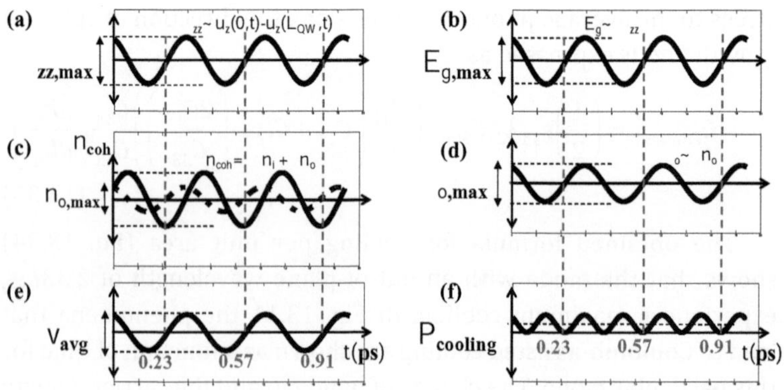

Figure 13.11 The time diagram of (a) the average out-of-plane strain, (b) the change of bandgap due to the change of piezoelectric field, and (c) the change of carrier density due to the change of bandgap. It is composed of an in-phase (Δn_i) and an out-of-phase (Δn_o) component. The amplitude of the out-of-phase component is 5.4×10^{-5} times lower than the total carrier density change for the phonon mode under consideration. (d) The piezostress due to Δn_o and (e) the average velocity. It is in phase with the piezostress shown in (d) and therefore results in a net cooling power, as shown in (f).

We can then simplify the cooling per unit area by defining the function $\mathcal{F}$:

$$\mathcal{F} = \iint_{k_{qz}=0,k_{||q}=0}^{k_{qz}=\frac{\pi}{a},k_{||q}=\left(\frac{\sqrt{2}\pi}{a}\right)} \frac{1}{1+\frac{v_s^2(k_{qz}^2+k_{||q}^2)}{R^2}}\left(\frac{1}{L_{e-h}k_{qz}}\right)$$

$$(1-\cos(k_{qz}L_{e-h}))\left(\frac{1}{\exp\left(\frac{\hbar v_s\left(\sqrt{k_{qz}^2+k_{||q}^2}\right)}{kT}\right)-1}+\frac{1}{2}\right)\hbar k_{||q}dk_{||q}dk_{zq} \tag{13.37}$$

and then express I_{cooling} as

$$I_{\text{cooling}} = \left(\frac{2}{a\pi^2\rho}\right)C_{\text{Coulomb}}\mathcal{F} \tag{13.38}$$

Therefore the cooling efficiency can be calculated as

$$\eta_{\text{cooling}} = \frac{I_{\text{cooling}}}{I_{\text{abs}}} = \left(\frac{2}{a\pi^2\rho I(1-\exp(-\alpha L_{\text{QW}})+\frac{\alpha_{\text{2PA}}L_{\text{QW}}I}{1+\alpha_{\text{2PA}}L_{\text{QW}}I})}\right)C_{\text{coulomb}}\mathcal{F} \tag{13.39}$$

A wurtzite c-GaN/InGaN QW, as shown in Fig. 13.12, was considered to explore the Coulomb effects. The *k.p* method in conjunction with Poisson equation can be used to calculate the self-consistent energy levels and the absorption of the MQW [52] in presence of Coulomb interaction. Effective masses and band line-ups were calculated according to Ref. [9]. The QW structure consists of 1 nm of $In_{0.15}Ga_{0.85}N$ sandwiched between 6 nm of GaN barriers. As can be seen in Fig. 13.12, the spatial separation of electron and hole wave functions due to the presence of a large piezoelectric field, leads to the screening of the built-in and piezoelectric electric field after photogeneration. As a result, significant blue shift of the interband transition energies is expected for a large photogenerated carrier density due to the QCSE.

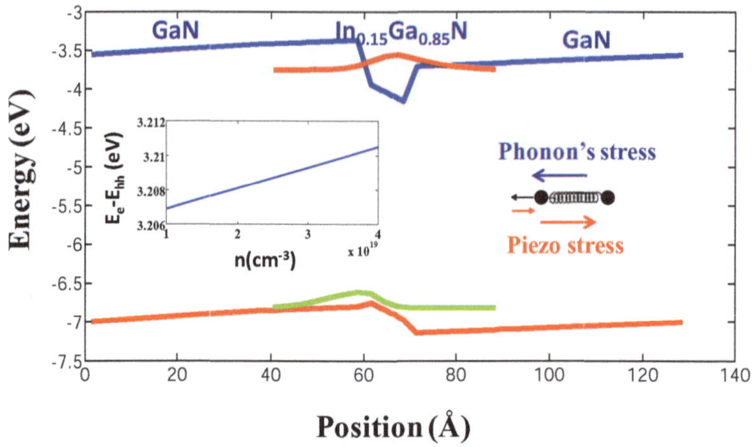

Figure 13.12 The band diagram of the designed MQW system. The electron wave function is in green, whereas the heavy-hole wave function is in red color. The inset shows the bandgap versus carrier density. Note that the direction of piezostress will be reversed for a phonon with opposite strain. This results in an effective phonon damping.

The inset of Fig. 13.12 shows the bandgap versus the carrier density with the slope equal to $L_{\text{e-h}}\lambda_{21}$. According to Fig. 13.13a, we can see that the net cooling efficiency increases significantly with the presence of the Coulomb interaction. We can see that the net cooling is possible for a laser intensity of 5 mW/μm^2 and for the detunings between 0–4 THz. The dependence of the net cooling efficiency on laser intensity can be explained by the fact that both the external quantum efficiency η_{ext} and the Coulomb cooling are dependent on laser intensity. At very high intensities, the probability of two-photon absorption increases, which is almost independent of detuning near the band edge [23] . As a result, in this regime, the change of carrier density with the change of bandgap reduces and the Coulomb cooling efficiency is suppressed. This behavior can be understood considering that the Coulomb cooling efficiency decreases due to the reduction of dn/dE_g. For extremely low excitation levels, the SRH recombination of the photogenerated carriers reduces the external quantum efficiency and the net cooling. The reduction of the internal quantum efficiency for large blue detunings is due to the increased carrier density and domination of the Auger recombination rate. In both cases, the reduction of the internal quantum efficiency effectively diminishes the positive effects of the Coulomb interaction on the net laser cooling. As a result, below and above certain detunings, the net cooling efficiencies for both cases (with and without the Coulomb effect) become essentially equal.

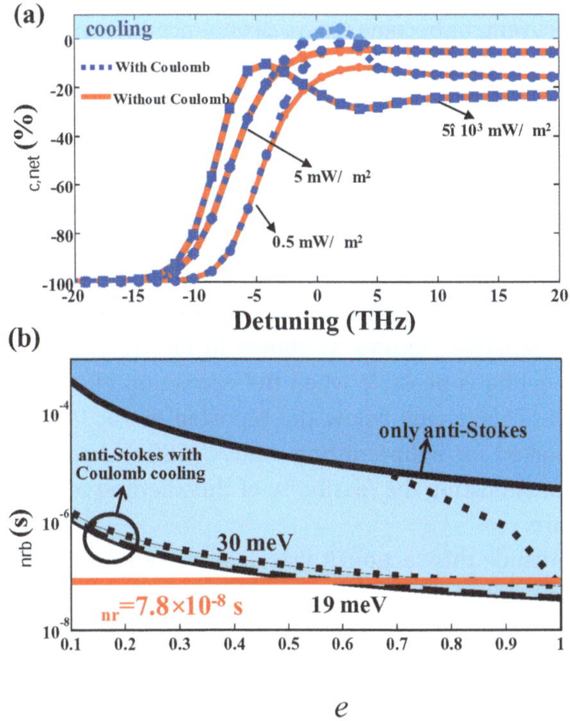

Figure 13.13 (a) The net cooling efficiency versus laser detuning for various laser intensities. Cases with both anti-Stokes and Coulomb cooling are shown by the dotted lines. The solid lines show only anti-Stokes cooling. These simulations have been performed for the structure that has an absorption broadening of 19 meV when the entire PL is extracted out from the material without reabsorption. (b) The break-even nonradiative recombination lifetime (τ_{nrb}) versus extraction efficiency for only anti-Stokes cooling (solid line) and both the anti-Stokes and Coulomb cooling (different values of absorption broadening has been considered). The 30 meV absorption broadening is a usual experimental value [19]. The lowest broadening is expected to be close to 19 meV (the value for GaN epilayers [15]). The red line represents the nonradiative recombination lifetime for a similar structure [53]. In all of the simulations, it is assumed that the temperature is 300 K. Reprinted with permission from Ref. [30]. Copyright 2014, AIP Publishing LLC.

The break-even nonradiative lifetime is defined as the shortest nonradiative lifetime that allows laser cooling (see Eqs. 13.1 and 13.2) [41]. The presence of the internal Coulomb cooling lowers the break-even nonradiative lifetime, as demonstrated in Fig. 13.13b. This improvement becomes more pronounced when the absorption broadening is lower since dn/dE_g increases as the absorption broadening decreases. By considering the nonradiative lifetime in a similar InGaN/GaN QW structure [53], we see that net cooling is possible with the presence of Coulomb cooling when the broadening is 30 meV and the extraction efficiency is better than 85% [53]. In this structure, the composition of the InGaN layer is close to the GaN barrier. Therefore the ultimate value of the broadening (19 meV) for the GaN epilayer [15], was also used to recalculate the break-even nonradiative lifetime, as shown in Fig. 13.13b . In this case, the net cooling is possible when the extraction efficiency is better than 55%. This is well below the reported extraction efficiency of 73% reported for light-emitting diodes (LEDs) made out of these materials, indicating the feasibility of this method of cooling in the near future.

To conclude this section, it was demonstrated that the Coulomb interaction could be utilized to facilitate the laser cooling of piezoelectric materials. The process is similar to optomechanical backaction cooling, where there is a backaction force due to displacement. Here the backaction force comes from the piezoelectric effect when the atoms move with phonons. The Coulomb cooling efficiency is highly dependent on laser intensity and detuning. It is shown that at optimum conditions the Coulomb interaction could make the net cooling possible for the structure under consideration. It has been also shown that the reduction of absorption broadening enhances the Coulomb cooling efficiency because of a higher change of the carrier density per change of carrier detuning. Increasing the laser intensity increases the Coulomb cooling efficiency until the point where the two-photon absorption dominates. However, the optimum laser intensity not only depends on the Coulomb cooling efficiency but also on the recombination processes. As for the laser detuning, the Coulomb cooling efficiency is maximized near the band edge where the slope of absorption versus detuning is the highest. Owing to the high efficiency of this mechanism, the next step is to investigate it in other strong piezoelectric semiconductors such as III–V quantum dots.

13.7 Comparison of Coulomb-Assisted Cooling in Piezoelectric Materials with Collisionally Aided Laser Cooling

In this section, we briefly introduce the laser cooling by means of collision redistribution of radiation in gaseous systems. Our description of this type of cooling is mainly based on the paper published by Berman and Stenholm [4] and the experimental demonstration of the cooling reported by Vogl and Weitz [49]. This topic is covered in detail in the Chapter 1 on collisional redistribution laser cooling in this book. We compare this method of laser cooling with the Coulomb-assisted laser cooling of piezoelectric materials. Consider a compressed gaseous system, as shown in Fig. 13.14a. Without any collision between the atoms and neglecting the Doppler shift of the radiation, the luminescence linewidth is very narrow and the energy of the radiated photons is essentially equal to the energy of the photons belonging to the excitation laser. A red-detuned excitation of an atom, far from its natural linewidth, results in excitation of virtual carriers [21]. However, when the atoms collide with each other, the "virtually excited" atoms can gain energy out of the collisions and can get *really* excited to the upper state. As a result the luminescence energy will be close to the bandgap of the atom and is spectrally broadened because of collisions as shown in Fig. 13.14b. The simple picture described in here neglects the linewidth broadening by the Doppler effect (which is equal to $\omega_D \approx \omega_{ex}/c\sqrt{kT/M}$, where ω_{ex} is the radial frequency of the excitation photons and M is the mass of the atom). Furthermore certain conditions need to be met in order to have an effective cooling/heating by this process. Assuming that the total thermal energy of the particle to be $ET \approx kT$, the duration of the collision and the collision rate are represented by τ_c and Γ, respectively, then for a certain detuning (Δ), we should have [4]:

$$\frac{E_T}{\hbar} > \tau_c^{-1} > |\Delta| > \Gamma \tag{13.40}$$

The cooling efficiency is related to the difference between the frequency of the emitted (ν_{fl}) and the absorbed photons(ν_{ab}):

$$\eta_{\text{cooling}} = \alpha(\nu_{\text{ab}})\frac{\nu_{\text{fl}} - \nu_{\text{ab}}}{\nu_{\text{ab}}} = \alpha(\nu_{\text{ab}})\frac{\Delta}{\nu_{\text{ab}}} \tag{13.41}$$

where $\alpha(\nu_{\text{ab}})$ denotes the absorbance of the pumped photons. Therefore to increase the cooling or heating efficiency, the detuning should be increased but not to the extent to result in a considerable decrease of absorbance. Also the condition of $\tau_{\text{c}}|\Delta| < 1$, as indicated in Eq. 13.40, allows one to analyze the system using "impact approximation" whereas the condition $\frac{E_T}{\hbar}\tau_{\text{c}} > 1$ ensures the thermal equilibrium of the atomic ensemble with the effective temperature of T. Each collision excites the atom to its upper state and results in an energy gain equal to the laser detuning. Considering that the spontaneous decay of the atoms is dominant (which is the case when the population inversion is very small), then the excitation rate can be obtained as

$$\dot{n}_2 = \frac{2\chi^2\Gamma}{\Delta^2 + (\Gamma + \gamma/2)^2} \cong \frac{2\chi^2\Gamma}{\Delta^2} \tag{13.42}$$

where γ is the radiative rate of the atoms. At the steady state, that is, $\dot{n}_2 = 0$, the occupation probability of the excited level is equal to

$$n_2 = \frac{2\chi^2\Gamma}{\Delta^2\gamma} \tag{13.43}$$

The condition of $|\Delta| > \Gamma$, as indicated in Eq. 13.40, is therefore necessary to have a small excitation and dominant radiative decay of the atoms ($\lambda >> \dot{n}_2$). The normalized energy change per unit time is expressed by

$$\frac{dE_T/E_T}{E_T} = \frac{\hbar\Delta}{E_T}\frac{2\chi^2\Gamma}{\Delta^2} \tag{13.44}$$

It is possible to find the cooling power and efficiency using the above equations. The cooling power can be found as follows:

$$P_{\text{cool}} = \gamma n_2(E_2 - E_1) - \dot{n}_2 E_{\text{ex}} = \dot{n}_2(\hbar\Delta) \tag{13.45}$$

where E_1, E_2, and E_{ex} are the energies of the first and second levels and the photons of the excitation laser. For the absorbed power, we can write

$$P_{\text{abs}} = \dot{n}_2 E_{\text{ex}} \tag{13.46}$$

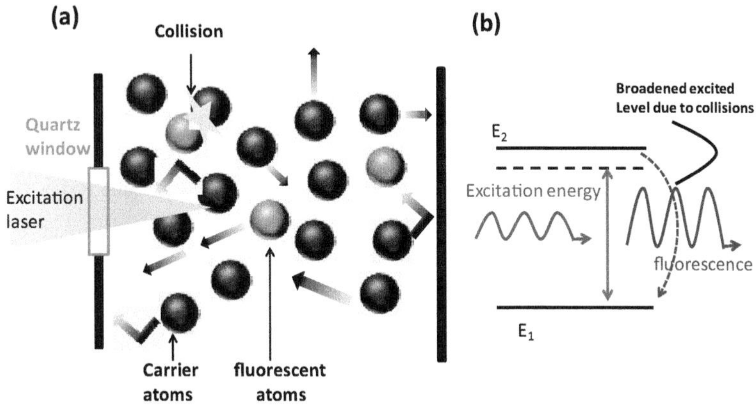

Figure 13.14 (a) The schematics of a gaseous system that consists of active florescent atoms and carrier atoms. The collision between the atoms results in broadening of the luminescence and upconversion of the luminescence energy with the respect to the absorbed photon energy. (b) The energy level diagram and the excitation and radiation of an atom in the system. With collisions the atoms can be excited to the upper state and emit photons with a center frequency equal to $(E_2 - E_1)/\hbar$. Therefore a red-detuned excitation results in cooling.

Therefore the cooling efficiency is equal to $P_{\text{cool}}/P_{abs} = \hbar\Delta/E_{\text{ex}}$. This method of cooling was experimentally demonstrated by Vogl and Weitz, who provided a simple explanation for the cooling mechanism, as shown by the graph of Fig. 13.15. the bandgap of the atoms depends on the interatomic distance, the bandgap has a minima and a maxima. Since the pump laser is red tuned with respect to the bandgap, therefore the absorption of the pump laser can only happen when the atoms get close enough to each other to have a reduced bandgap. However, on average the distance between the atoms corresponds to a higher bandgap. Since the fluorescence is a slower phenomenon compared to the collision events, atoms emit photons with a higher energy equal to their bandgap energy at the average interatomic distance. Neglecting the fluorescence trapping, in this process the energy of collisions gets extracted by the fluorescence.

It is worthwhile to compare this method of cooling with that of Coulomb-assisted cooling, as described in the previous section. In the latter approach the propagation of the phonon through the QW structure changes the average interatomic distance and results in a net strain. Similar to collisionally aided cooling, this effect results in an effective change of the MQW bandgap. The change of bandgap results in an immediate change of the photogenerated carrier density in phase with the average atoms' velocity. Therefore, the corresponding piezoelectric force on the atoms does a *net* work on them. This work can either coherently retard or amplify the atomic vibration by the phonon mode. In the case of cooling, the excess energy taken from the phonon bath is electrostatically transferred to the photogenerated electrons through the piezoelectric effect. Similar to the collisionally aided cooling, the fluorescence happens at a much slower timescale compared to cooling/heating cycle of the phonons and therefore the extra energy of the carriers can be extracted out from the material effectively.

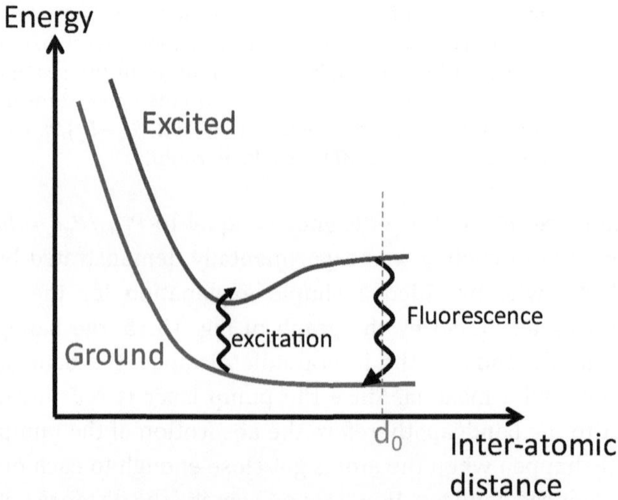

Figure 13.15 The energies of the ground state and the excited states versus the interatomic distance. The average interatomic distance is denoted by d_0. This graph is similar to the one indicated in Ref. [49]. This technique of cooling is similar to Coulomb-assisted cooling of piezoelectric materials in the sense that in both cases the photoexcitation happens as the interatomic distance decreases and the fluorescence is a slower process that extracts the excess energy of the carriers outside of the material. However, the mechanism of a shrinking bandgap and the excess energy gained by the carriers is different for both cases.

It should be noted that the laser is red-detuned from the bandgap to achieve Coulomb-assisted laser cooling. This is due to the change of absorption per change of the bandgap for a slightly red-detuned laser. As a result when atoms get closer to each other, the bandgap shrinks, as described by Eq. 13.25, and a significant photoexcitation happens. This is exactly similar to collisionally aided cooling, as shown in Fig. 13.15, where the reduction in interatomic distance results in photoexcitation. The excess photogenerated carriers gain an average energy by the described coherent piezoelectric interaction with the phonon. Therefore the average energy of the luminescence will be higher that the absorbed photons and extraction of the PL results in cooling.

13.8 Concluding Remarks

In this chapter we discussed the mechanisms that can potentially increase the laser cooling efficiency to make it practical for bulk semiconductors. Our focus was on Coulomb interaction in piezoelectric semiconductors. The equations regarding the anti-Stokes laser cooling efficiency and the external quantum efficiency were investigated to determine different absorption and recombination parameters and their roles in cooling of solids with lasers. These parameters include parasitic absorption by impurities, defects and surface states, extraction efficiency of the PL, and radiative, Auger, and SRH recombination rates. In addition the cooling efficiency depends on the bandgap of the material and the carrier density. It is concluded from the literature that laser cooling is very weak, even at optimum conditions, rendering it impossible for bulk semiconductors with the highest material quality achieved so far. Therefore other mechanisms that can initiate the laser cooling process should be investigated in order to assist anti-Stokes laser cooling. To establish the theoretical bases for understanding the Coulomb-assisted laser cooling, we described both the Coulomb interaction and the piezoelectricity separately. The Coulomb interaction here refers to the electrostatic attraction between the electron and hole pairs. The excitonic effect is a good example, which has been already shown to increase the laser cooling efficiency due to the enhancement in the excitonic radiative decay and its effect on steepening of the band-tail absorption. In general, the electric field between the photogenerated electron and hole pairs can either blue-shift or red-shift the transition energies. This phenomenon can be used for laser cooling purposes, as we have shown in this chapter. We provided a brief theory of piezoelectricity, particularly for zinc blende and wurtzite crystals that are currently at the center of attention for semiconductor laser cooling due to their high internal quantum efficiency. Our approach is inspired by the optomechanical cooling of micro- and nanomechanical systems, and therefore we explained the basics of backaction optomechanics and proceeded to establish the described theory by using the same mechanism since each phonon mode acts like a mechanical mode of vibration with low quality factor. The low quality factor of the phonon modes might be discouraging at first, however, there exist numerous phonon modes and the net cooling is the sum of the cooling powers of all those phonon modes. To validate the assumptions, we considered highly piezoelectric InGaN/GaN QW systems. Each acoustic phonon mode changes the bandgap by producing a net strain across the QW region. Since the material is piezoelectric, the strain is associated with an effective electric field that changes the bandgap by the QCSE. The change of the bandgap changes the photogeneration of the carriers. Now, since the electron and hole pairs are separated in space, the change in the photogenerated carrier density results in a change in the electrostatic field between the electrons and holes. The change in the electric field is associated with the piezoelectric force exerted on the atoms within the QW region. Therefore there is a closed loop that starts with the strain produced by the phonons and ends with a coherent, retarding piezoelectric force that prevents atoms from vibrating. Using the standard equations regarding photogeneration, the QCSE, and piezoelectricity, the laser cooling efficiency and the break-even nonradiative recombination rate can be found with and without the Coulomb effect. The results show a significant reduction of the break-even SRH lifetime and reveal the possibility of cooling in such systems. Finally, we compared this method to the collisionally aided laser cooling (for detailed information of this method, the reader is referred to the chapter on collisional redistribution laser cooling in this book) . In both methods, the change of interatomic distance results in a change of the bandgap and photoexcitation occurs with lower energy photons, while the slower radiative recombination extracts the excess energy of the material through emitting higher-energy photons.

It is worth mentioning that many of the proposed ideas can be further improved or investigated in other material systems. Furthermore, future experiments are needed to investigate each of the proposed mechanisms individually. And lastly, although the above theory applies to piezoelectric materials, the backaction optomechanical concepts can be used to damp the lattice vibration with any type of force associated with the photogenerated carriers. For these reasons, we believe that the extension of the optomechanical concepts to laser cooling of bulk materials merits further investigations.

Acknowledgments

The authors would like to acknowledge partial support from NSF awards nos. ECCS-1310620, ECCS-1206155, and ECCS-0901855, as well as ARO award nos. W911NF-13-1-0485 and W911NF-11-1-0390. The authors would also like to acknowledge the help of Travis Hamilton, Eric Dexheimer, and Abigail Turay in preparation of this chapter.

References

1. Adachi, S. (1994). *GaAs and Related Materials: Bulk Semiconducting and Superlattice Properties* (World Scientific).
2. Arnau, A. (2008). *Piezoelectric Transducers and Applications* (Springer).
3. Aspelmeyer, M., Meystre, P., Schwab, K. (2012). Quantum optomechanics, *Phys. Today,* **65**, pp. 29–35.
4. Berman, P. R., Stenholm, S. (1978). Heating or cooling collisionally aided fluorescence, *Opt. Commun.,* **24**, pp. 155–157.
5. Bernardini, F., Fiorentini, V. (2002). First-principles calculation of the piezoelectric tensor d of III-V nitrides, *Appl. Phys. Lett.,* **80**, p. 4145.
6. Chern, G.-W/, Lin, K.-H/, Huang, Y.-K., Sun, C.-K. (2003). Spectral analysis of high-harmonic coherent acoustic phonons in piezoelectric semiconductor multiple quantum wells, *Phys. Rev. B,* **67**, p. 121303.
7. Chern, G.-W., Lin, K.-H., Sun, C.-K. (2004). Transmission of light through quantum heterostructures modulated by coherent acoustic phonons, *J. Appl. Phys.,* **95**, pp. 1114–1121.

8. Christmas, U. M., Andreev, A., Faux, D. (2005). Calculation of electric field and optical transitions in InGaN/GaN quantum wells, *J. Appl. Phys.*, **98**, p. 073522.
9. Chuang, S. L. (1995). *Physics of Optoelectronic Devices* (Wiley).
10. Dowd, P., Braun, W., Smith, D. J., Ryu, C., Guo, C.-Z., Chen, S., Koelle, U., Johnson, S., Zhang, Y.-H. (1999). Long wavelength (1.3 and 1.5 μm) photoluminescence from InGaAs/GaPAsSb quantum wells grown on GaAs, *Appl. Phys. Lett.*, **75**, pp. 1267–1269.
11. Dyakonov, M., Kachorovskii, V. Y. (1994). Nonthreshold Auger recombination in quantum wells, *Phys. Rev. B*, **49**, p. 17130.
12. Epstein, R. I., Buchwald, M. I., Edwards, B. C., Gosnell, T. R., Mungan, C. E. (1995). Observation of laser-induced fluorescent cooling of a solid, *Nature*, **377**, pp. 500–503.
13. Fathipour, V., Jang, S., Hassaninia, I., Mohseni, H. (2014). Approaching high temperature photon counting with electron-injection detectors, *Proc. SPIE*, **9220**, p. 92200J.
14. Fathipour, V., Memis, O. G., Jang, S. J., Brown, R. L., Nia, I. H., Mohseni, H. (2014). Isolated electron injection detectors with high gain and record low dark current at telecom wavelength, *IEEE J. Sel. Topics Quantum Electron.*, **20**, pp. 65–70.
15. Fischer, A., Shan, W., Park, G., Song, J., Kim, D., Yee, D., Horning, R., Goldenberg, B. (1997). Femtosecond four-wave-mixing studies of nearly homogeneously broadened excitons in GaN, *Phys. Rev. B*, **56**, p. 1077.
16. Fuchs, G., Schiedel, C., Hangleiter, A., Härle, V., Scholz, F. (1993). Auger recombination in strained and unstrained InGaAs/InGaAsP multiple quantum-well lasers, *Appl. Phys. Lett.*, **62**, pp. 396–398.
17. Hausser, S., Fuchs, G., Hangleiter, A., Streubel, K., Tsang, W. (1990). Auger recombination in bulk and quantum well InGaAs, *Appl. Phys. Lett.*, **56**, pp. 913–915.
18. Hossein-Zadeh, M., Vahala, K. J. (2010). An optomechanical oscillator on a silicon chip, *IEEE J. Sel. Topics Quantum Electron.*, **16**, pp. 276–287.
19. Jiang, H., Singh, J. (1999). Linear electro-optic effect due to the built-in electric field in InGaN/GaN quantum wells, *Appl. Phys. Lett.*, **75**, pp. 1932–1934.
20. Johnson, S., Ding, D., Wang, J.-B., Yu, S.-Q., Zhang, Y.-H. (2007). Excitation dependent photoluminescence measurements of the nonradiative lifetime and quantum efficiency in GaAs, *J. Vac. Sci. Technol. B*, **25**, pp. 1077–1082.

21. Khurgin, J. (1992). Coulomb enhancement of ultrafast nonlinearities in quantum-well structures, *J. Opt. Soc. Am. B,* **9**, pp. 157–172.

22. Kippenberg, T. J., Vahala, K. J. (2007). Cavity opto-mechanics, *Opt. Express,* **15**, pp. 17172–17205.

23. Krishnamurthy, S., Nashold, K., Sher, A. (2000). Two-photon absorption in GaN, GaInN, and GaAlN alloys, *Appl. Phys. Lett.,* **77**, pp. 355–357.

24. Lin, K.-H., Lai, C.-M., Pan, C.-C., Chyi, J.-I., Shi, J.-W., Sun, S.-Z., Chang, C.-F., Sun, C.-K. (2007). Spatial manipulation of nanoacoustic waves with nanoscale spot sizes, *Nat. Nanotechnol.,* **2**, pp. 704–708.

25. Melgaard, S., Albrecht, A. R., Hehlen, M., Seletskiy, D., Sheik-Bahae, M. (2014). Optical refrigeration cools below 100 K, *Optical Society of America*, paper FTh4D.4.

26. Metzger, C. H., Karrai, K. (2004). Cavity cooling of a microlever, *Nature,* **432**, pp. 1002–1005.

27. Mohseni, H., Litvinov, V., Razeghi, M. (1998). Interface-induced suppression of the Auger recombination in type-II InAs/GaSb superlattices, *Phys. Rev. B,* **58**, p. 15378.

28. Nia, I. H., Memis, O. G., Kohoutek, J., Gelfand, R. M., Mohseni, H. (2012). Surface plasmon enhancement of photon extraction efficiency by silver nanoparticles: with applications in laser cooling of semiconductors, *Proc. SPIE,* **8457**, p. 84573V.

29. Nia, I. H., Mohseni, H. (2014). Exploring Coulomb interaction in piezoelectric materials for assisting the laser cooling of solids, *Proc. SPIE,* **9000**, p. 90000M.

30. Nia, I. H., Mohseni, H. (2014). A proposal for Coulomb assisted laser cooling of piezoelectric semiconductors, *Appl. Phys. Lett.,* **105**, p. 042102.

31. Nia, I. H., Mohseni, H. (2015). Photoluminescence study of suspended MQW structures for laser refrigeration purposes, *Proc. SPIE,* **9380**, p. 93800E.

32. Norris, A. N. (2006). Optimal orientation of anisotropic solids, *Q. J. Mech. Appl. Math.,* **59**, pp. 29–53.

33. Babilotte, P., Ruello, P., Vaudel, G., Pezeril, T., Mounier, D., Breteau, J.-M., Gusev, V. (2010). Picosecond acoustics in p-doped piezoelectric semiconductors, *Appl. Phys. Lett.,* **97**, p. 174103.

34. Pokatilov, E. P., Nika, D. L., Balandin, A. A. (2003). Phonon spectrum and group velocities in AlN/GaN/AlN and related heterostructures, *Superlattice Microst.,* **33**, pp. 155–171.

35. Pringsheim, P. (1929). Zwei Bemerkungen über den Unterschied von Lumineszenz-und Temperaturstrahlung, *Z. Phys.*, **57**, pp. 739–746.

36. Ruan, X., Kaviany, M. (2007). Advances in laser cooling of solids, *J. Heat Trans.*, **129**, pp. 3–10.

37. Rupper, G., Kwong, N., Binder, R. (2006). Large excitonic enhancement of optical refrigeration in semiconductors, *Phys. Rev. Lett.*, **97**, p. 117401.

38. Rupper, G., Kwong, N., Binder, R. (2007). Optical refrigeration of GaAs: theoretical study, *Phys. Rev. B*, **76**, p. 245203.

39. Sanders, G., Stanton, C., Kim, C. S. (2001). Theory of coherent acoustic phonons in $In_xGa_{1-x}N$/GaN multiple quantum wells, *Phys. Rev. B*, **64**, p. 235316.

40. Sermage, B., Heritage, J., Dutta, N. (1985). Temperature dependence of carrier lifetime and Auger recombination in 1.3 μm InGaAsP, *J. Appl. Phys.*, **57**, pp. 5443–5449.

41. Sheik-Bahae, M., Epstein, R. I. (2004). Can laser light cool semiconductors?, *Phys. Rev. Lett.*, **92**, p. 247403.

42. Sheik-Bahae, M., Imangholi, B., Hasselbeck, M., Epstein, R., Kurtz, S. (2006). Advances in laser cooling of semiconductors, *Proc. SPIE*, **6115**, p. 611518.

43. Shi, C., Asbeck, P. M., Edward, T. Y. (1999). Piezoelectric polarization associated with dislocations in wurtzite GaN, *Appl. Phys. Lett.*, **74**, pp. 573–575.

44. Shimada, K. (2006). First-principles determination of piezoelectric stress and strain constants of Wurtzite III–V nitrides, *Jpn. J. Appl. Phys.*, **45**, p. L358.

45. Smith, C., Abram, R., Burt, M. (1985). Theory of Auger recombination in a quantum well heterostructure, *Superlattice Microst.*, **1**, pp. 119–123.

46. Turchinovich, D., Jepsen, P. U., Monozon, B., Koch, M., Lahmann, S., Rossow, U., Hangleiter, A. (2003). Ultrafast polarization dynamics in biased quantum wells under strong femtosecond optical excitation, *Phys. Rev. B*, **68**, p. 241307.

47. Turchinovich, D., Monozon, B., Jepsen, P. U. (2006). Role of dynamical screening in excitation kinetics of biased quantum wells: nonlinear absorption and ultrabroadband terahertz emission, *J. Appl. Phys.*, **99**, p. 013510.

48. Van de Walle, C. G. (1989). Band lineups and deformation potentials in the model-solid theory, *Phys. Rev. B*, **39**, p. 1871.

49. Vogl, U., Weitz, M. (2009). Laser cooling by collisional redistribution of radiation, *Nature,* **461**, pp. 70–73.

50. Yoo, K., Albrecht, J., Ram-Mohan, L. (2010). Strain in layered zinc blende and wurtzite semiconductor structures grown along arbitrary crystallographic directions, *Am. J. Phys.,* **78**, pp. 589–597.

51. Zhang, J., Li, D., Chen, R., Xiong, Q. (2013). Laser cooling of a semiconductor by 40 kelvin, *Nature,* **493**, pp. 504–508.

52. Zhao, H., Arif, R., Ee, Y.-K., Tansu, N. (2009). Self-consistent analysis of strain-compensated InGaN–AlGaN quantum wells for lasers and light-emitting diodes, *IEEE J. Quantum Electron,* **45**, pp. 66–78.

53. Zhao, H., Liu, G., Zhang, J., Poplawsky, J. D., Dierolf, V., Tansu, N. (2011). Approaches for high internal quantum efficiency green InGaN light-emitting diodes with large overlap quantum wells, *Opt. Express,* **19**, pp. A991–A1007.

Index

* 9 7 8 9 8 1 4 7 4 5 0 4 8 *
#0001 - 010916 - C52 - 229/152/26 [28] - CB - 9789814745048